IUTAM Symposium on Recent Advances of Acoustic Waves in Solids

IUTAM BOOKSERIES

Volume 26

Aims and Scope of the Series

The IUTAM Bookseries publishes the proceedings of IUTAM symposia under the auspices of the IUTAM Board.

For other titles published in this series, go to
www.springer.com/series/7695

Tsung-Tsong Wu • Chien-Ching Ma
Editors

IUTAM Symposium on Recent Advances of Acoustic Waves in Solids

Proceedings of the IUTAM Symposium on Recent Advances of Acoustic Waves in Solids, Taipei, Taiwan, May 25-28, 2009

Springer

Editors
Tsung-Tsong Wu
Institute of Applied Mechanics
National Taiwan University
1, Section 4, Roosevelt Road
10617 Taipei
Taiwan R.O.C.
wutt@ntu.edu.tw

Chien-Ching Ma
Department of Mechanical Engineering
National Taiwan University
1, Section 4, Roosevelt Road
10617 Taipei
Taiwan R.O.C.
ccma@ntu.edu.tw

ISSN 1875-3507 e-ISSN 1875-3493
ISBN 978-90-481-9892-4 e-ISBN 978-90-481-9893-1
DOI 10.1007/978-90-481-9893-1
Springer Dordrecht Heidelberg London New York

Library of Congress Control Number: 2010934855

Cover design: SPI Publisher Services

Printed on acid-free paper

Springer is part of Springer Science+Business Media (www.springer.com)

Preface

This book is a collection of papers presented at the IUTAM symposium on "Recent Advances of Acoustic Waves in Solids" which was held in Taipei, Taiwan, May 25-28, 2009. Fifty invited participants from 16 countries were participated in this Symposium.

Acoustic wave in solids is not a new topic; instead, it has been a topic of importance in both engineering and science for several decades. However, started from the mid 90's, emerging needs in engineering applications have re-triggered new and important topics, such as phononic crystals, UHF acoustic filters in mobile communications, quantitative NDE of materials, etc. The purpose of this symposium was to bring renowned scholars and leading researchers from the areas of acoustic waves in solids together to discuss the latest advances, in particular, the emerging topics such as phononic band gap materials and acoustic metamaterials, acoustic filters in mobile communications, waves and quantitative nondestructive evaluation of materials, and waves in complex media.

We are grateful to all of the participants for their contribution and support to this symposium. The organizing of the symposium would not be possible without the valuable advices and guidance from all the scientific committee members, Jan D. Achenbach (USA), Arthur G. Every (South Africa), Ken-ya Hashimoto (Japan), Vincent Laude (France), Ioannis E. Psarobas (Greece), Chau-Shioung Yeh (Taiwan), Yook-Kong Yong (USA), Jüri Engelbrecht (Estonia, IUTAM Representative). The excellent works of the Local Organizing Committee and the Symposium Secretariat are gratefully acknowledged. Sincere thanks are also extended to Professor Jan D. Achenbach and Professor Yih-Hsing Pao for their encouragement and support of this symposium at the early planning stage.

During the symposium held in the Institute of Applied Mechanics, the scientific committee proposed and obtained unanimous consensus to dedicate this book to Professor Yih-Hsing Pao, in honor of his 80th birthday and profound contribution to the field of acoustic waves in solids.

Finally, but not the least, generous supports of the symposium from Taiwan government and related agencies are gratefully acknowledged, including National Science Council, Ministry of Education, National Applied Research Laboratories, National Taiwan University, IUTAM and the Society of Theoretical and Applied Mechanics in Taiwan.

Tsung-Tsong Wu
Chien-Ching Ma
Taipei, November, 2009

In Honor of Professor Yih-Hsing Pao's 80th Birthday

Prof. Yih-Hsing Pao was born in Nanking, China in 1930. After enrolling at National Chiao-Tung University in Shanghai, he transferred to National Taiwan University and received a Bachelor's Degree in civil engineering in 1952. He then pursued advanced studies in the United States, earning a Master Degree in mechanics from Rensselaer Polytechnic Institute in 1955, and a Ph.D. in applied mechanics from Columbia University in 1959.

Prof. Pao is a world-renowned scholar in applied mechanics. Since 1960, his innovative contributions have covered a broad range of theories and applications in modern engineering technology. Chief among them is a seminal treatise entitled "Diffraction and Scattering of Elastic Waves," co-authored with his former colleague Dr. C.C. Mow. This book laid the foundation for investigating soil-structure interaction under dynamic impact and blasting. The theory was later extended to Non-Destructive Evaluation of materials and structures, a modern technique employing ultrasonics to characterize a material or to detect defects in aircrafts, bridges, railways, as well as human bodies. His work on the theory of generalized rays for waves in layered media has been applied extensively to foundations of structures and to underwater acoustics. One of his papers has been selected by the International Union of Theoretical and Applied Mechanics as a landmark publication of the 20th century. In a research project with his student, Prof. Francis Moon of Cornell University, Prof. Pao pioneered the new field of electromagneto-mechanics, which is basic to study the stability of vehicles by magneto-levitation and of superconducting coil structures carrying strong current.

He has made contributions in building fundamental theories, numerical analysis, technical applications, and his research in earthquakes has been widely referenced by civil, mechanical, and aeronautical engineers.

During Prof. Pao's celebrated career in the US, he attained prestige matched by only a few in his field. This is evidenced by numerous leading positions in professional societies, such as the general chairman and the chairman of scientific committee of 9th US National Congress of Applied Mechanics. A special session was dedicated to honor his achievements at the 11th (1990) US National Congress of Applied Mechanics in Tucson, Arizona.

At the invitation of the Taiwan government in 1984, Prof. Pao made an admirable and far-sighted decision to return to Taiwan to found the Institute of Applied Mechanics at National Taiwan University. After more than two decades, his sustained and dedicated leadership has resulted in a first-class research center of Applied Mechanics. By creating an attractive research environment, he was able to recruit outstanding faculty members from Taiwan and abroad. Under his guidance, with depth of vision and breadth of knowledge, the Institute has set the standards of excellence in Taiwan. It emphasizes multidisciplinary efforts in both research and teaching, and in the balance of theory with experiments and applications. These efforts have established the Institute as the forerunner not only in Taiwan, but also in Asia and the world. As a leader in numerous professional societies and government advisory committees, Prof. Pao's influence has spread far beyond the university campus. Among other appointments, he has been elected a Member of the US National Academy of Engineering in 1985, and Academician of Academia Sinica in 1986. He is also the recipient of numerous prestigious awards, including the Senior Scientist Award from the Humboldt Foundation of Germany, the presidential award from the Republic of China and an honorary Doctorate Degree by National Chiao-Tung University. Prof. Pao has authored and co-authored several books and more than one hundred research articles published in various scientific and technical journals. He has been invited to contribute articles in *Physical Acoustics, Mechanics Today,* and *Applied Mechanics Reviews.*

Scientific Committee

Professor Tsung-Tsong Wu (Chair)
Institute of Applied Mechanics, National Taiwan University, Taipei, Taiwan

Professor Jan D. Achenbach
Department of Mechanical Engineering, Northwestern University, IL, USA

Professor Jüri Engelbrecht (IUTAM Representative)
Tallinn University of Technology, Tallinn, Estonia

Professor Arthur G. Every
School of Physics, University of Witwatersrand, Johannesburg, South Africa

Professor Ken-Ya Hashimoto
Department of Electrical and Mechanical Engineering, Chiba University, Chiba-shi, Japan

Professor Vincent Laude
LPMO department, FEMTO-ST institute, CNRS, Besancon, Cedex, France

Dr. Ioannis E. Psarobas
Department of Physics - Section of Solid State Physics, The University of Athens, Greece

Professor Chau-Shioung Yeh
Institute of Applied Mechanics, National Taiwan University, Taipei, Taiwan

Professor Yook-Kong Yong
Department of Civil Engineering, Rutgers University, NJ, USA

Local Committee

Professor Chien-Ching Ma (Chair)
Department of Mechanical Engineering, National Taiwan University, Taipei, Taiwan

Dr. Jia-Hong Sun (Secretariat General)
Institute of Applied Mechanics, National Taiwan University, Taipei, Taiwan

Ken-ya Hashimoto
Masahiko Hirao
José Sánchez-Dehesa
Hirotsugu Ogi
Yook-Kong Yong
Yung-Yu Chen
Anastasiya A. Bondarenko
Kuan-Chiun Weng
Chau-Shioung Yeh
Daniel Torrent
Victor P. Plessky
Da-Ming Zhu
Renata Troian
Thomas C.T. Ting
Jin-Chen Hsu
Benjamin Abbott
Arvi Ravasoo
Mahmoud I. Hussein
Tsung-Tsong Wu
Sean Wu
Vincent Laude
Igor Solodov
Badreddine Assouar
Yih-Hsing Pao
Ruyen Ro
Yue-Sheng Wang
Bahram Djafari-Rouhani
Abdelkrim Khelif
Shih-Yung Pao
Ji Wang
Jan D. Achenbach
Younkers Achaoui
Walter Steurer
C. S. Lam
Massimo Ruzzene
Mathias Fink
Ioannis E. Psarobas
Yukihiro Tanaka
Per-Ake Jansson
Ping-Hung Lee
Tanel Peets
Jüri Engelbrecht
Piotr Borejko
Chi-Kuang Sun
Sridhar Krishnaswamy
Mao-Kuen Kuo
Junhui Hu
Jia-Hong Sun
Arthur G. Every
Younho Cho
Horn Jiunn Sheen
Jaesun Lee
Yung-Chun Lee
Chin-Teh Sun
Chien-Ching Ma

IUTAM Symposium on Recent Advances of Acoustic Waves in Solids
May 25-28, 2009 Taipei, Taiwan

Table of Contents

Part I
Waves in General

Existence of Exceptional Body Waves and Subsonic Surface Waves in Monoclinic and Orthotropic Materials

T. C. T. Ting

Division of Mechanics and Computation, Durand 262, Stanford University, Stanford, CA 94305, USA

tting@uic.edu

Abstract. Explicit conditions are given for the existence of an exceptional body wave in monoclinic materials with the symmetry plane at $x_1 = 0$, $x_2 = 0$ or $x_3 = 0$ and in orthotropic materials with the symmetry planes coinciding with the coordinate planes. The non-existence of an exceptional body wave ensures the existence of a subsonic surface wave. If an exceptional body wave exists, explicit conditions are given for the existence of a subsonic surface wave in monoclinic and orthotropic materials except when $tr\mathbf{L}(\hat{v})$ needs to be computed.

1. Introduction

It is known [1-5] that a subsonic surface (Rayleigh) wave exists in an anisotropic elastic half-space if the limiting wave, which is a body wave, is not exceptional. If the limiting wave is exceptional but the transonic state is not of Type 1 (see Section 4), a subsonic surface wave exists. If the limiting wave is exceptional and the transonic state is of Type 1, a subsonic surface wave exists when $tr\mathbf{L}(\hat{v}) \leq 0$. $\mathbf{L}(\hat{v})$ is one of three Barnett-Lothe tensors [6,7] where $\hat{v}$ is the limiting wave speed. Explicit conditions for the existence of an exceptional body wave for a general anisotropic elastic material have been presented by Chadwick and Smith [8] in terms of the elastic stiffness $C_{\alpha\beta}$. Using a different approach Ting [9] recently obtained the conditions in terms of the reduced elastic compliance $s'_{\alpha\beta}$. The existence conditions in terms of $s'_{\alpha\beta}$ appear to have simpler expressions. We apply the conditions presented in [9] to monoclinic materials with the symmetry plane at one of the coordinate planes. We also consider the special case of orthotropic materials with the symmetry planes coinciding with the coordinate planes. If an

T.-T. Wu and C.-C. Ma (eds.), *IUTAM Symposium on Recent Advances of Acoustic Waves in Solids*, IUTAM Bookseries 26, DOI 10.1007/978-90-481-9893-1_1,

exceptional body wave does not exist, a subsonic surface wave exists. If an exceptional body wave exists, conditions are given for identifying whether the transonic state is of Type 1. If it is not of Type 1, a subsonic surface wave exists. If it is of Type 1, a subsonic surface wave exists when $tr\mathbf{L}(\hat{\upsilon}) \leq 0$. Hence, through the existence of an exceptional body wave we provide explicit conditions for the existence of a subsonic surface wave in monoclinic and orthotropic materials with the exception when $tr\mathbf{L}(\hat{\upsilon})$ needs to be computed.

2. Basic Equations

In a fixed rectangular coordinate system x_i (i=1, 2, 3), the equation of motion is

$$\sigma_{ij,j} = \rho \ddot{u}_i , \tag{2.1}$$

where σ_{ij} is the stress, u_i is the displacement, ρ is mass density, the dot denotes differentiation with time t and a comma denotes differentiation with x_i . The stress-strain relation is

$$\sigma_{ij} = C_{ijks} u_{k,s} , \tag{2.2}$$

$$C_{ijks} = C_{jiks} = C_{ksij} = C_{ijsk} , \tag{2.3}$$

in which C_{ijks} is the elastic stiffness. The C_{ijks} is positive definite and possesses the full symmetry shown in (2.3). The third equality in (2.3) is redundant because the first two imply the third ([10], *p*.32).

For a steady state motion in the x_1 -direction with a constant wave speed $\upsilon>0$ in the half-space $x_2 \geq 0$, a general solution for the displacement $\mathbf{u}$ in (2.1) and (2.2) is

$$\mathbf{u} = \mathbf{a} e^{ikz} , \quad z = x_1 + p x_2 - \upsilon t , \tag{2.4}$$

where k >0 is the real wave number, and p and $\mathbf{a}$ satisfy the equation [5,10,11]

$$[\mathbf{Q} - X\mathbf{I} + p(\mathbf{R} + \mathbf{R}^T) + p^2 \mathbf{T}]\,\mathbf{a} = \mathbf{0} , \quad X = \rho \upsilon^2 . \tag{2.5}$$

The superscript T denotes the transpose, $\mathbf{I}$ is the identity matrix and

$$Q_{ik} = C_{i1k1} , \quad R_{ik} = C_{i1k2} , \quad T_{ik} = C_{i2k2} . \tag{2.6}$$

Introducing the vector

$$\mathbf{b} = (\mathbf{R}^T + p\mathbf{T})\mathbf{a} = -p^{-1}(\mathbf{Q} - X\mathbf{I} + p\mathbf{R})\mathbf{a}, \tag{2.7}$$

in which the second equality follows from (2.5), the stress computed from (2.4) and (2.2) can be written as

$$\sigma_{i1} = Xu_{i,1} - \varphi_{i,2}, \sigma_{i2} = \varphi_{i,1}, \tag{2.8}$$

where the vector

$$\varphi = \mathbf{b}\, e^{ikz}, \tag{2.9}$$

is the stress function. The two equations in (2.7) can be written as

$$(\mathbf{N} - p\mathbf{I})\xi = \mathbf{0}, \tag{2.10}$$

in which [12]

$$\mathbf{N} = \begin{bmatrix} \mathbf{N}_1 & \mathbf{N}_2 \\ \mathbf{N}_3 + X\mathbf{I} & \mathbf{N}_1^T \end{bmatrix}, \; \xi = \begin{bmatrix} \mathbf{a} \\ \mathbf{b} \end{bmatrix}, \tag{2.11}$$

$$\mathbf{N}_1 = -\mathbf{T}^{-1}\mathbf{R}^T, \; \mathbf{N}_2 = \mathbf{T}^{-1}, \; \mathbf{N}_3 = \mathbf{R}\mathbf{T}^{-1}\mathbf{R}^T - \mathbf{Q}. \tag{2.12}$$

The matrix $\mathbf{N}_2$ is symmetric and positive definite while $-\mathbf{N}_3$ is symmetric and positive semi-definite [13, 10].

A steady wave is a body wave when p is real. A body wave is *exceptional* with respect to the surface $x_2 = 0$ if the wave produces no traction on the surface [14]. Otherwise the body wave is *normal.* It is shown in [8, 9, 15] that an exceptional body wave with respect to the surface $x_2 = 0$ is necessarily a transonic wave. The converse need not hold. A transonic wave is a body wave that need not be exceptional.

3. Exceptional Body Wave

The explicit expression of $\mathbf{N}_1$, $\mathbf{N}_2$, $\mathbf{N}_3$ in (2.12) was given by Ting [13] in terms of the elastic compliance $s_{\alpha\beta}$ and by Barnett and Chadwick [16] in terms of the elastic stiffness $C_{\alpha\beta}$. The $s_{\alpha\beta}$ and $C_{\alpha\beta}$ are the contracted notations of S_{ijks} and C_{ijks}, respectively. $\mathbf{N}_1$, $\mathbf{N}_2$, $\mathbf{N}_3$ have complicated expression in terms of $s_{\alpha\beta}$ or $C_{\alpha\beta}$. For the problems we will study here, an equation alternate to (2.10) is more convenient. Using a modified Lekhnitskii formalism Ting [17] (see also [9]) obtained the following alternate equation

$$\Gamma\,\xi = \mathbf{0}, \tag{3.1}$$

where

$$\Gamma = \begin{bmatrix} Xs'_{61}-p & -1 & Xs'_{65} & s'_{66}-ps'_{61} & s'_{62} & s'_{64}-ps'_{65} \\ Xs'_{21} & -p & Xs'_{25} & s'_{26}-ps'_{21} & s'_{22} & s'_{24}-ps'_{25} \\ Xs'_{41} & 0 & Xs'_{45}-p & s'_{46}-ps'_{41} & s'_{42} & s'_{44}-ps'_{45} \\ Xs'_{11}-1 & 0 & Xs'_{15} & s'_{16}-ps'_{11} & s'_{12} & s'_{14}-ps'_{15} \\ 0 & X & 0 & -1 & -p & 0 \\ Xs'_{51} & 0 & Xs'_{55}-1 & s'_{56}-ps'_{51} & s'_{52} & s'_{54}-ps'_{55} \end{bmatrix}, \tag{3.2}$$

$$s'_{\alpha\beta} = s_{\alpha\beta} - \frac{s_{\alpha 3}s_{3\beta}}{s_{33}}. \tag{3.3}$$

$s'_{\alpha\beta}$ is the *reduced elastic compliance*. It is symmetric so that $s'_{\alpha\beta} = s'_{\beta\alpha}$.
The fifth equation implicit in (3.1) gives

$$Xa_2 = b_1 + pb_2. \tag{3.4}$$

Elimination of a_2 in (3.1) using (3.4) yields

$$\hat{\Gamma}\,\hat{\xi} = \mathbf{0}, \tag{3.5}$$

where

$$\hat{\Gamma} = \begin{bmatrix} Xs'_{16}-p & Xs'_{56} & Xs'_{66}-1-p^2 & Xs'_{26}-p & Xs'_{46} \\ Xs'_{12} & Xs'_{25} & Xs'_{26}-p & Xs'_{22}-p^2 & Xs'_{24} \\ Xs'_{14} & Xs'_{45}-p & Xs'_{46} & Xs'_{24} & Xs'_{44}-p^2 \\ Xs'_{11}-1 & Xs'_{15} & Xs'_{16}-p & Xs'_{12} & Xs'_{14} \\ Xs'_{15} & Xs'_{55}-1 & Xs'_{56} & Xs'_{25} & Xs'_{45}-p \end{bmatrix}, \tag{3.6}$$

$$\hat{\xi}^T = [a_1 - pX^{-1}b_1,\ \ a_3 - pX^{-1}b_3,\ \ X^{-1}b_1,\ \ X^{-1}b_2,\ \ X^{-1}b_3]. \tag{3.7}$$

It can be shown that

$$|\Gamma| = -X^{-2}\left|\hat{\Gamma}\right|. \tag{3.8}$$

In (3.6), the 3×2 matrix on the upper left corner and the 2×3 matrix on the lower right corner are the transpose of each other. The 3×3 matrix on the upper right corner and the 2×2 matrix on the lower left corner are symmetric.

For an exceptional body wave with respect to the surface $x_2 = 0$,

$$b_1 = b_2 = b_3 = 0 . \tag{3.9}$$

Equation (3.4) and (3.5) reduce to

$$a_2 = 0 , \tag{3.10}$$

$$\frac{a_3}{a_1} = \frac{Xs'_{15}}{1 - Xs'_{55}} = \frac{1 - Xs'_{11}}{Xs'_{15}} = \frac{-s'_{12}}{s'_{25}} = \frac{(p / X) - s'_{16}}{s'_{56}} = \frac{s'_{14}}{(p / X) - s'_{45}} . \tag{3.11}$$

The second equality in (3.11) gives

$$X = \frac{(s'_{11} + s'_{55}) \pm \sqrt{\eta}}{2(s'_{11}s'_{55} - s'_{15}s'_{15})} , \tag{3.12}$$

$$\eta = (s'_{11} + s'_{55})^2 - 4(s'_{11}s'_{55} - s'_{15}s'_{15}) = (s'_{11} - s'_{55})^2 + (2s'_{15})^2 \geq 0. \tag{3.13}$$

Equation (3.12) provides two possible wave speeds for an exceptional body wave. The last equality in (3.11) leads to

$$\frac{p}{X} = \frac{1}{2}\left[(s'_{16} + s'_{45}) \pm \sqrt{\kappa} \right] , \tag{3.14}$$

$$\kappa = (s'_{16} - s'_{45})^2 + 4s'_{14}s'_{56} \geq 0. \tag{3.15}$$

κ cannot be negative because p has to be real for a body wave.

When (3.11) does not hold, an exceptional body wave does not exist. This means that a subsonic surface wave exists. If (3.11) holds, a subsonic surface wave exists if the first transonic state is not of Type 1.

4. First Transonic State of Type 1

The p in (2.4) for the inhomogeneous plane wave is complex when the wave speed υ is very small. As υ increases, p becomes real at the *limiting wave speed* $\hat{\upsilon}$. In the (x_1, x_2)-plane, a vertical line with a distance υ^{-1} to the x_2-axis does not intersect the slowness curves when υ^{-1} is very large. The vertical line touches the slowness curve for the first time when the distance to the x_2-axis is $\hat{\upsilon}^{-1}$. It is the *first transonic wave*, and is also called *the limiting wave*. Since p is real, it is a body wave. If we continue to move the vertical line towards the x_2 -axis, the line will touch the slowness curves for the second time. It is the second transonic wave. The number of transonic waves is at least three and at most fifteen [8]. An exceptional body wave is a transonic wave but a transonic wave need not be exceptional. If an exceptional body wave is the limiting wave, the limiting wave is *exceptional.*

At the limiting wave speed $\hat{\upsilon}$ there may be more than one limiting waves. All waves at the limiting wave speed $\hat{\upsilon}$ are referred as the *first transonic state*. According to Chadwick and Smith [5] (see also [9]), there are six types of first transonic state depending on whether the vertical line touches the slowness curves at one, two or three different points or at the same point.

The first transonic state is of Type 1 if the vertical line touches only one of the slowness curves at only one point. Let $\hat{p}_1$ be the p for this limiting wave which is real. The determinant of the 6×6 matrix $\mathbf{\Gamma}$ shown in (3.2) must have the expression

$$|\Gamma| = (p - \hat{p}_1)^2 K_4(p). \tag{4.1}$$

In the above $K_4(p)$ is a quartic equation in p that has no real roots. If $K_4(p)$ has a real root, it is not of Type 1. It can be of other types or a subsequent transonic state.

5. Monoclinic Materials with the Symmetry Plane at x_1=0

When the material has a symmetry plane at $x_1 = 0$ we have

$$s'_{15} = s'_{16} = s'_{25} = s'_{26} = s'_{45} = s'_{46} = 0. \tag{5.1}$$

Equation (3.11) reduces to

$$\frac{a_3}{a_1} = \frac{0}{1 - Xs'_{55}} = \frac{1 - Xs'_{11}}{0} = \frac{-s'_{12}}{0} = \frac{p/X}{s'_{56}} = \frac{s'_{14}}{p/X}. \tag{5.2}$$

By inspection, (5.2) does not hold if one of the following holds,

$$s'_{12}s'_{56} \neq 0, \tag{5.3a}$$

$$s'_{14}s'_{56} < 0, \tag{5.3b}$$

$$s'_{12} = 0,\ s'_{14}s'_{56} > 0,\ s'_{11} \neq s'_{55}. \tag{5.3c}$$

When (5.3a), (5.3b) or (5.3c) holds, an exceptional body wave does not exist so that a subsonic surface wave exists.

We next consider several cases for which an exceptional body wave exists.

Case I-1. When

$$s'_{56} = 0, \tag{5.4}$$

an exceptional body wave exists with $X = 1/s'_{55}$, $p = 0$ and $a_1 = 0$. Equation (3.8) simplifies to

$$|\Gamma| = p^2 K_4(p) \tag{5.5}$$

where

$$K_4(p) = \frac{1}{X^2}\begin{vmatrix} -p & Xs'_{66} - 1 - p^2 & -p \\ Xs'_{12} & -p & Xs'_{22} - p^2 \\ Xs'_{11} - 1 & -p & Xs'_{12} \end{vmatrix}, \quad X = 1/s'_{55}. \tag{5.6}$$

$K_4(p)$ is a quadratic equation in p^2 that can be written as

$$K_4(p) = \gamma[p^4 - 2\alpha p^2 + \beta], \tag{5.7}$$

where γ, α, β are constants. It has a real root for p when

$$\alpha^2 \geq \beta > 0, \ \alpha > 0 \tag{5.8a}$$

or

$$\beta \leq 0. \tag{5.8b}$$

Thus, when (5.8a) or (5.8b) holds, the transonic state is not of Type 1 so that a subsonic surface wave exists.

Case I-2. When

$$s'_{12} = s'_{14} = 0, \tag{5.9}$$

an exceptional body wave exists with $X = 1/s'_{11}$, $p = 0$ and $a_3 = 0$. Equation (3.8) simplifies to (5.5) in which

$$K_4(p)=\frac{1}{X^2}\begin{vmatrix} 0 & Xs'_{22}-p^2 & Xs'_{24} \\ -p & Xs'_{24} & Xs'_{44}-p^2 \\ Xs'_{55}-1 & 0 & -p \end{vmatrix}, \quad X=1/s'_{11}. \tag{5.10}$$

Again, $K_4(p)$ is a quadratic equation in p^2 shown in (5.7). If (5.8a) or (5.8b) holds, the transonic state is not of Type 1 so that a subsonic surface wave exists.

Case I-3. When

$$s'_{12}=s'_{11}-s'_{55}=0\,,\quad s'_{14}s'_{56}>0\,, \tag{5.11}$$

an exceptional body wave exists with $X=1/s'_{11}=1/s'_{55}$, $p/X=\pm\sqrt{s'_{14}s'_{56}}$ and $a_3/a_1=\pm\sqrt{s'_{14}/s'_{56}}$. Equation (3.8) reduces to

$$|\Gamma| = X^{-2}(p^2-X^2 s'_{14}s'_{56})^2(p^2-Xs'_{22})\,. \tag{5.12}$$

The transonic state is not of Type 1 so that a subsonic surface wave exists.

Case I-4. Consider the case

$$s'_{12}=s'_{14}=s'_{56}=s'_{11}-s'_{55}=0\,. \tag{5.13}$$

This is a degenerate case of Case I-3 for which $X=1/s'_{11}=1/s'_{55}$, $p=0$, and a_3/a_1 is arbitrary. Equation (5.12) simplifies to

$$|\Gamma| = X^{-2}p^4(p^2-Xs'_{22})\,. \tag{5.14}$$

A subsonic surface exists for this case.

6. Monoclinic Materials with the Symmetry Plane at x_2=0

When the material has a symmetry plane at $x_2=0$ we have

$$s'_{14}=s'_{16}=s'_{24}=s'_{26}=s'_{45}=s'_{56}=0\,. \tag{6.1}$$

Equation (3.11) reduces to

$$\frac{a_3}{a_1}=\frac{Xs'_{15}}{1-Xs'_{55}}=\frac{1-Xs'_{11}}{Xs'_{15}}=\frac{-s'_{12}}{s'_{25}}=\frac{p/X}{0}=\frac{0}{p/X}\,. \tag{6.2}$$

Equation (6.2) does not hold if one of the following holds,

$$s'_{25} = 0\,,\ s'_{12}s'_{15} \neq 0\,, \tag{6.3a}$$
$$s'_{12} = 0\,,\ s'_{15}s'_{25} \neq 0\,, \tag{6.3b}$$
$$s'_{12}s'_{25} \neq 0\,,\ s'_{12}(s'_{11}s'_{25} - s'_{12}s'_{15}) \neq s'_{25}(s'_{12}s'_{55} - s'_{15}s'_{25})\,. \tag{6.3c}$$

We will prove (6.3c) below. When (6.3a), (6.3b) or (6.3c) holds, an exceptional body wave does not exist so that a subsonic surface wave exists.

We next consider several cases for which an exceptional body wave exists.

Case II-1. When

$$s'_{15} = s'_{25} = 0\,, \tag{6.4}$$

an exceptional body wave exists with $X = 1/s'_{55}$, $p = 0$ and $a_1 = 0$. Equation (3.8) simplifies to (5.5) and (5.6). Thus the discussion following (5.7) applies here.

Case II-2. When

$$s'_{12} = s'_{15} = 0\,, \tag{6.5}$$

an exceptional body wave exists with $X = 1/s'_{11}$, $p = 0$ and $a_3 = 0$. Equation (3.8) simplifies to (5.5) in which

$$K_4(p) = \frac{1}{X^2}\begin{vmatrix} Xs'_{25} & Xs'_{22} - p^2 & 0 \\ -p & 0 & Xs'_{44} - p^2 \\ Xs'_{55} - 1 & Xs'_{25} & -p \end{vmatrix},\quad X = 1/s'_{11}\,. \tag{6.6}$$

Again, $K_4(p)$ is a quadratic equation in p^2 shown in (5.7) so that the discussion following (5.7) applies here.

Case II-3. When

$$s'_{12} = s'_{25} = s'_{15} = s'_{11} - s'_{55} = 0\,, \tag{6.7}$$

an exceptional body wave exists with $p = 0$, $X = 1/s'_{11} = 1/s'_{55}$ and a_3/a_1 is arbitrary. Equation (3.8) simplifies to (5.14) so that a subsonic surface wave exists.

Case II-4. When

$$s'_{12}s'_{25} \neq 0\,, \tag{6.8}$$

the second the third equalities in (6.2) can be solved for X as

$$\frac{1}{X} = \frac{1}{s'_{25}}(s'_{11}s'_{25} - s'_{12}s'_{15}) = \frac{1}{s'_{12}}(s'_{12}s'_{55} - s'_{15}s'_{25})\,. \tag{6.9}$$

Thus an exceptional body wave exists if (6.8) and

$$s'_{12}(s'_{11}s'_{25} - s'_{12}s'_{15}) = s'_{25}(s'_{12}s'_{55} - s'_{15}s'_{25}) \tag{6.10}$$

hold. This proves the earlier statement that an exceptional body wave does not exist if (6.3c) holds. When (6.8) and (6.10) hold, X is given by (6.9), $p=0$ and $a_3/a_1 = -s'_{12}/s'_{25}$.

The determinant in (3.8) reduces to

$$|\Gamma| = \frac{-1}{X^2}\begin{vmatrix} -p & 0 & Xs'_{66}-1-p^2 & -p & Xs'_{46} \\ Xs'_{12} & Xs'_{25} & -p & Xs'_{22}-p^2 & 0 \\ 0 & -p & Xs'_{46} & 0 & Xs'_{44}-p^2 \\ Xs'_{11}-1 & Xs'_{15} & -p & Xs'_{12} & 0 \\ Xs'_{15} & Xs'_{55}-1 & 0 & Xs'_{25} & -p \end{vmatrix}. \tag{6.11}$$

Let

$$\frac{Xs'_{15}}{1-Xs'_{55}} = \frac{1-Xs'_{11}}{Xs'_{15}} = \frac{-s'_{12}}{s'_{25}} = w\,,\ \text{say}. \tag{6.12}$$

This means

$$Xs'_{15} + w(Xs'_{55}-1) = 0\,,\ \ (Xs'_{11}-1) + wXs'_{15} = 0\,,\ \ s'_{12} + ws'_{25} = 0\,. \tag{6.13}$$

The determinant in (6.11) remains the same if we add to column one w times column two and add to row four w times row five. Using (6.13) we have

$$|\Gamma| = \frac{-1}{X^2}\begin{vmatrix} -p & 0 & E_{11} & E_{12} & E_{13} \\ 0 & Xs'_{25} & E_{12} & E_{22} & 0 \\ -pw & -p & E_{13} & 0 & E_{33} \\ 0 & 0 & -p & 0 & -pw \\ 0 & Xs'_{55}-1 & 0 & Xs'_{25} & -p \end{vmatrix}. \tag{6.14}$$

In the above, E_{ij} is the 3×3 matrix at the upper right corner of the determinant $|\Gamma|$ in (6.11), i.e.,

$$\mathbf{E} = \begin{bmatrix} Xs'_{66}-1-p^2 & -p & Xs'_{46} \\ -p & Xs'_{22}-p^2 & 0 \\ Xs'_{46} & 0 & Xs'_{44}-p^2 \end{bmatrix}. \tag{6.15}$$

The determinant $|\Gamma|$ in (6.14) can be written in the form of (5.5) in which

$$K_4(p) = X^{-2}\left((Xs'_{55}-1)[E_{22}E_{33}-2wE_{13}E_{22}+w^2(E_{11}E_{22}-E_{12}E_{12})]-\mathbf{y}^T\mathbf{E}\mathbf{y}\right), \tag{6.16a}$$

$$\mathbf{y}^T = [Xs'_{12},\ \ p,\ \ Xs'_{25}]. \tag{6.16b}$$

$K_4(p)$ in (6.16a) is a quadratic equation in p^2 in the form of (5.7) so that the discussion following (5.7) applies here.

Case II-5. When

$$s'_{12} = s'_{25} = 0, \tag{6.17}$$

an exceptional body wave exists in which X is given by (3.12) and $p = 0$. There are two solutions for the X. For each X, a_3 / a_1 is provided by the first equality in (6.2). The determinant in (3.8) reduces to (6.11) with $s'_{12} = s'_{25} = 0$. Following the derivation from (6.12) to (6.16), it can be shown that $|\Gamma|$ has the form of (5.5) in which $K_4(p)$ is shown in (6.16a). Since $s'_{12} = s'_{25} = 0$, $\mathbf{y}^T\mathbf{E}\mathbf{y}$ in (6.16b) simplifies to

$$\mathbf{y}^T\mathbf{E}\mathbf{y} = p^2 E_{22}. \tag{6.18}$$

7. Monoclinic Materials with the Symmetry Plane at x_3=0

When the material has a symmetry plane at $x_3 = 0$ we have

$$s'_{14} = s'_{15} = s'_{24} = s'_{25} = s'_{46} = s'_{56} = 0\,. \tag{7.1}$$

Equation (3.11) reduces to

$$\frac{a_3}{a_1} = \frac{0}{1 - Xs'_{55}} = \frac{1 - Xs'_{11}}{0} = \frac{-s'_{12}}{0} = \frac{(p/X) - s'_{16}}{0} = \frac{0}{(p/X) - s'_{45}}\,. \tag{7.2}$$

It is not difficult to see that (7.2) can be satisfied for any choice of $s'_{\alpha\beta}$. Hence an exceptional body wave always exists for monoclinic materials with the symmetry plane at $x_3 = 0$. We now consider several special cases below.

Case III-1. When

$$s'_{12} \neq 0, \tag{7.3}$$

we have $X = 1/s'_{55}$, $p/X = s'_{45}$ and $a_1 = 0$. Equation (3.8) reduces to

$$|\Gamma| = (p - Xs'_{45})^2 K_4(p)\,, \tag{7.4}$$

$$K_4(p) = \frac{1}{X^2}\begin{vmatrix} Xs'_{16} - p & Xs'_{66} - 1 - p^2 & Xs'_{26} - p \\ Xs'_{12} & Xs'_{26} - p & Xs'_{22} - p^2 \\ Xs'_{11} - 1 & Xs'_{16} - p & Xs'_{12} \end{vmatrix}, \quad X = 1/s'_{55}\,. \tag{7.5}$$

$K_4(p)$ is a quartic equation in *p*. If it has a real root, the transonic state is not of Type 1, and a subsonic surface wave exists.

Case III-2. When

$$s'_{12} = 0\,, \tag{7.6}$$

we have either $X = 1/s'_{55}$ or $X = 1/s'_{11}$. For $X = 1/s'_{55}$, the result is identical to Case III-1 with $s'_{12} = 0$ in (7.5). For $X = 1/s'_{11}$, we have $p/X = s'_{16}$ and $a_3 = 0$. Equation (3.8) reduces to

$$|\Gamma| = X^{-2}(p - Xs'_{16})^2 (p^2 - Xs'_{22})[(p^2 - Xs'_{44})(Xs'_{55} - 1) + (p - Xs'_{45})^2] \tag{7.7}$$

This is in the form of (4.1) in which $K_4(p)$ has a real root. Hence the transonic state is not of Type 1, and a subsonic surface wave exists.

Case III-3. When

$$s'_{12} = s'_{11} - s'_{55} = 0\,, \tag{7.8}$$

we have $X = 1/s'_{11} = 1/s'_{55}$ and either $p/X = s'_{45}$ and $a_1 = 0$ or $p/X = s'_{16}$ and $a_3 = 0$. Equation (3.8) reduces to

$$|\Gamma| = X^{-2}(p - Xs'_{16})^2(p - Xs'_{45})^2(p^2 - Xs'_{22})\,. \tag{7.9}$$

This is in the form of (4.1) in which $K_4(p)$ has a real root. Hence the transonic state is not of Type 1, and a subsonic surface wave exists.

Case III-4. When

$$s'_{12} = s'_{16} - s'_{45} = s'_{11} - s'_{55} = 0\,, \tag{7.10}$$

we have $X = 1/s'_{11} = 1/s'_{55}$, $p/X = s'_{16} = s'_{45}$ and a_3/a_1 is arbitrary. Equation (3.8) reduces to (7.9) with $s'_{16} = s'_{45}$ so that the statement following (7.9) applies here.

In summary, the existence of a subsonic surface wave is assured except when $s'_{12} \neq 0$ and the quartic equation (7.5) has no real roots.

8. Orthotropic Materials

Orthotropic materials are special monoclinic materials. We could specialize the results obtained in the previous section for monoclinic materials with the symmetry plane at $x_3 = 0$ by adding the condition

$$s'_{16} = s'_{26} = s'_{45} = 0\,. \tag{8.1}$$

Equation (7.2) reduces to

$$\frac{a_3}{a_1} = \frac{0}{1 - Xs'_{55}} = \frac{1 - Xs'_{11}}{0} = \frac{-s'_{12}}{0} = \frac{p/X}{0} = \frac{0}{p/X}\,. \tag{8.2}$$

Equation (8.2) can be satisfied for any choice of $s'_{\alpha\beta}$. Hence an exceptional body wave always exists for orthotropic materials with the symmetry planes coinciding with the coordinate planes. We now specialize Cases III-1 through III-5 below.

Case IV-1. When

$$s'_{12} \neq 0, \tag{8.3}$$

we have $X = 1/s'_{55}$, $p = 0$ and $a_1 = 0$. Equations (7.4) and (7.5) apply here with $s'_{45} = 0$ and $K_4(p)$ reduces to

$$K_4(p) = \frac{1}{X^2}\begin{vmatrix} -p & Xs'_{66} - 1 - p^2 & -p \\ Xs'_{12} & -p & Xs'_{22} - p^2 \\ Xs'_{11} - 1 & -p & Xs'_{12} \end{vmatrix}, \quad X = 1/s'_{55}. \tag{8.4}$$

$K_4(p)$ is a quadratic equation in p^2 in the form of (5.7) so that the discussion following (5.7) applies here.

Case IV-2. When

$$s'_{12} = 0, \tag{8.5}$$

we have either $X = 1/s'_{55}$ or $X = 1/s'_{11}$. For $X = 1/s'_{55}$, the result is identical to Case IV-1 with $s'_{12} = 0$ in (8.4). For $X = 1/s'_{11}$, we have $p = 0$ and $a_3 = 0$. Equation (7.7) reduces to

$$|\Gamma| = X^{-2} p^2 (p^2 - Xs'_{22})[(p^2 - Xs'_{44})(Xs'_{55} - 1) + p^2]. \tag{8.6}$$

This is in the form of (4.1) in which $K_4(p)$ has a real root. Hence the transonic state is not of Type 1, and a subsonic surface wave exists.

Case IV-3. When

$$s'_{12} = s'_{11} - s'_{55} = 0, \tag{8.7}$$

we have $X = 1/s'_{11} = 1/s'_{55}$, $p = 0$ and a_3/a_1 is arbitrary. Equation (7.9) with $s'_{16} = s'_{45} = 0$ applies here.

In summary, the existence of a subsonic surface wave is assured except when $s'_{12} \neq 0$ and the quadratic equation (8.4) has no real roots.

9. Conclusion

We have presented explicit conditions for the existence of an exceptional body wave in monoclinic and orthotropic materials. This approach is not necessarily the best way to determine if a subsonic surface wave exists. There have been studies on the explicit expression of the secular equation for the surface wave speed $\upsilon = \sqrt{X / \rho}$ of a surface wave in an anisotropic elastic material (see [19-22] and the references therein). The secular equation is a polynomial in X of degree three for cubic, hexagonal and orthotropic materials, of degree four for monoclinic materials with the symmetry plane at $x_3 = 0$ or at $x_1 = 0$ and of degree five for monoclinic materials with the symmetry plane at $x_2 = 0$. One could study if a real and positive root for X exists from the secular equation. The present approach is simplest in determining the existence of a subsonic surface wave when an exception body wave does not exist.

It should be noted that the analysis presented in Section 3 is related to the analysis of one-component waves in plates [22] and one-component Rayleigh waves, Stoneley waves, Love waves, slip waves and waves in a layered plate [23].

References

[1] Barnett, D. M., and Lothe, J.: Consideration of the existence of surface wave (Rayleigh wave) solutions in anisotropic elastic crystals. *J. Phys. F.* **4** 671-686 (1974).

[2] Barnett, D. M., and Lothe, J.: The existence of Rayleigh (surface) waves solutions in anisotropic elastic half-spaces. In J. Miklowitz and J. Achenbach, editors, *Modern Problems in Elastic Wave Propagation*, Wiley, New York, 445-457, (1978).

[3] Barnett, D. M., and Lothe, J.: Free surface (Rayleigh) waves in anisotropic elastic half-spaces, The surface impedance methods. *Proc R. Soc. Lond.* **A402** 135-152 (1985).

[4] Lothe, J., and Barnett, D. M.: On the existence of surface-wave solutions for anisotropic half-space with free surfaces. *J. Appl. Phys.* **47** 428-433 (1976).

[5] Chadwick, P., and Smith, G. D.: Foundations of the theory of surface waves in anisotropic elastic materials. *Adv. Appl. Mech.* **17** 303-376 (1977).

[6] Barnett, D. M., and Lothe, J.: Synthesis of the sextic and the integral formalism for dislocations, Greens function and surface waves in anisotropic elastic solids. *Phys. Norv.* **7** 13-19 (1973).

[7] Chadwick, P., and Ting, T. C. T.: On the structure and invariance of the Barnett-Lothe tensors. *Q. Appl. Math.* **45** 419-427 (1987).

[8] Chadwick, P.: A general analysis of transonic states in an anisotropic elastic body. *Proc. R. Soc. Lond.* **A401** 203-223 (1985).

[9] Ting, T. C. T.: Explicit conditions for the existence of exceptional body waves and subsonic surface waves in anisotropic elastic solids. *Wave Motion* **46** 323-335 (2009).

[10] Ting, T. C. T.: *Anisotropic Elasticity: Theory and Applications*, Oxford University Press, New York (1996).

[11] Stroh, A. N.: Steady state problems in anisotropic elasticity. *J. Math. Phys.* **4** 77-103 (1962).

[12] Ingebrigtsen, K. A., and Tonning, A.: Elastic surface waves in crystal. *Phys Rev.* **184** 942-951 (1969).

[13] Ting, T. C. T.: Some identities and the structure of $\mathbf{N}_i$ in the Stroh formalism of anisotropic elasticity. *Q. Appl. Math.* **46** 109-120 (1988).

[14] Alshits, V. I., and Lothe, J.: Elastic waves in triclinic crystals. III. the problem of exceptional surface waves and some of their general properties. *Sov. Phys. Crystallography*, **24**(6) 644-648 (1979).

[15] Chadwick, P., and Smith, G. D.: Surface waves in cubic elastic materials. In H. G. Hopkins and M. J. Sewell editors, *Mechanics of Solids, The Rodney Hill 60th Anniversary Volume*, Oxford, 47-100 (1982).

[16] Barnett, D. M., and Chadwick, P.: The existence of one-component surface waves and exceptional subsequent transonic states of types 2, 4, and E1 in anisotropic elastic media. In J. J. Wu, T. C. T. Ting and D. M. Barnett, editors, *Modern Theory of Anisotropic Elasticity and Applications*, SIAM, Philadelphia, 199-214 (1991).

[17] Ting, T. C. T.: A new modified Lekhnitskii formalism *à la* Stroh for steady-state waves in anisotropic elastic materials. *Wave Motion* **32** 125-140 (2000).

[18] Dongye, C., and Ting, T. C. T.: Explicit expressions of Barnett-Lothe tensors and their associated tensors for orthotropic materials. *Q. Appl. Math.* **47** 723-734 (1989).

[19] Ting, T. C. T.: An explicit secular equation for surface waves in an elastic material of general anisotropy. *Q. J. Mech. Math.* **55** 297-311 (2002).

[20] Ting, T. C. T.: Explicit secular equations for surface waves in an anisotropic elastic half-space – from Rayleigh to today. In Robert V. Goldstein and Gerard A. Maugin, editors, *Proc. NATO Workshop on Surface Waves in Anisotropic and Laminated Bodies and Defects Detection*, Kluwer, Dordrecht, Netherlands, 95-116 (2004).

[21] Ting, T. C. T.: The polarization vector and secular equation for surface waves in an anisotropic elastic half-space. *Int. J. Solids Structures* **41** 2065-2083 (2004).

[22] Ting, T. C. T.: Dispersion equations for steady waves in an anisotropic elastic plate or a layered plate. *Proc. Roy. S. London* **A464**(2091) 613-629 (2008).

[23] Ting, T. C. T.: Existence of one-component Rayleigh waves, Stoneley waves, Love waves, slip waves and one-component waves in a plate or layered plate. The George Herrmann Special Issue, *J. Mech. Materials and Structures* **4**(4) 631-647 (2009).

Deformation Waves in Microstructured Materials: Theory and Numerics

Jüri Engelbrecht, Arkadi Berezovski and Mihhail Berezovski

Centre for Nonlinear Studies, Institute of Cybernetics at Tallinn University of Technology, Tallinn, Estonia

je@ioc.ee

Abstract. A linear model of the microstructured continuum based on Mindlin theory is adopted which can be represented in the framework of the internal variable theory. Fully coupled systems of equations for macro-motion and microstructure evolution are represented in the form of conservation laws. A modification of wave propagation algorithm is used for numerical calculations. Results of direct numerical simulations of wave propagation in periodic medium are compared with similar results for the continuous media with the modelled microstructure. It is shown that the proper choice of material constants should be made to match the results obtained by both approaches.

1. Introduction

The classical theories of continua describe the behaviour of homogeneous materials. In reality, however, materials are always characterized by a certain microstructure at various scales. The character of a microstructure can be regular (like in laminated composites) or irregular (like in polycrystalline solids or alloys). Even more, regularity and irregularity may be combined like for some FGMs. The characteristic scale of a microstructure must always be compared with the spatial scale of excitation. The choice of proper mathematical models is extremely important in order to describe the wave fields with needed accuracy.

In general terms, the starting point for describing a microstructure could be either the discrete or the continuum approach. In the discrete approach the volume elements are treated as point masses with interaction [1]. Or, especially for laminated composites, the effective stiffness theory has been used [2]. The homogenization methods based on properties and geometry of constituents are widely used

T.-T. Wu and C.-C. Ma (eds.), *IUTAM Symposium on Recent Advances of Acoustic Waves in Solids*, IUTAM Bookseries 26, DOI 10.1007/978-90-481-9893-1_2,

for static and quasi-static problems [3]. From the viewpoint of continua, the straight-forward modelling leads to assigning all the physical properties to every volume element dV in a solid which means introducing the dependence on space coordinates. Thus, the governing equations are so complicated that can be solved only by numerical methods.

Another way is to separate macro- and microstructure in continua. Then the conservation laws for both structures should be formulated separately [4, 5] or in a more sophisticated way the microstructural quantities could be introduced into one set of conservation laws for the macrostructure [6]. Quite recently it has been shown that the generalization of such theories can be obtained by using the concept of dual internal variables [7].

To check the capabilities of the theory, it is useful to compare the theoretical predictions with results of direct numerical simulation of wave propagation through a certain known microstructure. In what follows, the derivation of a microstructure model is presented in the one-dimensional setting. The concept of dual internal variables is applied for the physical description of continua with microstructure. The finite volume wave propagation algorithm is used for both direct numerical simulation and the microstructure modeling. Results of direct numerical simulations of wave propagation in a periodically layered medium are compared with similar results for the homogeneous medium with a modelled microstructure.

2. Governing Equations

The governing equations of thermoelasticity are local balance laws for linear momentum and energy [8]. In the one-dimensional case these governing equations are reduced to (no body forces)

$$\frac{\partial}{\partial t}(\rho_0 v) - \frac{\partial \sigma}{\partial x} = 0, \tag{2.1}$$

$$\frac{\partial}{\partial t}\left(\frac{1}{2}\rho_0 v^2 + E\right) - \frac{\partial}{\partial x}(\sigma v - Q) = 0, \tag{2.2}$$

complemented by the second law of thermodynamics

$$\frac{\partial S}{\partial t} - \frac{\partial}{\partial x}(Q/\theta + K) \geq 0. \tag{2.3}$$

Here t is time, ρ_0 is the matter density, v is the physical velocity, σ is the Cauchy stress, E is the internal energy per unit volume, S is the entropy per unit volume, θ is temperature, Q is the material heat flux, and the "extra entropy flux" K vanishes in most cases, but this is not a basic requirement.

3. Internal Variables

Up to now the microstructure was not specified. In the framework of the phenomenological continuum theory it is assumed that the influence of the microstructure on the overall macroscopic behaviour can be taken into account by the introduction of an internal variable φ, which we associate with the integral distributed effect of the microstructure, and a certain dual internal variable ψ. We suppose that the free energy depends on the internal variables φ, ψ and their space derivatives $W = W^*(u_x, \varphi, \varphi_x, \psi, \psi_x)$. Then the constitutive equations follow

$$\sigma := \frac{\partial W^*}{\partial u_x}, \quad \tau := -\frac{\partial W^*}{\partial \phi}, \quad \eta := -\frac{\partial W^*}{\partial \phi_x}, \quad \xi := -\frac{\partial W^*}{\partial \psi}, \quad \varsigma := -\frac{\partial W^*}{\partial \psi_x}. \tag{3.1}$$

We include into consideration the non-zero extra entropy flux [9]

$$K = -\theta^{-1}\eta\dot{\phi} - \theta^{-1}\varsigma\dot{\xi}. \tag{3.2}$$

It can be checked that the dissipation inequality in the isothermal case reduces to

$$(\tau - \eta_x)\,\dot{\phi} + (\xi - \varsigma_x)\,\dot{\psi} \geq 0. \tag{3.3}$$

In the non-dissipative case the dissipation inequality can be satisfied by the choice

$$\dot{\phi} = m(\xi - \varsigma_x), \quad \dot{\psi} = -m(\tau - \eta_x), \tag{3.4}$$

where m is a coefficient. The latter two evolution equations express the duality between internal variables: one internal variable is driven by another one and vice versa.

The simplest free energy dependence is a quadratic function [10]

$$W^* = \frac{\rho_0 c^2}{2} u_x^2 + A\phi u_x + \frac{1}{2}B\phi^2 + \frac{1}{2}C\phi_x^2 + \frac{1}{2}D\psi^2, \tag{3.5}$$

where A, B C, D, and c are material constants.

Here we include only the contribution of the second internal variable itself. In this case, the evolution equation for the internal variable φ is a hyperbolic equation [7]

$$\ddot{\phi} = m^2 D(\tau - \eta_x) . \tag{3.6}$$

As a result, we can represent the equations of motion in the form

$$\rho_0 u_{tt} = \rho_0 c^2 u_{xx} + A\phi_x, \tag{3.7}$$
$$I\phi_{tt} = C\phi_{xx} - Au_x - B\phi, \tag{3.8}$$

where $I = 1/(m^2 D)$ is an internal inertia measure. In terms of stresses introduced by Eq. (3.1), the same system of equations is represented as

$$\rho_0 \frac{\partial^2 u}{\partial t^2} = \frac{\partial \sigma}{\partial x}, \quad I\frac{\partial^2 \phi}{\partial t^2} = -\frac{\partial \eta}{\partial x} + \tau. \tag{3.9}$$

It is worth to note that same equations are derived in [11] but based on different considerations.

3.1 Single Wave Equation

The governing equations (3.7) and (3.8) can be reduced to one equation. We can determine the first space derivative of the internal variable from Eq. (3.8) and its third derivatives from Eq. (3.7). Inserting the results into the balance of linear momentum (3.7), we obtain a higher order equation [9] with clearly separated wave operators which describe the influence of the microstructure

$$u_{tt} = \left(c^2 - \frac{A}{\rho_0 B}\right) u_{xx} + \frac{C}{B}(u_{tt} - c^2 u_{xx})_{xx} - \frac{I}{B}(u_{tt} - c^2 u_{xx})_{tt}. \tag{3.10}$$

3.2 System of Equations

At the same time, in terms of strain and velocity, Eq. (3.7) is rewritten as

$$\rho_0 v_t = \rho_0 c^2 \varepsilon_x + A\phi_x. \tag{3.11}$$

The particle velocity and the strain are related by the compatibility condition

$$\varepsilon_t = v_x, \tag{3.12}$$

which form the system of equations for these two variables.

Similarly, introducing a microvelocity w as follows:

$$w_x := -D\psi, \tag{3.13}$$

and using Eq. (3.6) with $m = 1$, we have

$$\phi_t = w_x, \tag{3.14}$$

that is nothing else but the compatibility condition at micro-level. It follows from Eqs. (3.14) and (3.8) that

$$I\dot{w}_x = C\phi_{xx} - A\varepsilon - B\phi. \tag{3.15}$$

Integrating the latter equation over x, we arrive at

$$Iw_t = C\phi_x - \int (A\varepsilon + B\phi)dx \tag{3.16}$$

Thus, we have two coupled systems of equations (3.11), (3.12) and (3.14), (3.16) for the determination of four unknowns: ε, v, φ, and w. These two systems of equations are solved numerically to describe the microstructure dynamics.

4. Numerical Simulations

4.1 Algorithm Description

There are many computational methods used to describe wave propagation phenomena (see, e.g. [12]). In our computations we apply a modification of the wave propagation algorithm [13] that was successfully applied to the simulation of wave propagation in inhomogeneous media with rapidly-varying properties [14]. In simulations of wave propagation in a layered medium with known location of inhomogeneities, the numerical scheme is the same as described in [14]. However, the wave propagation algorithm is modified in order to solve the coupled systems of equations in the modelling of the microstructure. This modification is needed to treat the source terms which appeared in equations due to their coupling.

4.2 Linear Waves

As an example, the propagation of a pulse in an one-dimensional medium which can be represented as an elastic bar is analysed. This bar is assumed homogeneous except of a region of length d, where periodically alternating layers of size l are inserted. The density and longitudinal velocity in the bar are chosen as $\rho_0 = 4510$ kg/m^3 and $c = 5240$ m/s, respectively. The corresponding parameters for the material of the inhomogeneity layers are $\rho_l = 2703$ kg/m^3 and $c_l = 5020$ m/s, respectively. The shape of the pulse before the crossing of the inhomogeneity region is formed by an excitation of the strain at the boundary for an limited dimensionless time period ($0 < t < 100$)

$$\varepsilon(0,t) = (1 + \cos(\pi(t-50)/50)) . \tag{4.1}$$

The time step used in calculations is by definition a unit. The length of the pulse $L = 100\,\Delta x$ is comparable with the size of inhomogeneity ($l = 128\,\Delta x$). Using the notion of the bar, it must be stressed that l and L are much smaller than the diameter of the bar [15].

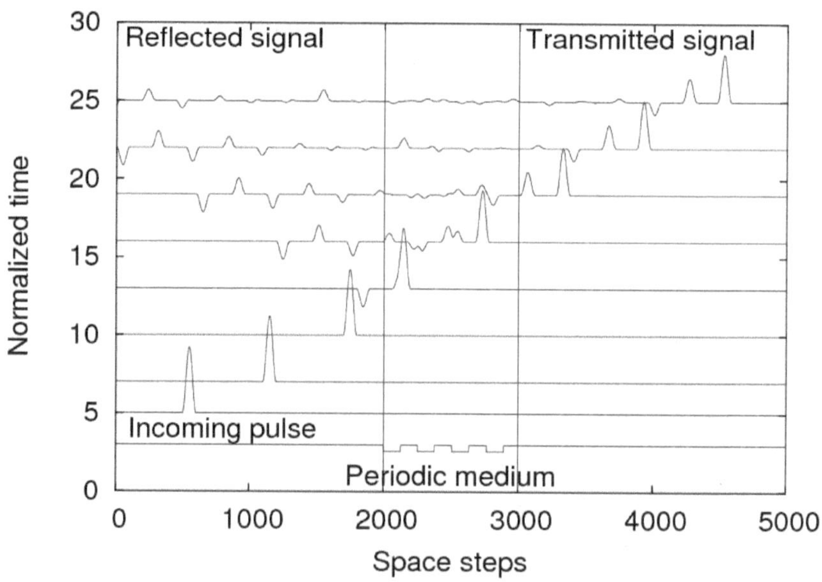

Fig. 4.1 Scattering of a pulse by a periodic multilayer.

Direct numerical simulation of linear elastic wave propagation in the medium with variable properties shows that the pulse holds its shape up to the entering into the inhomogeneity region. After the interaction with the periodic multilayer, the single pulse is separated into many reflected and transmitted parts as it can be seen in Fig. 4.1. Normalized time shown in Fig. 4.1 is measured in hundreds of time steps. During the propagation in the periodic medium, the amplitude of the pulse is diminished due to multiple reflections.

The same pulse propagation was simulated by the microstructured model described above with the following choice of material parameters: $A = 49\ \rho_0 c^2$, $I = \rho_1$, $C = I c_1^2$, $B = 24.6\ A^2 \rho_0 c^2$. In this case, there is no assumption of periodicity of microstructure, however, in calculations of the pulse propagation, the internal length l for the microstructure is kept the same as in the case of periodic multilayer. The ratio of scales l and L together with the value of the parameter A determines the contribution of the microstructure to the macromotion.

Here the coupled systems of equations (3.11), (3.12) and (3.14), (3.16) are solved simultaneously. It should be noted that no boundary conditions for the internal variable are prescribed. A non-zero solution for the microstructure is induced due to the coupling.

Results of numerical simulation are presented in Fig.4.2, where the corresponding transmitted pulses from the solution of the problem with periodic multilayer are also shown.

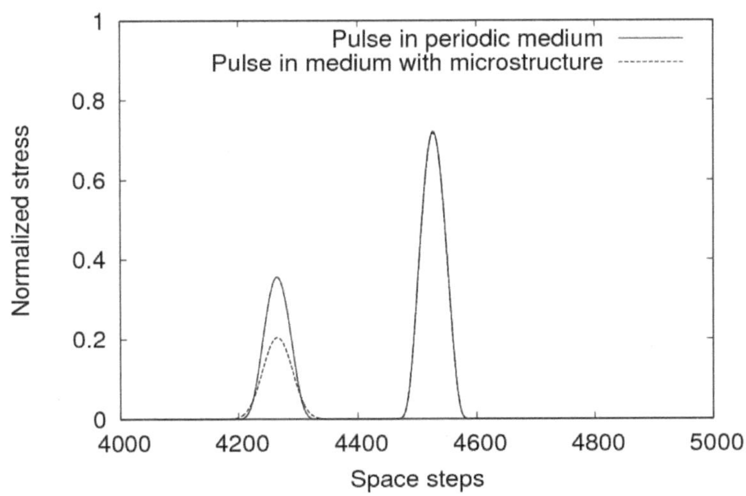

Fig. 4.2 Transmitted pulses.

As one can see, the adjustment of material parameters in the microstructure model allows us to reproduce the first pulse with perfect accuracy while the second one is essentially smaller in amplitude, because of the absence of a reflected trail in the case of the microstructure model.

4.3 Weakly Nonlinear Waves

We consider again the propagation of a pulse in a layered 1D medium (elastic bar) where the length of inhomogeneity $l = 4\,\Delta x$ is much smaller than the length of the pulse $L = 100\,\Delta x$. The properties of materials are the same as previously with a weak nonlinearity for the less stiff material (cf. [16])

$$\sigma = \rho_0 c^2 u_x (1 + N u_x) , \tag{4.2}$$

where N is a parameter of nonlinearity.

Direct computations in this weakly nonlinear case ($N = 0.04$) show that the initial bell-shaped pulse is transformed in a train of soliton-like pulses propagating with amplitude-dependent speeds (Fig. 4.3) like for the celebrated KdV case.

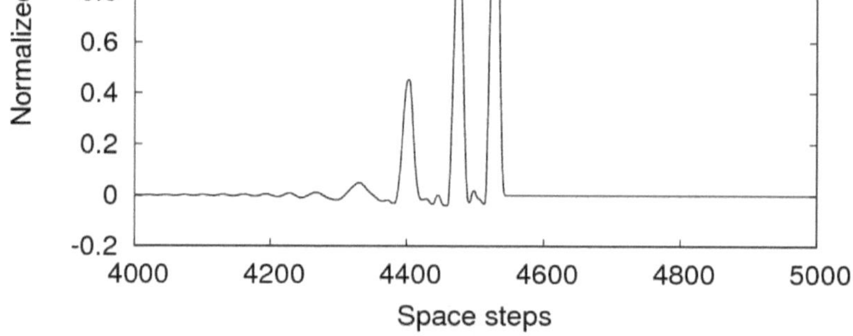

Fig. 4.3 Transformation of a bell-shaped pulse in a weakly nonlinear periodic medium (after 4600 time steps).

If we return to the microstructure model then the linear governing equations (3.7), (3.8) must be modified. Instead of the free energy function (3.5), a cubic function is used:

$$W^* = \frac{\rho_0 c^2}{2} u_x^2 + A\phi u_x + \frac{1}{2} B\phi^2 + \frac{1}{2} C\phi_x^2 + \frac{1}{2} D\psi^2 + \frac{\rho_0 c^2}{6} M\phi^3 + \frac{\rho_0 c^2}{6} N u_x^3, \tag{4.3}$$

where M and N are new material constants (see [17]).

Now the governing equations yield (cf. (3.7), (3.8))

$$\rho_0 u_{tt} = \rho_0 c^2 u_{xx} + \rho_0 c^2 N u_x u_{xx} + A\phi_x, \tag{4.4}$$

$$I\phi_{tt} = C\phi_{xx} + M\phi_x\phi_{xx} - A u_x - B\phi. \tag{4.5}$$

Besides dispersive effects (see [10]), the governing equations (4.4) and (4.5) include also nonlinear effects in macro- and microscale. The dispersive effects are analysed in [10] while the influence of nonlinearities is described in [18]. It is not surprising that the balance between the dispersive and nonlinear effects can occur resulting in emergence of solitons.

For numerical simulation, the system of equations (4.4), (4.5) can be represented in the form of a single (4th order) equation (like Eq. (3.10)). The initial value problem for such a model nonlinear equation is solved by the pseudospectral method [18]. The initial pulse-type excitation leads to the train of solitons similar to that shown in Fig. 4.3.

5. Conclusions

If we know all the details of a given microstructure, namely, size, shape, composition, location, and properties of inclusions as well as properties of a carrier medium, the classical wave theory is sufficient for the description of wave propagation. Usually our knowledge about the microstructure is limited – we know only the characteristic scale of microstructure and its physical properties. Then the accuracy of classical theories is not sufficient and the more advanced theories of continua should be used.

In the paper, we have compared results of direct numerical simulations of wave propagation in given layered media with the corresponding results obtained by a continuous model of the microstructure. The presented model looks like a promising variant of the theory, complicated enough to describe various effects of the microstructure. This model can be naturally extended to include non-linear effects and dissipation [19]. However, numerical simulations demonstrate that the straight-forward numerics and the modelling on the basis of continuum theories need a careful matching of material coefficients.

Some general remarks should be made in addition. The concept of dual internal variables introduced in [7] permits to model consistently microstructure(s) for both dissipative (not analysed here) and non-dissipative processes (see above). Such an approach gives an excellent basis to clarify the structure of generalised continuum theories such like linear Cosserat, micromorphic, and second gradient elasticity theories. This will be a subject of our forthcoming publications.

Once the wavefields in microstructured materials are described with needed accuracy, the respective mathematical models can also be used for solving the inverse problems. In linear cases, the dependence of phase velocities on the microstructure can be used for determining the material properties. In nonlinear cases, when the balance between dispersive and nonlinear effects supports the propagation of solitary waves, the algorithms for solving the inverse problems can be based on the analysis of shapes of solitary waves. It has been shown namely [17] that the nonlinearity of the microstructure leads to asymmetric solitary waves. This property can be used for constructing an algorithm which determines the parameters of the microstructure from measured asymmetry (see [17]).

Acknowledgements: Support of the Estonian Science Foundation is gratefully acknowledged.

References

[1] Maugin, G.A.: *Nonlinear Waves in Elastic Crystals*, Oxford University Press (1999).
[2] Sun, C.T., Achenbach, J.D., Herrmann, G.: Continuum theory for a laminated medium. *J. Appl. Mech.* **35** 467–475 (1968).
[3] Nemat-Nasser, S., Hori, M.: *Micromechanics: Overall Properties of Heterogeneous Materials*, Elsevier, Amsterdam (1993).
[4] Mindlin, R.D.: Microstructure in linear elasticity. *Arch. Rat. Mech. Anal* **16** 51-78 (1964).
[5] Eringen, A.C., Suhubi, E.S.: Nonlinear theory of micro-elastic solids II. *Int. J. Eng. Sci.* **2** 189-203 (1964).
[6] Maugin, G.A.: On the thermomechanics of continuous media with diffusion and/or weak nonlocality. *Arch. Appl. Mech.* **75** 723-738 (2006).
[7] Ván, P., Berezovski, A., Engelbrecht, J.: Internal variables and dynamic degrees of freedom. *J. Non-Equilib. Thermodyn.* **33** 235-254 (2008).
[8] Nowacki, W.: *Thermoelasticity*. Pergamon/PWN, Oxford/Warszawa (1962).
[9] Maugin, G.A.: Internal variables and dissipative structures. *J. Non-Equilib. Thermodyn.* **15** 173-192 (1990).
[10] Engelbrecht, J., Berezovski, A., Pastrone, F., Braun, M.: Waves in microstructured materials and dispersion. *Phil. Mag.* **85** 4127-4141 (2005).

[11] Engelbrecht, J., Cermelli, P., Pastrone, F.: Wave hierarchy in microstructured solids. In: Maugin, G.A. (ed.) Geometry, *Continua and Microstructure*, pp. 99-111. Hermann Publ., Paris (1999).

[12] *Effective Computational Methods for Wave Propagation*. Kampanis, N. A., Dougalis, V.A., Ekaterinaris, J.A. (eds.) , Chapman & Hall/CRC, Boca Raton (2008).

[13] LeVeque, R.J.: *Finite Volume Methods for Hyperbolic Problems*. Cambridge University Press (2002).

[14] Berezovski, A., Engelbrecht, J., Maugin, G.A.: *Numerical Simulation of Waves and Fronts in Inhomogeneous Solids*. World Scientific, Singapore (2008).

[15] Lakes, R.S.: Experimental method for study of Cosserat elastic solids and other generalized continua. In: Mühlhaus, H.-B. (ed.) *Continuum Models for Materials with Microstructure*, pp.1-22. Wiley, New York (1995).

[16] Berezovski, A., Berezovski, M. and Engelbrecht, J.: Numerical simulation of nonlinear elastic wave propagation in piecewise homogeneous media. *Mater. Sci. Engng. A* **418** 364-369 (2006).

[17] Janno, J., Engelbrecht J.: An inverse solitary wave problem related to microstructural materials. *Inverse Problems*. **21** 2019-2034 (2005).

[18] Engelbrecht, J. Berezovski, A., Salupere, A.: Nonlinear deformation waves in solids and dispersion. *Wave Motion*. **44** 493-500 (2007).

[19] Pastrone, F., Cermelli, P. and Porubov, A.: Nonlinear waves in 1-D solids with microstructure. *Mater. Phys. Mech.* **7** 9-16 (2004).

Part II
Waves and NDE

Generation and Propagation of Anti-Plane Surface Waves on a Body with Depth-Dependent Properties

Jan D. Achenbach and Oluwaseyi Balogun

Department of Mechanical Engineering, Northwestern University, Evanston IL 60208, USA

achenbach@northwestern.edu

Abstract. The generation and propagation of anti-plane surface waves on an inhomogeneous half-space of depth dependent shear modulus and mass density, is discussed in this paper. The radiation of surface waves an anti-plane line load is analyzed by an application of the reciprocity theorem. Next the governing equation for free surface waves is reformulated in a form that is amenable to a surface wave solution in the high frequency range. The boundary condition on the free surface yields an equation for the velocity of surface waves, in terms of the wave number and derivatives of the functions defining the depth dependence of the shear modules and the mass density. This equation does not always have a real-valued solution, and when it does the amplitude of the corresponding wave motion does not always display the decrease with depth that would define a surface wave. Numerical examples are presented to illustrate these observations.

1. Introduction

In recent years there has been a renewed interest in horizontally polarized (anti-plane) surface waves along the flat surface of an elastic body, whose shear modules and mass density depend on the distance from the flat surface.

It is well known that anti-plane surface waves do no exist on a homogenous body with a free surface. A boundary condition that represents some constraint on the surface is required to support anti-plane surface waves. The best known example are Love waves which can exist due to the presence of an elastic layer.

In this paper we consider the existence of anti-plane surface waves on a half space, but one with smoothly varying depth dependent properties. It is shown that there are

T.-T. Wu and C.-C. Ma (eds.), *IUTAM Symposium on Recent Advances of Acoustic Waves in Solids*, IUTAM Bookseries 26, DOI 10.1007/978-90-481-9893-1_3,

limitations on the depth dependence of the shear modules and the mass density for surface waves to exist. For the case that anti-plane surface waves do exist, we will determine, by a novel method based on the reciprocity theorem, the amplitudes of the surface waves radiated by a time-harmonic line load applied to the free surface of the half-space.

For applications in seismology, the dependence of μ and ρ on depth has been dealt with numerically by replacing the continuous inhomogeneity by a representative layering which may, however, produce effects on the interfaces, or by a formulation which is amenable to the Runge-Kutta technique, see [1, p. 267]. It appears that no analytical solution for surface waves, are available for general depth dependence of μ and ρ. For in-plane surface (Rayleigh) waves, a special depth dependence, exp (αz), when taken for both the elastic constants and the mass density, does give a solution for the Rayleigh waves on a halfspace with a free boundary, see [2,3]. However, this approximation does not support anti-plane surface waves on a free half-space.

In [4] the governing equation for the anti-plane displacement has been formulated in general terms, and an appropriate substitution has been used to reduce the governing equation to one that is amenable to a WKBJ solution see [5, p.25]. The corresponding high frequency surface-wave solution was subsequently used to satisfy the condition on the traction-free boundary. The resulting equation relates the velocity of the anti-plane surface waves to the wave number of such waves. The equation shows that the velocity of surface waves also depends on the value of the shear modulus and its first three derivatives at $z = 0$, as well as on the value of the mass density and its first order derivative at $z = 0$. A WKBJ solution for a layer of inhomogeneous material discussed by Brekhovskikh, was applied in [6] for the determination of the reflection coefficient of acoustic and electromagnetic waves in inhomogeneous layered media. Some interesting papers on antiplane shear waves by Shuvalov *et al.* [7,8] that address the determination of antiplane wave solutions, follow the model presented by Brekhovskikh [6]. Reference [7] is concerned with general properties of dispersion spectra in a monoclinic plate. Reference [8] deals with the propagation of antiplane surface waves in semi-infinite elastic media with vertically periodic continuous and/or discrete variation of material properties. In contrast to these references, the model presented in this work directly addresses the determination of high frequency antiplane surface wave solutions in a general inhomogeneous halfspace without equivalent layering.

In [4] it has been shown that the existence of a surface wave speed does not necessarily mean that a corresponding surface wave whose amplitude decays with depth also exists. The conditions for a surface wave are briefly reviewed in this paper. The results are illustrated for a specific example of depth dependence of the shear modulus and the mass density.

2. Governing Equations

The relevant material properties are the shear modulus $\mu(z)$ and the mass density $\rho(z)$, where z is the distance from the free surface of a half space. We have

$$\mu(z) = \mu_0 g(z) \text{ and } \rho(z) = \rho_0 h(z)\,, \tag{1}$$

where

$$g(0) = g_0 = 1 \quad \text{and} \quad h(0) = h_0 = 1 \tag{2}$$

The analytical form of the surface wave is taken as

$$v(x,z,t) = V(z)e^{i(kx-\omega t)}\,, \tag{3}$$

Thus the wave propagates in the *x*-direction, the displacement is in the *y*-direction, and *V*(*z*) is assumed to be real-valued, and

$$\lim_{z\to\infty} V(z) = 0\,, \tag{4}$$

in order that Eq. (3) represents a surface wave. Also, *V(z)* must satisfy

$$\mu(z)\frac{d^2V}{dz^2} + \frac{d\mu}{dz}\frac{dV}{dz} - k^2\mu(z)V(z) + \rho(z)\omega^2 V(z) = 0 \tag{5}$$

with boundary condition

$$\mu(0)\left.\frac{dV}{dz}\right|_{z=0} = 0 \tag{6}$$

3. Generation of Anti-Plane Surface Waves

The generation of anti-plane surface waves requires the application of a mechanical excitation to the surface of the body. A prototype mechanical excitation is that due to a time-harmonic antiplane line-load. Such a load does not only generate surface waves, but also body waves. It may, however, be expected that in the two-dimensional configuration of a line load applied to the free surface of a half-space, the body waves will decay as $1/r^{1/2}$, where *r* is the distance from the point of application of the line load, while the surface waves do not display geometrical decay.

For an inhomogeneous solid, the usual mathematical techniques for determination of the radiated field, such as Fourier transform methods, do not work. A new technique for the determination of the radiated surface waves is presented in this paper, in the application of the reciprocity theorem [9].

For an elastic body of volume V and boundary S, the reciprocity theorem relates two elastodynamic states, State A and State B, defined by forces, f_i, stress tensors τ_{ij}, and displacements u_i, by integrals over the volume V and the boundary S. For the steady-state time-harmonic case, where

$$f_i(\mathbf{x},t) = f_i(\mathbf{x})e^{-i\omega t}\ ,\ u_i(\mathbf{x},t) = u_i(\mathbf{x})e^{-i\omega t}\ ,\ \tau_{ij}(\mathbf{x},t) = \tau_{ij}(\mathbf{x})e^{-i\omega t}\ , \tag{7}$$

The reciprocity theorem has been derived in [9] in the following form

$$\int_V (f_j^A u_j^B - f_j^B u_j^A)dV = \int_S (\tau_{ij}^B u_j^A - \tau_{ij}^A u_j^B)n_i dS\ , \tag{8}$$

where $\boldsymbol{n}$ is the unit vector along the outward normal to S. It is shown in [9] that the reciprocity theorem as stated by Eq. (8) applies to isotropic, anisotropic as well as inhomogeneous solids. Extensions to linear viscoelasticity have also been worked out.

For the two-dimensional configuration shown in Fig 1, elastodynamic State A is defined by the line load F_o and the surface waves that radiate in both the positive and negative x-directions. We have

$$f_y^A = F\delta(x)\delta(z) \tag{9}$$

Using the general expression given by Eq.(3), the force radiates surface waves of the following form

$$x > 0\ \ v^A = RV(z)e^{ikx} \tag{10}$$

$$x < 0\ \ v^A = RV(z)e^{-ikx}\ , \tag{11}$$

where R is the radiation constant that will be determined by reciprocity considerations. Note that the term exp($-i\omega t$) has been omitted.

For the configuration shown in Fig 1, the reciprocity relation reduces to the form

$$\int_v (f_y^A v^B - f_y^B v^A)dV = \int_S (\tau_{xy}^B v^A - \tau_{xy}^A v^B)n_x dS = 0 \tag{12}$$

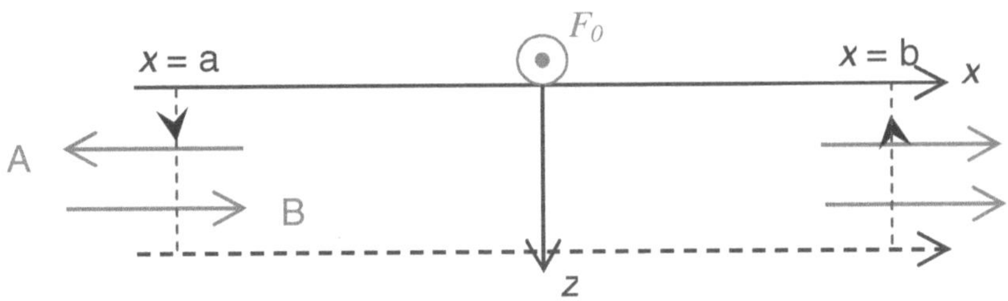

Fig. 1 Contour for the application of the reciprocity theorem.

As shown in Fig 1, the contour for integration along S is chosen as a rectangle with integrations at $x = b$, $\infty > z \geq 0$ and $x = a$, $0 \leq z < \infty$. The part of the contour at $z \to \infty$ is left out since the surface waves have exponentially vanished as $z \to \infty$. By taking the line load slightly inside the half-space the contribution from the traction-free surface vanishes, and we are left with a contribution from the line load in the left-hand side of Eq. (12).

For State B, which is the virtual state, we choose a free surface wave propagating in the positive x-direction.

$$v^B = V(z)e^{ikx} \tag{13}$$

$$\tau_{xy}^B = ik\mu(z)V(z)e^{ikx} \tag{14}$$

Substitution of the expressions for States A and B into Eq.(12) yields

$$F = ikR\left\{\int_\infty^0 \mu(z)V(z)^2\,(e^{2ikb} - e^{2ikb})dz - 2\int_0^\infty \mu(z)V(z)^2 dz\right\} \tag{15}$$

It is noted that waves propagating in the same direction in the integral along x=b cancel each other, while the counter-propagating waves in the integral along x=a produce a contribution that allows the calculation of R. The result is

$$R = \frac{1}{2}\frac{i}{k}\frac{F}{I}\,, \tag{16}$$

where

$$I = \int_0^\infty \mu(z)V(z)^2 dz \tag{17}$$

For given $\mu(z)$ we have to determine $V(z)$ to calculate the integral in Eq. (17), and to obtain the final result.

4. Determination of *V*(*z*)

Equation (5) was considered in some detail in Ref. [4]. To render the equation amenable to solution the authors of [4] eliminated the second term by the substitution

$$V(z)=\frac{1}{\sqrt{g(z)}}\overline{V}(z) \tag{18}$$

After some manipulation Eq. (5) then simplifies to

$$\frac{d^2\overline{V}}{dz^2}-\omega^2 q(z,\omega)\overline{V}=0, \tag{19}$$

where

$$q(z,\omega)=\frac{1}{c_S{}^2}-\frac{1}{c^2}+\frac{f(z)}{\omega^2} \tag{20}$$

$$c^2=c(z)^2=\frac{\mu(z)}{\rho(z)} \tag{21}$$

$$f(z)=\frac{1}{2}\frac{1}{g}\frac{d^2g}{dz^2}-\frac{1}{4}\frac{1}{g^2}\left(\frac{dg}{dz}\right)^2 \tag{22}$$

Next a postulated solution of Eq. (19) in the form

$$\overline{V}(z)=A\exp\left\{-\int_0^z\phi dz\right\} \tag{23}$$

was considered. Substitution of Eq. (23) into (18) and the result in Eq. (19) yields

$$\phi^2-\frac{d\phi}{dz}-\omega^2 q=0 \tag{24}$$

Following Heading [5] a solution was sought of the form

$$\phi=\phi_0(z,\omega)\omega+\phi_1(z,\omega)+\phi_2(z,\omega)\frac{1}{\omega}+O\left(\frac{1}{\omega^2}\right), \tag{25}$$

for the case that ω is large and positive and q(z, ω) tends to a limit as $\omega\to\infty$. Substitution in Eq.(24), only keeping terms of order ω^2 and ω, yields

$$\varphi_0 = q^{1/2} \tag{26}$$
$$\varphi_1 = (dq^{1/2} / dz) / 2q^{1/2} \tag{27}$$

The high-frequency approximation to $\overline{V}(z)$ then becomes

$$\overline{V}(z) = A\frac{q_0^{1/4}}{q^{1/4}}\exp\left\{-\omega\int_0^z q^{1/2}dz\right\}, \tag{28}$$

where $q(z,\omega)$ is defined by Eq.(20), and q_o is

$$q_0 = \frac{1}{c_s^2} - \frac{1}{c_o^2} + \frac{f(0)}{\omega^2} \tag{29}$$

For details we refer to Ref. [4]. The expression for the actual displacement follows from Eqs. (18) and (28) as

$$V(z) = \frac{A}{g^{1/2}}\frac{q_0^{1/4}}{q^{1/4}(z,\omega)}\exp\left(-\omega\int_0^z q^{1/2}(z,\omega)dz\right) \tag{30}$$

It may be checked that Eq. (30) reduces to the right result for the homogenous case. The stress τ_{xz} corresponding to Eq. (30) is

$$\tau_{xz}(z) = ik\mu(z)V(z)\,, \tag{31}$$

while τ_{yz} becomes

$$\tau_{yz} = \mu(z)\frac{dV}{dz} = -\frac{A}{g^{1/2}}\frac{q_o^{1/4}}{q^{5/4}}\left\{\frac{1}{2}\frac{q}{g}\frac{dg}{dz} + \frac{1}{4}\frac{dq}{dz} + \omega q^{3/2}\right\} \tag{32}$$

It follows from Eq. (20) that

$$\frac{dq}{dz} = \frac{2}{c^3}\frac{dc}{dz} + \frac{df}{dz}\frac{1}{\omega^2}, \tag{33}$$

where $c(z)$ and $f(z)$ are defined by Eqs. (20) and (21), respectively. Also

$$\frac{dc}{dz} = c_0\frac{d}{dz}\left[\frac{g(z)}{h(z)}\right]^{1/2} = \frac{1}{2}c_0\left[\frac{1}{g^{1/2}}\frac{dg}{dz}\frac{1}{h^{1/2}} - \frac{g^{1/2}}{h^{3/2}}\frac{dh}{dz}\right] \tag{34}$$

$$\frac{df}{dz} = \frac{1}{2g^3}\left(\frac{dg}{dz}\right)^3 - \frac{1}{g^2}\frac{dg}{dz}\frac{d^2g}{dz^2} + \frac{1}{2g}\frac{d^3g}{dz^3} \tag{35}$$

For a traction-free surface at $z = 0$, τ_{yz} as given by Eq. (32) must vanish at $z = 0$. At $z = 0$, we have

$$g = g_0 = 1, \qquad h = h_0 = 1, \tag{36}$$

and we define for simplicity of notation

$$\frac{dg}{dz} = g_1, \quad \frac{dh}{dz} = h_1, \quad \frac{d^2 g}{dz^2} = g_2, \quad \frac{d^3 g}{dz^3} = g_3 \tag{37}$$

By virtue of Eq. (22) and Eqs. (34)-(35), we obtain at $z = 0$

$$f_0 = f(0) = \frac{1}{2} g_2 - \frac{1}{4} g_1^{\,2}, \quad f_1 = \left.\frac{df}{dz}\right|_{z=0} = \frac{1}{2} g_1^{\,3} - g_1 g_2 + \frac{1}{2} g_3 \tag{38}$$

$$c_1 = \left.\frac{dc}{dz}\right|_{z=0} = \frac{1}{2} c_0 (g_1 - h_1) \tag{39}$$

By introducing Eqs. (36)-(39) into Eq. (32), and by multiplying the result by c_s, taking into account that $\omega = kc_s$, and by introducing a characteristic length L, the condition $\tau_{yz} = 0$ at $z = 0$ yields

$$(kL)\left(1 - \frac{c_S^{\,2}}{c_0^{\,2}} + \frac{f_0 L^2}{(kL)^2}\right)^{3/2} + \frac{1}{2} g_1 L \left(1 - \frac{c_S^{\,2}}{c_0^{\,2}} + \frac{f_0 L^2}{(kL)^2}\right) + \frac{1}{4}\frac{c_S^{\,2}}{c_0^{\,2}}(g_1 L - h_1 L) + \frac{1}{4}\frac{f_1 L^3}{(kL)^2} = 0 \tag{40}$$

It should be noted that in addition to depending on kL, i.e the surface wave is dispersive, c_s also depends on the first three derivatives of $g(z)$ and the first derivative of $h(z)$ at $z = 0$. Equation (40) is the anti-plane surface waves dispersion relation.

5. Results and Discussion

In accordance with Eq. (3), the functional form of the displacement of the anti-place surface wave is of the form

$$v(x, z, t) = V(z) \exp\{ik(x - c_s t)\} \tag{41}$$

where we have used $\omega = kc_s$, $V(z)$ is given by Eq. (30), and c_s is the solution of Eq. (40). In order to compute the radiation constant R from Eq. (16), we must have an expression for V(z) which is real-valued and represents a surface wave, while c_s should also be real valued.

For $V(z)$ to be real valued and represent a surface wave, we must have $q>0$, where q is defined by Eq.(20). Thus

$$\frac{c_s^{\ 2}}{c^2} < 1 + \frac{f(z)}{k^2} \tag{42}$$

On the other hand Eq. (42) indicates that for a real value of c_s the first term of (42) should be real-valued, i.e,

$$\frac{c_s^{\ 2}}{c^2} \le 1 + \frac{f(0)}{k^2} \tag{43}$$

These conditions are not necessarily contradictory but they must be satisfied at the same time. A real-valued solution of Eq. (40) is also restricted by conditions on other terms, for example we should probably have $g_1 \le 0$. These restrictions have been illustrated by an example in Ref [4]. For that example it is possible to obtain a real-valued solution for c_s, but it is found that $q(z,\omega)$ becomes negative at a certain depth, and hence $V(z)$ would not be real-valued and decaying with depth. Hence it is recommended that for selected profiles $g(z)$ and $h(z)$ it should first be checked that Eq. (40) produces a real valued c_s. If c_s is indeed real-valued, it should be checked that $q > 0$, to determine the surface wave displacement and stress profiles.

As a sample example let us consider

$$g(z) = 1 + \left\{1 - \exp[\alpha(z/L)^2]\right\}\exp[-\beta(z/L)], \quad h(z) = 1 \tag{44}$$

It may be checked that

$$g_1 = 0, \quad g_2 = 2\frac{\alpha}{L^2}, \quad g_3 = -6\frac{\alpha\beta}{L^3}, \quad h_1 = 0 \tag{45}$$

The profile for $\mu(z)/\mu_o = g(z)$ is shown in Fig 2. Using Eq. (45), c_s follows from (40) as

$$\left(\frac{c_s}{c_0}\right)^2 = 1 + \frac{\alpha}{(kL)^2} - \frac{1}{(kL)^2}\left(\frac{3}{4}\alpha\beta\right)^{2/3} \tag{46}$$

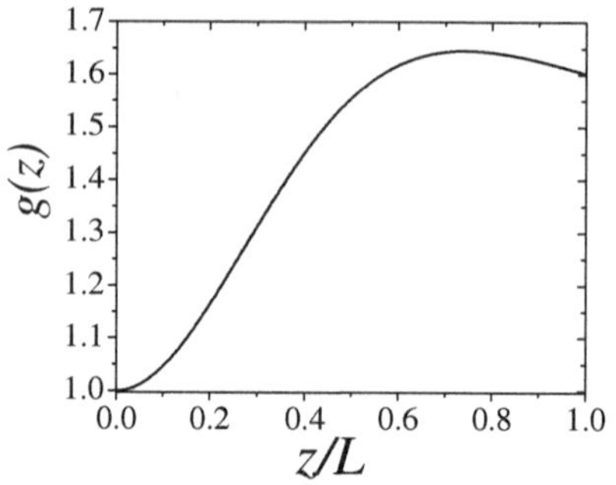

Fig. 2 Shear modulus profile given by Eq. (44) for $\alpha = 5$ and $\beta = 0.5$.

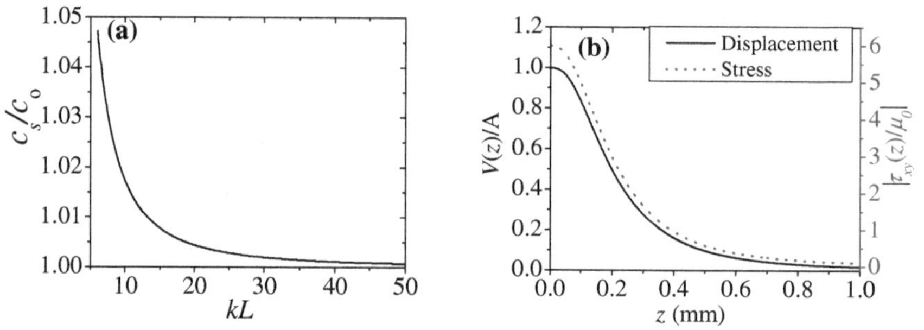

Fig. 3 (a) Calculated surface wave dispersion curve and (b) displacement and stress profiles.

To check that for the example of Eq. (44), the condition (42) is satisfied, we write

$$\frac{c_s^{\ 2}}{c^2} = \frac{c_s^{\ 2}}{c_0^{\ 2}}\frac{c_0^{\ 2}}{c^2} = \left\{1 + \frac{1}{k^2}\left[f_0 - \left(-\frac{1}{4}f_1\right)^{2/3}\right]\right\}\frac{c_0^{\ 2}}{c^2}, \tag{47}$$

where Eq. (46) has been used. According to Eq. (44), we should then have

$$\frac{c_0^{\ 2}}{c^2} < \frac{1 + \dfrac{f(z)}{k^2}}{1 + \dfrac{1}{k^2}\left[f_0 - \left(-\dfrac{1}{4}f_1\right)^{2/3}\right]}, \tag{48}$$

within the penetration depth of the surface wave. For the example given by Eq. (44) we clearly have that $c_0^{\ 2}/c^2 < 1$. Hence the condition (48) implies that Eq. (42) will be satisfied if

$$f(z) > f_0 - \left(-\frac{1}{4}f_1\right)^{2/3} \tag{49}$$

For the example of Eq. (44) we have $f_0 = 2\alpha/L^2$ and $f_1 = -3\alpha\beta/L^3$, and hence the condition (49) depends on the values of α and β. Figure 3(a) shows the velocity of surface waves. The dimensionless anti-plane displacement $V(z)/A$, where $V(z)$ is defined by Eq. (30) and the dimensionless stress $|\tau_{xz}(z)/\mu_o|$, which follows from Eq. (31) are shown in Fig. 3(b). For the results of Fig. 3, the surface wave velocity is obtained from Eq. (46) as $c_s/c_o = 1.0472$ at $kL = 6$, where $c_0 = 3040$ m/s and $L = 1$ mm. The calculated surface wave velocity is seen to decrease with kL and it approaches the transverse wave velocity at $z = 0$. The stress and displacement profiles are seen to decrease with z as expected for a surface wave.

Acknowledgements: This work was supported by the National Science Foundation under grant number OISE-0730259 and the Northwestern University Infrastructure Technology Institute under grant number DTRT06-G-0015.

References

[1] Aki K. and Richards P.G.: *Quantitative Seismology*, Vol. **I** p. 267, W.H. Freeman and Company, San Fransisco (1980).

[2] Destrade M.: Seismic Rayleigh waves on an exponentially graded, orthotropic elastic half-space. *Proc. Roy. Soc. A* **463** 495-502 (2007).

[3] Kulkarni S.S. and Achenbach J.D.: Application of reciprocity theorem to determine line-loaded generated surface waves on an inhomogeneous transversely isotropic half-space. *Wave Motion* **45** 350-360 (2008).

[4] Achenbach J.D. and Balogun O.: Anti-Plane surface waves on a half-space with depth-dependent properties. *Wave Motion*, to appear.

[5] Heading J.: *An Introduction to Phase-Integral Methods*, p. 25, Methuen and Co. Ltd., London (1962).

[6] Brekhovskikh L.M.: *Waves in Layered Media*, p. 193-199, Academic Press, New York, (1960).

[7] Shuvalov A.L.: Ponchet O., and Kiselev A. P., Shear horizontal waves in transversely inhomogeneous plates. *Wave Motion* **45** 605-615 (2008).

[8] Shuvalov A.L, Ponchet O., and Gockin S.V.: Existence and spectral properties of shear horizontal surface acoustic waves in vertically periodic half spaces. *Proc. Roy. Soc. A* **465** 1489-1511 (2009).

[9] Achenbach J.D.: *Reciprocity of Elastodynamics*, Cambridge University Press, Cambridge, UK (2003).

Resonance Ultrasound Microscopy for Imaging Young's Modulus of Solids

Masahiko Hirao and Hirotsugu Ogi

Graduate School of Engineering Science, Osaka University, Toyonaka, Osaka 560-8531, Japan

hirao@me.es.osaka-u.ac.jp

Abstract. Resonance ultrasound microscopy is developed for quantitatively measuring local Young's modulus of solid surfaces, using monocrystal langasite ($La_3Ga_5SiO_{14}$) as a probing oscillator. The langasite oscillator is acoustically isolated so that the contact with the specimen only affects the vibration, making the quantitative measurement possible. The vibrational analysis is proposed to calculate the local stiffness and local damping from the resonance frequency and internal friction of the oscillator, respectively. This method is applied to a polycrystalline copper and a SiC fiber as illustrative examples.

1. Introduction

Young's modulus is the most important engineering stiffness. It is required in designing any structures, including micro/nano electromechanical systems. Usually, it is obtained by a tensile test and by ultrasonic velocity measurements, for which specimens must be large enough and have simple shapes such as plate, sphere, rectangular parallelepiped, and so on. It has never been straightforward to measure Young's modulus in a localized area. Since the local Young's modulus is more sensitivity to the presence of small defects, it can also be useful for assess-ing the integrity of small structures.

Surface probing techniques have been developed for measuring Young's modulus in a local area. Yamanaka *et al.* [1, 2] developed ultrasound atomic-force microscopy (UAFM), where an AFM cantilever contacts with the specimen surface. The resonance frequency of the bending vibration of the cantilever varies depending on the elastic stiffness of the contacting material, providing an elastic image with high spatial resolution. Rabe *et al.* [3, 4] also presented a similar

T.-T. Wu and C.-C. Ma (eds.), *IUTAM Symposium on Recent Advances of Acoustic Waves in Solids*, IUTAM Bookseries 26, DOI 10.1007/978-90-481-9893-1_4,

methodology called atomic-force acoustic microscopy. These methods with an AFM cantilever achieved high-resolution elastic-stiffness microscopy (~ 50 nm). However, quantitative evaluation of the local stiffness is still difficult because the cantilever is not isolated acoustically. Namely, many ambiguous components participate in the resonant system, not only the cantilever and the specimen but also the piezoelectric transducer attached to it to excite the vibration, the mechanical properties of the material at the gripping end, and the gripping condition.

We propose an alternative Young's modulus microscopy, *resonant ultrasound microscopy* (RUM). It uses a bar-shape langasite oscillator. The vibrations are excited and detected by a noncontacting line antenna [5], and the oscillator is acoustically isolated from any other materials except for the specimen. Thus, the elastic properties of the specimen are extracted unambiguously, realizing the quantitative measurement. The isolated oscillator significantly contributes to the modeling to deduce Young's modulus from the resonance frequency change. Furthermore, the internal-friction change of the oscillator indicates the local damping property of the specimen. We can thus simultaneously construct elastic and anelastic images of solids. This paper presents the principle of RUM and shows examples of elastic and anelastic imaging for polycrystalline copper and SiC fiber.

2. Isolated Langasite Oscillator

Figure 2.1 shows the schematic of the RUM probe, consisting of the langasite oscillator, monocrystal diamond tip attached at the end of the oscillator, a fixture gripping the oscillator, a line antenna embedded in the fixture, and a cylindrical guide surrounding the fixture. The langasite oscillator is a rectangular-column-shape single crystal, and its longitudinal direction is along the [100] direction of the trigonal system. The cross-sectional area is A=0.6x0.6 mm^2 and the length L=7.0 mm. We selected the [100] direction to be aligned in the longitudinal direction of the oscillator because Young's modulus in this direction is small, providing higher sensitivity to the material's stiffness [5]. A conical monocrystal diamond tip is attached at the center of the bottom surface of the

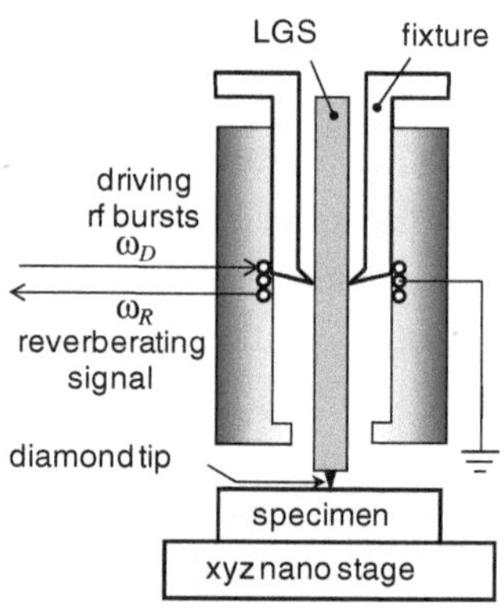

Fig. 2.1 RUM probe with a monocrystal langasite oscillator.

oscillator, which contacts the specimen. The fixture holds the oscillator at the center nodal point. The line antenna measures the resonance frequency contactlessly. Therefore, the langasite oscillator is acoustically isolated except for the contact with the specimen, so that the change in the resonance frequency solely and directly gives the elastic stiffness of the specimen. The biasing force for the contact is kept unchanged, being independent of surface roughness; it principally equals the oscillator weight plus the fixture (0.59 mN). This configuration is an important advantage of RUM, because the stable biasing force allows us to analytically calculate the effective Young modulus [6].

The resonance frequency is measured by the superheterodyne spectroscopy method [7, 8]. We apply a burst signal of frequency ω_D (~ 50 V_{pp} amplitude and 100 μs duration) to the generation wire of the antenna. The quasistatic electric field is induced in the longitudinal direction near the center area of the oscillator bar, owing to the converse piezoelectric effect, where the maximum stress occurs for the fundamental resonant mode. After the excitation, the oscillator vibrates with the resonance frequency ω_R, and the vibrational amplitude decays. The detection wire of the antenna receives the ringing signal via the piezoelectric effect, which is mixed with the reference signal of the driving burst to produce the beating signal with the differential frequency of $\omega_R - \omega_D$. The amplitude is integrated over a long interval by the analog integrator. When the driving frequency equals the resonance frequency, the integrated amplitude takes a large value. Thus, by scanning the driving frequency and acquiring the integrated amplitude, we obtain a spectrum with a sharp peak, from which the resonance frequency is determined by fitting the Lorentzian function.

The contacting point is moved using the 3-axis stage. At a measuring point, the resonance frequency is first measured in a noncontacting state. Then, the stage lifts the specimen up to make the contact and the resonance frequency in the con tacting condition is measured; the frequency change between them is recorded. The probe is separated from the specimen, and the stage moves the specimen to the next measuring point. The probe never scratches the specimen surface.

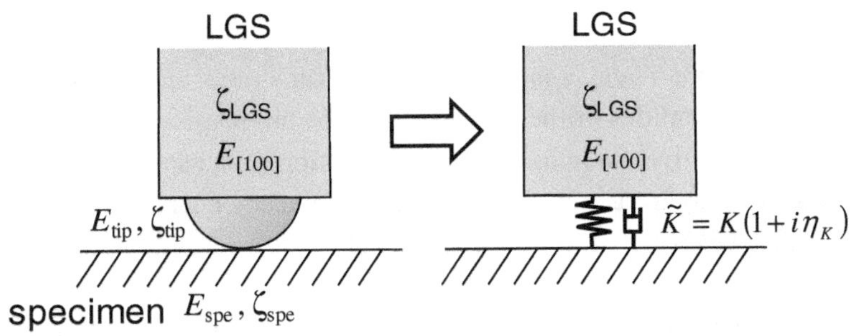

Fig. 3.1 Modeling the contact interface by a spring with complex spring constant, $\tilde{K}$.

3. Vibrational Analysis

The contact problem of the isolated oscillator can be modeled as a resonator of a bar oscillator connected with the rigid wall through a spring and a damper as shown in Fig. 3.1. The spring constant K and the damping constant η_K are associated with Young's modulus E_{spe} and internal friction ζ_{spe} of the specimen at the local contacting area as well as those of the oscillator ($E_{[100]}$ and ζ_{LGS}) and the tip (E_{tip} and ζ_{tip}). Considering the boundary conditions at the free end (top) and the spring end (bottom), we derive the frequency and damping equations as follows [5]:

$$K = K_{osc}\beta L \tan(\beta L), \tag{3.1}$$

$$\eta_K = \frac{1}{p}\left[2(\varsigma_m - \varsigma_0)\left((\beta L)^2 + p\right) - \frac{\varsigma_{LGS}}{2}\left((\beta L)^2 - p\right)\right]. \tag{3.2}$$

Here, β denotes the wavenumber in the longitudinal direction and $K_{osc}=E_{[100]}A/L$ is the spring constant of the oscillator for a static load.$p=K/K_{osc}$ indicates the contribution of the contact stiffness to the resonator system; a larger p causes a larger frequency change. ζ_{LGS}, ζ_m, and ζ_0 are internal friction of the oscillator, measured internal friction at the contact condition, and the loss due to the propagation of sound wave into the specimen, respectively. Assuming a Hertzian contact with isotropic bodies [9], the effective Young's modulus E_{Hertz} and internal friction ζ_{spe} are obtained by [5]

$$E_{Hertz} = \sqrt{\frac{K^3}{6FR}} = \left(\frac{1-\nu_{spe}^2}{E_{spe}} + \frac{1-\nu_{tip}^2}{E_{tip}}\right)^{-1}, \tag{3.3}$$

$$\varsigma_{spe} = \frac{3}{2}\left[1 + \frac{E_{spe}}{E_{tip}} \cdot \frac{1-\nu_{tip}^2}{1-\nu_{spe}^2}\right]\eta_K - \frac{E_{spe}}{E_{tip}} \cdot \frac{1-\nu_{tip}^2}{1-\nu_{spe}^2}\varsigma_{tip}. \tag{3.4}$$

Here, E and ν are Young's modulus and Poisson's ratio, and the subscripts *spe* and *tip* indicate quantities of the specimen and the tip, respectively. Thus, we can determine the effective modulus and internal friction of the specimen at the localized contact area by measuring the resonance frequency change and the internal-friction change between before and after the contact.

Equations (3.3) and (3.4) are applicable to contact with isotropic materials. However, the materials show elastic anisotropy at a localized area because of crystal symmetry, oriented defects, segregation of precipitates, and so on. It is then necessary to discuss the applicability of the isotropic approximation. Willis [10] derived the elastic field caused by a contact between an indenter and an anisotropic half-space solid using Fourier transformation. Following his analytical method, we derived the contact stiffness for anisotropic solids:

$$E_{Willis} = \sqrt[3]{\frac{8I_1}{I_0^3}}, \tag{3.5}$$

Here, I_0 and I_1 are given in detail elsewhere [11] and they depend on all independent elastic constants of the specimen and the tip. Figure 3.2 compares the effective Young's modulus calculated using anisotropic elastic constants (E_{Willis}) and that calculated assuming elastic isotropy (E_{Hertz}) for contact between the monocrystal diamond and hexagonal and cubic materials. These two moduli prin cipally show a good correlation. Materials with high elastic anisotropy cause larger difference between them, although the difference is smaller than ~10%. Therefore, the stiffness evaluation with RUM could involve 10% error associated with the elastic anisotropy.

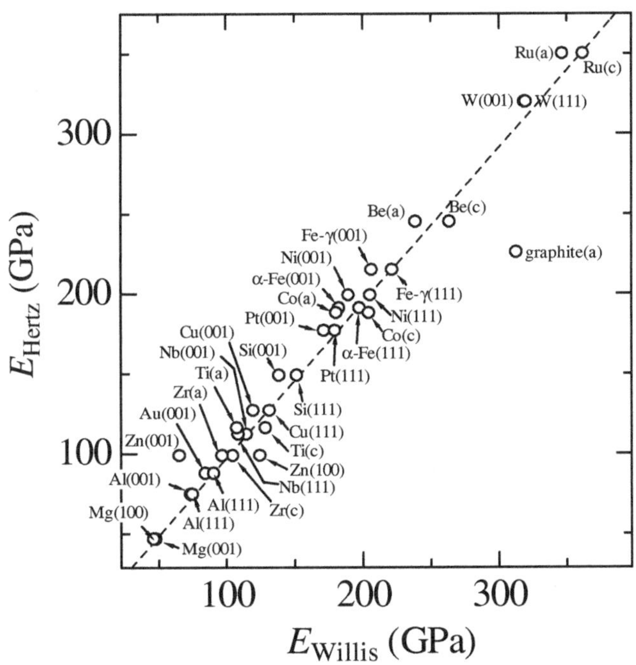

Fig. 3.2 Comparison between the effective Young's modulus determined by considering elastic anisotropy of the contacting materials (E_{Willis}) and that determined by assuming elastic isotropy (E_{Hertz}). Various cubic and hexagonal crystals are involved.

4. Elastic and Anelastic Imaging

Figure 4.1 shows the elastic and anelastic images of a polycrystalline copper. We first measured the crystallographic orientations of individual grains using electron backscatter-diffraction (EBSD) method and calculated the anisotropic effective Young's modulus (Fig. 4.1(a)). Then, we measured the resonance frequency change and internal friction of the same area to obtain the stiffness and internal-friction images as shown in Figs. 4.1(b) and (c), respectively. Young's modulus images by EBSD and RUM show a good agreement, indicating high capability of quantitative stiffness measurement with RUM. The RUM modulus shows a low stiffness at the grain boundary indicated by the arrow in Fig. 4.1(b). The grain boundary shows higher internal friction (Fig. 4.1(c)). The softening at grain boundary is possible due to highly distorted lattice structure, where dislocations are densely formed and absorbs the acoustic energy.

We applied this method to a silicon-carbide fiber (SCS-6) embedded in Ti-alloy matrix as shown in Fig. 4.2. This fiber has an annular structure; carbon core, inner carbon coating, chemical-vapor-deposited β-SiC, and outer carbon coating. The RUM images in Fig. 4.3 clearly show the difference of the stiffness among the individual components. The microstructure of the fiber was studied by transmission-electron microscopy in detail by Ning and Pirouz [12]. The inside region of the core consists of blocks of turbostratic carbon (TC) with 1–50 nm size, whose chemical bonds are similar to those of graphite. They are randomly oriented. The inner and outer carbon coatings also consist of TC blocks with 30-50 nm size, but their *c* axes are predominantly aligned in the radial direction. The basal planes of TC grains are in alignment with the longitudinal direction of the fiber, along which the covalent bonds appear. Therefore, the carbon coating layers show higher modulus than in the carbon core.

Young's modulus in the SiC region is smaller than the expected value of 327 GPa calculated using reported values for the bulk β-SiC (Young's modulus of β-SiC is 447 GPa). This will be caused by microdefects introduced during the CVD deposition process. The RUM measurement provides the stiffness in a localized region, where the stress field increases the degree of the stress concentration near defects if any and lowers the apparent stiffness more significantly than in the case of the application of a uniform stress field.

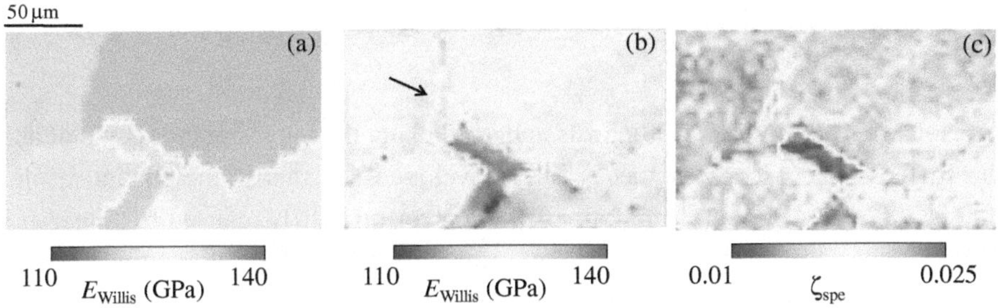

Fig. 4.1 Elastic and anelastic images. (a) The effective Young's modulus image calculated by crystallographic orientation determined by the EBSD method. (b) Effective image modulus measured by RUM. (c) Internal friction image measured by RUM.

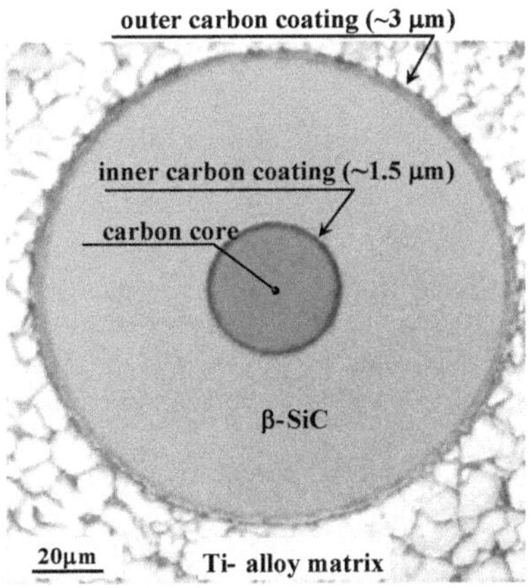

Fig. 4.2 Scanning-electron microscopy image of SCS-14 fiber embedded in a Ti-6Al-4V matrix.

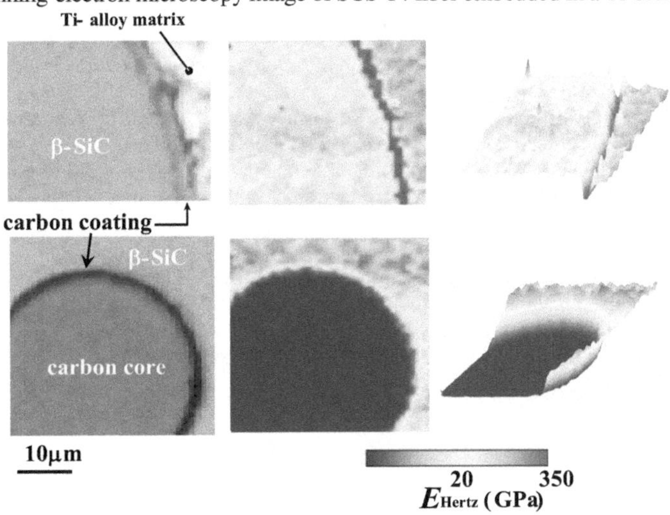

Fig. 4.3 Images by SEM (left), stiffness images by RUM (center), and three-dimensional stiffness images by RUM (right).

5. Conclusions

An isolation of vibrating oscillator is achieved using the langasite monocrystal and the stiffness microscopy is successfully developed. Currently, the spatial resolution is 200 nm. Although it may involve the error up to 10% caused by material's elastic anisotropy, it can provide the local stiffness and local internal friction quantitatively as has been confirmed by the stiffness image obtained by electro-backscattering diffraction measurement. This microscopy is applied to a single SiC fiber and clearly shows the stiffness difference among the components. The stiffness in the SiC region is much lower than that expectation, which is explained by microdefects introduced during the deposition of the SiC phase.

References

[1] Yamanaka, K., Ogiso, H., and Kolosov, O.: Ultrasonic force microscopy for nanometer resolution subsurface imaging. *Appl. Phys. Lett.* **64** 178-180 (1994).

[2] Yamanaka, K., Tsuji, T., Noguchi, A., Koike, T., and Mihara, T.: Nanoscale elasticity measurement with in situ tip shape estimation in atomic force microscopy. *Rev. Sci. Instrum.* **71** 2403-2408 (2000).

[3] Rabe, U. and Arnold, W.: Acoustic microscopy by atomic force microscopy. *Appl. Phys. Lett.* **64** 1493-1495 (1994).

[4] Rabe, U., Janser, U.K., and Arnold, W.: Vibrations of free and surface-coupled atomic force microscope cantilevers: Theory and experiment. *Rev. Sci. Instrum.* **67** 3281-3293 (1996).

[5] Ogi, H, Niho, H., and Hirao, M.: Internal-friction mapping on solids by resonance ultrasound microscopy. *Appl. Phys. Lett.* **88** 141110 (2006).

[6] Ogi, H., Hirao, M., Tada, T., and Tian, J.: Elastic-stiffness distribution on polycrystalline copper studied by resonance ultrasound microscopy: Young's modulus microscopy. *Phys. Rev. B* **73** 174107 (2006).

[7] Hirao, M., Ogi, H., and Fukuoka, H.: Resonance EMAT system for acoustoelastic stress evaluation in sheet metals. *Rev. Sci. Instrum.* **64** 3198-3205 (1993).

[8] Petersen, G.L., Chick, B.B., Fortunko, C.M., and Hirao, M.: Resonance techniques and apparatus for elastic-wave velocity determinations in thin metal plates. *Rev. Sci. Instrum.* **65** 192-198 (1994).

[9] Johnson, K.: *Contact Mechanics*, Cambridge Univ. Press, Cambridge (1985).

[10] Willis, J.R.: Hertzian contact of anisotropic bodies. *J. Mech. Phys. Solids* **14** 163-176 (1966).

[11] Ogi, H., Inoue, T., Nagai, H., and Hirao, M.: Quantitative imaging of Young's modulus of solids: A contact-mechanics study. *Rev. Sci. Instrum.* **79** 053701 (2008).

[12] Ning, X.J. and Pirouz, P.: The microstructure of SCS-6 SiC fiber. *J. Mater. Res.* **6** 2234-2248 (1991).

Non-Classical Nonlinearity in Solids for Defect-Selective Imaging and NDE

Igor Solodov

Department of Nondestructive Testing, Institute of Polymer Technology (IKT-ZFP), University of Stuttgart, 70569 Stuttgart, Germany

igor.solodov@ikt.uni-stuttgart.de

Abstract. The non-classical nonlinear phenomena are based on anomalously high local nonlinearity of micro- and macro-scale defects and thus are defect-selective, i.e. appeal directly to the vulnerable (faulty) areas within material or a product. The paper suggests an overview of the mechanisms and manifestations of the non-classical acoustic phenomena which lay the foundations of the new approaches to ultrasonic NDE and defect-selective imaging.

1. Introduction

A traditional view on nonlinear ultrasonics is concerned with classical idea of elastic wave distortion due to material nonlinearity: waveform deformation caused by a local velocity variation accumulates with propagation distance and provides progressive transition of a harmonic wave into sawtooth- or N-type waves. As a result, the spectrum acquires higher (ultra-) harmonics of the fundamental frequency which deliver information on the matter. In classical (crystalline or free from defects isotropic) media, the material nonlinearity is quite low and normally only few harmonics are observable so that classical nonlinear non-destructive evaluation (NDE) is basically "second harmonic" NDE.

In imperfect (damaged) materials, nonlinear response is provided by the higher-order Contact Acoustic Nonlinearity (CAN) [1]: strongly nonlinear local vibrations of defects due to mechanical constraint of their fragments which efficiently generate multiple ultra-harmonics and support multi-wave interactions. Another contribution to non-classical nonlinear spectrum comes from resonance properties of planar defects [2]. Similar to resonance behaviour of an air bubble in liquid, vibrations of a certain mass of material around a cracked defect are managed by reduced stiffness which provides a specific characteristic frequency of the defect and brings the nonlinear resonance scenario into elastic wave-defect interaction.

T.-T. Wu and C.-C. Ma (eds.), *IUTAM Symposium on Recent Advances of Acoustic Waves in Solids*, IUTAM Bookseries 26, DOI 10.1007/978-90-481-9893-1_5,

In this paper, basic mechanisms responsible for frequency conversion by nonlinear defects are discussed and major features of CAN spectra derived. Experimental methodologies of nonlinear laser vibrometry (NLV) and nonlinear air-coupled emission (NACE) are developed and used to study non-classical spectra in elastic wave-defect interaction. Applications for defect-selective imaging and NDE are demonstrated for a series of hi-tech materials and industrial components.

2. Phenomenology of Harmonic Generation via CAN

2.1 "Clapping" Mechanism

Consider a pre-stressed crack (static stress σ^0) driven with longitudinal acoustic traction $\sigma_\sim$ which is strong enough to provide clapping of the crack interface. The clapping nonlinearity comes from asymmetrical dynamics of the contact stiffness: the latter is higher in a compression phase (due to clapping) than that for tensile stress when the crack is assumed to be supported only by edge-stresses.

Such behaviour of a clapping interface can be approximated by a piece-wise stress (σ)-strain (ε) relation [3]:

$$\sigma = C[1 - H(\varepsilon)(\Delta C / C)]\varepsilon \,, \tag{2.1.1}$$

where $H(\varepsilon)$ is the Heaviside unit step function; $\Delta C = [C - (d\sigma / d\varepsilon)_{\varepsilon>0}]$, and C is the intact material (linear) stiffness.

The bi-modular pre-stressed contact driven by a harmonic acoustic strain $\varepsilon(t) = \varepsilon_0 \cos \nu t$ is similar to a "mechanical diode" and results in a pulse-type modulation of its stiffness $C(t)$ (Fig. 2.1). It also provides an unconventional nonlinear waveform distortion: a half-period rectified output (Fig. 2.1) instead of the saw-tooth like profile in classical materials. Since $C(t)$ is a pulse-type periodic function of the driving frequency ν (Fig. 2.1, right), the nonlinear part of the spectrum induced in the damaged area ($\sigma^{NL}(t) = \Delta C(t) \cdot \varepsilon(t)$) contains a number of its higher harmonics $n\nu$ (both odd and even orders) whose amplitudes are modulated by the *sinc*-envelope function [3]:

$$A_n = \Delta C \Delta \tau \varepsilon_0 \left[sinc\left((n+1)\Delta t\right) - 2cos\left(\pi \Delta t\right) sinc\left(n \Delta t\right) + sinc\left((n-1)\Delta t\right) \right], \tag{2.1.2}$$

where $\Delta \tau = \tau / T$ ($\tau = (T / \pi) Arc\cos(\varepsilon^0 / \varepsilon_0)$) is the normalized modulation pulse length.

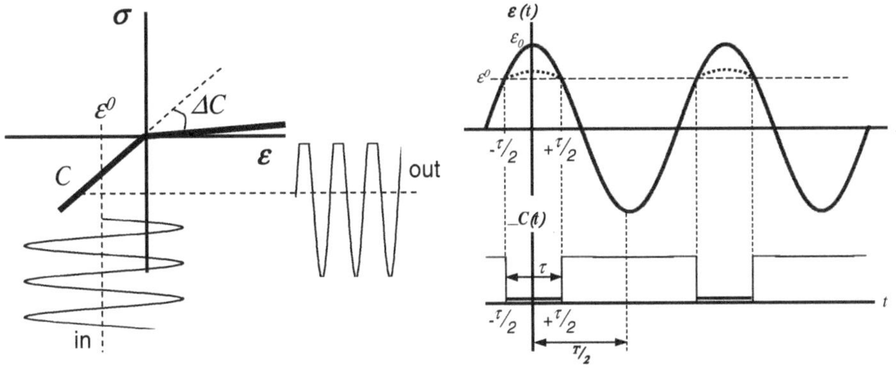

Fig. 2.1 Mechanical diode model (left); stiffness modulation and waveform distortion (right).

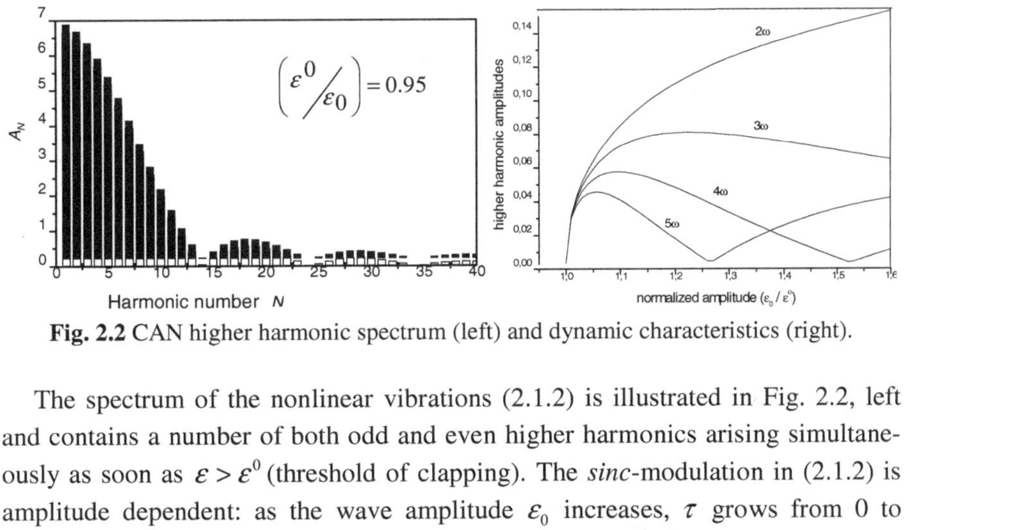

Fig. 2.2 CAN higher harmonic spectrum (left) and dynamic characteristics (right).

The spectrum of the nonlinear vibrations (2.1.2) is illustrated in Fig. 2.2, left and contains a number of both odd and even higher harmonics arising simultaneously as soon as $\varepsilon > \varepsilon^0$ (threshold of clapping). The *sinc*-modulation in (2.1.2) is amplitude dependent: as the wave amplitude ε_0 increases, τ grows from 0 to $T/2$. This affects dynamic characteristics of the higher harmonics (Fig. 2.2, right) and provides their oscillations due to the spectrum "compression" effect.

2.2 Nonlinear Friction Mechanisms

For a shear wave drive, the surfaces of the contact interface are mechanically coupled by the friction force caused by the interaction between asperities. If the driving amplitude is small enough, the interface shear motion is constrained by the interaction between neighbouring asperities that prevents the contact surfaces from sliding (micro-slip mode). The mechanical diode model for the micro-slip motion is shown in

Fig. 2.3, left and demonstrates a step-wise increase in tangential stiffness as the neighbouring asperities interact. This interaction is independent of the direction of shear motion and causes stiffness variation twice for the input signal period (Fig. 2.3, right). Such a constraint introduces a symmetrical nonlinearity and provides only odd harmonic generation. Like for the clapping mechanism, their amplitudes are *sinc*-modulated due to pulse-type stiffness variation (Fig. 2.4):

$$A_{2N+1} = 2\Delta C\varepsilon_0\left(\frac{\tau}{T}\right)\left\{\mathrm{sinc}\left(\frac{2N\tau}{T}\right)+\mathrm{sinc}\frac{2(N+1)\tau}{T}\right\} \tag{2.2.1}$$

and exhibit similar non-power dynamics.

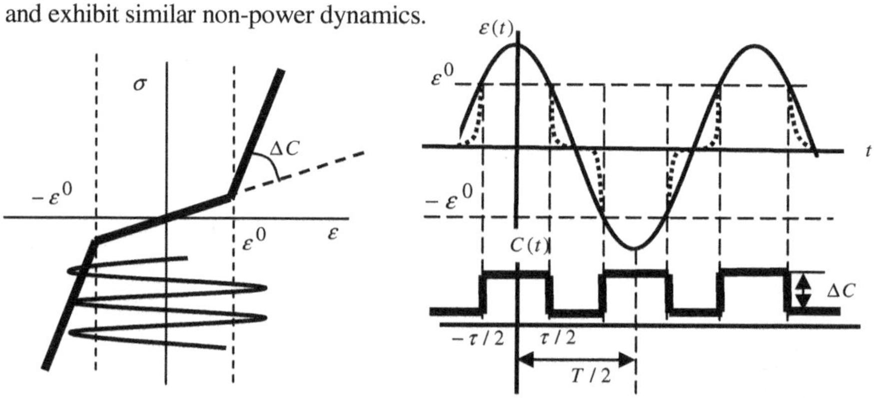

Fig. 2.3 Mechanical diode model (left), stiffness modulation and waveform distortion in micro-slip mode (right).

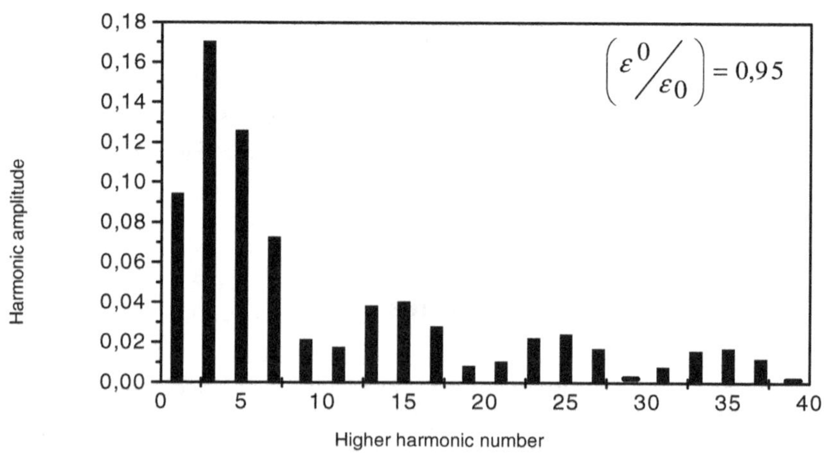

Fig. 2.4 Sinc-modulated odd higher harmonic CAN spectrum in micro-slip mode.

When the amplitude of tangential traction is greater than the contact static friction force the micro-slip motion changes for sliding. An oscillating shear wave drive is accompanied by a cyclic transition between static and kinematic friction (stick-and-slide mode) so that the contact stress-strain relation follows a hysteric loop [3]. The contact tangential stiffness changes symmetrically (independent of the direction of motion) between the static (for a stick phase) and dynamic values (in slide phase) twice over the input strain period that provides odd higher harmonics generation. Similar to the above, the CAN features *sinc*-spectrum modulation and non-power dynamics.

3. CAN Nonlinear Resonance Modes

Besides the higher harmonic generation, the experiments [2, 4] also revealed different scenarios of CAN dynamics which expand considerably nonlinear spectra of cracked defects. These scenarios exhibit the forms of dynamic instability, i.e. an abrupt change of the output for a slight variation of the input parameters. To illustrate the feasibility of the new nonlinear vibration modes and ascertain their basic spectral patterns, we assume that the crack exhibits both resonance and nonlinear properties and thus is identified as a nonlinear oscillator [2]. Its characteristic frequency (ω_0) is determined by a linear stiffness and an associated mass of the material inside the damaged area. The contact nonlinearity is introduced as displacement (*X*) dependent nonlinear interaction force $F^{NL}(X)$. The driven vibrations (driving force $f(t) = f_0 \cos \nu t$) of the nonlinear oscillator are found as a solution to the nonlinear equation:

$$\ddot{X} + \omega_0^2 X = f(t) + F^{NL}(X) \tag{3.1}$$

In the second order of the perturbation approach $F^{NL} \sim \cos(\nu - \omega_0)$ that accounts for the interaction between driving and natural frequency vibrations. If $\nu - \omega_0 \approx \omega_0$, the resonance increase in the output at $\omega_0 \approx \nu / 2$ is observed (subharmonic generation). The higher-order terms in the interaction correspond to the frequency relation $m\nu - n\omega_0$ that provides resonance output at $\omega_0 \approx m\nu / (n+1)$. For $n = 1$, the crack generates ultra-subharmonics (USB) of the second order $m\nu / 2$; the higher order USB correspond to the higher values of *n*.

In reality, a damaged area has a more complicated structure that can be conceived as a set of coupled nonlinear oscillators. If the frequency of the driving acoustic wave is $\nu \approx \omega_\alpha + \omega_\beta$, the difference frequency components $\nu - \omega_\alpha \approx \omega_\beta$ and $\nu - \omega_\beta \approx \omega_\alpha$

provide cross excitation of the coupled oscillators. It results in a resonant generation of the frequency pair ω_α, ω_β centred around the subharmonic position. The higher-order nonlinear terms in (3.1) expand the CAN spectrum which comprises multiple ultra-frequency pairs (UFP) centred around the higher harmonics and USB [2].

The USB and UFP belong to the class of the instability modes and can be interpreted, respectively, as a half-frequency and combination frequency decay of a high-frequency phonon (driving frequency signal). The resonance instability manifests in the avalanche-like amplitude growth beyond the input threshold [5]. The reverse amplitude excursion results in bistability [4]: the input amplitudes for the up and down transitions are different (amplitude hysteresis). Such dynamics is a distinctive signature of the nonlinear acoustic phenomena associated with nonlinear resonance.

4. Experimental Study of Non-Classical CAN Spectra

The examples of the USB and UFP spectra observed in damaged materials by using nonlinear laser vibrometry (NLV) [6] are shown in Figs. 4.1, 4.2.

The USB spectrum in Fig. 4.1 is measured in a cracked area of a polystyrene plate driven at ~1.3.kHz with a shaker. The higher harmonic pattern changes abruptly for the USB spectrum as the driving amplitude grows beyond a certain threshold value. The "wavy" amplitude modulation in Fig. 4.1 indicates involvement of the CAN mechanisms into USB generation.

Fig. 4.2 shows a section of the nonlinear spectrum measured in a glass-fiber reinforced composite (GFRP) with an impact damage for a 20-kHz excitation beyond the UFP-threshold. One can identify the positions of the second (40 kHz) and third (60 kHz) harmonics as well as ultra-subharmonics (50 & 70 kHz). The UFP lines are centred around the USB positions and distanced by $\Delta \cong 1.2$ kHz. The UFP signals with larger Δ and smaller amplitude are also seen in Fig. 4.2.

The experimental results on the nonlinear dynamics obtained with fractured flaws are summarized schematically in Fig. 4.3 for a defect represented by a pair of coupled oscillators (normal frequencies ω_1 and ω_2) [6]. At low amplitude of the driving excitation (frequency ν), the nonlinear spectrum follows the non-resonant scenario of the previous section and results in the higher harmonic generation. As the input amplitude exceeds the threshold value, resonance instability generally leads to activation of the USB components first.

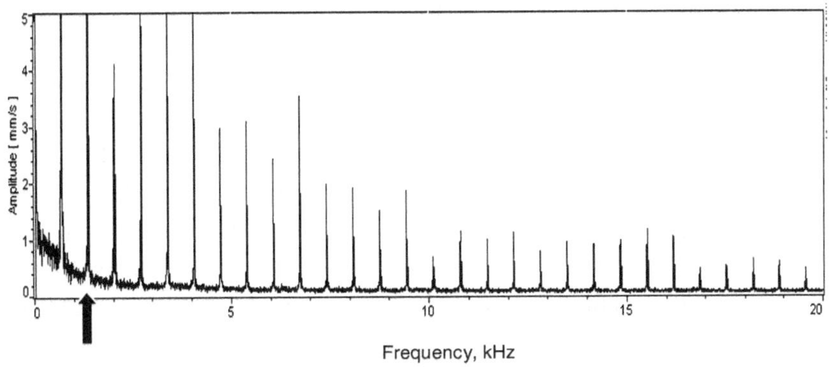

Fig. 4.1 USB spectrum in cracked area of polystyrene plate. The arrow indicates driving frequency.

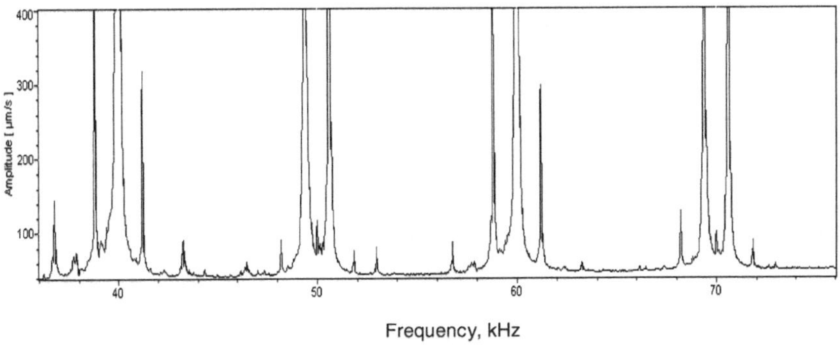

Fig. 4.2 UFP spectrum in impact damaged area of GFRP-specimen. Driving frequency is 20 kHz.

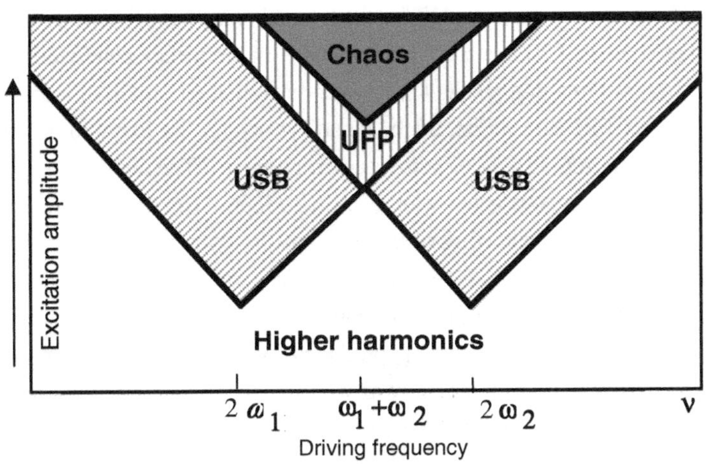

Fig. 4.3 Schematic of nonlinear dynamics of a cracked defect presented as a pair of coupled nonlinear oscillators.

The threshold amplitude depends on the driving frequency: a minimal threshold requires frequency matching to the main subharmonic resonance ($\nu = 2\omega$). The frequency zones for the USB generation expand readily as the excitation amplitude increases. Further increase of acoustic excitation above a given threshold gives rise to the UFP spectra. A direct transition to the UFP-instability is also possible when the sum-frequency resonance matching conditions are satisfied. The *V*-shaped zones in Fig. 4.3 are typical for parametric resonance modes [5] and indicate that the frequency matching is not required for high driving amplitudes. Finally, the multiple UFP bring the system to a quasi-continuous spectrum which indicates a build-up of chaotic vibrations.

5. CAN Application for Nonlinear Imaging and NDE

The nonlinear spectra discussed above are produced locally in the damaged area while an intact part of material outside the defects vibrates linearly, i.e. with no frequency variation in the output spectrum. Thus, nonlinear defects are ***active*** sources of new frequency components rather than ***passive*** scatterers in conventional ultrasonic testing. This makes nonlinearity a ***defect-selective*** indicator of damage presence and development. The high localization of nonlinear spectral components around the origin is a basis for nonlinear imaging of damage.

The NLV [6] uses a sensitive scanning laser interferometer for detecting nonlinear vibrations of defects. The excitation system includes piezo-stack transducers operating at 20 and 40 kHz. After a 2D-scan and FFT of the signal received, the C-scan images of the sample area are obtained for any spectral line within the frequency bandwidth of 1 MHz.

Figure 5.1 shows imaging results for an oval delamination on top of a piezo-actuator embedded into a GFRP composite. Such “smart” structures are likely to be used for active structural health monitoring of aerospace components. The actuator itself was used as an internal excitation source fed with a few volt input. The higher harmonic images selectively reveal the boundary ring of the delamination where clapping and rubbing of the contact surfaces are, apparently, expected. On the contrary, the driving frequency (50 kHz) image indicates only a standing wave pattern over the area of the actuator.

Fig. 5.2, left shows fatigue cracking produced by cyclic loading in Ni-base superalloy. Such a crack of ~ 1.5 mm length, with average distance between the edges of only ≈ 5μm, is clearly visualized in the USB-image (Fig. 5.2, right) whereas linear NDE using slanted ultrasonic reflection failed to detect such small cracks.

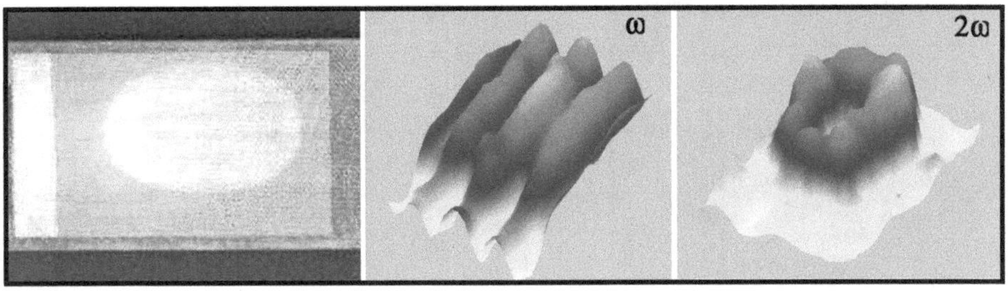

Fig. 5.1 Fundamental frequency (ω) and higher harmonic imaging of a delamination in a "smart" structure.

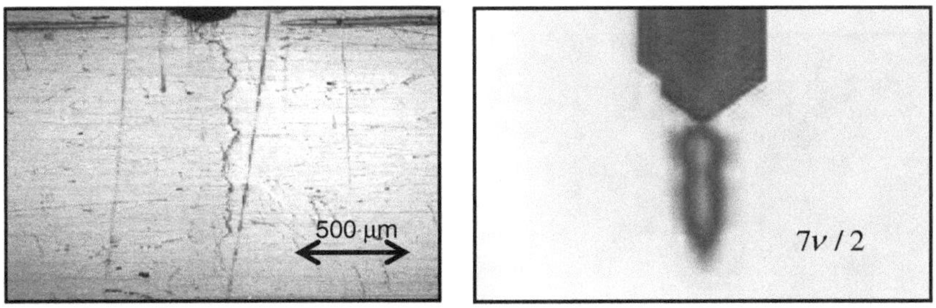

Fig. 5.2 Right: USB-image of 5µm-wide fatigue crack in Ni-base super-alloy; left: crack photo.

Similarly to all nonlinear modes discussed, the UFP-components generally display a strong spatial localization around the defects and are applicable for the detection of damage. The benefit of the UFP-mode is illustrated in Fig. 5.3 for a 14-ply epoxy based GFRP composite with a 9.5J-impact damage in the central part. The linear image at driving frequency of 20 kHz reveals only a standing wave pattern over the whole sample (Fig. 5.3, left). The image at the first UFP-side-lobe of the 10th harmonic of the driving frequency (198.8 kHz) yields a clear indication of the damaged area (Fig. 5.3, right).

The scanning laser vibrometer suffers from variation of optical reflectivity; e.g. the measurements fail in the damaged areas with particularly strong scattering of laser light. Our experiments demonstrated that planar defects as localized sources of nonlinear vibrations efficiently radiate nonlinear airborne ultrasound. Such a nonlinear air-coupled emission (NACE) is proposed as an alternative (and in many cases superior) methodology to locate and visualize the defects in NDE [7].

A practical version of the NACE for nonlinear imaging of defects uses a high-frequency focused air-coupled (AC-) ultrasonic transducer as a receiver [8].

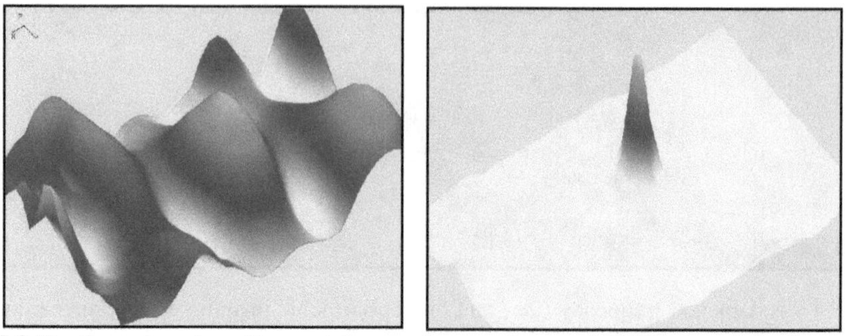

Fig. 5.3 Nonlinear imaging of impact damage in central part of GFRP plate: left - linear (20 kHz-image); right - UFP – image.

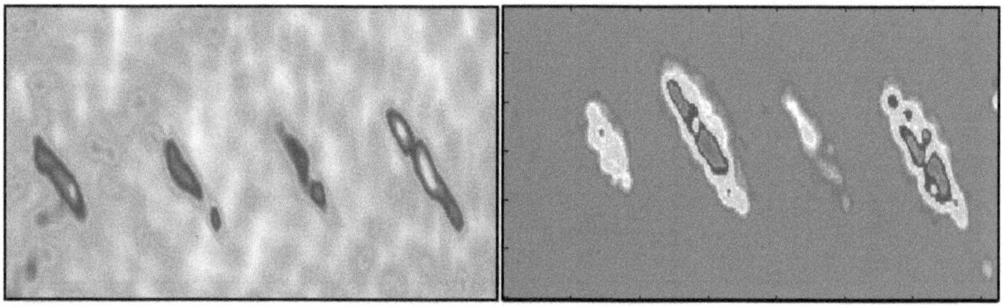

Fig. 5.4 Nonlinear imaging of an impact induced damage in multi-ply ($+45^0$; -45^0) GFR-plate: NLV (left); NACE (9th-11th) higher harmonic image (right).

Fig. 5.5 NACE imaging in steel specimens: (9-11th) harmonic imaging of 50 μm-wide fatigue crack (left); (5x40 *mm*) hammer peening area in steel plate (right).

In Fig. 5.4, the NACE imaging results are compared with NLV of multiple impact damage on a reverse side of a carbon fibre-reinforced (CFR-) multi-ply ($+45^0$; -45^0) composite plate (175x100x1 mm). Both techniques reliably visualize the defects with similar sensitivity.

Fig. 5.5 (left) shows the (9-11th) harmonic NACE image of the 50μm-wide fatigue crack in a steel plate (150x75x5 mm) with two horizontally located grip holes for cyclic loading at some distance from the crack. The image reveals that the NACE detects not only the crack itself but also the fatigue structural damage in the plasticity areas between the crack and the grip holes.

To verify the NACE sensitivity to micro-damage induced by plastic deformation, the NACE inspection was implemented for a steel specimen with a cold work area (5x40 mm) produced by hammer peening. The image in Fig. 5.5, right confirms that the NACE develops even without serious cracked defects and clearly discerns the micro-damage induced by plastic deformation.

References

[1] Solodov, I.Yu.: Ultrasonics of nonlinear contacts: propagation, reflection and NDE-applications. *Ultrasonics* **36** 383-390 (1998).

[2] Solodov, I., Wackerl, J., Pfleiderer, K., Busse, G.: Nonlinear self-modulation and subharmonic acoustic spectroscopy for damage detection and location. *Appl. Phys. Lett.* **84** 5386-5388 (2004).

[3] Pecorary, C., Solodov, I.: Non-classical nonlinear dynamics of solid interfaces in partial contact for NDE applications. In: Delsanto, P. (ed.) *Universality of Non-Classical Nonlinearity with Application to NDE and Ultrasonics*, Ch. 19, pp. 307-324, Springer, New York, USA (2006).

[4] Solodov, I., Korshak, B.: Instability, chaos, and "memory" in acoustic wave-crack interaction. *Phys. Rev. Lett.* **88** 014303, 1-3 (2002).

[5] Minorsky, N.: *Nonlinear Oscillations*. D. Van Nostrand Co. Inc., Princeton (1962).

[6] Solodov, I., Pfleiderer, K., Busse, G.: Nonlinear acoustic NDE: Inherent potential of complete nonclassical spectra. In: Delsanto, P. (ed) *Universality of Non-Classical Nonlinearity with Application to NDE and Ultrasonics*, Ch. 29, pp. 465-484, Springer, New York, USA (2006).

[7] Solodov, I., Busse, G.: Nonlinear air-coupled emission: the signature to reveal and image micro-damage in solid materials. *Appl. Phys. Lett.* **91** 251910 (2007).

[8] Solodov, I., Busse, G.: Listening for nonlinear defects: a new methodology for nonlinear NDE. In: Enflo, B.O. et al. (ed.) *Nonlinear Acoustics - Fundamentals ands Applications,* AIP, pp. 569-573 (2008).

Counterpropagating Ultrasonic Waves for Inhomogeneous Materials Characterization

Arvi Ravasoo

Centre for Nonlinear Studies, Institute of Cybernetics at Tallinn University of Technology, Akadeemia tee 21, 12618 Tallinn, Estonia

arvi@ioc.ee

Abstract. The interaction of two counterpropagating waves in an inhomogeneously prestressed material (a structural element) and in a physically nonlinear material with essentially changing continuous properties is studied theoretically. The influence of the parameters of a two-parametric prestressed state and the exponentially changing material properties on the profiles of boundary oscillations, evoked in the specimen with two parallel boundaries by the counterpropagating waves is clarified. Algorithms for qualitative and quantitative ultrasonic nondestructive characterization of two-parametric prestress and for qualitative characterization of essentially and exponentially changing properties of the elastic material with quadratic nonlinearity are proposed. Utilization of these algorithms in practical applications of ultrasonic nondestructive characterization of inhomogeneous materials is illustrated on the basis of numerical simulation data.

1. Introduction

The progress in material science and technology requires very precise determination of the physical properties and states of materials. The utilization of ultrasonic waves is one of the very promising methods for this purpose [1-3]. Ultrasonic wave speed measurements are widely used to infer the homogeneous properties and states of solids. The inhomogeneity in materials leads to the necessity of extracting more information from the data of ultrasonic nondestructive testing [4-5]. This motivates to develop new ultrasonic techniques based on interaction of waves in the material, on nonlinear effects of wave propagation, on the data about distortion of wave profiles by propagation, etc.

T.-T. Wu and C.-C. Ma (eds.), *IUTAM Symposium on Recent Advances of Acoustic Waves in Solids*, IUTAM Bookseries 26, DOI 10.1007/978-90-481-9893-1_6,

In this paper counter-propagation, reflection and interaction of two ultrasonic waves in an inhomogeneously prestressed material (structural element) and in a material with essentially changing continuous physical properties is investigated. It is shown that material inhomogeneity generates distortion in the profiles of interacting waves. This phenomenon enables to propose enhanced algorithms for qualitative and quantitative ultrasonic nondestructive characterization of inhomogeneous materials. Established ultrasonic methods are illustrated by two model problems.

The first model problem is: the ultrasonic nondestructive characterization of inhomogeneous two-parametric prestress in the physically nonlinear elastic material (structural element).

A two-dimensional specimen or a structural element with two parallel boundaries is considered. In the civil engineering applications such objects may be, for example, the plates with different types of supports, thin-walled open sections, etc. The specimen is undergoing two-dimensional prestress that corresponds to plane strain. Two counter-propagating longitudinal one-dimensional harmonic waves are excited simultaneously on opposite boundaries of the specimen. Wave motion is studied theoretically using nonlinear theory of elasticity and the perturbation technique. The nonlinear part of the obtained solution describes nonlinear effects of wave motion that are sensitive to the nonlinear physical properties of the material and to the parameters of prestress. Analyses of the influence of prestress on the nonlinear distortion of the profiles of boundary oscillations enables to pose the problems of qualitative and quantitative nondestructive characterization of prestress in the material.

The second model problem is the ultrasonic nondestructive characterization of the nonlinear elastic material with essentially changing continuous physical properties.

Man-made materials like composite materials called functionally graded materials (FGMs) [7-8] are considered. The possibility for qualitative nondestructive characterization of the exponentially graded nonlinear elastic FGMs is studied in detail. This group of FGMs is widely used not only in the thermal-protection systems of space planes but also in electrical, chemical and many other technologies.

The ultrasonic counterpropagating waves are excited in a specimen simultaneously on two parallel boundaries. The one-dimensional problem is governed theoretically by the equation of motion in the form of the second order partial differential equation with variable space coefficients. The analytical solution to this equation is unknown and therefore it is solved numerically using the programme package Maple.

Analyses of the results of numerical simulations verifies the fact that the profiles of the recorded boundary oscillations are informative about the variable material properties. This enables to propose a method for qualitative nondestructive characterization of FGMs with essentially changing continuous properties.

2. Problem Formulation

The intention is to excite simultaneously two ultrasonic longitudinal waves in a material (specimen, structural element, etc.) with inhomogeneous properties. This means a certain geometrical constraint - a specimen must have two parallel surfaces. Objects with parallel surfaces are widely used in civil engineering and machinery. Two different specimens are considered: (i) the physically nonlinear elastic material (structural element) undergoing inhomogeneous prestress and (ii) the physically nonlinear elastic FGM (structural element) with essentially changing continuous properties. In both cases the small but finite deformations of the material are described by the nonlinear theory of elasticity [9] in Lagrangian rectangular coordinates X_J, $J = 1, 2, 3$. The second Piola-Kirchhoff stress tensor $T_{KL}(X_J,t)$, $K,L = 1, 2, 3$, where t denotes the time is used for stress characterization. The physical and geometrical nonlinearity of the problem is taken into account.

2.1 Prestressed Material

The material of most structures is itself isotropic and homogeneous but the inhomogeneity may be introduced by prestress. This process is studied theoretically by the assumption that the components of the displacement vector at this present state $U^*_K(X_J,t)$ are expressed by the formula

$$U^*_K(X_J,t) = U^0_K(X_J) + U_K(X_J,t), \tag{2.1.1}$$

where $U^0_K(X_J)$ and $U_K(X_J,t)$ are displacements evoked by prestress and wave motion, respectively.

The equilibrium of the specimen undergoing static plane strain ($U^*_3(X_J,t) = U^0_3(X_J) = 0$) is described by a system of two elliptic second-order partial differential equations [10]

$$\begin{aligned}
&\left(1+k_1U^0_{I,I}+k_2U^0_{J,J}\right)U^0_{I,I}+\left(2+k_3U^0_{I,J}+2k_4U^0_{J,I}\right)U^0_{I,IJ}\\
&+\left(k_7+k_3U^0_{I,I}+k_3U^0_{J,J}\right)U^0_{I,JJ}+\left(k_4U^0_{I,J}+k_3U^0_{J,I}\right)U^0_{J,II}\\
&+\left(k_3U^0_{I,J}+k_4U^0_{J,I}\right)U^0_{J,JJ}+\left(k_6+k_5U^0_{I,I}+k_5U^0_{J,J}\right)U^0_{J,JI}=0.
\end{aligned} \tag{2.1.2}$$

Here indices $I = 1$, $J = 2$ specify the first equation and $I = 2$, $J = 1$ the second equation. The coefficients k_n, $n = 1, 2, \ldots, 7$ are certain functions of the Lamé constants λ and μ, the elastic constants of the third order ν_1, ν_2 and ν_3 and the density ρ_0 [10]. Indices I and J after a comma indicate differentiation with respect to X_I and X_J, respectively.

The quasi-one-dimensional problem of longitudinal wave propagation in two-dimensional specimen is governed by the equation of motion [10]

$$(1+f_1)U_{1,11}+f_2U_{1,1}+f_3U_{1,1}U_{1,11}-c^{-2}U_{1,tt}=0. \tag{2.1.3}$$

The coefficients of Eq. (2.1.3) are functions of prestress and material properties [10].

2.2 Functionally Graded Material

Considerable interest is generated in recent years for man-made materials like composite materials called functionally graded materials (FGMs) with essentially changing continuous physical properties. Here, FGMs with physical properties that are described by the theory of elasticity with quadratic nonlinearity [9] are considered. The variables in space material properties are the density ρ, the second order elastic Lamé coefficients λ and μ and the third order elastic coefficients ν_1, ν_2 and ν_3. In the case of one-dimensional deformations the elastic coefficients are grouped to the linear elastic coefficient $\alpha = \lambda + 2\mu$ and to the nonlinear elastic coefficient $\beta = 2(\nu_1 + \nu_2 + \nu_3)$.

The one-dimensional response of the material to the external loading is governed by the equation of motion [11]

$$(1+k_8U_{,X})U_{,XX}+k_9U_{,X}+k_{10}U_{,X}^2-k_{11}U_{,tt}=0, \tag{2.2.1}$$

where U denotes the displacement, t the time and the indices after the comma indicate differentiation with respect to the coordinate X or the time t, accordingly. The coefficients of the Eq. (2.2.1) are functions of the variable in space material properties ρ, α and β [11].

The specimen of the exponentially graded material is considered. The thickness of the specimen with two parallel boundaries is denoted as h and the variation of essentially changing continuous properties of the specimen are determined by the formula

$$\gamma(X)=\gamma_0\left[1+\gamma_{11}\exp\left(-\gamma_{12}X\right)+\gamma_{21}\exp\left(-\gamma_{22}\left(X-h\right)\right)\right], \tag{2.2.2}$$

where $\gamma=\rho$, α, β, respectively.

Three different schemes of inhomogeneity are studied. The scheme A – a symmetric exponential variation of material properties nearby both boundaries, the scheme B - exponential variation of material properties just nearby the boundary $X = 0$ and the scheme C - the same just nearby the boundary $X = h$.

3. Counterpropagating Waves

The wave process in the prestressed material is governed by Eq. (2.1.3). To solve this equation it is necessary to determine the coefficients, i.e., to have some preliminary information about the prestressed state of the specimen. This information may be obtained from the observation data of the loading scheme of the specimen (structural element). Here the problem is solved by assumption that the type of prestressed state and the physical properties of the material are known. Theoretically, unknown parameters of the prestressed state are determined by the solution to the set of Eqs. (2.1.2). As it was mentioned above, the small but finite deformations of the specimen are described by the nonlinear theory of elasticity. Consequently, the strain is small. This enables to introduce a small parameter ε and to solve both governing Eqs. (2.1.2) and (2.1.3) making use of the perturbation technique.

The prestressed state is described by the perturbative analytical solution to the Eqs. (2.1.2) in the form of series

$$U_K^0\left(X_1,X_2\right)=\sum_{m=1}^{\infty}\varepsilon^m U_K^{0(m)}\left(X_1,X_2\right) \tag{3.1}$$

with a small parameter $|\varepsilon| \ll 1$. Solutions to Eqs. (2.1.2) are sought in the form of polynomials with respect to coordinates X_1 and X_2 for a special case of prestress that corresponds to the pure bending with compression or tension. The result is that the problem of ultrasonic nondestructive characterization of the prestressed state turns to determination of the constant coefficients of polynomials.

Now it is possible to solve Eq. (2.1.3) with known space dependent variable coefficients. Again, the perturbation technique is used and the solution to equation (2.1.3) is sought in the form

$$U_1(X_1, X_2, t) = \sum_{n=1}^{\infty} \varepsilon^n U_1^{(n)}(X_1, X_2, t). \tag{3.2}$$

Equation (2.1.3) is solved following the perturbation procedure under the initial conditions equal to zero and the boundary conditions in terms of particle velocity.

The obtained analytical solution, which is too cumbersome to be presented here, describes the initial stage of counter-propagation of waves with arbitrary smooth initial profiles in a specimen undergoing pure bending with tension or compression.

The wave motion in FGMs with essentially changing continuous properties is governed by the Eq. (2.2.1). The analytical solution of this equation is unknown. Counter-propagation of longitudinal harmonic waves in exponentially graded FGMs is studied numerically making use of the symbolic manipulation software Maple.

4. Nondestructive Material Characterization

The effects of wave-wave, wave-material and wave-prestress interaction accompany counter-propagation of waves in materials. All these effects may be successfully employed by ultrasonic nondestructive characterization of materials with complex properties. Wave-wave interaction is on principle a nonlinear phenomenon. In the linear case the superposition of waves occurs. Another nonlinear phenomenon is the wave-prestress interaction. Theoretically, deformation fields evoked in the material by the wave motion and prestress are bounded, provided the geometrical nonlinearity of the problem is taken into account. The wave-material interaction is a fundamental phenomenon and this takes place regardless the models (linear or nonlinear) of wave propagation in the material.

Ultrasonic nondestructive characterization of inhomogeneous materials leads to the necessity of extraction of more information from the wave propagation data. Therefore the interest may be directed to utilization of different interaction phenomena and accompanied nonlinear effects for enhancement of nondestructive testing techniques. One possible way in this direction is illustrated below by two examples.

4.1 Inhomogeneous Prestress

The idea is to solve the problem of nondestructive determination of material inhomogeneity (prestress) on the basis of the data about counter-propagation and interaction of waves in the specimen. It is assumed that the geometry of the specimen (structural element) and the physical properties of the material are known. The inspection confirms the fact that the specimen is undergoing prestress that corresponds to the pure bending with tension or compression (Fig. 4.1). The solution (3.1) to Eq. (2.1.2) describes this case as a two-parametric prestressed state characterized by the main domain of the component of the second Piola-Kirchhoff stress tensor $T_{22} = a + bX_1$. The purpose is to evaluate constants a and b on the basis of wave interaction data.

Harmonic waves with the same amplitude and frequency are excited in terms of particle velocity on opposite boundaries of the specimen undergoing two-parametric prestress. The evolution of wave profile is recorded on the same boundaries in terms of stress.

The linear part of the solution (the first term in solution (3.2)) describes simultaneous propagation of two harmonic waves in the prestress free physically linear material where the interaction of waves is determined by superposition of wave profiles.

The nonlinear effects that accompany counter-propagation of harmonic waves are sensitive to the nonlinear physical properties of the material and to the parameters of prestress. The main part of these effects is described by the second term in solution (3.2). Nonlinear effects include evolution of the second harmonic, influence of the prestress to the evolution of the first harmonic, nonlinear interaction between two first harmonics and influence of the nonlinear physical properties of the material to the wave propagation. Evolution of the third and the higher harmonics are neglected here as the higher orders are small phenomena.

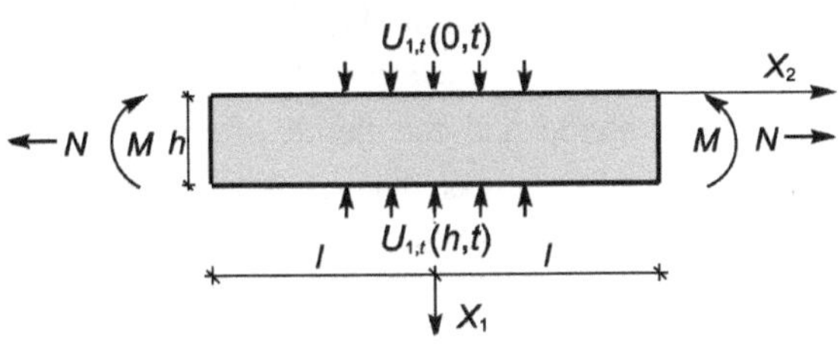

Fig. 4.1 Loading scheme of the prestressed specimen (N – normal force, M – couple).

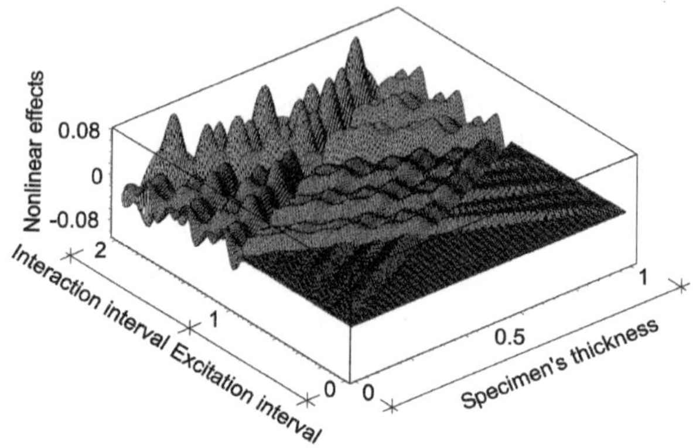

Fig. 4.2 Evolution of nonlinear effects in a prestressed material.

The informative nonlinear part of boundary oscillations evoked by the counterpropagating harmonic waves with equal amplitudes in the prestressed physically nonlinear elastic material (duralumin, thickness $h = 0.1$ m) is illustrated in Fig. 4.2. Oscillations coincide on both boundaries. It is possible to distinguish two intervals on the time axis – the interval of excitation and the interval of interaction. Essential is, that the nonlinear wave interaction amplifies the boundary oscillation amplitude in the interval of interaction. This phenomenon facilitates utilization of nonlinear effects of wave interaction in nondestructive characterization of prestress.

4.1.1 Qualitative prestress characterization

Nonlinear parts of boundary oscillations in the prestress free material are characterized by constant but different values of the amplitudes in the interval of excitation and interaction. These amplitudes are sensitive to the linear physical properties of the material (density, Lamé constants) and less sensitive to the value of the excitation frequency. Homogeneous prestress ($T_{22} = a$) modulates the boundary oscillation (Figs. 4.2). The shape and the depth of modulation are informative about the sign and the value of prestress. The oscillation profiles on both boundaries coincide. The inhomogeneous prestress ($T_{22} = a + b\,X$) modulates oscillations on different boundaries in different way.

The analysis of the influence of prestress on the boundary oscillations leads to the conclusion that nonlinear effects caused by counter-propagation of two harmonic waves enable to pose the problem of qualitative nondestructive characterization of prestress in the material. It is possible to determine qualitatively the presence and the nature of prestress and to distinguish (i) prestress free material, (ii) homogeneously prestressed material, (iii) material undergoing pure bending and (iv) material undergoing arbitrary prestress.

4.1.2 Quantitative prestress characterization

The dependence of the depth of modulation of nonlinear boundary oscillations provoked by counter-propagation of two harmonic waves in the material enables to pose a problem of quantitative characterization of inhomogeneous prestress.

The plots of nonlinear oscillations on the boundaries of the prestress free material are composed on the basis of the analytical solution (3.2). Two first local maxima of the boundary oscillation amplitudes on both boundaries are determined and characterized by the values of instants τ_1 and τ_2.

The next step is to compose plots of boundary oscillation amplitudes versus prestress parameters a and b for both instants τ_1 (Fig. 4.3) and τ_2.

In order to evaluate the unknown real values of prestress parameters a and b in a physical experiment, the counterpropagating harmonic waves are excited in a prestressed specimen and the oscillation profiles are recorded on both boundaries. The difference of the values of oscillation amplitudes on opposite boundaries is determined for both instants τ_1 and τ_2. Resorting to the corresponding plots, two possible values of the parameter b are determined making use of the calculated differences. The value of the parameter a and the final value of the parameter b are determined making use of the value of the recorded oscillation amplitude on one of the boundaries at the instant τ_1 or τ_2.

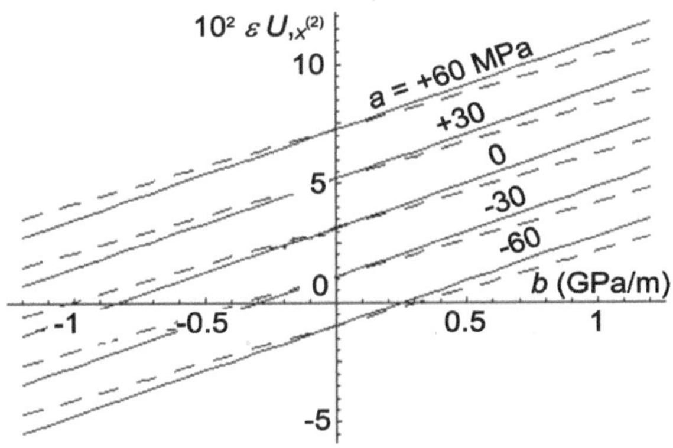

Fig. 4.3 Boundary oscillation amplitude variation at the instant τ_1 versus prestress parameters a and b (solid line – $X = 0$, dashed line – $X = h$).

4.2 Functionally Graded Material

The problem of ultrasonic nondestructive characterization of functionally graded materials (FGMs) with essentially changing continuous physical properties is solved on the basis of the equation of motion of the material (2.2.1). It is assumed that the geometry of a specimen and the type of physical properties of the material are known. Here, the exponentially graded nonlinear elastic material is studied. Due to analytical complexity the problem of counter-propagation of harmonic waves in the material is treated numerically using the symbolic manipulation software Maple.

The loading scheme of the prestress free specimen is similar to that in Fig. 4.1 (no N, no M). Material properties vary into the depth of the specimen (along the axis $X \equiv X_1$) in accordance with three different schemes described above. One-dimensional wave motion is excited simultaneously on boundaries $X = 0$ and $X = h$ in terms of stress and the evoked boundary oscillations are recorded in terms of particle displacement.

Numerous numerical experiments are posed and analysed. The material is characterized by the density $\rho = 6000$ kg/m, the linear elasticity $\alpha = 400$ GPa and the nonlinear elasticity $\beta = -1000$ GPa. Variation of material properties is determined by the values of constants $\gamma_{i1} = 1$, $\gamma_{i2} = 150$ m^{-1} ($i = 1, 2$) in Eq. (2.2.2). The excitation frequency equals to $\omega = 1.5391 \cdot 10^6$ rad/s.

Analyses of the results of numerical simulations (Fig. 4.4 et al.) lead to the conclusion that the changes of boundary oscillation profiles caused by variable density and linear elasticity are of the same order while changes caused by nonlinear elasticity are the higher order small phenomena. The changes in material properties are clearly reflected in changes of boundary oscillation profiles.

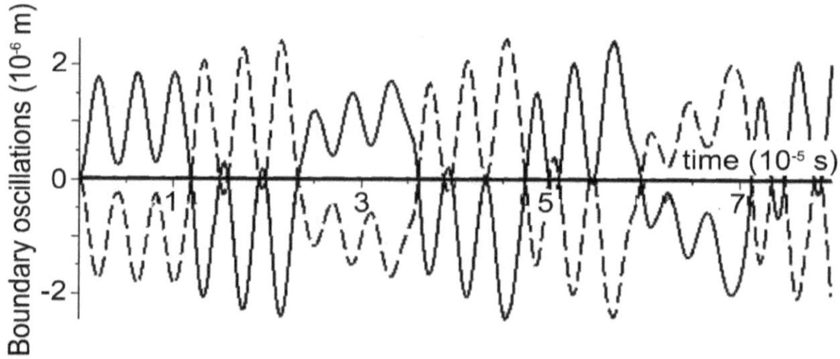

Fig. 4.4 Oscillations on opposite boundaries of the specimen with exponentially graded properties (solid line – $X = 0$, dashed line – $X = h$).

Consequently, on the basis of boundary oscillation profile data, it is possible to propose a method for qualitative nondestructive characterization of FGMs with essentially changing continuous properties. This qualitative method enables to distinguish specimens made by (i) homogeneous material, (ii) symmetrically distributed material properties, (iii) asymmetrically distributed material properties and also distinguish the most relevant property of the material responsible for inhomogeneity.

5. Conclusions

In this paper a relatively simple method is proposed for nondestructive ultrasonic characterization of materials with continuously variable properties. The method is based on utilization of the data about nonlinear counter-propagation, interaction and reflection of ultrasonic waves in the material (specimen, structural element).

In comparison with single wave experiments, the amount of information available from multi-wave propagation, reflection and interaction data increases essentially. This is demonstrated by model problems of counter-propagation of two harmonic waves in the physically nonlinear elastic material (structural element) undergoing two-parametric prestress and in the exponentially graded physically nonlinear elastic material.

The information embedded in nonlinear effects of boundary oscillations evoked by counterpropagating waves is sufficient to solve qualitative and quantitative problems of nondestructive characterization of two-parametric inhomogeneous prestress.

Qualitative nondestructive characterization, i.e. identification of the type of inhomogeneity (symmetric, asymmetric, etc.) of physically nonlinear material with essentially changing continuous properties (FGMs) may be implemented on the basis of data of boundary oscillation profile distortions.

Acknowledgements: The research was supported by the Estonian Science Foundation through the Grant No. 7728.

References

[1] Krautkrämer, J., Krautkrämer, H.: *Ultrasonic Testing of Materials.* 3rd Edition, Springer, Berlin (1990).

[2] Achenbach, J.D.: *The Evaluation of Materials and Structures by Quantitative Ultrasonics.* Springer, NY (1993).

[3] Si-Chaib, M.O., Menad, S., Djelouah, H., Bocquet, M.: An ultrasound method for the acoustoelastic evaluation of simple bending stresses. *NDT&E International* **34** 521-529 (2001).
[4] Baganas, K.: Wave propagation and profile reconstruction in inhomogeneous elastic media. *Wave Motion* **42** 261-273 (2005).
[5] Liu, G.R., Han, X., Lam, K.Y.: Stress waves in functionally gradient materials and its use for material characterization. *Composites: Part B* **30** 383-394 (1999).
[6] Ravasoo, A.: Non-linear interaction of waves in prestressed material. *Int. J. Non-Linear Mech.* **42** 1162-1169 (2007).
[7] Suresh, S., Mortensen, A.: *Fundamentals of Functionally Graded Materials*. IOM Comm. Ltd, London (1998).
[8] Chin, E.S.C.: Army focused research team on functionally graded armour composites. *Material Science and Engineering* **A259** 155-161 (1999).
[9] Eringen, A.C.: *Nonlinear Theory of Continuous Media*. McGraw-Hill, NY (1962).
[10] Ravasoo, A., Lundberg, B.: Nonlinear interaction of longitudinal waves in an inhomogeneously predeformed elastic medium. *Wave Motion* **34** 225-237 (2001).
[11] Ravasoo, A.: Nonlinear Waves in Characterization of Inhomogeneous Elastic Material. *Mechanics of Materials* **31** 205-213 (1999).

Ultrasonic Characterization of the Mechanical Properties of Thin Films and Coatings

Sridhar Krishnaswamy

Department of Mechanical Engineering, Northwestern University, Evanston, IL 60208, USA

s-krishnaswamy@northwestern.edu

Abstract. This paper presents a review of recent work at Northwestern University in the area of ultrasonic characterization of thin film material properties. Contact and non-contact ultrasonic methods that are used to characterize the mechanical properties of micro- and nano-structured thin films are briefly described. Specifically, the methods used are: Line-focus Acoustic Microscopy, narrowband and broadband Photoacoustic guided wave methods, and Photoacoustic bulk-wave pump-probe techniques. Applications to several free-standing thin-films and thin-film on substrate systems are discussed.

1. Introduction

Characterization of the thickness and mechanical properties of thin films and coatings is of great industrial importance. The microelectronics industry uses thin films and multi-layered components in semiconductor chips and surface acoustic wave devices. In many MEMS applications, the mechanical properties of thin membranes are critical to the proper performance of micro-devices. Thin film coatings are also used on conventional engineering materials such as steel to obtain enhanced surface hardness properties and to act as thermal barriers in engine components, for instance.

The mechanical properties of thin films can be investigated using tensile tests, bulge tests, and nano-indentation tests, but these are intrusive and not nondestructive. Nondestructive ultrasonic techniques such as acoustic microscopy using high-frequency contact piezoelectric transducers can be used to characterize both the thickness and the mechanical properties of films and coatings. Photoacoustic techniques, where high frequency ultrasound is launched via rapid thermal expansion in

T.-T. Wu and C.-C. Ma (eds.), *IUTAM Symposium on Recent Advances of Acoustic Waves in Solids*, IUTAM Bookseries 26, DOI 10.1007/978-90-481-9893-1_7,

materials, are nondestructive and also noncontact. The high temporal and spatial resolution of photo-acoustic techniques enables *in-situ* nondestructive evaluation of mechanical properties of ultra-thin films.

At Northwestern University, we use a suite of ultrasonic methods to characterize the mechanical properties of thin films. The techniques include (i) contact acoustic microscopy using a line-focus acoustic microscope (LFAM); (ii) broadband and narrowband photoacoustic (PA) guided-wave acoustic microscopy, and (iii) bulk-wave pump-probe techniques. In all these techniques, the velocities of bulk and/or surface acoustic waves are measured and are related to the elastic properties of the thin films. In this paper, we will briefly introduce the basic aspects of the experimental and analytical methods. Applications to several micro- and nano-structured films are described and discussed.

2. Line-Focus Acoustic Microscopy

Line-focus acoustic microscopy (LFAM) has proved to be an accurate method to determine the mechanical constants of isotropic or anisotropic films that are on an elastic substrate. An acoustic microscope was first used for the quantitative characterization of materials by Weglein [1]. He proposed a method that uses a point-focus acoustic beam to determine the film thickness by measuring the velocity variation of the leaky Rayleigh wave propagating in a thin film/substrate structure. Compared to point-focus beams, a line-focus beam generates wave modes propagating along the normal to the line, and thus it is particularly useful for measuring velocities and elastic constants in anisotropic materials. Kushibiki and Chubachi first systematically established the theoretical and experimental methods for LFAM, and applied it to characterize a variety of materials [2]. The LFAM has since then been used for characterization of mechanical properties of various thin films with thicknesses in the micrometer range [3-6].

LFAM experiments are generally based on measuring the so-called $V(z)$ curve, which is an output of the measured voltage as a function of the distance between the focal plane of the acoustic transducer and the specimen surface denoted by the z-distance. The measurement and analysis of the $V(z)$ curves provides information about the velocities of the transducer generated leaky surface acoustic waves (LSAW) on the sample surface. A simple interference model by ray theory has been established for the determination of characteristics of the LSAW that propagate along the interface between the coupling water and the solid sample.

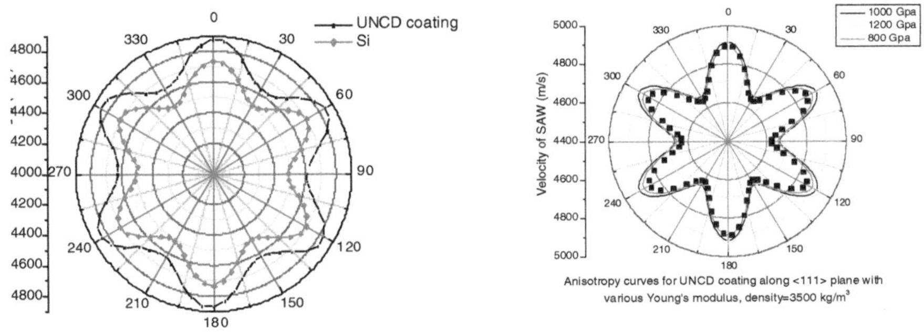

(a) (b)

Fig. 1 (a) Polar plot of anisotropic SAW velocity for plain and UNCD-coated silicon wafer. (b) Comparison with analytical fit for various values of UNCD Young's modulus.

The LSAW velocity is calculated from the periodicity Δz of the interference signal appearing in the $V(z)$ curves and is given by:

$$v_{LSAW} = v_w \Big/ \sqrt{1-(1-v_w/2f\Delta z)^2} \quad (1)$$

where v_w is the acoustic velocity in water, and f is the operating frequency of the transducer. The experimental procedure of LFAM is as follows: first the $V(z)$ curves are obtained by adjusting the distance between the transducer and sample surface; secondly the fringe interval Δz is determined by filtering and fast Fourier transform; thirdly v_{LSAW} is measured at different acoustic frequencies so as to get a dispersion curve of the LSAW. The mechanical properties of the thin films are then evaluated by least-square fitting the measured dispersion curves to an appropriate theoretical model by iterating over the material properties.

Figure 1 shows a representative result for UNCD-coated and plain Silicon wafer [7]. It is seen that the UNCD layer stiffens the system in comparison to the uncoated silicon wafer. The six8fold symmetry is of the underlying silicon wafer (the UNCD grain structure being on the order of 3-5nm, it is essentially isotropic). A two-layer model was used to obtain the effective Young's modulus of the UNCD coating, which is estimated at 900GPa from Fig. 1(b).

3. Photoacoustic Methods

Photoacoustic methods can be broadly classified as bulk-wave and guided-wave techniques. The technique using a femtosecond laser to launch longitudinal acoustic waves with extremely high frequency (usually higher than gigahertz) resulting from transient photothermal reaction is sometimes called picosecond ultrasonics or femtosecond transient pump-probe technique [8]. Of particular interest in this work are the femtosecond laser generated bulk acoustic waves that propagate perpendicular to the film and are reflected from the film and substrate interface. Guided acoustic waves including surface acoustic wave (SAW) and Lamb waves are also intrinsically related to the elastic properties of layered materials through their dispersive nature. Photoacoustic guided waves have been applied to characterize the mechanical properties thin films and hard coatings with thicknesses ranging from hundreds of nanometers to tens of microns [9-12]. In the following, we describe bulk-wave and guided-wave PA methods of materials characterization.

3.1 Broadband Guided-Wave PA Method

Similar to the LFAM, the PA laser ultrasonic method can also be used to obtain the dispersion curves of surface acoustic waves (SAW) in thin films. It is a direct and convenient non-contact way of thin-film characterization and many applications have been introduced. To launch a broadband SAW on the sample, a pulsed high power pump laser is line focused on the surface to generate a SAW pulse with a frequency bandwidth that is mainly determined by the laser pulse duration and the laser focal spot size (Fig. 2). The shorter the pulse is and the smaller the spot size, the higher the frequency of the generated acoustic wave. In our work, a regen amplified femtosecond laser is used as the generation laser. A stabilized balanced Michelson interferometer is used for the optical detection of the acoustic pulses. The broadband ultrasonic wave disperses as it propagates along the surface. By measuring the ultrasonic wave train at two different source to receiver distances, the dispersion curve can be calculated after phase deconvolution of the signals. After fast Fourier transform of the two signals, discontinuities for the phase spectra are corrected by phase unwrapping. The phase velocity spectrum is then given by:

$$v_{SAW}(f) = 2\pi f \Delta L / \Delta\varphi(f) \tag{2}$$

where ΔL is the spacing between the two measurement positions and $\Delta\varphi(f)$ is the phase difference after phase unwrapping. Since a single broadband measurement is adequate to generate the entire dispersion curve, the broadband PA method is a faster technique for thin film characterization as compared to the LFAM.

Pulsed laser generated broadband guided-waves in free-standing thin plates have been used in the past to measure the dispersion relation of Lamb wave modes, and in turn to determine the elastic properties of the materials. In particular, the antisymmetric A0 lamb mode, which is quite easy to detect with photoacoustic methods, can be correlated with small-deflection plate theory, when the acoustic wavelength is much larger than the thickness of the film. A simple expression for the acoustic phase velocity of A0 mode in thin films can be derived as [12]:

$$v_{A0} = \frac{\omega}{k} = \sqrt{\frac{D}{\rho h}k^2 + \frac{\sigma}{\rho}} \qquad (3)$$

where ω is the frequency, k the wavenumber, h is the plate thickness, D is the flexural rigidity, σ is the in-plane stress (typically residual stress) and ρ is the density. It is known that the above simplified model gives good results in determining the phase velocity when $kh < 0.5$.

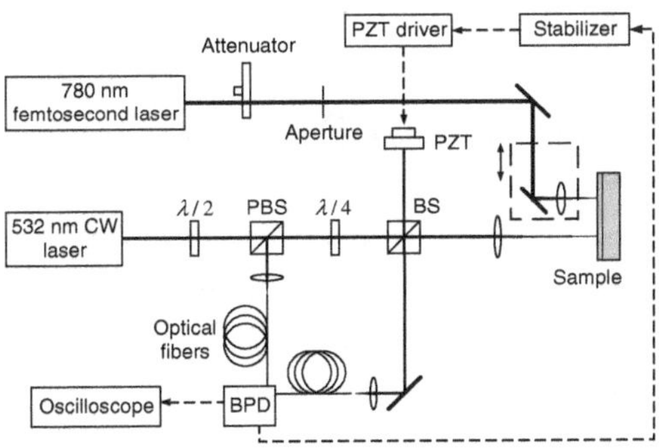

Fig. 2 Broadband guided-wave photoacoustic setup.

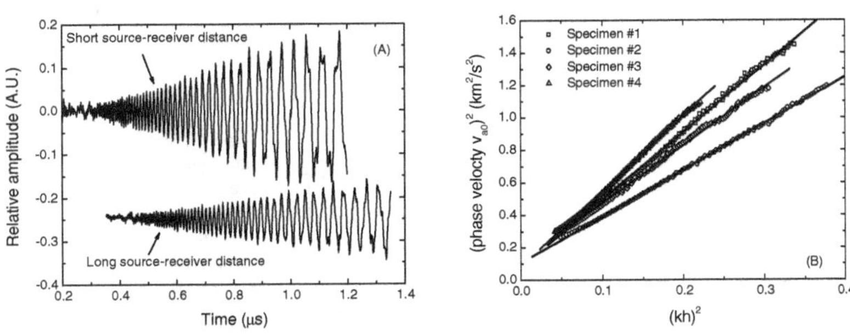

(a) (b)

Fig. 3 (a) Dispersive guided wave signal for thin film sample #1; (b) Measured dispersion curves (dots) and linear fitted curves (lines) for specimens #1-#4. The measured flexural rigidities are 4.82,5.82,3.64,1.76×10^{-9} Nm; and the residual stresses are 235, 299, 334, 242 MPa, respectively.

Several free-standing layered thin film structures with various thicknesses and in-plane extent were fabricated using standard microfabrication processes. A photomask was designed to create several rectangular free-standing films on (100) Si wafers. Silicon nitride was grown on both sides of the wafer with Low Pressure Chemical Vapor Deposition (LPCVD). The wafer was patterned and the silicon was etched to leave a free-standing film of silicon nitride of thickness ranging from 200-400nm and lateral dimensions ranging from 100-2000 μm. Aluminum films ranging from 300-500nm in thickness were subsequently e-beam evaporated onto the silicon nitride.

Figure 3(a) shows the measured time traces of the broadband A_0 mode for specimen #1 at two source-to-receiver positions, from which the strong dispersive nature is distinguished. Fig. 3(b) shows the experimentally determined A_0 mode dispersion curves for four different specimens [13]. The solid lines in Fig. 3(b) correspond to the linear fitting of experimental data, and give the flexural rigidity and residual stresses of the films. The derived mechanical properties are listed in the figure. The measured tensile residual stresses range from 230 to 330 MPa.

3.2 Narrowband Guided-Wave PA Method

In some cases, it is better to use a narrowband guided-wave PA method. Unlike the broadband approach, where data have to be collected at multiple source-to-rec eiver distances, the narrowband PA technique can be used on a much smaller footprint. Narrowband acoustic waves can be generated using lasers in a number of ways [11,12].

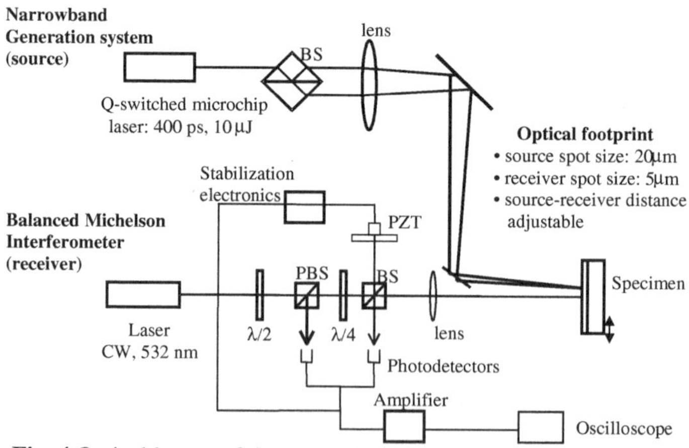

Fig. 4 Optical layout of the narrowband photo-acoustic microscope.

The technique used here (Fig. 4) requires the coherent interference of two crossed pulsed laser beams that are obtained from a microchip laser (parameters: 480 ps pulse width, 13.2 μJ energy, and 1064 nm wavelength). The beams are focused onto the surface of the thin film to be tested resulting in an intensity grating due to coherent interference. The rapid thermal heating that ensues in the thin film leads to generation of acoustic waves with a fixed wavenumber 'k' given by [12]:

$$k = \frac{4\pi}{\lambda_e}\sin(\theta/2) \tag{4}$$

where, λe is the generation laser beam wavelength and θ is the angle between the two crossed laser beams. The resulting guided waves were detected with a Michelson interferometer. By varying the angle, θ, the film can be probed over a range of wavenumbers. Dispersion curves can thus be obtained for the films.
Photo-acoustic data were obtained for several two-layer Al/Si3N4 thin film samples. Dispersion curves of guided acoustic modes over a range of *kh* values were obtained by adjusting the angle between the two crossed generation beams. Figure 5a shows a representative time trace of the So mode obtained for a wavenumber of 159,000 1/m (corresponding to a grating spacing of 39.5 μm). As expected, the frequency content of the Ao mode is significantly lower than that of the So mode. The Ao mode signal has a peak frequency of 8.6 MHz corresponding to a phase velocity of 339 m/s. The So mode waveform, on the other hand, has a peak frequency of 176 MHz, corresponding to a velocity of 6952 m/s.

Figure 6 shows the experimentally measured dispersion curves for the So and the Ao modes respectively. As expected from small *kh* values, the former is non-dispersive and the latter is dispersive. Up to three thin film parameters can be extracted by minimizing the least-squares error between the experimentally measured dispersion data and the asymptotic expressions. Assuming that the Poisson's ratios for the two materials are the same as the bulk values, and independently measuring the layer thicknesses using ellipsometry and profilometry, it is possible to obtain the Young's moduli of the two materials and the thickness-average residual stress. For the two-layer 420 nm Al/236 nm Si3N4 thin film sample, the measured values are: Young's modulus of Aluminum equal to 80MPa, Young's modulus of silicon nitride equal to 196 MPa, and residual stress of 188MPa. Results obtained for additional film ratios of 510 nm Al/ 250 nm Si3N4 and 510 nm Al/ 397 nm Si3N4 ranged from 47-82 GPa for the Aluminum Young's modulus and 192-268 GPa for the Si3N4

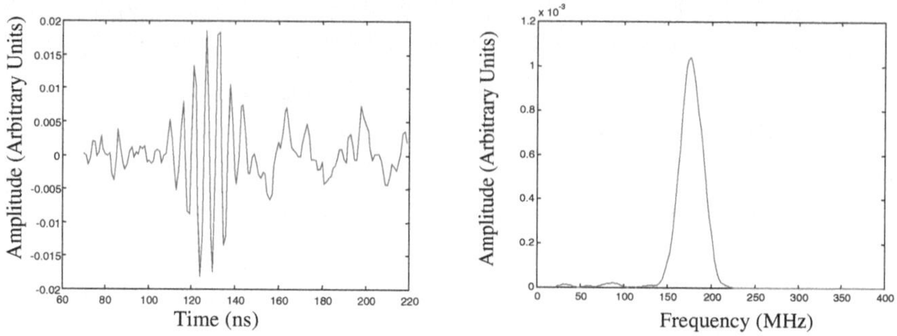

(a) (b)

Fig. 5 So mode propagating on 420 nm Al/236 nm silicon nitride. (a) Left: displacement vs time (ns) and (b) Right: spectrum in Mhz.

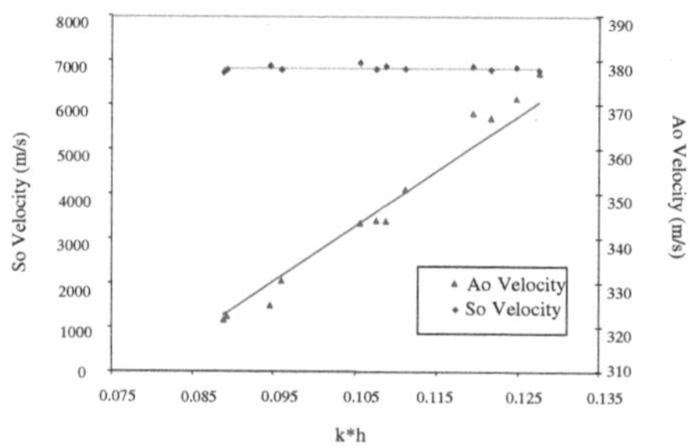

Fig. 6 Dispersion curve for 420nm-Al/236nm-silicon nitride: Velocity (m/s) vs (kh).

The measured modulus for aluminum is within the range of 50-90 GPa obtained by Lim [14] on evaporated films. The measured modulus of the silicon nitride is also within the range of 250 GPa that is cited for LPCVD grown films, and values of 210 GPa and 290 GPa for Plasma Enhanced Chemical Vapor Deposition films and Low Pressure Chemical Vapor Deposition films respectively [15]. The literature values for the modulus of aluminum and silicon nitride on the nanometer scale appear to vary greatly. However, the results of the modulus for aluminum and silicon nitride obtained from the photo-acoustic experiments are in the same range as published values.

3.3 Bulk-Wave PA Method

Another photoacoustic technique using an ultrashort femtosecond laser pump-probe method has been developed to quantitatively characterize the mechanical and thermal properties of thin films [8]. Basically, the pump-probe technique involves the excitation of the sample by a strong pump pulse and monitoring the subsequent relaxation processes by a weaker probe pulse, which is delayed with respect to the pump pulse by means of a variable optical delay line. Typically, the intensity of the reflected probe beam is monitored. The relative variations of the surface reflection coefficient due to the thermal and acoustic response are typically on the order of 10^{-7} to 10^{-5}. Highly sensitive lock-in detection is therefore applied by setting its reference frequency to the modulation frequency of the pump pulse. In our experiments, the laser pulses with a pulse duration of ~100 fs and a repetition rate of 80 MHz are generated from a Ti:sapphire laser. The laser beam is split into two beams of unequal intensity. The intense pump beam is used as a heating source while the lower power beam is used as the detection beam to monitor the change in reflectivity on the sample surface. The probe beam passes through a gold-coated retroreflector mounted on a micro-positioning motorized stage to vary the optical delay length. As the delay path length of the probe beam increases, a time delay is achieved between the arrivals of the pump and the probe pulses. The reflected probe beam, which contains a snapshot of the transient information of the sample surface, is sent into a balanced photodetector and the weak signal is amplified by the lock-in amplifier. By moving the delay stage, the laser induced transient reflectance change on the sample surface is recorded at various times. The ultrashort laser pulse launches bulk acoustic waves of very high frequency normally into the thin films, which then reflect off boundaries. These high frequency acoustic echoes are typically found embedded within a large transient thermal signal. Since non-dispersive bulk acoustic waves are measured, the pump-probe technique provides a direct measure of the properties of the thin film.

Cr/Si Thin Films: Next, magnetron-sputtered chromium thin films deposited on the <100> silicon wafers were investigated using broadband PA guided wave and bulk-wave pump-probe techniques [7]. The dotted line in Fig. 7(a) shows the measured SAW phase velocity dispersion curve for a 308nm Cr thin film on silicon wafer substrate using the guided-wave PA approach.

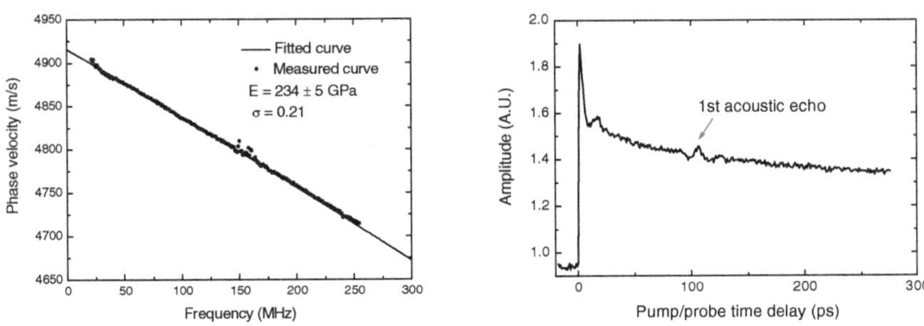

(a) (b)

Fig. 7 Cr-Si thin film on substrate: (a) Guided-wave PA and (b) bulk-wave PA results.

From the experimental dispersion curve measurements, the Young's modulus for the Cr film is calculated as 234±5 GPa and its Poisson's ratio is 0.21 (using the bulk value for its density).

Bulk-wave pump-probe experiments were also conducted on the Cr thin films. Fig.7(b) shows the thermo-reflectance variation as a function of time (pump-probe delay). An obvious acoustic echo can be clearly distinguished from the thermal decay. The measured longitudinal wave velocity in Cr is estimated around 6620 m/s. The estimated Young's modulus is about 275 GPa.

Acknowledgements: This paper presents an overview of the work done in collaboration with several of the author's students and post-docs: Feifei Zhang, Carmen Lilley, and Todd Murray. Financial support was provided in part by the US National Science Foundation.

References

[1] Weglein, R.D.: Acoustic microscopy applied to SAW dispersion and film thickness measurement. *IEEE Trans. Sonics. Ultrason.* **SU-27** 82 (1980).

[2] Kushibiki, J., Chubachi, N.: Material characterization by line-focus-beam acoustic microscope. *IEEE Trans. Sonics. Ultrason.* **SU-32** 189 (1985).

[3] Achenbach, J.D., Kim, J.O., Lee, Y.C. In: Briggs, A. (Ed.): *Advances in Acoustic Microscopy, Plenum Press*, New York, p.153, (1995).

[4] Briggs, A.: *Acoustic Microscopy,* Oxford University Press, Oxford, (1992).

[5] Takanaga, I., Kushibiki, J.: A method of determining acoustical physical constants for piezoelectric materials by line-focus-beam acoustic microscopy. *IEEE Trans. Ultrason. Ferroelect. Freq. Contr.* **49** 893 (2002).

[6] Zinin, P., Lefeuvre, O., Briggs A. et al.: Determination of density and elastic constants of a thin phosphoric acid-anodized oxide film by acoustic microscopy. *J. Acoust. Soc. Am.* **106** 2560 (1999).

[7] Zhang, F.: *Photoacoustic Guided-wave and Bulk-wave Characterization of the Mechanical Properties of Thin Film Materials*. PhD Dissertation, Northwestern University, (2006).

[8] Thomsen C., Grahn H.T., Maris, H.J., Tauc, J.: Surface generation detection of phonons by picosecond light-pulses. *Phys. Rev. B* **34** 4129 (1986).

[9] Rogers, J.A., Maznev, A.A., Banet, M.J., Nelson, K.A.: Optical generation and characterization of acoustic waves in thin films: Fundamentals and applications. *Annu. Rev. Mater. Sci.* **30** 117 (2000).

[10] Hess, P.: Surface acoustic waves in materials science. *Phys. Today* **55** 42 (2002).

[11] Duggal, A.R., Rogers, J.A., Nelson, K.A.: Real-time optical characterization of surface acoustic modes of polyimide thin-film coatings. *J. Appl. Phys.* **72** 2823 (1992).

[12] Hernandez, C.M., Murray, T.W., Krishnaswamy, S.: Photoacoustic characterization of the mechanical properties of thin films. *Appl. Phys. Lett.* **80** 691 (2002).

[13] Zhang, F., Lilley, C., Krishnaswamy, S.: Bulk-wave and Guided-wave Photoacoustic Evaluation of the Mechanical Properties of Aluminum / Silicon Nitride Double-layer Thin Films. *Ultrasonics* **45** 66-76 (2006)

[14] Lim, Y.Y., Chaudhri, M.M., Enomoto, Y.: *Mater. Res.* **14**(6) 2314 (1999).

[15] Tabuta, O., Kawahata, K. and Sugiyama S.: *Sensors and Actuators* **20** 135 (1989).

Evaluation of Corrosion in Carbon Steel Pipes by Laser-Generated Guided Wave

Do-Youn Kim[1], Joon-Hyun Lee[2], Younho Cho[2], Jaesun Lee[1] and Jan D. Achenbach[3]

[1]Graduate School of Mechanical Engineering, Pusan National University, Jangjeon-dong, Gumjeong-gu, Busan, 609-735, Korea

[2]School of Mechanical Engineering, Pusan National University, Jangjeon-dong, Gumjeong-gu, Busan, 609-735, Korea

[3]Walter P. Murphy and Distinguished McCormick School Professor, Center for Quality, Prevention and Failure Prevention, Northwestern University, U.S

Abstract. The objective of this research is to locate and evaluate wall thinning in pipe elbow by a non-contact guided wave technique with laser source as a transmitter and air-bone transducer as a receiver, respectively. Wall thinning of carbon steel pipe is one of the most serious problems in nuclear industry; especially the one in carbon steel pipe elbow caused by FAC (Flow-Accelerated Corrosion). Therefore, development of a robust NDE technique for the pipe elbows is essential for safe operation of nuclear power plants. Specimens used in this study were carbon steel which is widely used in real nuclear power plants. The geometry of wall thinning was given as 120mm extent, 80mm-length and 5mm-depth. The L(0,1) and L(0,2) dominant modes group shows a promising variation in the ultrasound guided wave data analysis based on the response obtained by the laser generation/air-coupled detection system. The trends of these characteristics and subsequent signal processing were used to estimate the size and location of wall thinning.

1. Introduction

Carbon steel is one of the principal structural materials in power plants. Since local wall thinning caused by FAC occurs inside the elbows by flowing high temperature and high pressure water with high velocity, it can make a big disaster if this defect grows without any regular inspection. Therefore structural evaluation of elbows with local wall thinning becomes more important in order to maintain the integrity of coolant piping systems [1, 2].

T.-T. Wu and C.-C. Ma (eds.), *IUTAM Symposium on Recent Advances of Acoustic Waves in Solids*, IUTAM Bookseries 26, DOI 10.1007/978-90-481-9893-1_8,

Most inspections were carried out using point by point examination. This method is inefficient and also takes a lot of time to inspect a large area structures. In these situations, the developed method for solving the drawbacks compare to conventional technique is the inspection using guided waves.

An ultrasonic guided wave technique is one of powerful tools to detect the reduction and has lots of benefit over conventional ultrasonic methods due to such features as lower cost, ease of operation, and testing speed. Moreover, broad-band, multi-mode guided waves, such as those generated by a laser system, have the potential for detecting flaws in various sizes [3-5]. For the purpose of this study, the characteristic of the guided wave will be shown, when it passed through the elliptical defect of elbow, a laser generation/ air-coupled transducer ultrasonic hybrid systems were employed as a way of detecting the defect of elbow by using guided wave. In addition, linear slit array were used for the directivity of laser-generated guided wave and the determination of wavelength. Air-coupled transducer as guided waves detector was controlled to detect specific mode among guided waves by adjusting its receiving angle to a leak direction of selected mode.

2. Selective Generation and Reception of Guided Wave Mode

The problem in laser based guided wave testing is the difficulty to generate a desired mode due to the dispersive nature of Lamb waves [6]. While the excitation of a particular mode is made by a laser pulse, the different components of the wave will travel with different speeds and at least two modes are present even at low frequency range. This could make the evaluation of defect difficult due to interpretation of received signal. In this study, the selective generation and reception of guided wave modes are achieved by the technique that used the relation of dispersion curves and linear slit array [7]. Figures 1(a) show the process of selective generation using this linear slit array. The elements gap (Δs) in Figure 1(a) is equal to the wavelength of generated modes and illustrated as the diagonal line with a slope of $\Delta s/d$ in Figure 1(a).

The active modes lie on at the intersection points between the line and the phase velocity of dispersion curves, and therefore it is possible to generate specific modes selectively by adjust the elements gap. The method to receive the modes generated by the above-mentioned technique is to rotate the air-coupled transducer by the angle based on Snell's law for the propagation velocity in air (Cair) and the phase velocity of the specific mode (Cp) as shown in Figure 1(b). In this study, the velocity of wave in air is 340 m/s and the phase velocity of modes is obtained in Figure 1.

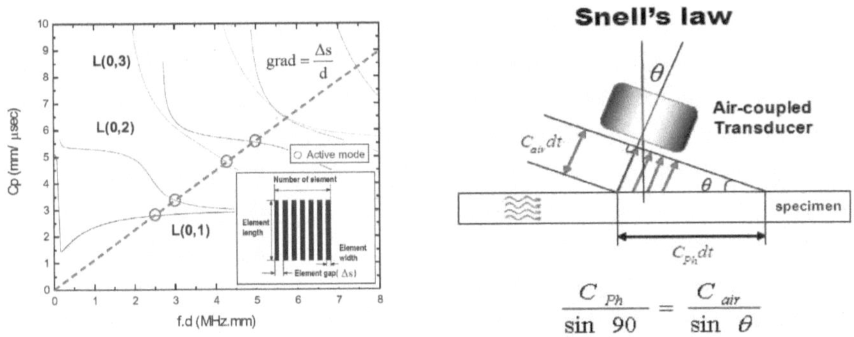

(a) Phase velocity of selected mode in dispersion curve. (b) Determination of receipt angle for specific mode.

Fig. 1 Optional receipt of guided wave mode.

Table 1 Theoretical values of L(0,1) and L(0,2) modes at 8mm wavelength on each defect.

Wavelength [mm]	Mode	Frequency [kHz]	Phase velocity [mm/μsec]	Receiving angle [θ°]
8mm	L (0,1)	305	3.2	6.09
	L (0,2)	382	3.6	5.4

This study adopted the modes of L(0,1) and L(0,2) as the suitable modes for experiments due to readily excited, received experimentally at low frequency-thickness and only slightly dispersive [8]. Table 1 shows frequencies, phase velocities, reception angles of L(0,1) and L(0,2) modes at 8mm slit spacing. In the process of this calculation, the velocity of wave in air was 340 m/s and the phase velocity of modes was obtained from the dispersion curves in Figure 1.

3. Specimen and Experimental Setup

The specimen used in the test was 8.5mm thick carbon steel elbow. To evaluate the guided wave interaction with defect in elbow, through compared with defect region and defect-free region. An elliptical defect with a constant width 120 mm and depth 5 mm was machined on the inner surface of 8.5 mm thick elbow having diameter of 218 mm. Figure 2 shows the shape of the side and front mentioned defect on carbon steel elbow.

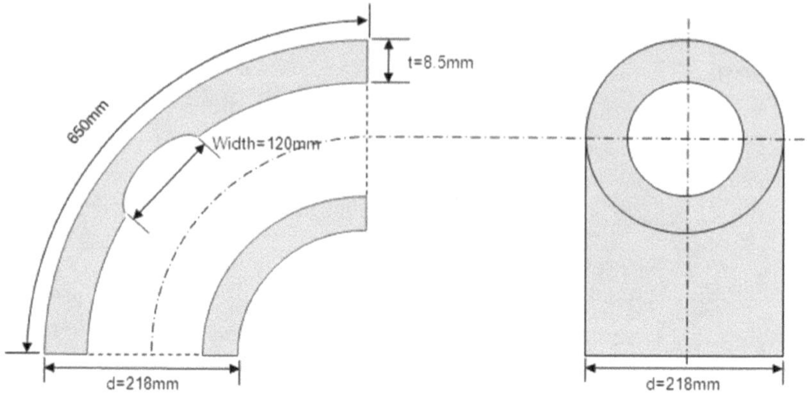

Fig. 2 Shape of defects in 8.5mm thick elbows.

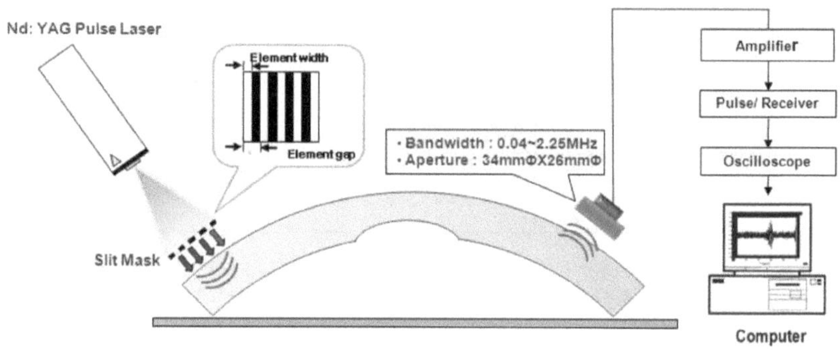

Fig. 3 Schematic diagram of experimental setup.

A schematic diagram of the apparatus used to perform experiment is shown in Figure 3. As shown in this figure, the laser and air-coupled transducer were positioned on the same side of the test elbows and act as the generator and detector of the guided wave signal scanning at 10mm steps along the longitudinal direction. A wavelength of fiberized Nd:YAG pulse laser system was used to generate ultrasonic waves is 532 nm and this pulse laser system emited energy of 32mJ at one pulse. The beam of this laser illuminated a linear array slit and transmitted beam act as line source on the elbow. The guided wave generated by this source propagated separation distance start at 160mm to end at 380mm between the source to the receiver, perpendicular to the surface of the elbow, and was subsequently detected using the air coupled transducer with a standoff from 5 mm the outer surface of elbow. In addition, the received signals from the air-coupled transducer were magnified by the amplifier and displayed through the signal averaging scheme with 1000 sampling data on the screen of oscilloscope. Here, the interval between slits, the width and the number of slits were fabricated 8 mm, 4 mm and 7 respectively.

4. Experimental Results

4.1 Characteristic of L(0.1), L(0.2) Modes in Defect Region of Elbow

Figure 4 shows the variation of amplitude in defect region. After scanning to right and left direction as shown in this figure, it is possible to evaluate defect by the variation of L(0,1) and L(0,2) modes. The signal of L(0,1) mode is appeared 16cm to 40cm in regardless of defect region, but amplitude of L(0,2) mode is disappeared 23cm and 33cm to start defect region. This result indicates that the signal on defect region and passed by defect region is affected by defect.

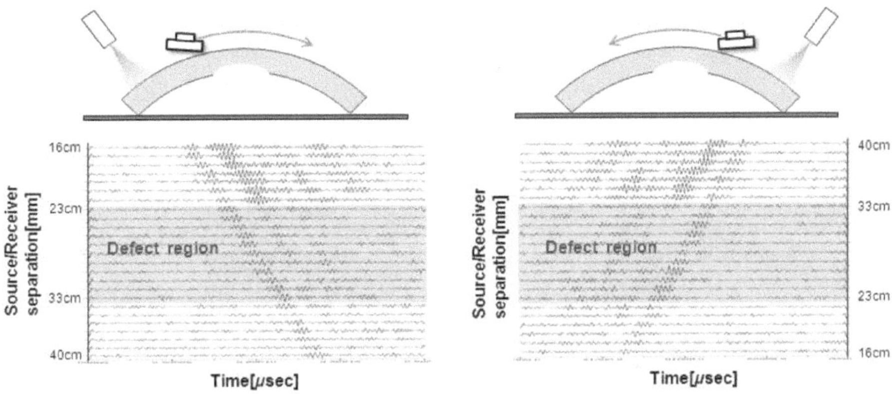

(a) Scanning to right direction. (b) Scanning to left direction.

Fig. 4 Guided wave signals of L(0,1), L(0,2) mode in elbow with defect region.

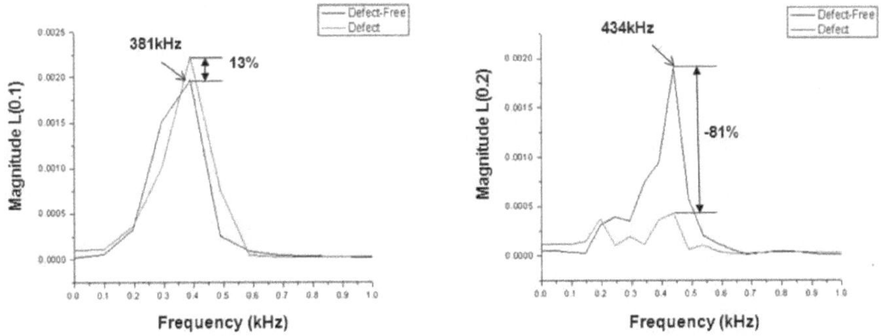

(a) Frequency spectrum of L(0,1). (b) Frequency spectrum of L(0,2).

Fig. 5 Signal characteristics of L(0,1) and L(0,2) mode on frequency spectrum.

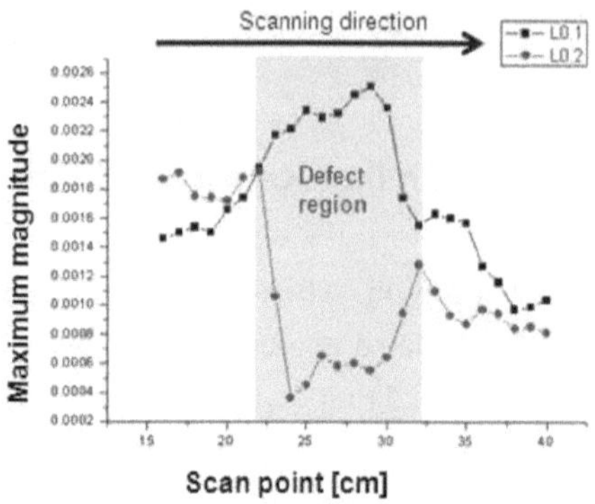

(a) Magnitude of right direction.

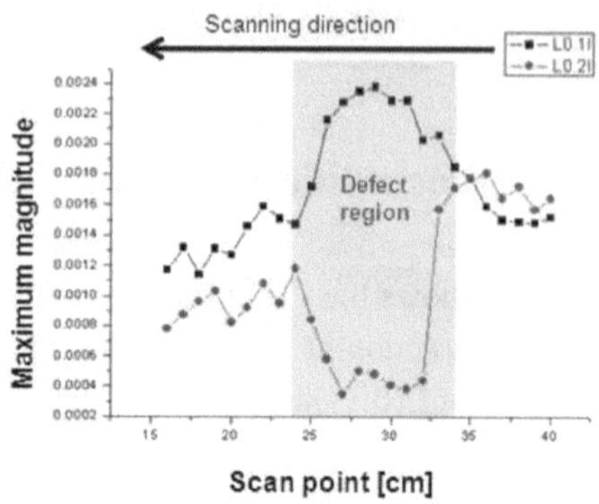

(b) Magnitude of left direction.

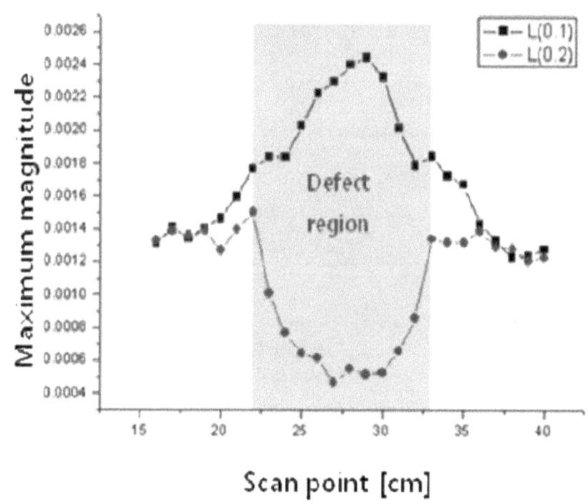

(c) Synthesis of right and left maximum magnitudes.

Fig. 6 Maximum magnitude using line scan technique on defect region.

Figure 5 shows the characteristics of the frequency spectra of L(0,1) mode with the frequency of 381kHz (Theoretical value is 376kHz) and L(0,2) mode with 434kHz (Theoretical value is 423kHz) distinctly. The plots in Figures 5 (a)-(b) are frequency spectra of these modes in the 0~1MHz range.

In analysis of the frequency spectra were obtained by performing a Fast-Fourier Transform (FFT) of the time-domain waveforms, the magnitude of L(0,1) mode with the center frequency of 381kHz in the defected region is increased by 13%.

However, the L(0,2) mode of waves propagating in the defected region suffers a dramatic attenuation. The maximum decrease in center peak magnitude of the signal with frequency of 434 kHz is 81%

4.2 Defect Localization along the Longitudinal Direction

As shown in a experimental setup of figure 3, guided waves are received with a constant source/receiver separation along the longitudinal direction. The air-coupled transducer is passed by the center of defect for obtaining clear characteristics of defect. Figures 6 shows the results from line scan of right and left direction, synthesis of right and left maximum magnitudes using pitch-catch method in defect region. The maximum magnitude of the frequency spectrum in L(0,1) and L(0,2) modes were plotted as a function of the scan position by scanning at 10mm steps along the longitudinal direction respectively.

The defect region in figure 6(a)-(b), as a depth of elliptical defect in increased, magnitude of L(0,1) mode is increased linearly but it isn't distinguished clearly from start point of defect region. Maximum magnitude of the L(0,2) mode is decreased rapidly in the defect region, so is the factor that can distinguish defect region. Figure 6(c) is synthesized from right and left maximum magnitude. The L(0,1), L(0,2) mode show inverse characteristic and the L(0,2) mode has relation more accurate between depth of defect and variation of signal. As a result of maximum magnitude, we estimate the relative depth of elliptical defect.

5. Conclusions

The possibility for estimating thickness reduction using the group velocity of guided wave was applied to the elbow. As a result, to evaluate the thickness reduction using the group velocity, mode identification was conducted by time-frequency analysis. In the elbow, L(0,1) and L(0,2) modes were appeared in defect-free region, but amplitude of L(0,2) mode were disappeared in wall-thinning and characteristic the maximum magnitude of frequency spectrum of L(0,2) mode is varied in the defect region. So we could know that it is possible to evaluate wall-thinning of elbow by using the ratio of L(0,2) to L(0,1) for the magnitude quantitatively.

Acknowledgements: This work was supported by the BK21 program and WCU (World Class University) program through the Korea Science and Engineering Foundation funded by the Korea Ministry of Education and Science in the period between 2008 and 2009.

References

[1] Dooley, R. B., and Chexal, V.K.: Flow-accelerated corrosion of pressure vessels in fossil plants. *International Journal of Pressure Vessels and Piping* **77** 85-90 (2000).

[2] Demma,A., Cawley, P., Lowe, M., Roosenbrand, A. G., Pavlakovic, B.: The relection of guided waves form notches in pipes: a guide for interpreting corrosion measurements. *NDT&E International* **37** 167-180 (2004).

[3] Lowe, M. J. S., Alleyne, D. N., Cawely, P.: Defect detection in pipes using guided waves. *Ultrasonics* **36** 147-154 (1996).

[4] Jhang, K., Kim, H., Kim, H., and Ha, J.: Laser Generation of Focused Lamb Waves. *Journal of the Korean Society for Nondestructive Testing* **22** (6) 637-642 (2002).

[5] Wright, W. M. D., Hutchins, D. A.: Air-coupled ultrasonic testing of metals using broadband pulses in through-transmission. *Ultrasonic* **37** 19-22 (1999).

[6] Alleyne, D. N., and Cawley, P.: The Interaction of Lamb Waves with Defect. *IEEE Trans. Ultrason. Ferroelec. Freq. Contr.* **39A**(3) 381-397 (1992).

[7] Kin, HM, Lee, TH and Jhang, KY.: Non-Contact Guided Wave Technique with Enhanced Mode-Selectivity. *J Korean Soc NDT* **26**(6) 597-602 (2004).

[8] Hutchins, David A., and William, M. D.: Wright and Gordon Hayward: Air-Coupled Piezoelectric Detection of Laser-Generated Ultrasound. *IEEE Trans. Ultrason. Ferroelec. Freq. Contr.* **41A**(6) 796-805 (1994).

Measurement of Surface and Lamb Waves with Application on Acoustic Sensing and Non-Destructive Evaluation

Yung-Chun Lee, Shi Hoa Kuo and Cheng-Hsien Chung

Department of Mechanical Engineering, National Cheng Kung University, Tainan 701, Taiwan

yunglee@mail.ncku.edu.tw

Abstract. This paper summarizes the design and fabrication of several novel ultrasound transducers for measuring surface and Lamb waves. The experimental setup and measurement method for each developed transducer are also addressed. According to piezoelectric materials being used, two types of ultrasound transducers are discussed which are PVDF focusing transducers and micro-machined PZT transducers. For the PVDF transducers, a defocusing measurement along with a *V(f,z)* waveform processing technique are developed which allow accurately determination of dispersion curves of either surface waves of a layered medium or Lamb waves of a thin plate. Elastic properties of coating layers and thin plates are therefore non-destructive determined. For the micro-machined PZT transducers, a differential type wave velocity measurement is developed which is extremely sensitive to small wave velocity variation. Hence, the mechanical and dielectric loading effects of a fluid to the Lamb waves of a piezoelectric plate are experimentally and theoretically characterized, which can be used for acoustic sensing of fluids.

1. Introduction

Ultrasonic transducers are key components in the engineering applications of all kinds of ultrasounds and elastic waves. The dominant way of constructing ultrasound transducers is still using piezoelectric materials for electrically generating and receiving acoustic waves in a wide range of frequency as well as measurement methods. We will discuss the using of two different piezoelectric materials, the polyvinylidene fluoride (PVDF) films and the PZT ceramics, for fabricate a number of ultrasound transducers of different purposes

T.-T. Wu and C.-C. Ma (eds.), *IUTAM Symposium on Recent Advances of Acoustic Waves in Solids*, IUTAM Bookseries 26, DOI 10.1007/978-90-481-9893-1_9,

and applications. Corresponding experimental setup and measurement method for each developed transducer are also established so that the potentials of the transducers can be explored for the purpose of non-destructive evaluation or acoustic sensing.

2. PVDF Focusing Transducers

2.1 Fabrication of Focusing PVDF Transducers

PVDF and its copolymer Poly(vinylidene fluoride-trifluoroethylene) [P(VDF-TrFE)] have been widely used in making ultrasound transducers since their acoustic impedance is much closer to that of human tissues and water. It is expected to have better impulse response and broad-band characteristics. There are basically two approaches in fabricating PVDF-based focusing ultrasound transducers: to permanently deform a piezoelectric polymer film into a concave shape [1] and to form a piezoelectric polymer film by spin-coating on a concave surface of a substrate [2]. The latter is favorable for transducers of higher operating frequency and/or lager aperture angle. In this work, we proposed a modified method which can spin-coat a P(VDF-TrFE) copolymer film from its precursor solution and simultaneously heats up the film by an infrared lamp to achieve thermal crystallization. In this way, both point-focused and line-focused PVDF transducers are obtained with a wide range of transducer frequencies and aperture angles.

Figure 1(a) and (b) show one SEM image of a spin-coated piezoelectric film and its measured ferroelectric hysteresis loop, respectively. The film has a thickness of 6 μm and shows excellent piezoelectric characteristics. The construction of a point-focused PVDF transducer is shown in Fig. 2. Two transducers have been made; one has a radius of curvature of 5 mm and a full opening aperture angle of 30°. The pulse/echo reflected waveform and the frequency spectrum of this focusing transducer are shown in Fig. 3(a) and (b), respectively. The central frequency is 52 MHz and the -6dB bandwidth is 44 MHz. The second transducer has a radius of curvature of 11 mm and a full opening aperture angle of 95°. The central frequency and -6 dB bandwidth are 14 MHz and 20 MHz, respectively.

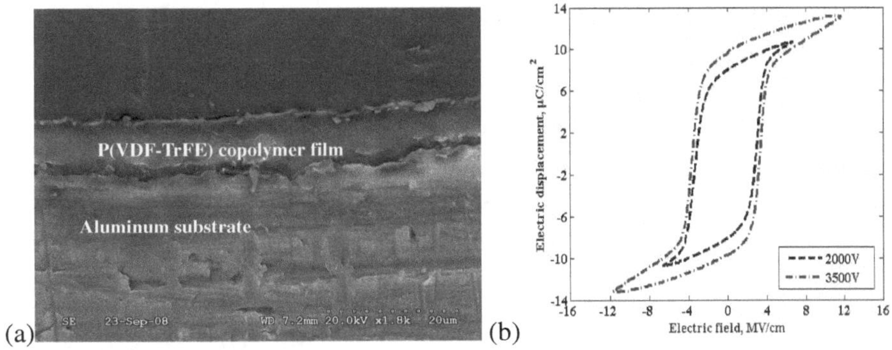

Fig. 1 (a) Spin-coated P(VDF-TrFE) film on an aluminum baking substrate and (b) its ferroelectric hysteresis loop.

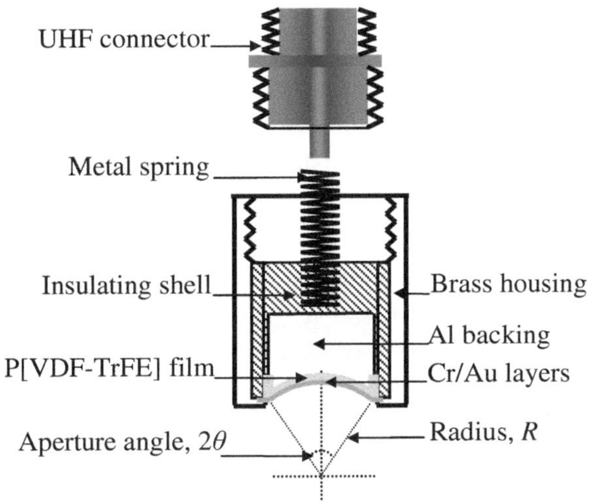

Fig. 2 Construction of a P(VDF-TrFE) focusing transducer.

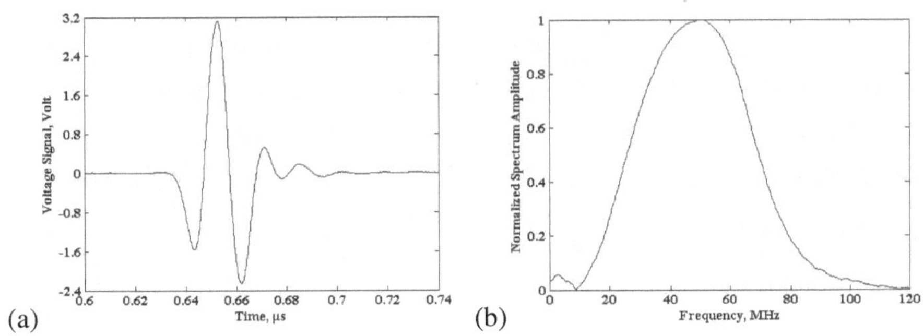

Fig. 3 A (a) pulse/echo reflected waveform and (b) its frequency spectrum of a point-focus PVDF transducer.

2.2 Defocusing Measurement and *V(f,z)* Waveform Processing

A defocusing measurement method for Lamb wave dispersion curves are shown in Fig. 4. Figure 4(a) shows schematically the experimental setup where a sample is placed in a water tank and the P(VDF-TrFE) point-focused transducer can perform standard pulse/echo reflective waveform measurements at different defocusing distance. Figure 4(b) shows few of the defocusing waveforms measured from a 250 μm stainless steel plate sample. The defocusing measurements are carried out in a much dense step of distance interval of 10 μm. Following the *V(f,z)* waveform processing method [3], the measured waveforms are first Fourier transformed with respect to time (*t*) into frequency (*f*) domain to yield the *V(f,z)*. The *V(f,z)* can be viewed as a function of two variables, *f* and *z*. For a specific and fixed frequency, *f*, the *V(f,z)* can be considered as a standard *V(z)* curve in conventional acoustic microscopy. The oscillation period of a *V(f,z)* curve, Δz, can be determined by doing Fourier transform again with respect to z and hence transformed into 1/z domain. An image plot of double Fourier transform data is shown in Fig. 4(c). A peak coordinate searching algorithm can be applied to trace continuous modes in the entire (*f* - 1/Δ*z*) image plane.

As indicated in Fig. 4(c), both A0 and S0 modes as well as other high order modes are observed. Figure 4(d) shows the measured dispersion curves for S0 and A0 modes and their comparison with their theoretical counterparts. Notice that the Rayleigh wave velocity of the plate can be easily obtained from the overlapping of S0 and A0 modes.

However, in order to accurately determine both two elastic constants without any ambiguity due to inversely curve-fitting on the dispersion curves, it is better to measure either longitudinal or transverse bulk waves in a more direct approach. Therefore, the high-frequency focusing transducer with a small aperture angle of 30o is used for directly measuring longitudinal wave in the thickness direction of a thin plate. The pulse/echo waveform is shown in Fig. 5. Longitudinal wave velocity can be accurately determined from the time-of-flight. Finally, both Young's modulus and shear modulus can be determined in a more accurate and straightforward way. A series of measurements have been carried out on a number of stainless steel and glass plates with thickness from 125 μm to 250 μm, and the measurement accuracy in both Young's modulus and shear modulus is within 5 %.

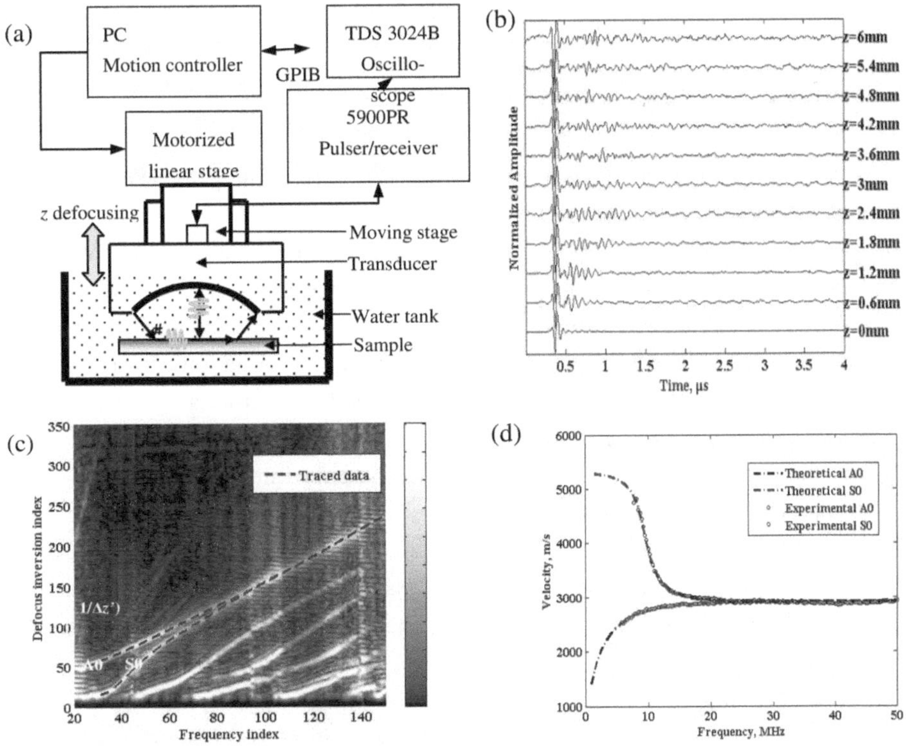

Fig. 4 (a) Experimental setup of a defocusing measurement method and (b) measured waveforms, (c) an image plot from *V(f,z)* signal processing, and dispersion curves of Lamb waves.

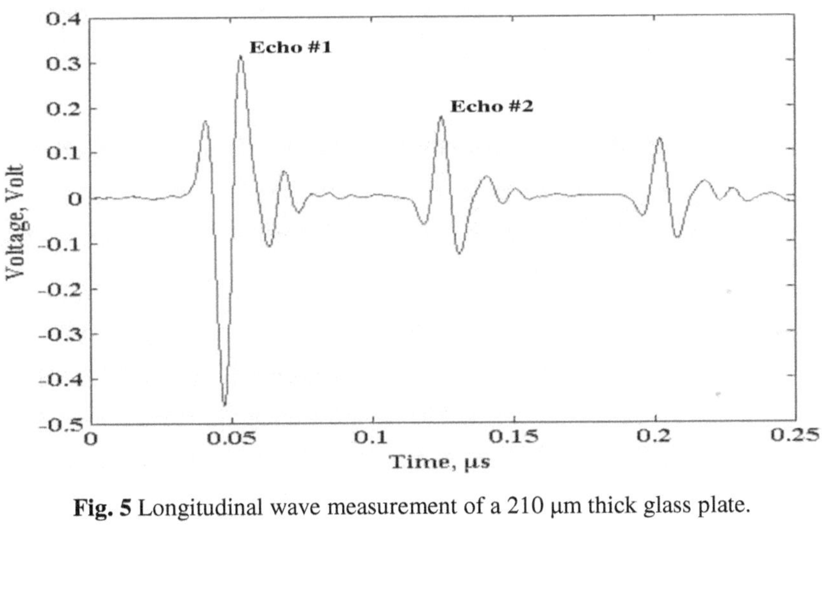

Fig. 5 Longitudinal wave measurement of a 210 μm thick glass plate.

3. Micro-machined PZT Transducer and Measurement

3.1 Fluid Loading Effects of Leaky Lamb Waves

Lamb waves are 2D guided plate waves propagating in a thin plate. There exist a number of modes of Lamb waves and each mode has a specific frequency-wavelength or frequency-velocity relationship called dispersion curve. When a plate is immersed in a fluid, the Lamb waves become leaky Lamb waves which have slightly different wave velocities and wave attenuation along propagation direction due to shedding energy to the adjacent fluid. The variation of wave velocity and the attenuation of a leaky Lamb wave are called fluid-loading effect which is mostly due to mechanical coupling at the solid/fluid interfaces and dominated by the fluid's elastic properties. For a piezoelectric plate immersed in a fluid, additional loading effects are induced from the electrical coupling between electrical fields in the plate and the fluid, and are related to the dielectric and conductive properties of the fluid. Inversely, if one can characterize the nature of the loading effects, leaky Lamb waves can be applied to acoustic sensing of both mechanical and electrical properties of a fluid.

Theoretical analysis on leaky Lamb waves of a piezoelectric plate to reveal the quantitative dependence of wave velocity and attenuation as functions of the elastic and electrical properties of a loading fluid can be carried out with rigorous partial wave analysis [4] or approximate perturbation analysis [5]. However, since the velocity variation and wave attenuation are usually small, it is still a grand challenge to the precision and accuracy of an acoustic wave measurement system.

3.2 Miniature PZT Transducer and Differential Measurement

To precisely measure the fluid loading effects of leaky Lamb waves, a new type of differential measurement method and new PZT line transducers are developed. Figure 6 shows schematically the measurement method which utilizes four PZT transducers in contact with a piezoelectric plate. Two of them are for wave generation and the other two for wave receiving. These transducers are laboratory-made from a PZT-5H piezoelectric ceramic. The original PZT plate has a thickness of 0.5 mm and both surfaces coated with Au/Cr electrodes. A precision wafer dicing saw machine cuts the PZT plate into long slender bars of a width, height, and length of 0.2 mm, 0.5 mm, and 12.7 mm, respectively. The long PZT bar is glued to a brass block for electrically conducting and mechanically backing. The brass block is then embedded into an epoxy matrix contained by an aluminum case. A SMA connector is mounted on the aluminum case and its signal line is connected to the brass block and the inner electrode of the long PZT bar. A minimal amount of epoxy is applied on the side walls of the PZT bar to strengthen its position as well as to electrically isolate the brass block. Finally, thin metal films of chromium (50 nm) and gold (200 nm) are thermally evaporated on the surface of PZT transducer. The Au/Cr film covers the outer electrode of the PZT element and connects it to the aluminum case or the grounding metal of SMA connector.

The differential measurement as shown in Fig. 1 first fixes the sample plate in a fixture and then partially immerses the plate in a fluid. The transmitting transducer pair then launches Lamb waves which are then received by the receiving transducer pair. The Lamb waves from the transmitters to the receivers can be divided into two parts. From the transmitters to the air/liquid interface, the acoustic wave is a free mode Lamb wave since there is no fluid loading. From the air/fluid interface to the wave receivers, however, the waves become leaky Lamb waves subjected to fluid loadings. In Fig. 6, L_o denotes the distance between transmitters and receivers, and x the distance from receivers to the air/fluid interface. The time-of-flight of Lamb waves traveling from transmitters to receivers is,

$$\tau = \frac{L_o - x}{c_o} + \frac{x}{c}, \tag{1}$$

where c_o and c are the phase velocities of free-mode and leaky Lamb waves, respectively. During the measurements, the assembly of transmitters, receivers, and sample plate are always kept in the same configuration but can move vertically by a servo-controlled linear stage. The waveform measurements starts from a small value of x and then gradually pushing the sample-plate/transducers assembly into the fluid for wave measurements. In Eq. (1), taking differentiation of τ with respect to x gives,

$$c = \left[\frac{d\tau}{dx} + \frac{1}{c_o} \right]^{-1}. \tag{2}$$

Therefore, with given c_o the leaky Lamb wave velocity, c, can be determined from the linear correlation between change in time-of-flight ($\Delta\tau$) and change in distance x (Δx). It is essentially a differential type of measurement which is very sensitive to the change in wave velocity and much less vulnerable to measurement errors and environmental noises.

Figure 8 shows the measurement results on an X-cut $LiNbO_3$ plate and the wave is propagating along 40° azimuthal angle measured from the z-axis of the plate. Figure 8(a) shows the experimental and theoretical leaky Lamb wave velocities of S0 mode at 3 MHz as a function of the dielectric constant of permittivity of the fluid, which is a mixture of water and ethanol with different concentration ratio. As predicted by the partial wave analysis, the leaky Lamb wave velocity is decreasing for about 400 m/sec with increase of fluid permittivity from 20 to 80. Figure 8(b) and 8(c) show the functional dependence of leaky Lamb wave velocity and attenuation on the electric conductivity of a salt water of different concentration. One can see the wave velocity is almost monotonously decreased for increasing fluid's conductivity by the wave attenuation is increasing first and then drop back to lower value. This is known as conductivity shielding effect.

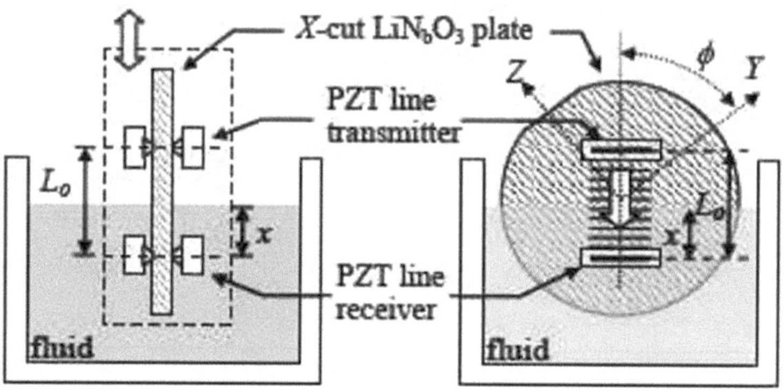

Fig. 6 A differential measurement method of leaky Lamb waves.

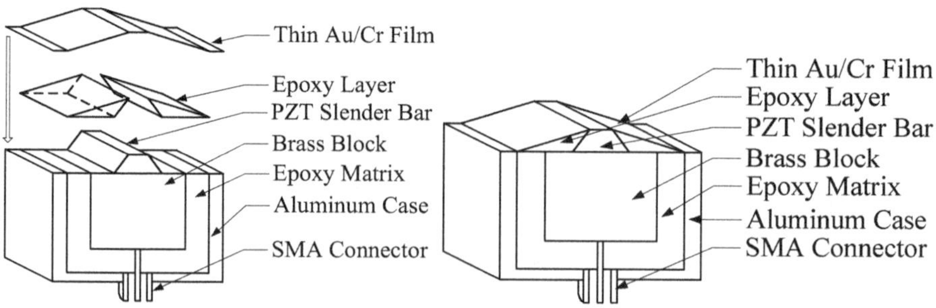

Fig. 7 Design and construction of line-PZT transducer.

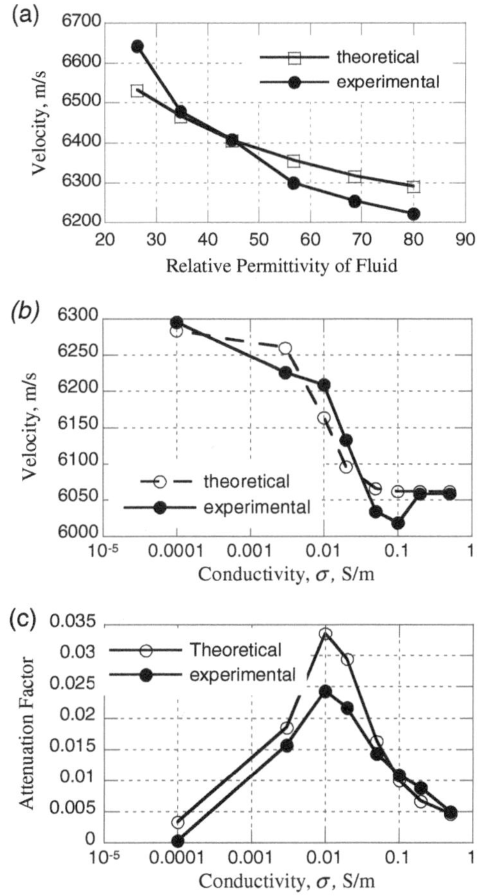

Fig. 8 Electrical loading effects on S0 mode of leaky Lamb waves at 3 MHz of on a *X*-cut $LiNbO_3$.

References

[1] Xiang, D., Hsu, N. N., Blessing, G. V.: Design, construction and application of a large aperture lens-less line-focus PVDF transducer. *Ultrasonics* **34** 641-647 (1996).

[2] Robert, M., Molingou, G., Snook, K., Cannata, J., Shung, K. K.: Fabrication of focused poly(vinylidene fluoride-trifluoroethylene) P(VDF-TrFE) copolymer 40-50 MHz ultrasound transducers on curved surfaces. *J. Appl. Phys.* **96** 252-256 (2004).

[3] Lee, Y.-C., Cheng, S.-W.: Measuring lamb wave dispersion curves of a bi-layered plate and its application on material characterization of coating. *IEEE Trans. Ultrason. Ferroelec. Freq. Contr.* **48** 830-837 (2001).

[4] Nayfeh, A. H.: *Wave Propagation in Layered Anisotropic Media. Elsevier*, New York (1995).

[5] Josse, F., Shana, Z. A.: On the use of ZX-LiNbO3 acoustic plate mode devices as detectors for dilute electrolytes. *Sensors & Actuators* **B9** 97-112 (1992).

Part III
Phononic Crystals

Guided Elastic Waves at Periodically Structured Surfaces and Interfaces

A. G. Every and A. A. Maznev

School of Physics, University of the Witwatersrand, PO Wits 2050, Johannesburg, South Africa

Arthur.every@wits.ac.za

Abstract. A brief overview is provided of guided elastic waves at periodically structured surfaces, with particular attention given to their folded dispersion relations, the occurrence of band gaps, their coupling to bulk elastic waves, and their interaction with light. Two examples of physical situations and the dynamic behavior they give rise to are used to illustrate these ideas. The first pertains to the dispersion relation for guided waves at a periodic array of thin coplanar slits in a solid, and the role that they play in the scattering of bulk waves. Next, behavior for laser generated dispersive Rayleigh and Sezawa modes in a patterned thin film structure on a substrate is discussed.

1. Introduction

Acoustic waves at periodically structured surfaces and interfaces are of importance in geophysics, materials science, microelectronic and many other areas. The periodic structuring induces coupling between modes whose wave vectors differ by a reciprocal lattice vector, which renders surface and interfacial waves dispersive, and gives rise to band gaps. While this phenomenon is common for bulk waves in periodic structures, surface and interface waves exhibit specific effects not encountered with bulk waves. Most importantly, periodicity induces interaction between surface and bulk modes. As a result, true surface waves only exist within a limited frequency range. Above a certain frequency threshold, surface modes disappear or turn into so-called "leaky-" or "pseudo-" surface modes which are attenuated via radiation of bulk waves.

The periodicity-induced attenuation of surface waves at a sinusoidally corrugated surface was predicted by Brekhovskikh [1] in what was perhaps the first study of surface waves in periodic structures. Subsequently, surface waves at periodically

T.-T. Wu and C.-C. Ma (eds.), *IUTAM Symposium on Recent Advances of Acoustic Waves in Solids*, IUTAM Bookseries 26, DOI 10.1007/978-90-481-9893-1_10,

corrugated surfaces have attracted considerable attention from both theoreticians [2-4] and experimentalists [5-7]. In particular, Brillouin scattering measurements on small amplitude gratings on Si [5,6] revealed the expected bandgaps at the Brillouin zone (BZ) boundary, as well as quasi-bandgaps in the leaky region inside the BZ caused by the hybridization between the zone-folded leaky Rayleigh mode and the longitudinal resonance

The past few years have seen an explosion of activity in the study of guided acoustic waves in periodic structures, largely related to the general progress in the field of phononic crystals [8]. Researchers have studied a wide range of structures more complex than an elastic half-space with a corrugated surface and have elucidated a number of new and interesting phenomena.

In this paper we consider two examples illustrating both general concepts and specific phenomena particular to the structures studied. We start by discussing guided interfacial waves at a periodic arrangement of coplanar slits in a solid [9,10], and the role that a leaky branch of these waves plays in the scattering of bulk waves at the interface. This problem has attracted considerable attention over the years, but a clear picture of the dispersion relation of these guided modes has only emerged recently. Next, we consider surface acoustic waves (SAWs) in a periodically patterned thin film structure on a silicon substrate [11,12]. An "active" light scattering technique referred to as laser-induced transient gratings has allowed us to produce good quality dispersion curves and elucidate a number of features such as large bandgaps inside the BZ resulting from the hybridization of zone-folded Rayleigh and Sezawa surface modes [12]. We will highlight the common features shared by the surface and interfacial waves in the two structures studied, and also point out some of their differences in behavior. While in the first part of the paper the discussion is based on the results of numerical modeling, the second part presents experimental data accompanied by a largely heuristic interpretation. It should be mentioned, however, that the plain wave expansion method presented in the first part can be easily adapted for the numerical analysis of the structure studied in the second part of the paper, which presents a subject for future work.

The purpose here is not to summarize a completed body of research, or to provide a broad survey of the field. Indeed, some important areas of research such as guided waves in structures with 2D periodicity and in periodically structured plates are left outside the scope of this article. Our intention is merely to sketch some aspects of a rich and exciting field, and hopefully stimulate further investigations aimed at filling in the bigger picture.

2. Guided Waves at a Periodic Array of Coplanar Slits

There is an extensive literature on the scattering of elastic waves by a periodic arrangement of coplanar slits in a solid, see e.g. references [9,10,13-15]. This is a model which can be taken to represent e.g. fracture, mine stoping or the dry contact between solids. Figure 1 depicts the physical situation we consider. Two identical semi infinite elastically isotropic solids on opposite sides of the $z=0$ plane are bonded to each other within regularly spaced strips parallel to the y axis, each of width b, and between the strips, the surfaces of the two solids are unattached. The repeat distance of this arrangement in the x direction is D.

The translational symmetry of the system is broken by this interface. The abrupt change in the z-direction gives rise to reflection, mode conversion and interfacial waves. In the x-direction, the translational symmetry of the interface is limited to integrals multiples of D, and as a result, in the scattering of waves at the interface, the wave vector component k_x is preserved only to within an integral multiple of the reciprocal lattice vector $G=2\pi/D$. As is common practice in such situations, one restricts k_x to the first BZ $|k_x|<\pi/D$, and "folds" the regions of the dispersion relation lying beyond the zone boundaries into the first zone, to constitute a stack of higher lying branches to the dispersion relation. For some purposes, though, an extended zone description is also useful.

The situation has reflection symmetry through the $z=0$ plane, and this allows one to treat separately wave fields which are symmetric and anti-symmetric with respect to reflection through this plane. An incoming plane wave from one side and its resultant scattered field is then taken as the superposition of symmetrically and anti-symmetrically disposed pairs of incoming waves and their scattered fields. We assume that the sagittal plane for the waves is perpendicular to the slits, and we are thus dealing with a plane strain problem of L and sagittally (SV) polarized T waves, in which the displacement field is confined to the xz-plane and independent of y. Fig. 1 depicts a pair of symmetrically incident T waves of wave vectors $\mathbf{k}=(k_x,-k_z)$ and $\mathbf{k}=(k_x,k_z)$, and angular frequency ω incident on the interface at an angle θ and giving rise to scattering, with the displacement field being $\mathbf{u}(x,z)\exp(-i\omega t)$.

The solids are of density ρ and their L and T acoustic slownesses are respectively $\alpha=1/V_L$ and $\beta=1/V_T$. The scattered field conforms to the Sommerfeld radiation conditions, and at the interface the combined incident and scattered field satisfies the boundary conditions (BC) of continuity of displacements u_x, u_z and traction forces within the joined regions, and the vanishing of traction forces within the unattached regions. In our calculations we employ a coupling parameter $0 \le \eta(x) \le 1$, with $\eta(x)=1$ in the joined regions and $\eta(x)=0$ in the unattached regions. Allowing $\eta(x)$ to vary continuously rather than abruptly at the junctions between attached and unattached regions, avoids unphysical singular behavior at the junctions, and gives one control over the convergence of the Fourier series representation of the scattered field.

The scattered field is a superposition of an infinite number of Bloch harmonics, i.e. outgoing L and T waves of the same frequency and having wave vectors $\left(k_x^n, \pm k_z^{\alpha n}\right)$ and $\left(k_x^n, \pm k_z^{\beta n}\right)$ respectively, where

$$k_x^n = k_x + nG\,;\; n = 0, \pm 1, \pm 2, \ldots$$

$$k_z^{\alpha n} = \sqrt{\omega^2 \alpha^2 - \left(k_x^n\right)^2}\;,\quad k_z^{\beta n} = \sqrt{\omega^2 \beta^2 - \left(k_x^n\right)^2} \tag{1}$$

The amplitudes of these outgoing L and T waves are respectively A_n, and B_n depending on the value of ω and k_x, some of the $k_z^{\alpha n}$ and $k_z^{\beta n}$ are real, and the remainder imaginary. The choice of sign ($\pm$) is dictated by the Sommerfeld radiation conditions. For small ω, there is only one outgoing homogeneous T wave in each half space, the rest of the partial waves being evanescent. As ω is increased, successive critical values or thresholds are crossed, where evanescent waves change to homogeneous. These are represented by the lines labeled L0, T0 and T-1 in Fig.2

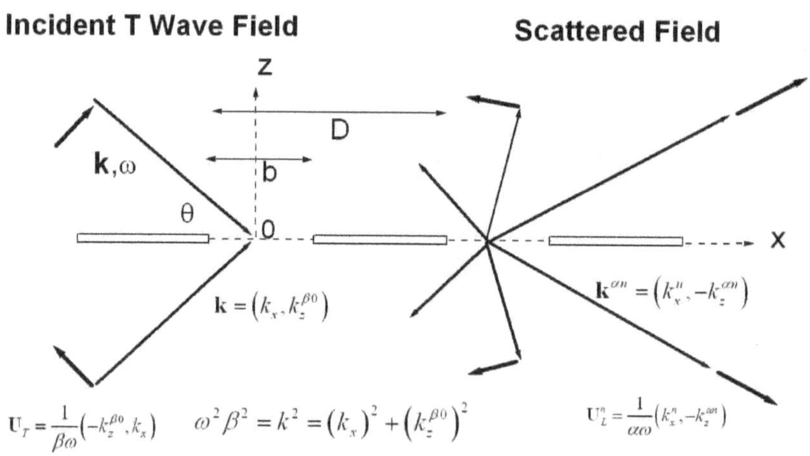

Fig. 1 Scattering of a symmetrically incident T wave at a periodic set of slits.

A numerical solution to this problem is obtained by imposing a finite cut off on the expansion, limiting n to the range $-N \leq n \leq N$, where $N = 47$ in our calculations. The values of the $2N+1$ coefficients A_n, and equal number of B_n are determined by imposing the boundary conditions at a discrete set of $2N+1$ equally spaced points x_p in the interval $-D/2 < x_p < D/2$. This yields

$$B_n = \delta_{n0} + 2A_n k_x^n k_z^{\alpha n} / \left(\omega^2 \beta^2 - 2\left(k_x^n\right)^2\right) \tag{2}$$

$$\sum_{n=-N}^{N} M_{pn} A_n = m_p \; ; \; p = 0, \pm 1, \ldots, \pm N \tag{3}$$

where M_{pn} and m_p are functions of $x_p, \eta\left(x_p\right), k_x^n, k_z^{\alpha n}, k_z^{\beta n}, \omega, \beta, n, D$. The solution of (3),

$$B_n = \delta_{n0} + 2A_n k_x^n k_z^{\alpha n} / \left(\omega^2 \beta^2 - 2\left(k_x^n\right)^2\right) \tag{4}$$

yields the A_n, and substitution into (2) yields the B_n for the symmetrical field. Similar analyses yield the scattering matrix $\tilde{M}$ and scattering amplitudes $\tilde{A}_n$ and $\tilde{B}_n$ for anti-symmetrically incident T waves, and incident L waves.

2.1 Interfacial Waves

The matrices M and $\tilde{M}$ become singular under certain conditions, leading to the existence of interfacial waves (IW), i.e. guided waves in the subsonic spectral range $0 < \omega < k_x / \beta$, and contained within the first BZ, which consist entirely of evanescent partial wave components. In the supersonic or radiation domain, $\omega > k / \beta$, there occur near-singular features associated with pseudo-interfacial waves (PIW). These consist predominantly of evanescent partial wave components, but through a weak coupling to the $n = 0$ T bulk wave continuum, are leaky rather than perfect guided modes. They feature prominently in the scattering of bulk waves at the interface.

Figure 2 depicts the dispersion relation, in terms of dimensionless frequency $\hat{\omega} = \omega \beta D / \pi$ and wave vector $\hat{k}_x = k_x D / \pi$, for Poisson's ratio $\nu = 0.3$, and $b / D = 0.16$. The anti-symmetrical field interfacial wave A persists down to $\hat{k}_x = 0$, approaching asymptotically and degenerating with the T_0 threshold at this lower limit. Its dispersion curve is horizontal at the BZ boundary $\hat{k}_x = 1$, corresponding to zero group velocity, and phase velocity there approaches the Rayleigh velocity in the limit $b / D \rightarrow 0$.

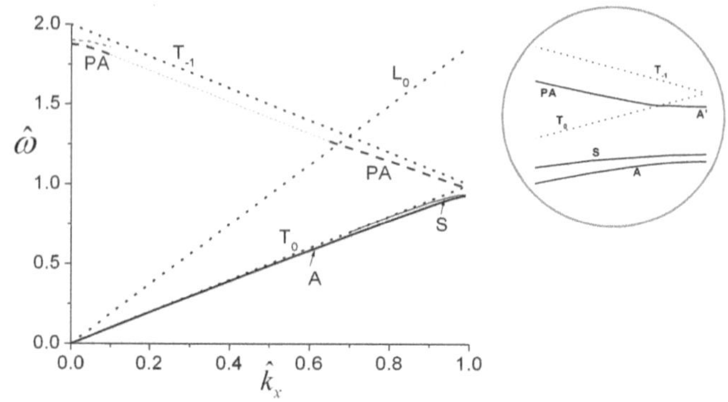

Fig. 2 Dispersion relation for $\nu = 0.3$ and $b/D = 0.16$. The dispersion relation is symmetric about the zone centre $\hat{k}_x = 0$ and zone boundary $\hat{k}_x = 1$. The inset is a blow up of the region near the BZ boundary.

At the BZ boundary there is a second anti-symmetrical IW A', separated from A by a band gap. A' penetrates through the T_0 threshold, and becomes the pseudo-interfacial wave PA, which lies a little below the T_{-1} threshold. PA is able to absorb from and radiate into the T wave continuum, and it exists therefore as a resonance rather than a true singularity. At the L0 wave threshold, a second channel opens up for PA to radiate into, and it becomes a much broader resonance, sharpening up again only near the BZ centre.

2.2 Pseudo Interfacial Waves in Scattering

The pseudo interfacial wave PA has a pronounced effect on the frequency dependence of the transmission and reflection of bulk waves at the interface, mainly resulting from the rapid variation of the phase of $\tilde{A}_0$ and $\tilde{B}_0$ for anti-symmetrically incident L and T waves in the vicinity of this mode. Figure 3 shows the frequency dependence, near resonance scattering, of the reflection amplitude $|R_T^T| = |B_0 + \tilde{B}_0|/2$, for a T wave incident from one direction on the interface at an angle $\theta = 7\pi/24$. It has been calculated by decomposing this wave into symmetrically and anti-symmetrically incident T waves. There is a sharp kink at $\hat{\omega}/2 = 0.6215$, which corresponds to the T_{-1} threshold. Below this value both B_0 and $\tilde{B}_0$ are in magnitude unity, since there is only one outgoing channel on each side of the interface into which to radiate, the $n = 0$ L mode being evanescent. At $\hat{\omega}/2 = 0.6145$ there is a sharp resonance due to the fact that near PA the phase of $\tilde{B}_0$ undergoes a rapid increase through approximately 2π. This causes $|R_T^T|$ to drop sharply to zero as the relative phase passes through π, rise rapidly to unity as the relative phase passes through zero, and then level off.

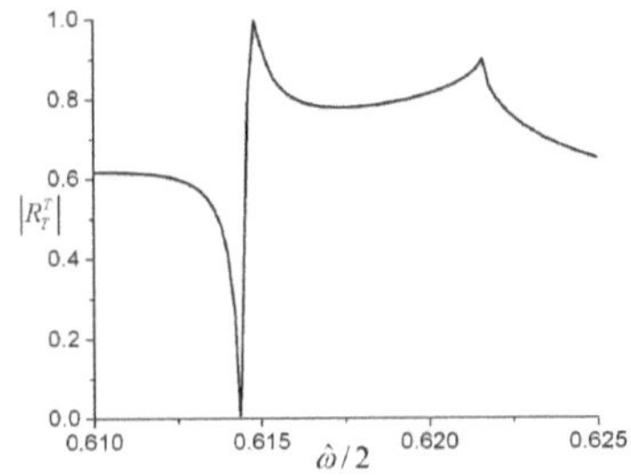

Fig. 3 Reflection amplitude $\left|R_T^T\right|$ for a T wave incident at angle $\theta = 7\pi/24$ near the PA resonance, for $\nu = 0.3$, and $b/D = 0.5$.

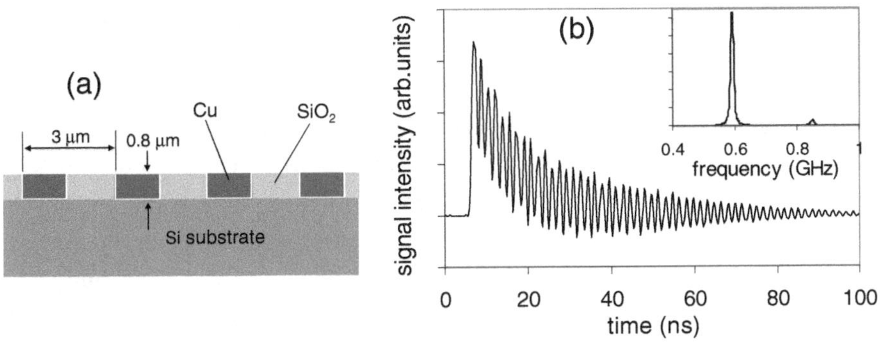

Fig. 4 (a) Schematic cross-section of the sample (not to scale); (b) An example of the signal waveform with the Fourier spectrum of acoustic oscillations shown in the inset.

3. SAW's in a Periodically Patterned Thin Film Structure on Silicon

In this section, we discuss measurements of surface acoustic modes in a structure depicted in Fig. 4(a), comprised of alternating copper and oxide strips on a silicon wafer. Acoustic waves in periodically patterned layered structures exhibit somewhat different behavior than interfacial waves considered in the previous section, or SAWs at periodically corrugated surfaces. Most importantly, guided modes in layered structures are dispersive even in the absence of the periodicity. Moreover, layered structures often support multiple acoustic modes. Periodicity induces coupling between the modes leading to "avoided crossing" bandgaps inside the BZ. This behavior differs from that of symmetric and anti-symmetric interfacial modes that remain uncoupled because the periodic structuring of the interface does not break the symmetry.

3.1 Experiment

The laser-induced transient grating technique also referred to as Impulsive Stimulated Thermal Scattering [16] is particularly well suited for studying surface wave propagation in periodic structures, as it allow one to measure waves inside the structure of interest as opposed to measuring transmission through the structure as is more common in traditional acoustics.

The transient grating set-up with optical heterodyne detection has been described in details elsewhere [16,17]. In short, two excitation pulses derived from a single laser source are crossed at the sample surface to form a spatially periodic intensity pattern. Absorption of the excitation light followed by rapid thermal expansion generates counter-propagating acoustic modes at the wavelength defined by the periodicity of the excitation grating. Detection of the acoustic waves is performed via diffraction of the quasi-cw probe beam focused at the center of the excitation pattern. Measurements were performed in two configurations, with the excitation grating wavevector parallel and perpendicular to the Cu and SiO2 strips.

A typical signal waveform for the grating wavevector parallel to the Cu lines is presented in Fig. 4(b). High frequency oscillations are due to surface acoustic modes while the slowly decaying component is the contribution of the "thermal grating" associated with the periodic temperature profile [16]. The decay of the acoustic oscillations is caused mainly by the finite size of the excitation spot, as the counter-propagating SAW eventually leave the probing area. The Fourier-spectrum of acoustic oscillations reveals the presence of two surface acoustic modes as is not uncommon for film/substrate structures: the fundamental mode and the weak second-order mode, often referred to as Rayleigh and Sezawa waves, respectively.

Acoustic dispersion curves presented in Fig. 5(a) are, again, quite typical for a structure comprising a "slow" film on a "fast" substrate, with the Sezawa mode emerging from under the "cut-off " determined by the velocity of the sagitally polarized bulk transverse wave T. Fig. 5(a) also shows calculated dispersion curves for a homogeneous layer with effective elastic properties.

3.2 Surface Wave Dispersion and Bandgaps Inside the BZ

Measurements with the grating wavevector perpendicular to the strips reveal a totally different picture. Acoustic dispersion curves measured in this configuration are shown in Fig. 5(b). As discussed in Section 2, in this case it is instructive to present the data in terms of the reduced wavenumber. As expected, the dispersion curves form a bandgap at the BZ boundary. It should be noted, however, that this is not a true bandgap as it overlaps with the left side of the middle dispersion branch. A much larger bandgap is formed inside the BZ. In order to elucidate the origin of this bandgap, dispersion curves for acoustic wavevector parallel to the copper lines from Fig. 5(a) are re-plotted in Fig. 5(b) versus the reduced wavenumber. It is evident that the bandgap arises as a result of the "avoided crossing" of the Sezawa mode and the zone-folded Rayleigh mode.

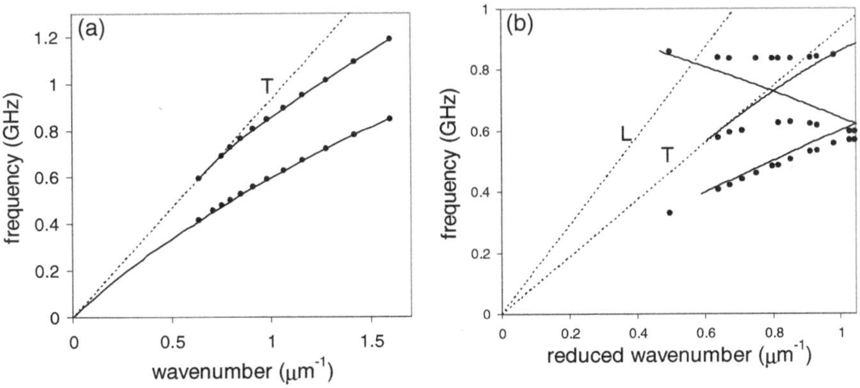

Fig. 5 SAW dispersion curves (symbols) measured with wavevector (a) along copper strips and (b) perpendicular to the strips, i.e. in the direction of periodicity. Dotted lines correspond to bulk wave thresholds. Solid lines in (a) were obtained by "effective medium" calculations. Solid lines in (b) correspond to the dispersion curves from (a) replotted vs. reduced wavenumber.

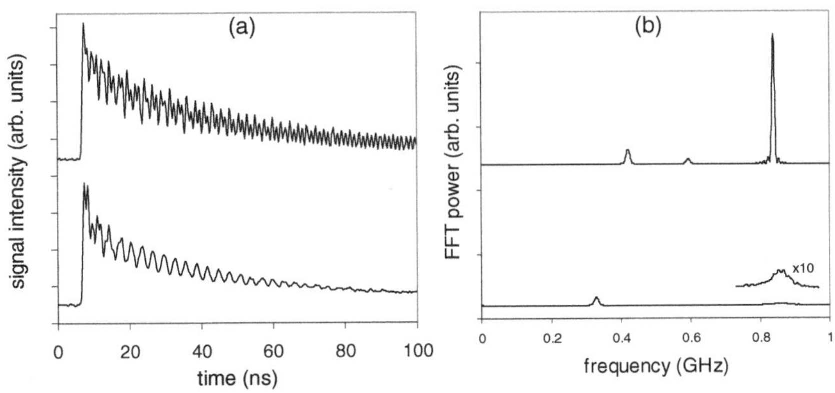

Fig. 6 (a) Signal waveforms and (b) corresponding spectra measured at wavenumbers 1.42 μm^{-1} (top) and 1.60 μm^{-1} (bottom). Corresponding reduced wavenumbers are 0.67 and 0.50 μm^{-1}, respectively. Inset in (b) shows the weak 3rd mode peak magnified by a factor of 10.

Why is one bandgap so much larger than the other? Ref. [12] offers a qualitative explanation based on the fact that in counter-propagating Rayleigh waves the direction of the particle motion is opposite (i.e. clockwise vs. counterclockwise) while in the counter-propagating Rayleigh and Sezawa waves it is the same. The conjecture is that the interaction between the modes is stronger in the latter case. However, the issue still awaits detailed numerical modelling to be resolved decisively.

3.3 Pseudo Surface Mode

As can be seen from Fig. 5(b), the upper dispersion branch mostly lies above the T line i.e. in the "leaky" region. However, in the experiment significant attenuation caused by the coupling to bulk waves is only observed above the L threshold. This point is illustrated by Fig. 6 showing two signal waveforms and corresponding spectra. The high frequency mode corresponding to the top waveform lies above the T threshold but below the L threshold and one can see that this mode has a long lifetime exceeding the time window of the measurement. As mentioned above, in the absence of attenuation the lifetime of the acoustic oscillations is determined by the "walk-out" of the acoustic wavepackets which is controlled by the group velocity. Since the dispersion curve of the upper branch is almost flat the group velocity is very low thus leading to a long lifetime. However, as soon as the upper branch crosses the longitudinal threshold, labeled "L" in Fig. 5(b), it becomes severely attenuated and decays almost completely within 10-15 ns following the excitation (see the lower waveform in Fig. 6). In view of the discussion in Section 2.1, the fact that attenuation increases sharply upon crossing the L threshold while a mode situated between T and L thresholds may be fairly long-lived does not appear entirely surprising. Again, detailed numerical modelling is desirable for the decisive interpretation of the results.

Acknowledgments: This work is based partly on research supported by the National Research Foundation.

References

[1] Brekhovskikh, L.M.: Propagation of surface Rayleigh waves along the uneven boundary of an elastic body. *Sov. Phys. Acoust.* **5** 288 (1960).
[2] Glass, N.E., Maradudin, A.A.: Leaky surface-elastic waves on both flat and strongly corrugated surfaces for isotropic, nondissipative media. *J. Appl. Phys.* **54** 796 (1983).
[3] Giovannini, L., Nizzoli, F., Marvin, A.M.: Theory of surface acoustic phonon normal modes and light scattering cross section in a periodically corrugated surface. *Phys. Rev. Lett.* **69** 1572 (1992).

[4] Maradudin, A.A., Zierau, W.: Surface acoustic waves of sagittal and shear-horizontal polarizations on large-amplitude gratings. *Geophys. J. Int.* **118** 325 (1994).

[5] Dutcher, J.R., Lee, S., Hillebrands, B., McLaughin, G.J., Nickel, B.G., Stegeman, G.I.: *Phys. Rev. Lett.* **68** 2464 (1992).

[6] Lee, S., Giovannini, L., Dutcher, J.R., Nizzoli, F., Stegeman, G.I., Marvin, A.M., Wang. Z., Ross, J.D., Amoddeo, A., Caputi, A.S.: Light-scattering observation of surface acoustic modes in high-order Brillouin zones of a Si(001) grating. *Phys. Rev. B* **49** 2273 (1994).

[7] Dhar, L., Rogers, J.A.: High Frequency Phononic Crystals Characterized With a Picosecond Transient Grating Photoacoustic Technique. *Appl. Phys. Lett.* **77** 1402 (2000).

[8] Sigalas, M., Kushwaha, M.S., Economou, E.N., Kafesaki, M., Psarobas, I.E. and Steurer, W.: Classical vibrational modes in phononic lattices: theory and experiment. *Z. Kristallogr.* **220** 765 (2005).

[9] Every, A.G.: Guided elastic waves at a periodic array of thin coplanar cavities in a solid. *Phys. Rev. B* **78** 174104 (2008).

[10] Danicki, E.J.: Resonant phenomena in bulk-wave scattering by in-plane periodic cracks. *J. Acoustic Soc, Am.* **105** 84 (1999).

[11] Profunser, D.M., Wright, O.B., Matsuda, O.: Imaging ripples on phononic crystals reveals acoustic band structure and Bloch harmonics. *Phys. Rev. Lett.* **97** 055502 (2006).

[12] Maznev, A.A.: Bandgaps and Brekhovskikh attenuation of laser-generated surface acoustic waves in a patterned thin film structure on silicon. *Phus. Rev. B* **78** 155323 (2008).

[13] Angel, Y.C., Achenbach, J.D: Reflection and transmission of elastic waves by a periodic array of cracks: oblique incidence. *Wave Motion* **7** 375 (1985).

[14] Mikata, Y.: Reflection and transmission by a periodic array of coplanar cracks: normal and oblique incidence. *J. Appl. Mech.* **60** 911 (1993).

[15] Maugin, G. A., Chevalier, Y., Louzar, M.: Interfacial waves in the presence of areas of slip. *Geophys. J. Int.* **118** 305 (1994).

[16] Rogers, J.A., Maznev, A.A., Banet, M.J., Nelson, K.A.: Optical generation and characterization of acoustic waves in thin films: fundamentals and applications. *Annu. Rev. Mater. Sci.* **30** 117 (2000).

[17] Maznev, A.A., Mazurenko, A., Zhuoyun, L., Gostein, M.: Laser-based surface acoustic wave spectrometer for industrial applications. *Rev. Sci. Instrum.* **74** 667 (2003).

Acoustic Cloaking by Metafluids

José Sánchez-Dehesa and Daniel Torrent

Wave Phenomena Group, Department of Electronics Engineering, Polytechnic University of Valencia, Valencia, Spain

jsdehesa@upvnet.upv.es

Abstract. In this work acoustic cloaking shells are explained and analyzed. Although the materials requirements for building these devices are materials with anisotropy in the mass density, it is explained how realize them with layers of isotropic metafluids based on sonic crystals. The cloaking efficiency as a function of the number of the number of layers is also studied.

1. Introduction

In a work by Cummer and Schurig [1] it was predicted that acoustic cloaking is possible in a two-dimensional (2D) geometry by means of a cloak made of an acoustic material having a strong unnatural mass anisotropy. This result in acoustics follows a previous analogous result discovered by Pendry et al. [2] in electromagnetism by using a material with equivalent requirements for the permittivity and permeability tensor components. However, while electromagnetic (EM) cloaking has been experimentally demonstrated by using a metamaterial specially designed [6], its acoustic counterpart has not been demonstrated yet. Moreover, the demonstration of acoustic cloaking is still waiting for some proposal of engineered material (metamaterial) that accomplishes the requirements on mass anisotropy predicted [1].

In this regards, the work by Milton et al. [7] describes conceptually how the mass anisotropy could be possible by spring loaded masses. Besides, a recent advance in the physical realization of metamaterial with mass density anisotropy has been performed by these authors by demonstrating that such property can be made possible by using non-symmetric lattices of solid cylinders [8].

In this work an acoustic cloak that could be physically realizable is presented. In brief, the proposed cloak is based on a multilayered structure consisting of two layers with the same thickness and made of two different acoustic isotropic metamaterials.

T.-T. Wu and C.-C. Ma (eds.), *IUTAM Symposium on Recent Advances of Acoustic Waves in Solids*, IUTAM Bookseries 26, DOI 10.1007/978-90-481-9893-1_11,

These metamaterials are built with sonic crystals (i.e., periodic arrays of sonic scatterers) based on two types of elastic cylinders that have to accomplish certain requirements on their mass density and effective sound speed. Numerical experiments based on multiple scattering method are present to support the exact performance of the proposed cloak.

2. Acoustic Cloaking in 2D

The solution reported by Cummer and Schurig for the acoustic cloaking in [1] requires a fluid material with an anisotropic density and a scalar bulk modulus. Moreover, these parameters must be dependent on the radial distance to the hidden object. The predicted functional form is

$$\begin{aligned} \rho_r(r) &= \frac{r}{r-R_1}\rho_b \\ \rho_\vartheta(r) &= \frac{r-R_1}{r}\rho_b \\ B(r) &= \left(\frac{R_2-R_1}{R_2}\right)^2 \frac{r}{r-R_1}B_b \end{aligned} \tag{2.1}$$

where R_1 and R_2 are the inner and outer radii of the cloaking shell, B is the bulk modulus of the shell, ρ_r and ρ_ϑ are the components of the diagonal mass density tensor, and the quantities with subscript b are those of the surrounding background that is a fluid or a gas. Materials with such uncommon properties do not exist in nature and, therefore, some engineered material must be introduced to accomplish them.

In previous work [8] it has been shown that, in the low frequency limit, arrangements of cylinders in non-symmetric lattices leads to acoustic metamaterials with anisotropic mass density and scalar bulk modulus, as required by the previous equations. The lattices considered in the mentioned work are single-cylinder lattices and, as a consequence, when the mass density of the cylinder is larger (smaller) than that of the background, the effective mass density tensor is always larger (smaller) than that of the background. This is an important drawback because in equations (1) one component of the mass density tensor is the reciprocal of the other and consequently the radial (angular) component of the mass tensor is always larger (smaller) than that of the background.

Therefore, a material having certain mass density in the radial direction and its reciprocal along the tangential direction cannot be engineered by using the theory developed so far. However, we suggest below a path to get the actual realization of such property by using periodic structures.

3. Multilayered Design

We arrive at the solution to this problem by exploring the possibility of building anisotropic materials based on sonic crystals with two types of materials cylinders, following a combination of two approaches previously introduced. Unfortunately, the practical realization of conditions (1) is impossible to achieve because of the limitation imposed by the close packing condition of the lattice. Therefore, in a natural way, we conclude that a multilayered composite structure made of two materials could overcome such problem and give a solution. It is interesting to note that an approach similar to this was also proposed to get EM cloaking [10]. However, while the EM cloak only verifies a reduced set of the conditions imposed for EM cloaking, the one reported here exactly matches the conditions for acoustic cloaking.

Let us consider a cloaking shell consisting of a multilayered structure that is made of alternating layers of materials of type 1 and 2, as shown in Figure 1. For any periodic system the bulk modulus (in the homogenization limit) does not depend on the type (isotropic or anisotropic) of lattice; in fact, it has been shown in [8] that the effective bulk modulus at large wavelength can be determined by performing a volume average of its reciprocal. This volume average, for a one-dimensional multilayered system of materials 1 and 2, becomes

$$\frac{1}{B_{eff}}=\frac{d_1}{B_1}+\frac{d_2}{B_2} \tag{3.1}$$

where B_1 (B_2) is the bulk modulus of material 1 (2) and d_1 (d_2) is the length of layer 1 (2).

To obtain the tensor associated with the effective speed of sound we need to first calculate the dispersion relation $K = K(\omega)$ of the system; i.e., the wavenumber as a function of the frequency. This calculation follows, for example, a procedure explained in textbooks like [12],

$$\cos K_x d = \cos k_{1x} d_1 \cos k_{2x} d_2 - \frac{1}{2}\left[\frac{\rho_1 k_{2x}}{\rho_2 k_{1x}} - \frac{\rho_2 k_{1x}}{\rho_1 k_{2x}}\right] \sin k_{1x} d_1 \sin k_{2x} d_2 \tag{3.2}$$

where

$$k_{ix}^2 = \frac{\omega^2}{c_i^2} - k_y^2, \text{ for i=1, 2.} \tag{3.4}$$

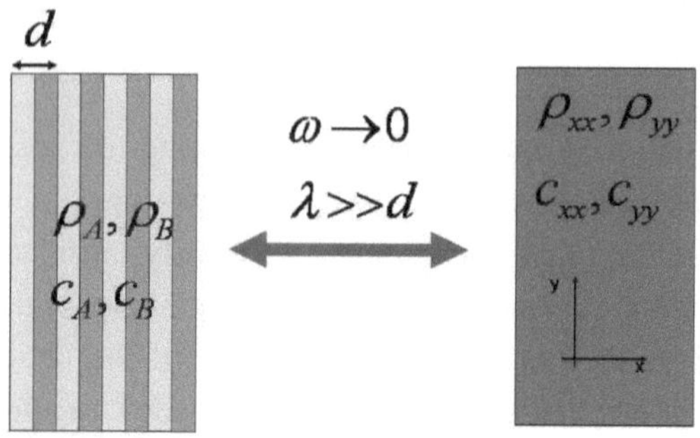

Fig. 1 Behavior of a one-dimensional periodic system in the low frequency limit. The effective medium is an anisotropic fluid-like material.

The effective speed of sound is defined (in the low frequency limit) as the ratio between the angular frequency ω and the wave number K. This ratio can be obtained by making a power series expansion of the trigonometric functions up to second order in their arguments. It is easy to show that the effective speed of sound tensor that follows is

$$c_\perp^2 = \frac{B_{eff}}{\rho_\perp}, c_\parallel^2 = \frac{B_{eff}}{\rho_\parallel} \tag{3.5}$$

Where the quantities $\rho_\perp$ and $\rho_\parallel$ are the components of the density tensor and are

$$\begin{aligned} \rho_\perp &= \frac{1}{d_1 + d_2}(d_1\rho_1 + d_2\rho_2) \\ \frac{1}{\rho_\parallel} &= \frac{1}{d_1 + d_2}\left(\frac{d_1}{\rho_1} + \frac{d_2}{\rho_2}\right) \end{aligned} \tag{3.6}$$

The components $c_\perp$ and $c_\parallel$ define the diagonal components of the speed tensor for the propagation along the perpendicular and parallel directions, respectively, to the layered system. These relations define a way to design a multilayered shell satisfying the conditions (2.1), as is explained in detail in [22]. The solution reported there is analyzed here.

The proposed cloak is schematically shown in Figure 2, where the 1D structure is transformed into a circular-shaped shell that it is expected to cloak a rigid core placed in its interior. To check the functionality of the multilayered cloak we have performed multiple scattering simulations by using the method developed in [14, 15].

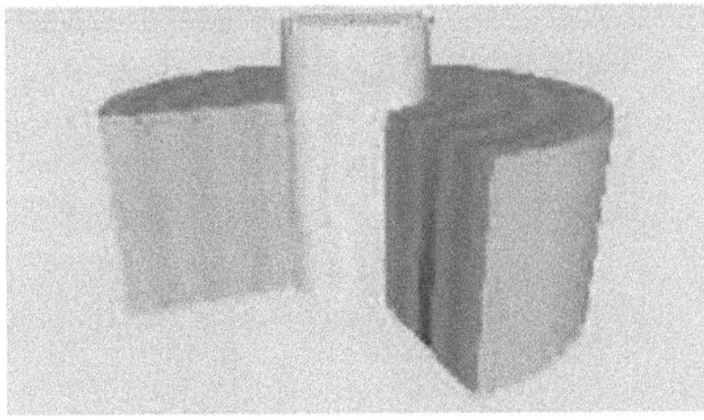

Fig. 2 Schematic view of the cloaking shell. It consists of a circular-shaped multilayered structure made up of two different materials of the same thicknesses.

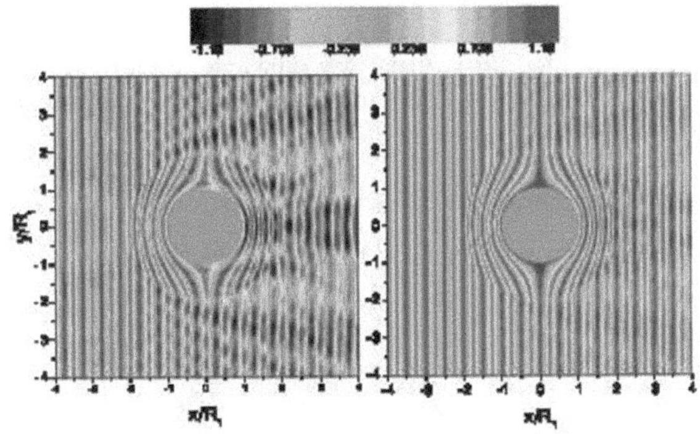

Fig. 3 Pressure map for a planar wave incident on a rigid cylindrical scatterer surrounded by a multilayered acoustic shell made up of 50 layers (left panel) and 200 layers (right panel). The radius of the shell is twice that of the core (R2 = 2R1).

Maps of the acoustic pressure are represented by the real part of the complex amplitude P and are shown in Figure 3 for the case of a rigid core of radius R_1 that is placed inside a multilayered shell of radius $R_2 = 2R_1$. The full structure is submitted to an acoustic field of wavelength $\lambda = R_1 / 2$.

The performance of two different shells are depicted in Figure 3 where the left panel corresponds to a shell made of 50 layers and the right panel to one composed of 200 layers, where each layer of thickness d is composed of two alternative layers of thickness of material 1 and 2. The cloaking effect is evident in both representations, but that corresponding to 200 layers is considerably more effective.

These results can be compared with the case of the rigid cylinder with no cloak that is represented in the left panel of Figure 4, where the incident wave is strongly scattered by the cylinder. On the other hand, in the right panel of Figure 4 is depicted the acoustic cloaking by a extremely thin cloak, its thickness being two order of magnitude smaller than the hidden cylinder, but it is also made of 200 layers. This result is very promising because it indicates the possibility of building cloaks as thinner as the available technology allows.

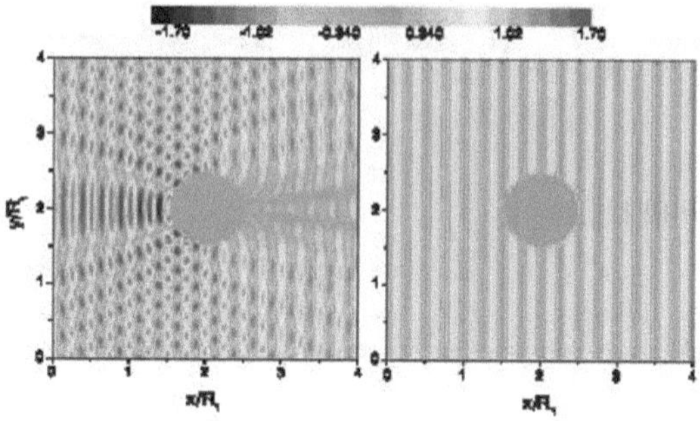

Fig. 4 Left panel: pressure map for a planar wave incident on a rigid cylindrical scatterer of radius R1. Right panel: map corresponding to the same scatterer surrounded by an extremely thin cloak shell (R2-R1 = 0.01R1) made of 200 layers.

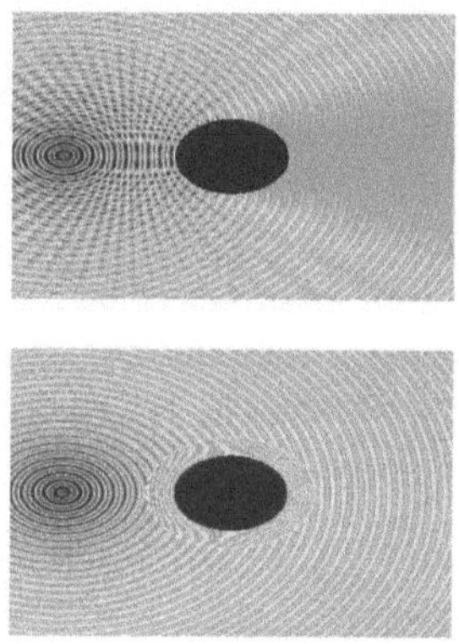

Fig. 5 Upper panel Interaction of a point source with a rigid cylinder. Lower panel Cloaking of the rigid cylinder from the point source.

In Figure 5 the interaction of a rigid cylinder with a point source is shown in the upper panel. It is clear that there is a strong reflected field due to the short wavelength of the incident field. In the lower panel the rigid object is surrounded by the cloaking shell and then the interaction with the point source disappears. This is a demonstration of the fact that the cloaking effect is independent of the nature of the incident field.

Now, it is also interesting to analyze the cloaking effect as a function of the number of layers employed in the fabrication of the cloak. The resulting behavior is important in order to simplify as much as possible the fabrication of the cloaking shell. We have studied the backscattered field as a typical parameter characterizing the cloak's performance and it is represented (in a logarithmic scale) in Figure 6 as a function of the frequency for different number of layers.

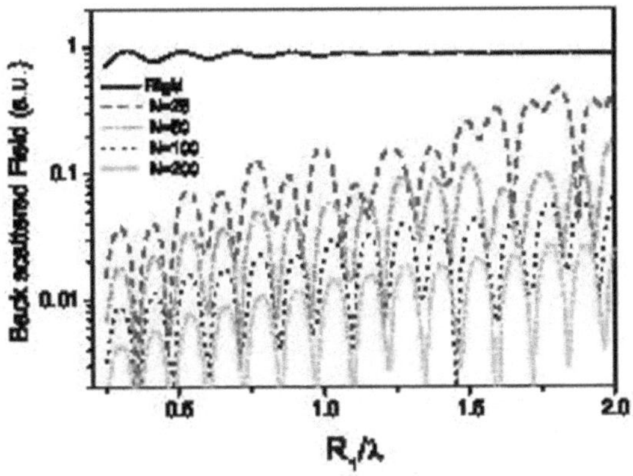

Fig. 6 Frequency response of the backscattered pressure for a rigid cylinder surrounded by a cloaking shell consisting of a multilayered structure as described in Figure 2. N is the number of layers in the structure. The case of the bare rigid cylinder without the shell is also represented.

It is remarkable in Figure 6 that only 50 layers are able to reduce in more than one order of magnitude (for a wide range of frequencies) the back scattered field in comparison with that for the corresponding naked rigid cylinder. Other interesting cases like a penetrable and a void regions are not reported here but we expect results analogous to those already published [13].

4. Acoustic Cloaking in 3D

In three dimensions the cloaking conditions for the acoustic parameters was derived in [19,20], yielding

$$\begin{aligned}
\rho_r(r) &= \frac{R_2 - R_1}{R_2}\frac{r^2}{(r-R_1)^2}\rho_b \\
\rho_\vartheta = \rho_\varphi &= \frac{R_2 - R_1}{R_2}\rho_b \\
B(r) &= \left(\frac{R_2 - R_1}{R_2}\right)^3 \frac{r^2}{(r-R_1)^2} B_b
\end{aligned} \tag{4.1}$$

The same approach used previously for the two dimensional case could be applied here, however now the cloaking shell should be a multilayered spherical fluid-like structure which, obviously, is more difficult to achieve in practice. Each of the fluids used in the structure can be made of lattice of elastic spheres.

The acoustic parameters $\rho_r(r)$ and $B(r)$ have a similar radial dependence in two and three dimensions, however the angular components of the density tensor $\rho_\vartheta = \rho_\varphi$ are now constant. This difference is not important for designing a multilayered cloaking shell and it can be achieved in a similar way as it was done in [22].

References

[1] Cummer, S. A., and Schurig, D.: One path to acoustic cloaking *New J. Phys.* **9** 45 (2007).
[2] Pendry, J. B., Schurig, D. and Smith , D. R.: Controlling electromagnetic fields. *Science* **312** 1780 (2006).
[3] Leonhardt, U. *Science* **312** 1777 (2006).
[4] Greenleaf A., Lassas, M. and Uhlman, G.: Anisotropic conductivities that cannot be detected by EIT. *Physiol. Meas.* **24** 413 (2003).
[5] Greenleaf, A., Lassas, M., and Uhlman, G.: On nonuniqueness for Calderon's inverse problem. *Math. Res. Lett.* **10** 685 (2003).
[6] Schurig, D., Mock, J. J., Justice, B. J., Cummer, S. A., Pendry, J., Starr, A. and Smith, D. R.: Metamaterial electromagnetic cloak at microwave frequencies. *Science* **314** 977(2006).
[7] Milton, G. W., Briane, M., and Willis, J. R.: On cloaking for elasticity and physical equations with a transformation invariant. *New J. Phys.* **8** 248 (2006).
[8] Torrent, D., and Sánchez-Dehesam J.: Anisotropic mass density by two dimensional acoustic metamaterials. *New J. Phys.* **10** 023004 (2008).
[9] Torrent, D., and Sánchez-Dehesa, J.: Acoustic metamaterials for new two dimensional sonic devices. *New J. Phys.* **9** 323 (2007).
[10] Huang, Y., Feng, Y., and Jiang, T.: Electromagnetic structure by layered structure of homogeneous isotropic materials. *Opt. Express.* **15** 11133 (2007).
[11] Schoenberg, M., and Sen, P. N.: Properties of a periodic stratified acoustic half-space and its relation to a Biot fluid. *J. Acoust. Soc.* **73** 61 (1983).
[12] Tretyakov, S.: *Analytical Modeling in Applied Electromagnetism.* Norwood, MA: Artech House, (2000).
[13] Cai, L.-W., and Sánchez-Dehesa, J.: Analysis of Cummer–Schurig acoustic cloaking. *New J. Phys.* **9** 450 (2007).
[14] Cai, L.-W.: Multiple scattering in single scatterers. *J. Acoust. Soc.* **115** 986 (2004).
[15] Cai, L.-W., and Sánchez-Dehesa, J.: Acoustical scattering by radially stratified scatterers *J. Acoust. Soc. Am.* (to be published).
[16] Milton, G. M.: *The Theory of Composites*. Cambridge, Cambridge University Press, 2002.
[17] Torrent, D., Håkansson, A., Cervera, F., and Sánchez-Dehesa, J.: Homogenization of two dimensional clusters of rigid rods in air. *Phys. Rev. Lett.* **96** 204302 (2006).
[18] Torrent, D., and Sánchez-Dehesa, J.: Effective parameters of clusters of cylinders embedded in a non viscous fluid or gas. *Phys. Rev. B* **74** 224305 (2006).
[19] Chen, H., and Chan, C. T.: Acoustic cloaking in three dimensions using acoustic metamaterials. *Appl. Phys. Lett.* **91** 183518 (2007).
[20] Cummer, S. A., Popa, B.-I., Schurig, D., Smith, D. R., Pendry, J., Rahm, M. and Starr, A.: Scattering derivation of a 3D acoustic cloaking shell. *Phys. Rev. Lett.* **100** 024301 (2008).
[21] Greenleaf, A., Kurylev, Y., Lassas, M., and Uhlmann, G.: Full-invisibility of active devices at all frequencies Commun. *Math. Phys.* **275** 749 (2007).
[22] Torrent, D., and Sánchez-Dehesa, J.: Acoustic cloaking in two dimensions: A feasible approach. *New J. Phys.* **10** 063015 (2008).

Band Structure and Phonon Transport in a Phononic Crystal Made of a Periodic Array of Dots on a Membrane

B. Djafari-Rouhani, Y. Pennec and H. Larabi

Institut d'Electronique, Microélectronique et Nanotechnologie, UMR CNRS 8520, Avenue Poincaré, Université de Lille1 Sciences et Technologies, 59652 Villeneuve d'Ascq, France

bahram.djafari-rouhani@univ-lille1.fr

Abstract. Using the finite difference time domain method, we investigate theoretically the band structure and phonon transport in a new type of phononic crystal constituted by a periodic array of cylindrical dots deposited on a homogeneous membrane. One new finding is the possibility of an absolute low frequency gap (as compared to the Bragg gap), similarly to the case of locally resonant structures. The existence of the low frequency gap requires very appropriate geometrical parameters, whereas it persists for various combinations of the materials constituting the plate and the dots. Besides, the band structure can exhibit one or more higher gaps when increasing the height of the cylinders. The results are discussed for different shapes of the cylinders such as circular, square or rotated square. The band structure can also display an isolated branch with a negative slope, useful for the purpose of negative refraction phenomena. We discuss the condition for wave guiding through different types of linear defects inside the phononic crystal. Finally, we investigate phonon transport between two substrates connected by a periodic array of particles and discuss different features appearing in the transmission spectrum.

1. Introduction

The study of phononic crystals [1-2], constituted by a periodical repetition of inclusions in a matrix background, has received a great deal of attention during the last decade [3]. Associated with the possibility of absolute band gaps in their band structure, these materials have found several potential

T.-T. Wu and C.-C. Ma (eds.), *IUTAM Symposium on Recent Advances of Acoustic Waves in Solids*, IUTAM Bookseries 26, DOI 10.1007/978-90-481-9893-1_12,

applications, in particular in the field of wave guiding and filtering as well as in the field of sound isolation. The band gaps may originate from the Bragg reflections resulting from the periodicity of the structure or be due to the existence of local resonances in each unit cell. In addition to infinite phononic crystals, several recent works have also been devoted to the study of acoustic waves in free or supported plates of 2D phononic crystals [4-9]. The existence of band gaps in such geometries may be useful for the purpose of introducing functionalities such as wave guiding and filtering in integrated high frequency devices. It is also worth noticing that, in the case of non metallic nanostructured materials, the thermal transport is mediated by phonons and, therefore, the knowledge and engineering of phononic band structure is a necessary step for the purpose of heat management. In particular, the existence of band gaps and/or the lowering of group velocities due to the bending of the dispersion curves are detrimental for phonon transmission and hence thermal conductivity.

In this paper, we are dealing with a new type of finite thickness phononic crystal constituted by a square array of cylindrical dots deposited on a thin homogeneous plate. One new finding of this work is the possibility of a low frequency gap where the wavelengths in the constituting materials are much larger than the typical lengths in the structure such as the period of the lattice or the thickness of the plate. The existence and evolution of this gap as a function of the geometrical and material parameters of the structure have been thoroughly discussed in a recent paper [10]. The band structure can also display one or more higher gaps depending on the height of the cylinders [10]. These results will be briefly recalled in this paper. Let us mention a recent paper by Wu et al [11] where a study on a similar structure has been carried out independently. Besides, we show the possibility of an isolated dispersion curve useful for negative refraction phenomena. The evolution of the dispersion curves and the band gaps will be discussed for different shapes of the cylinder's section, namely circular, square, and rotated square. We also study the possibility of confinement and wave guiding when a guide is created inside the phononic crystal by removing or modifying a row of dots. Finally, we calculate the phonon transmission between two substrates connected by a periodic array of particles and discuss the features appearing in the transmission spectra [12]. The calculations presented in this paper are based on the Finite Difference Time Domain (FDTD) method.

2. Dispersion Curves and Band Gaps

2.1 Cylindrical Dots of Circular Shape

The physical model considered is a square lattice of cylindrical dots deposited on a plate (Fig. 1a) with lattice parameter *a*. The filling factor is defined as $\beta = S/a^2$, *S* being the section area of the dots. The height of the cylinders is denoted by *h* and the thickness of the plate by *e*. The materials constituting the dots and the plate are assumed to be isotropic or of cubic symmetry with their crystallographic axes oriented along the coordinate axes *x*, *y* and *z* (Fig. 1a). In the following illustrations, the lattice parameter *a* is taken equal to $1\ \mu m$, unless stated otherwise. However, since the frequencies scale inversely with the lengths, the results can be transposed to different frequency domains.

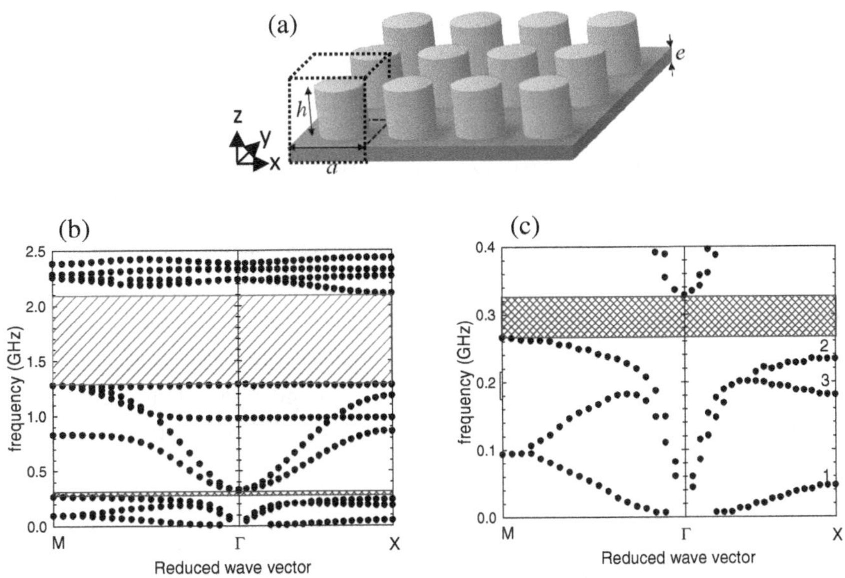

Fig. 1 (a) Phononic crystal made of a square lattice of dots deposited on a homogeneous plate. The dashed cube represents one unit cell of the periodic structure. (b) Band structure in the frequency range [0, 2.5] GHz for steel cylinders on a silicon plate, calculated in the first irreducible Brillouin zone of the phononic crystal. The parameters are $a=1\mu m$, $h=0.6\mu m$ and $e=0.1\mu m$. (c) Magnification of (b) in the frequency range [0, 0.4] GHz.

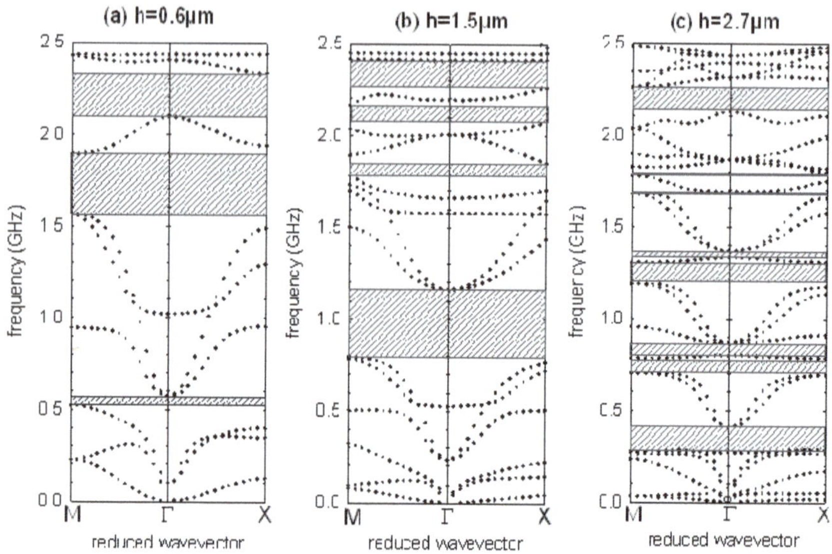

Fig. 2 Evolution of the band structure for different values of the height of the dots: (a) $h=0.6\mu m$, (b) $h=1.5\mu m$ and (c) $h=2.7\mu m$. The other geometrical parameters are $e=0.2\mu m$ and $\beta=56.4\%$.

Figure 1b gives an example of the band structure for propagation in the *(x, y)* plane, along the high symmetry axes of the first Brillouin zone, for a phononic crystal constituted by an array of circular steel cylinders on a silicon (Si) plate and the following geometrical parameters: filling factor $\beta=0.564$, height of the cylinders $h=0.6\ \mu m$ and thickness of the plate $e=0.1\ \mu m$. The band structure is displayed in the frequency range *[0, 2.5] GHz* and magnified for its lowest part (*[0, 0.4] GHz*) in Fig.1c. A new feature with respect to usual phononic crystals is the existence of a low frequency gap, extending from 0.*265 GHz* to *0.327 GHz*, where the acoustic wavelengths in all constituting materials are more than 10 times larger than the size of the unit cell. This result resembles the low frequency gap in the so-called locally resonant materials [13-14] where the opening of the gap results from the crossing of the normal acoustic branches with a flat band associated with a local resonance of the structure rather than from the Bragg reflections due to the periodicity of the structure. Actually, the occurrence of the low frequency gap in our structure is closely related to the relative motion of the dispersion curves and, in particular, to the relative bending of the shear horizontal (branch #2) and symmetric Lamb (branch #3) modes when changing the geometrical parameters of the structure. From a detailed study [10] of the shift and bending of both branches #2 and #3 with these parameters (in particular with the thickness of the plate and the height of the dots), it was found that the existence of this low frequency gap requires a very appropriate choice of the geometrical parameters such as those chosen in figure 1 (see also figure 2). On the contrary, we have shown [10] the persistence of this gap against very ifferent combinations of the materials constituting the dot and the plate among a set of five materials (tungsten, steel, silicon, aluminum and epoxy). One can notice the persistence of this gap even if the constituting materials are identical. This supports the origin of the gap as being related to the geometrical rather than physical parameters of the structure. On the other hand, the central frequency of the gap is very dependent upon the choice of the materials and happens at lower frequencies when we combine a high density material (tungsten) in the cylinders with a low density material (epoxy) in the plate.

The band structure in Fig. 1 displays also a higher Bragg gap, around *2 GHz*, which is in accordance with the period of the structure as usual. We have investigated the behavior of this and other higher gaps in the band structure as a function of the geometrical parameters h, e and β, along the high symmetry axes ΓX and ΓM of the irreducible Brillouin zone. In figure 2, we show the evolution of these gaps as a function of the height of the dots while keeping the other parameters constant ($\beta=0.564$ and and $e=0.2\mu m$). For $h=0.6\mu m$ (Fig.2a), we note the existence of three gaps. The lowest one at [0.*5193, 0.5717GHz*] closes for $h>1.0\mu m$. Besides, the band structure exhibits two higher gaps respectively in the frequency ranges [*1.560, 1.887GHz*] and [*2.092, 2.328GHz*]. When increasing h to $1.5\mu m$ (Fig. 2b) and then to $2.7\mu m$ (Fig. 2c), the central frequencies of these gaps move downwards together with the dispersion curves, whereas new absolute band gaps appear at higher frequencies. It is interesting to remark that, up to a certain frequency range, the opening of the gaps results from the bending of the dispersion curves rather than their folding, similarly to the case of locally resonant materials where the opening of the gaps results from the crossing of the normal acoustic branches with almost flat bands.

2.2 Negative Slope Branch

An interesting feature of the band structure in figure 2a (reproduced in Fig. 3a) is the existence of an isolated branch with a negative slope around the frequency 2.18 GHz. This branch can be useful for investigating the phenomenon of negative refraction which has been widely studied in phononic crystals of infinite extent [15-16] but not yet in phononic crystals of finite thickness. In particular, this branch satisfies some of the conditions to perform sound focusing by a flat lens. Indeed, as shown in figure 3b, the equifrequency surfaces (EFS) of the phononic crystal in the vicinity of 2.18 GHz are almost circular around the Γ point of the Brillouin zone. These EFS display decreasing frequency as one increases the wave vector, which means that the wave vector and group velocity are antiparallel whatever the direction of propagation of the acoustic wave.

Moreover, we present in Fig. 3a the dispersion curves of the Lamb modes for a homogeneous plate of thickness $e=0.2\mu m$. One can see that the sound line associated with the symmetric Lamb mode is almost linear in the frequency range of interest which means that its EFS are also circular. This line intersects the negative slope branch of the phononic crystal at 2.089 GHz. At this frequency one obtains a superposition of the quasi-circular EFS of both the homogeneous plate (representing the incident medium) and the phononic crystal and both media have the same phase velocities (or indices of refraction). This means that if an incident wave is generated at this frequency by a source situated in the plate in front of the phononic crystal, it will undergo the phenomenon of negative refraction and focusing at the back of the phononic crystal. Nevertheless, the conditions for realizing a satisfactory focusing still need to be investigated, since for instance the waves impinging the phononic crystal at different incident angles will undergo different transmission coefficients.

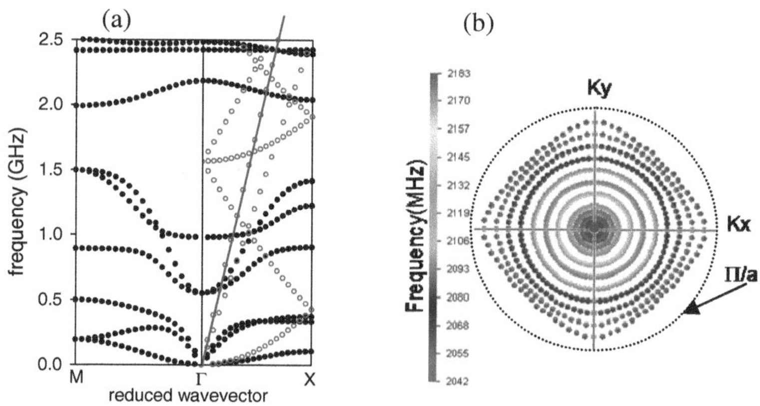

Fig. 3 (a) Band structure of Fig. 2a represented with black dots. The red open dots correspond to the band structure of a homogeneous silicon plate of thickness $e=0.2\mu m$. The red solid line is the sound line associated to the symmetric Lamb mode. (b) Equifrequency surfaces (EFS) around the Γ point of the Brillouin zone.

2.3 Confined Phonons in a Waveguide

In this section, we study the possibility of propagating confined modes in a linear waveguide inside the phononic crystal. The geometrical parameters are the same as in Fig. 1, i.e. β=*56.4%*, *h*=*0.6μm* and *e*=*0.1μm* that ensures the existence of the largest forbidden gaps. The FDTD calculation is performed by using a super cell containing five unit cells in the *y* direction. The guide is created by removing one row of dots in the third unit cell, thus constituting a linear waveguide in the *x* direction.

The width of the waveguide, δ, has been chosen as a variable parameter to investigate the existence and number of localized modes in the band gap. Figure 4a shows the band structure in the *ΓX* direction for the waveguide structure with δ=*0.55a*. The dispersion curve exhibits three additional branches inside the higher gap (*[1.287, 2.106GHz]*) while no supplementary branches appear inside the lowest gap. Increasing the width of the waveguide leads to the lowering of the frequencies of the confined modes. Fig. 4b shows the band structure for a waveguide with δ=*1.05a* which presents one new mode in the lowest band gap (*[0.2652, 0.3279GHz]*). To show the confinement of such modes inside the waveguide, we focus on the points C and D of the dispersion curves. The maps of the displacement fields associated with both modes are respectively sketched in Figs. 4c and d. In both cases, the acoustic field is essentially confined in the area of the waveguide and does not leak out into the rest of the structure.

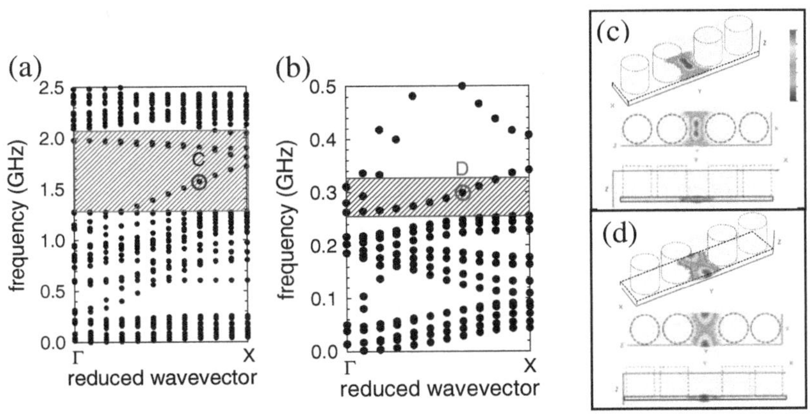

Fig. 4 (a) Band structure along the ΓX direction of the Brillouin zone calculated with a supercell containing (1x5) unit cells. The waveguide is formed by removing one row of dots in the third unit cell. The red hatched areas indicate the location of the absolute band gap of the perfect structure. The dispersion curves are calculated for a waveguide width of δ=*1.05a*. (b) Same as (a) for a waveguide of width δ=*0.55a*. (c) and (d) Maps of the modulus of the elastic displacement field for the modes denoted as C and D. The confinement is shown in three-quarter, top, and lateral views. The red color corresponds to the maximum displacement, whereas the blue color corresponds to the minimum.

2.4 Cylindrical Dots of Circular Shape

The dots in the phononic crystal are now assumed to be parallelepipeds of square section. Thus, the band structure can be modified by rotating the dots along the z axis. In the illustrations given in figure 5, the phononic crystal is constituted by steel rods on a silicon plate and the following geometrical parameters have been used: $a=1\mu m$, $h= 0.6\ \mu m$, $e=0.1\mu m$ and the side of the square $c=0.7\mu m$. The filling fraction which is chosen equal to 49% allows us to rotate the rods between 0 and 45°. Figure 5a presents the dispersion curves when the sides of the square are parallel to the x and y axes.

The band structure is very similar to the one presented in figure 1 for cylindrical dots of circular shape. It exhibits a low frequency gap that occurs between 0.233 and 0.285 GHz, and a high frequency gap in the frequency range [1.327, 2.120GHz], the latter being divided in two parts by an almost flat band at 1.610 GHz. The evolution of these gaps when the rods are rotated by an angle of 15, 30 and 45° are respectively presented in figures 5b and c. The lower gap (Fig. 5b) shifts towards higher frequencies and widens (from 0.052 to 0.079GHz) when the rods are rotated from 0 to 45°. The higher gap (Fig. 5c) remains almost unchanged except for the fact that the flat band moves downwards and leaves the band gap when the rotation angle reaches 30°. Therefore, the position of the flat band can be tuned inside the gap by changing the rotation angle and may be used in view of a selective filtering transmission through such a structure.

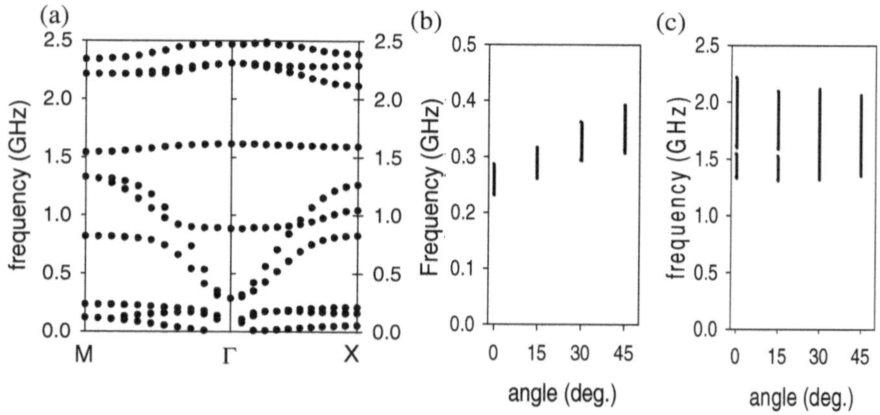

Fig. 5 (a) Band structure in the frequency range [0, 2.5] GHz for square rods of steel on a silicon plate. The parameters used for the calculation are $a=1\mu m$, $c=0.7\mu m$, $h=0.6\mu m$ and $e=0.1\mu m$. (b) and (c) Evolution of the low and high frequency gaps as a function of the angle of rotation of the rods.

3. Phonon Transport between two Substrates across a Periodic Array of Particles

This section is devoted to some preliminary results about the phonon transmission between two silicon substrates through a set of particles [12]. A few simulations have been performed when the substrates are connected by a two-dimensional array of dots. However, to avoid a huge computational time, most of the simulations are done in the geometry of Figure 6a where the dots are replaced by a one-dimensional set of plates. Figure 6b presents an example of the transmission coefficient in a case which is well-adapted for a physical discussion. The parameters are $a=1\mu m$, $h=1.4\mu m$ and $d=0.2\mu m$. The regular oscillations appearing in the whole spectrum, and in particular in the low frequency part of the spectrum, can be associated with Fabry-Perot resonances along the height of the plates. The frequencies of the peaks and their separation are closely related to the height h and the nature of the plates, but almost independent of the nature of the substrates and the period a. The separation between these peaks would decrease if one increases the height of the plates. Two examples of the displacement field, respectively for a peak and a dip in the transmission coefficient, are presented in Fig. 6c and d.

The spectrum in Fig. 6b displays also regular transmission gaps, around 5, 10 and 15 GHz, which can be associated to the period of the structure and the nature of the substrates, but are almost independent of the material constituting the plates. In addition, close to a zero of transmission, one may find a narrow peak for which the displacement field (Fig. 6e) shows an enhancement both inside the particles and in the vicinity of the interfaces between the substrates and the particles. Therefore, this selective transmission is believed to be the result of coherent coupling between the diffracted waves excited on both interfaces and the Fabry-Perot resonant modes inside the junctions.

Finally, above some threshold frequency, the transmission spectrum displays randomly fast oscillations. It is likely that this behavior happens when the plates can support transversely exited modes whose number increases when going to higher frequencies.

In figure 7, we give another example of the transmission coefficient for thicker plates, namely $d=0.4\mu m$. The main effect with respect to the previous case is to reduce the contrast in the Fabry-Perot oscillations and, more importantly, to decrease the threshold frequency above which the spectrum contains fast oscillations.

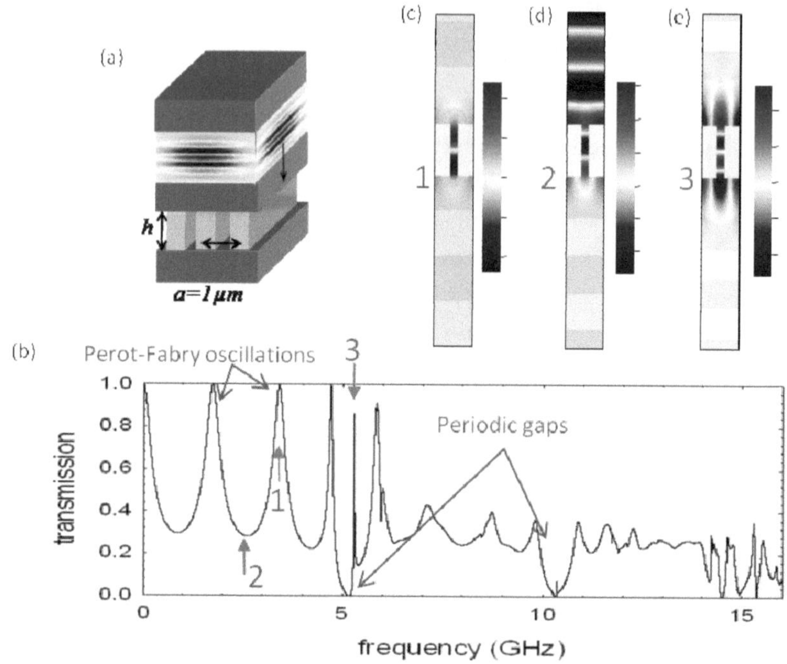

Fig. 6 (a) Two Silicon substrates connected by a one-dimensional array of Al plates. (b) Phonon transmission coefficient calculated with the following parameters: *a=1 µm, h=1.4 µm* and *d=0.2 µm*. (c), (d) and (e) Maps of the displacement field at three specific frequencies 3.412GHz, 2.604GHz and 5.294GHz, labeled 1, 2 and 3 in figure b.

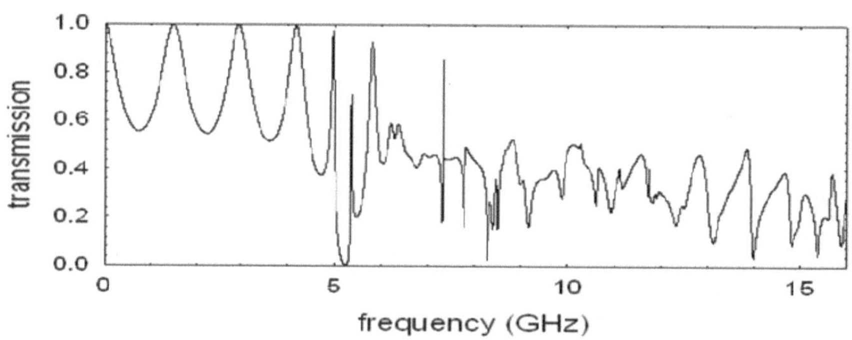

Fig. 7 Same as in figure 10b for *h=1.4 µm* and *d =0.4 µm.*

4. Conclusions

Using a FDTD method, we have studied the band structure of a phononic crystal constituted by a periodic array of dots deposited on a thin membrane. One major finding of this work is the possibility of a low frequency absolute band gap in which the acoustic wavelength in any constituent material is at least ten times larger than the characteristic lengths in the structure. The occurrence of this gap requires an appropriate choice of the geometrical parameters whereas it exists for a large variety of materials constituting the phononic crystal. However, the central frequency and the width of this gap are dependent upon the material properties. More particularly, in view of acoustic isolation, it would be suitable to choose a high density material for the cylinders and a low density material for the plate. The band structure can also display one or more higher gaps depending on the height of the cylinders. The band structure has been discussed for cylindrical dots of circular or square shapes. In the latter case, the band gaps can be tuned by rotating the squares with respect to the crystallographic axes of the crystal. Depending on the parameters, the band structure can also display an isolated dispersion curve with a negative slope which may be useful for studying the phenomena of negative refraction in such a finite thickness phononic crystal. In presence of a linear waveguide created by removing a row of cylinders, we have shown the possibility in any band gap of well confined modes without a significant leakage in the rest of the structure. In a forthcoming work, we have studied in detail the existence of confined modes as well as their ability to contribute to transmission for different types of waveguides obtained by changing the geometrical or material parameters of the dots along a row. Finally, in the last part of the paper, we studied the phonon transmission between two substrates across a periodic array of particles and discussed the different features appearing in the spectra. In particular, we showed the existence of a series of zero of transmission associated to the excitation of a surface wave at the substrate boundary, as well as the possibility of a narrow transmission peak next to the zero resulting from the coupling of the surface waves at both boundaries with a Perot-Fabry resonance inside the particles connecting the substrates

Acknowledgments: This work was supported in part by the European Commission (EC) 7th Framework Program (FP7), under the (IP) project reference No. 216176 (NANOPACK, www.nanopack.org).

References

[1] Kushwaha, M.S., Halevi, P., Dobrzynski, L., Djafari-Rouhani, B.: Acoustic band structure of periodic elastic composites. *Phys. Rev. Lett.* **71** 2022 (1993).

[2] Sigalas, M.M., Economou, E.N.: Band structure of elastic waves in two dimensional systems. *Solid State Commun.* **86** 141 (1993).

[3] For a comprehensive list of references on phononic crystals, see the phononic database at http://www.phys.uoa.gr/phononics/PhononicDatabase1.html

[4] Vasseur, J.O., Deymier, P.A., Djafari-Rouhani, B., Pennec, Y.: Absolute band gaps in two-dimensional phononic crystal plates. *Proceedings of IMECE*, ASME International Mechanical Engineering Congress and Exposition, Chicago, Illinois (2006).

[5] Hsu, J.C., Wu, T.-T.: Efficient formulation for band-structure calculations of two-dimensional phononic-crystal plate. *Phys. Rev. B* **74** 144303 (2006).

[6] Khelif, A., Aoubiza, B., Mohammadi, S., Adibi, A., Laude, V.: Complete band gaps in two-dimensional phononic crystal slab. *Phys. Rev. E* **74** 046610 (2006).

[7] Vasseur, J.O., Hladky-Hennion, A.C., Djafari-Rouhani, B., Duval, F., Dubus, B., Pennec, Y.: Waveguiding in two-dimensional piezoelectric phononic crystal plate. *J. Appl. Phys.* **101** 114904 (2007).

[8] Hsu, J.C., Wu, T.-T.: Lamb waves in binary locally resonant phononic plates with two-dimensional lattices. *Appl. Phys. Lett.* **90** 201904 (2007).

[9] Vasseur, J.O., Deymier, P.A., Djafari-Rouhani, B., Pennec, Y., Hladky-Hennion, A.C.: Absolute forbidden bands and waveguiding in two-dimensional phononic crystal plates. *Phys. Rev. B* **77** 085415 (2008).

[10] Pennec, Y., Djafari-Rouhani, B., Larabi, H., Vasseur, J. O., Hladky-Hennion, A. C.: Low-frequency gaps in a phononic crystal constituted of cylindrical dots deposited on a thin homogeneous plate. *Phys. Rev. B* **78** 104105 2008).

[11] Wu, T.-T., Huang, Z.-G., Tsai, T.-C., Wu, T.-C.: Evidence of complete band gap and resonances in a plate with periodic stubbed surface. *Appl. Phys. Lett.* **93** 111902 (2008).

[12] Djafari-Rouhani, B., Pennec, Y., Larabi, H.: Band structure and wave guiding in a phononic crystal constituted by a periodic array of dots deposited on a homogeneous plate. *Proceedings of SPIE on Nanotechnologies in Photonics*, San Jose 24-29 January 2009, Volume **7223** on Photonic and phononic crystal materials and devices IX (2009) .

[13] Liu, Z., Zhang, X., Mao, Y., Zhu, Y.Y., Yang, Z., Chan, C.T., Sheng P.: Locally Resonant Sonic Materials. *Science* **289** 1734 (2000).

[14] Goffaux, C., Sanchez-Dehesa, J., Levy Yeyati, A., Lambin, Ph., Khelif, A., Vasseur, J.O., Djafari-Rouhani, B.: Evidence of Fano-Like Interference Phenomena in Locally Resonant Materials. *Phys. Rev. Lett.* **88** 22 (2002).

[15] Sukhovich, A., Jing, L., Page, J. H.: Negative refraction and focusing of ultrasound in two-dimensional phononic crystals. *Phys. Rev. B* **77** 014301 (2008) and references therein.

[16] Hladky-Hennion, A., Vasseur, J., Dubus, B., Djafari-Rouhani, B., Ekeom, D., Morvan, B.: Numerical analysis of negative refraction of transverse waves in an elastic material. *J. Appl. Phys.* **104** 64906 (2008).

Material Parameters Determining the Band Gaps of Solid-Solid Phononic Crystals

Xiao-Zhou Zhou[1], Yue-Sheng Wang[1] and Chuanzeng Zhang[2]

[1]Institute of Engineering Mechanics, Beijing Jiaotong University, Beijing, 100044, China

[2]Department of Civil Engineering, University of Siegen, Siegen, D-57076, Germany

yswang@bjtu.edu.cn

Abstract. Material parameters directly determining band gaps for general three-dimensional phononic crystals are derived from the basic wave equations. These parameters include the mass density ratio, shear modulus ratio and Poisson's ratios of the scatterer and matrix materials. The effects of these parameters on phononic band gaps are discussed by computing two-dimensional systems with different filling fractions and lattice forms for both anti-plane and in-plane wave modes. The results show that the mass density ratio predominantly determines the band gap for the anti-plane mode, while that both mass density ratio and shear modulus ratio play equally important roles in controlling the band gaps for the in-plane mode. The maximum band gap will appear at both large density ratio and shear modulus ratio for either anti-plane or in-plane wave mode; but band gaps may appear in other situations depending on the filling fraction and lattice forms. Unlike one-dimensional phononic crystals, neither acoustic impedance ratio nor wave velocity ratio of the two-dimensional systems can determine the band gap independently. The analysis of the paper is relevant to the tuning of band gaps.

1. Introduction

Phononic crystals, a kind of periodic composites which possibly exhibit complete band gaps in their transmission spectra, have received considerable attention in recent years because of their unique physical properties and promising applications [1]. From the view of the application, tuning of the band gaps of phononic crystals is of importance. It has been found that the gap-width is strongly dependent upon the structural geometries of phononic crystals, including the scatterer's shape, lattice form, filling fraction, et al.[2-7]. For instance, a

T.-T. Wu and C.-C. Ma (eds.), *IUTAM Symposium on Recent Advances of Acoustic Waves in Solids*, IUTAM Bookseries 26, DOI 10.1007/978-90-481-9893-1_13,

triangle lattice and/or a square-section of scatterers are favorable to yield a wider band gap [5-7].

On the other hand, phononic band gaps are also determined by material combinations. Kushwaha et al [5, 6] showed that the band gaps become wider with the increase of the differences in both shear modulus and density when the values in the scatterer are higher than those in the matrix. They and Sigalas et al [8] indicated that a low-velocity material surrounded by a high-velocity material is more favorable to generate band gaps. Based on the comparison of some specific phononic crystals, Kee et al [9] concluded that the material parameters related to band gaps are the impedance and velocity, and that the contrast of the impedance is more important than that of the velocity. Recently Liu et al [10] argued that the density ratio rather than the contrast of the elastic constants is the dominant factor. It is noted that all these investigations were based on the calculation of band structures for the systems with particular material combinations. Some conclusions are even somewhat conflicting. In this paper, we will begin with the basic wave equations and derive the material parameters directly determining band gaps for general three-dimensional (3D) cases. Then the effects of these parameters on phononic band gaps are discussed in details by computing the two-dimensional (2D) cases with different filling fractions and lattice forms for both anti-plane and in-plane wave modes. For comparison, the results of the one-dimensional (1D) cases are also presented.

2. Theory

Consider a general periodic structure consisting of isotropic elastic scatterers embedded in an isotropic elastic matrix. Introduce the coordinate normalized by the lattice constant a, i.e. $\mathbf{r} = a^{-1}(x, y, z)$. Then harmonic wave equations may be written as

$$-\rho(\mathbf{r})\omega^2 a^2 u_i(\mathbf{r}) = \left[\lambda(\mathbf{r}) u_{j,j}(\mathbf{r})\right]_{,i} + \left[\mu(\mathbf{r})\left(u_{i,j}(\mathbf{r}) + u_{j,i}(\mathbf{r})\right)\right]_{,j}, \tag{2.1}$$

where $i, j = x, y, z$; $u_i(\mathbf{r})$ is the displacement components; $\rho(\mathbf{r})$ is the mass density; $\mu(\mathbf{r})$ and $\lambda(\mathbf{r})$ are the Lamé's constants; and the double subscripts imply the summation from 1 to 3. Let $h(\mathbf{r})$ represent $\rho(\mathbf{r})$, $\mu(\mathbf{r})$ or $\lambda(\mathbf{r})$. We can write $h(\mathbf{r}) = h_2 \bar{h}(\mathbf{r})$ with $\bar{h}(\mathbf{r}) = 1$ in the scatterers and $\bar{h}(\mathbf{r}) = h_1 / h_2$ in the matrix, where the subscripts 1 and 2 refer to the scatterers and matrix, respectively. Then (2.1) can be rewritten as

$$-\bar{\omega}^2\bar{\rho}(\mathbf{r})u_i(\mathbf{r})=\left[\kappa_2\bar{\lambda}(\mathbf{r})u_{j,j}(\mathbf{r})\right]_{,i}+\left[\bar{\mu}(\mathbf{r})\left(u_{i,j}(\mathbf{r})+u_{j,i}(\mathbf{r})\right)\right]_{,j}, \tag{2.2}$$

where $\kappa_2=2\nu_2/(1-2\nu_2)$ with ν_2 being the Poisson's ratio; and $\bar{\omega}=\omega a/c_{t2}$ is the normalized frequency with $c_{t2}=\sqrt{\mu_2/\rho_2}$ being the transverse wave velocity.

Equation (2.2) can be solved by using the plane wave expansion (PWE) method or wavelet method [11]. Here we will introduce the PWE method. Expand $\bar{h}(\mathbf{r})$ as

$$\bar{h}(\mathbf{r})=\sum_{\mathbf{G}}\mathrm{e}^{\mathrm{i}\mathbf{G}\cdot\mathbf{r}}\bar{h}_{\mathbf{G}}, \tag{2.3}$$

where $\mathrm{i}=\sqrt{-1}$ and $\bar{h}_{\mathbf{G}}$ is the Fourier coefficient which is given by

$$\bar{h}_{\mathbf{G}}=\begin{cases}(h_1/h_2)f+(1-f), & \mathbf{G}=0\\ (h_1/h_2-1)P(\mathbf{G}), & \mathbf{G}\neq 0\end{cases}, \tag{2.4}$$

in which f is the filling fraction; and $P(\mathbf{G})$ is the structural function related to the scatterer's shape. According to Bloch-theorem, the displacement field can be expressed as

$$u_i(\mathbf{r})=\mathrm{e}^{\mathrm{i}\mathbf{k}\cdot\mathbf{r}}u_{i\mathbf{k}}(\mathbf{r}),\quad i=x,y,z, \tag{2.5}$$

where $\mathbf{k}$ is the Bloch's wave vector. Since $u_{i\mathbf{k}}(\mathbf{r})$ is also a periodic function, it can be expanded in Fourier series as (2.3). Then we have

$$u_i(\mathbf{r})=\sum_{\mathbf{G}}u_{i\mathbf{k}+\mathbf{G}}\mathrm{e}^{\mathrm{i}(\mathbf{k}+\mathbf{G})\cdot\mathbf{r}},\quad i=x,y,z. \tag{2.6}$$

Substituting (2.3) and (2.6) into (2.2), we obtain

$$\begin{aligned}\bar{\omega}^2\sum_{\mathbf{G}_2}\bar{\rho}_{\mathbf{G}_1}u_{i\mathbf{k}+\mathbf{G}_2}=&\left[\sum_{\mathbf{G}_2}\bar{\mu}_{\mathbf{G}_1}(\mathbf{k}+\mathbf{G}_2)_j(\mathbf{k}+\mathbf{G}_1+\mathbf{G}_2)_j\right]u_{i\mathbf{k}+\mathbf{G}_2}\\ &+\left[\kappa_2\sum_{\mathbf{G}_2}\bar{\lambda}_{\mathbf{G}_1}(\mathbf{k}+\mathbf{G}_2)_j(\mathbf{k}+\mathbf{G}_1+\mathbf{G}_2)_i\right.\\ &\left.+\sum_{\mathbf{G}_2}\bar{\mu}_{\mathbf{G}_1}(\mathbf{k}+\mathbf{G}_2)_i(\mathbf{k}+\mathbf{G}_1+\mathbf{G}_2)_j\right]u_{j\mathbf{k}+\mathbf{G}_2}\end{aligned}. \tag{2.7}$$

Equation (2.4) shows that $\bar{\rho}_{\mathbf{G}_1}$, $\bar{\mu}_{\mathbf{G}_1}$ and $\bar{\lambda}_{\mathbf{G}_1}$ involve the material parameters (ρ_1/ρ_2 , μ_1/μ_2 and λ_1/λ_2), filling fraction f and structural function $P(\mathbf{G})$ which is related to the scatterer's shape and lattice form (**G**). λ_1/λ_2 may be re-written as $\lambda_1/\lambda_2 = (\mu_1/\mu_2)(\kappa_1/\kappa_2)$. Consequently, we can conclude that the band gaps are determined by two kinds of parameters: (*a*) structural parameters, including the filling fraction *f*, the scatterer's shape and the lattice form (**G**), and (*b*) material parameters, including the mass density ratio ρ_1/ρ_2 , the shear modulus ratio μ_1/μ_2 , and Poisson's ratios ν_1 and ν_2 . In this paper we will examine the influences of the material parameters on the band gaps.

If we consider a 2D system consisting of infinite cylinders parallel to the *z*-axis embedded in a matrix with the wave vector lying in *x*-*y* plane, then the anti-plane wave mode can be separated from the in-plane one. For the anti-plane mode, we have

$$-\bar{\omega}^2 \bar{\rho}(\mathbf{r}) u_z(\mathbf{r}) = \nabla\left[\bar{\mu}(\mathbf{r}) \nabla u_z(\mathbf{r})\right]. \tag{2.8}$$

Obviously, the material parameters for the anti-plane mode are only the mass density ratio ρ_1/ρ_2 and the shear modulus ratio μ_1/μ_2 .

Equation (2.7) can be rewritten in an eigenvalue equation. Then solution to this eigenvalue equation yields the dispersion curves from which we can get the band gaps.

3. Results and Discussion

The band gaps have been calculated for 2D phononic crystals with cylindrical scatterers in a square or triangle lattice. For the anti-plane mode, the band gap between the lowest two frequency bands is considered. This is the first potential gap for all phononic crystals. And for the in-plane mode, the band gap determined by the third and forth lowest bands is considered. This is the second but the most applicable potential band gap for most phononic crystals. The normalized gap-width $\Delta\bar{\omega}/\bar{\omega}_c$ (gap-width to mid-gap frequency ratio), which represents the more valuable band gap is calculated for both modes. We illustrate the variation of the gap-width in the logarithmic plane of ρ_1/ρ_2 and μ_1/μ_2 for some selected values of the filling fraction and Poisson's ratios. The 3D figures presented in this paper can provide a guide for the tuning of phononic band gaps by appropriate choices of the material combinations.

3.1 Anti-Plane Mode

Fig. 3.1 shows the normalized gap-width of the first band gap varying with ρ_1/ρ_2 and μ_1/μ_2 in a square lattice (Fig.3.1a) and a triangle lattice (Fig.3.1b) with *f*=0.4. The maximum band gap appears at both large ρ_1/ρ_2 and μ_1/μ_2 and becomes wider as these two parameters both increase. However, the band gaps mainly appear at large values of ρ_1/ρ_2 , that is to say, the density ratio predominantly determines the band gaps. This result is similar to that of Kushwaha et al [5, 6]. But the analysis in the last section indicates that band gaps are determined by the ratios of the material parameters rather than their differences as mentioned by Kushwaha et al [5, 6].

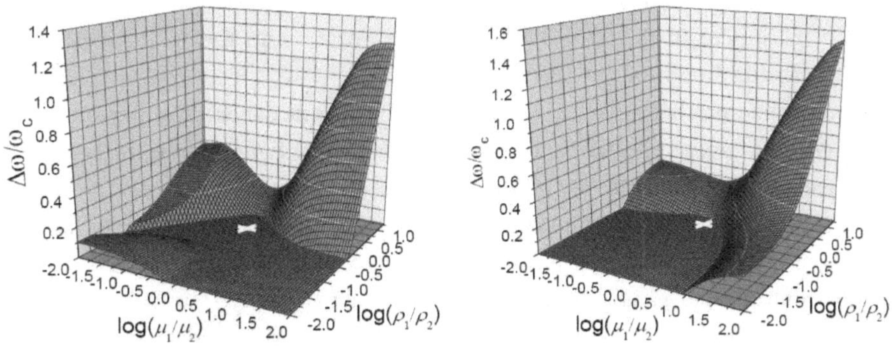

(a) (b)

Fig. 3.1 Variation of normalized gap-width in the logarithmic plane of ρ_1/ρ_2 and μ_1/μ_2 for the anti-plane mode, *f*=0.4. (*a*) square lattice; (*b*) triangle lattice.

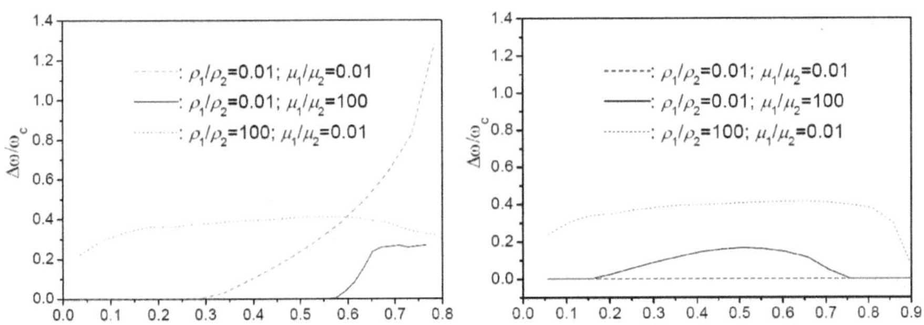

(a) (b)

Fig. 3.2 Variation of normalized gap-width with the filling fraction *f* for the anti-plane mode. (*a*) square lattice; (*b*) triangle lattice.

Band gaps also appear at large ρ_1/ρ_2 and small μ_1/μ_2 for both square and triangle lattices. For the square lattice, band gaps appear at both small ρ_1/ρ_2 and μ_1/μ_2 but not at small ρ_1/ρ_2 and large μ_1/μ_2; while it is contrary for the triangle lattice. To show the existence and width of the band gaps appearing at both small ρ_1/ρ_2 and μ_1/μ_2 as well as at small/large ρ_1/ρ_2 and large/small μ_1/μ_2 for various filling fractions, we plot the normalized gap-width varying with the filling fraction in Fig. 3.2 for three cases: (i) $\rho_1/\rho_2 = 0.01$, $\mu_1/\mu_2 = 0.01$ (dashed lines); (ii) $\rho_1/\rho_2 = 0.01$, $\mu_1/\mu_2 = 100$ (solid lines); and (iii) $\rho_1/\rho_2 = 100$, $\mu_1/\mu_2 = 0.01$ (dotted lines). It is shown that band gaps may appear in case (iii) for most values of the filling fraction and in case (ii) for only some particular values of the filling fraction. The gap is very narrow in case (ii) and thus is of little significance. No band gap appears in case (i) for the triangle lattice; while band gaps appear in this case for the square lattice with the width increasing with the filling fraction.

The above analysis also shows that neither the impedance ratio nor the wave velocity ratio can determine band gaps independently. For instance, in Fig. 3.1a, band gaps appear at large ρ_1/ρ_2 and small μ_1/μ_2, but not at small ρ_1/ρ_2 and large μ_1/μ_2. However, the impedance ratios $\sqrt{(\rho_2\mu_1)/(\rho_1\mu_2)}$ in these two situations could be close and even the same. Therefore, only the impedance ratio alone cannot control band gaps. Similarly, only the wave velocity ratio cannot control the band gaps as well. As shown in Figs. 3.1, the band gaps appear at both larger ρ_1/ρ_2 and μ_1/μ_2, where the transverse wave velocity ratio $\sqrt{(\rho_2\mu_1)/(\rho_1\mu_2)}$ may be either larger or smaller than 1. This result is different from that of Kushwaha et al [6] and Sigalas et al [8].

3.2 In-Plane Mode

Fig. 3.3 shows the normalized gap-width for $\nu_1 = \nu_2 = 0.2$ with the filling fraction f=0.4. The maximum band gap appears at both large ρ_1/ρ_2 and μ_1/μ_2, and becomes wider with the two parameters increasing. However, unlike the anti-plane mode, the band gap for the in-plane mode at large ρ_1/ρ_2 and small μ_1/μ_2 is very narrow and thus is of little importance. So we can conclude that the density ratio ρ_1/ρ_2 and the modulus ratio μ_1/μ_2 may play almost equally important roles in controlling the band gaps of the in-plane mode. This result is different from that of Kee et al [9].

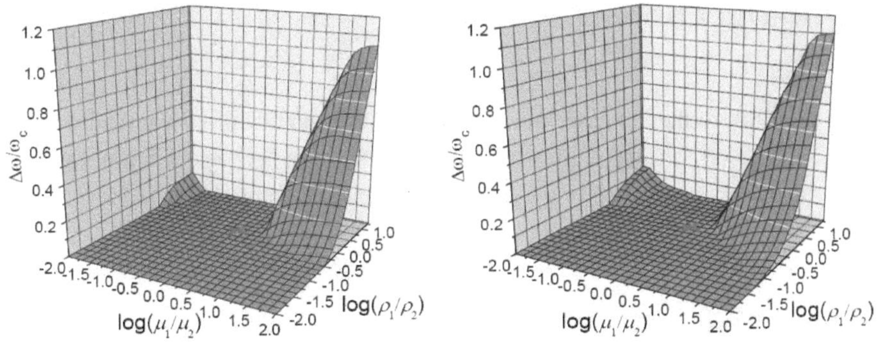

(a) (b)

Fig. 3.3 Variation of normalized gap-width in the logarithmic plane of ρ_1/ρ_2 and μ_1/μ_2 for the in-plane mode, $\nu_1 = \nu_2$ =0.2, *f*=0.4. (*a*) square lattice; (*b*) triangle lattice.

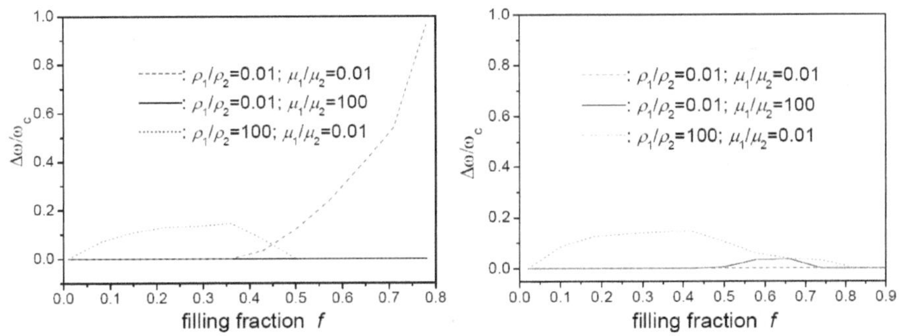

(a) (b)

Fig. 3.4 Variation of normalized gap-width with the filling fraction *f* for the in-plane mode, $\nu_1 = \nu_2$ =0.2. (*a*) square lattice; (*b*) triangle lattice.

To examine the existence of the band gaps in other cases, we take $\nu_1 = \nu_2$ =0.2 and present the normalized band-gap width versus the filling fraction in Fig. 3.4 for three cases (i), (ii) and (iii) as in Fig. 3.2. Narrow band gaps may appear in case (iii) (the dotted lines) for both square and triangle lattices with lower or mod erate filling fractions. Similar to Fig. 3.2a, the width of the band gap appearing in case (i) for the square lattice increases with the filling fraction, see the dashed line in Fig.3.4a. No band gap is found for other situations.

Detailed calculations also show that the influences of the Poisson's ratios on the band gaps are slight in a square lattice but not negligible in a triangle lattice. And the influence of the Poisson's ratio of the matrix material is more pronounced.

4. One-Dimensional System

For comparison to the 2D case, we consider a 1D phononic crystal consisting of two elastic solid layers stacked alternatively. An elastic wave propagates normally to the layer. The dispersion relation of the system can be written in the following form [12]

$$\cos(kh) = \cos(\omega t_1)\cos(\omega t_2) - b\sin(\omega t_1)\sin(\omega t_2) \equiv g(\omega), \tag{2.9}$$

where k is the wave number; $h = h_1 + h_2$ with h_j being the thickness of the component layers; $t_j = h_j/c_j$ with c_j being the transverse or longitudinal wave velocities of the component materials; $b = (1+Z^2)/2Z$ with $Z = (c_2\rho_2)/(c_1\rho_1)$ being the impedance ratio for a transverse or longitudinal wave. One can easily rewrite (2.9) in a form which is explicitly dependent on the mass density ratio ρ_1/ρ_2, shear modulus ratio μ_1/μ_2 and Poisson's ratios ν_1 and ν_2 (Poisson's ratios involved only for the longitudinal mode). Solution of $|g(\omega)| > 1$ yields the band gaps.

Figs. 4.1 and 4.2 illustrate the normalized band-gap width versus the mass density ratio ρ_1/ρ_2 and the shear modulus ratio μ_1/μ_2 for the transverse and longitudinal modes with the filling fraction *f*=0.4 and some selected values of the Poisson's ratios. It is observed that the band gaps can be determined independently by the impedance ratio $\sqrt{(\rho_1\mu_1)/(\rho_2\mu_2)}$. Wide band gaps can be obtained at either a large or a small impedance ratio. The band gaps of the longitudinal wave are somewhat dependent on the Poisson's ratios, while those of the transverse wave are not. In fact, the band structure for a 1D system determined by (2.9) is mainly dependent on the value of the parameter *b*, which is a function of the impedance ratio *Z* for either transverse or longitudinal wave. Therefore the band gaps in 1D case are basically determined by the impedance ratio. This behavior is quite different from the 2D case considered before.

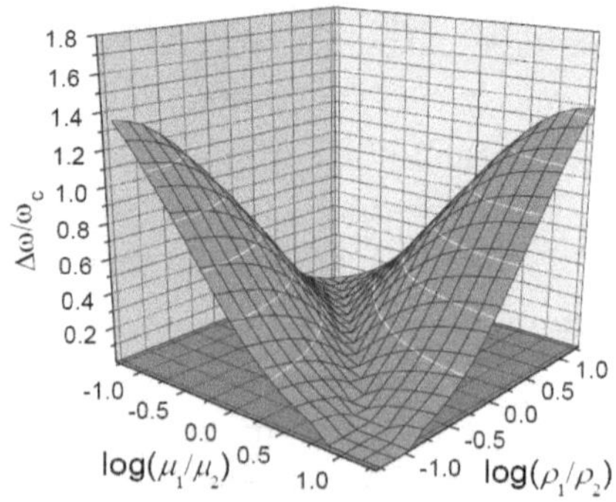

Fig. 4.1 Variation of normalized gap-width in the logarithmic plane of ρ_1/ρ_2 and μ_1/μ_2 for the transverse mode in 1D case with *f*=0.4.

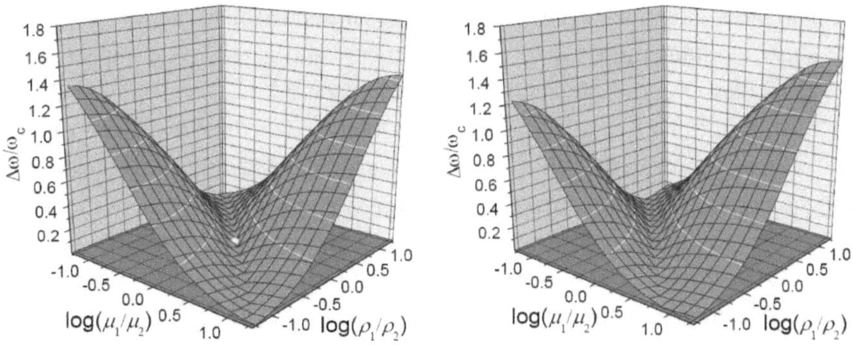

(a) (b)

Fig. 4.2 Variation of normalized gap-width in the logarithmic plane of ρ_1/ρ_2 and μ_1/μ_2 for the longitudinal mode in 1D case with *f*=0.4. (*a*) $\nu_1=\nu_2=0.2$; (*b*) $\nu_1=0.4$, $\nu_2=0.2$.

5. Concluding Remarks

In this paper, the material parameters determining band gaps for general 3D phononic crystals are derived from the basic wave equations. The influences of these parameters on phononic band gaps are discussed in details by computing 2D phononic crystals with different filling fractions and lattice forms for both anti-plane and in-plane wave modes. For comparison, the results for 1D phononic crystals are also presented. From the results we have following conclusions:

For the general 3D case, the material parameters determining phononic band gaps include the mass density ratio, shear modulus ratio, and Poisson's ratios of both scatterer and matrix materials. In particular, only the mass density ratio and shear modulus ratio are involved for the anti-plane mode in the 2D case.

For the anti-plane mode the mass density ratio predominantly determines the band gap; while for the in-plane mode both the mass density ratio and the shear modulus ratio may play equally important roles in controlling the band gaps.

In all considered cases, the band gaps easily appear at both large mass density ratio and shear modulus ratio and become wider with both of these two parameters increasing. That is to say, to get a wide band gap, it is favorable to embed a large-density stiff material into a small-density soft material. No band gap can appear when both of these two parameters are close to 1.

Band gaps may also appear in other situations depending on the filling fraction, lattice forms and Poisson's ratios. For example, besides the case of both large mass density ratio and shear modulus ratio, band gaps in a square lattice can also appear in other situations except the situation of a small mass density ratio and a large shear modulus ratio. For a triangle lattice, band gaps do not appear at both small mass density ratio and small shear modulus ratio.

For both anti-plane and in-plane modes, neither the impedance ratio nor the wave velocity ratio can determine the band gaps independently. In this aspect, 2D phononic crystals are quite different from 1D systems where the acoustic impedance ratio can determine the band gaps almost independently. A wide band gap can be obtained in the 1D phononic system with a larger acoustic mismatch (i.e. large or small impedance ratios).

Acknowledgments: The authors are grateful to the support provided by the National Natural Science Foundation of China (No. 10632020), the German Research Foundation (No. ZH 15/11-1), and jointly by the China Scholarship Council and the German Academic Exchange Service (No. D/08/01795).

References

[1] http://www.phys.uoa.gr/phononics/PhononicDatabase.html

[2] Vasseur, J. O., Djafari-Rouhaniy, B., Dobrzynskiy, L., Deymierz, P. A.: Acoustic band gaps in fibre composite materials of boron nitride structure. *J. Phy. Condens. Matter* **9** 7327-7341 (1997).

[3] Caballero, D., Sa´nchez-Dehesa, J., Rubio, C., Ma´rtinez-Sala, R., Sa´nchez-Pe´rez, J. V., Meseguer, F., Llinares, J.: Large two-dimensional sonic band gaps. *Phys. Rev. E* **60** R6316 (1999).

[4] Sainidou, R., Stefanou, N.: Formation of absolute frequency gaps in three-dimensional solid phononic crystals. *Phys. Rev. B* **66** 212301(2002).

[5] Kushwaha, M. S., Halevi, P., Martinez, G.: Theory of acoustic band structure of periodic elastic composites. *Phys. Rev. Lett.* **49** 2313-2322 (1994).

[6] Kushwaha, M. S., Halevi, P.: Band-gap engineering in periodic elastic composites. *Appl. Phys. Lett.* **64** 1085-1087 (1994).

[7] Vasseur, J. O., Djafari-Rouhani, B., Dobrzynski, L., Kushwaha, M. S., Halevi, P.: Complete acoustic band gaps in periodic fibre reinforced composite materials: the carbon/ epoxy composite and some metallic systems. *J. Phy.: Condens. Matter* **6** 8759-8770 (1994).

[8] Sigalas, M., Economou, E. N.: Band structure of elastic waves in two dimensional systems. *Solid State Commu.* **86** 141-143 (1993).

[9] Kee, C. S., Kim, J. E., Park, H. Y., Chang, K. J.: Essential role of impedance in the formation of acoustic band gaps. *J. Appl. Phys.* **87** 1593-1596 (2000).

[10] Liu, Y. H., Chang, C. C., Chern, R. L., Chang, C. C.: Phononic band gaps of elastic periodic structures: A homogenization theory study. *Phys. Rev. B* **75** 054104 (2007).

[11] Yan, Z. Z., Wang, Y. S.: Wavelet- based method for calculating elastic band gaps of two-dimensional phononic crystals. *Phys. Rev. B* **74** 224303 (2006).

[12] Guz, A. N., Shulga, A. N.: Dynamics of laminated and fibrous composites. *Appl. Mech. Rev.* **45** 35-60 (1992).

Behavior of Wave Motion in an Acoustic Metamaterial with Anisotropic Mass Density

C. T. Sun and H. H. Huang

School of Aeronautics and Astronautics, Purdue University, W. Lafayette, IN 47907, USA

sun@purdue.edu

Abstract. An elastic solid with frequency-dependent anisotropic mass density was developed to represent an acoustic metamaterial to study wave propagation. The band gaps of the material were related to the frequencies for which the mass density of the equivalent elastic solid becomes negative. Reflection and transmission of a pressure wave impinging on the metamaterial with a finite width were investigated using the equivalent elastic solid. It was found that the impinging pressure wave can induce a strong shear-dominated wave mode accompanied by a weak extension-dominated wave in the metamaterial oriented in some directions. Since the fluid-like medium cannot transmit shear waves, the shear-dominated mode in the metamaterial may be trapped inside the metamaterial, and, thus, the fluid-like material behind the metamaterial can remain basically undisturbed.

1. Introduction

Metamaterials have recently generated much excitement among physicists and engineers. A metamaterial is usually regarded as a material that contains man-made microstructures that give rise to unusual properties which are not found in natural materials. Extensive references on electromagnetic and acoustic metamaterials can be easily found, for example, in [1-8].

A one-dimensional lattice system that consists of periodically distributed local resonators was employed by Huang et al [9, 10] as an acoustic metamaterial. When the lattice was represented by a classical elastic solid, the effective mass density of the equivalent solid was found to become negative in certain frequency range that is near local resonance frequency of the resonator. It was found that the frequencies associated with negative effective mass densities are in the band gap of the metamaterial.

T.-T. Wu and C.-C. Ma (eds.), *IUTAM Symposium on Recent Advances of Acoustic Waves in Solids*, IUTAM Bookseries 26, DOI 10.1007/978-90-481-9893-1_14,

Recently, Huang and Sun [11] have considered a 2D metamaterial that possesses an anisotropic effective mass density. Their design of the metamaterial is shown in Fig. 1.1a. It is basically a 2D composite material with a specific microstructure in the form of a cavity with an internal mass connected in two directions by two different springs to the host matrix. For simplicity, this metacomposite is represented by the lattice model as shown in Fig. 1.1b. The longitudinal stiffness of the matrix is represented by the spring constant K_1 and the shear stiffness is represented by the shearing spring constant G, *and the mass of the surrounding matrix is lumped into a single mass* m_1. This model was first used by Milton et al. [12] to demonstrate the existence of anisotropic mass concept. It should be noted that another type of acoustic metamaterial with anisotropic mass density is a solid containing periodic microstructures formed with a fluid-like material considered by Torrent and Sánchez-Dehesa [13].

If the 2D lattice model shown in Fig. 1.1b is modeled as a homogeneous classical elastic solid, then the effective mass density of the representative elastic solid becomes frequency-dependent in order to match the dispersion curves of the original metamaterial. Further, the effective mass density can be shown to follow the property of second-order tensors. If in the original configuration, the two springs (k_{21} and k_{22}) connecting the internal mass to the host medium are different in stiffness then the representative homogenous elastic solid would possess an anisotropic mass density which may assume negative values for wave frequencies that are close to the local resonance frequencies of the internal mass [11]. In general, two modes of wave motion are excited in the metamaterial; one is dominated by extensional deformation and the other by in-plane shear deformation. For each mode, there are two branches representing the acoustical and optical modes, respectively.

In the present study, we employ the representative elastic solid with a frequency dependent anisotropic mass density to study the wave reflection and transmission into and out of the metamaterial. Attention is focused on how the anisotropic mass density can be used to alter the wave motion in the metamaterial.

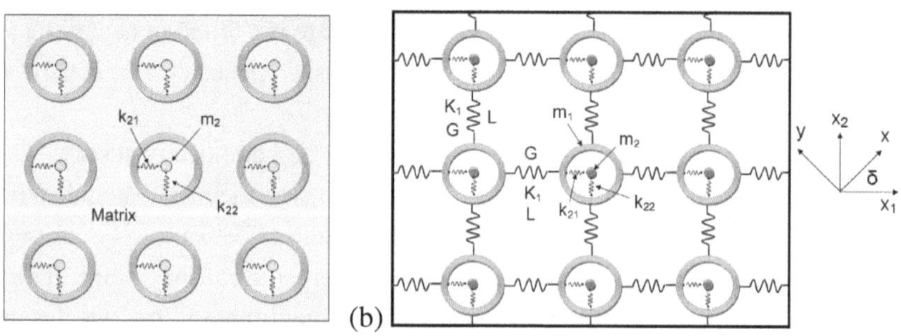

(a) (b)

Fig. 1.1 (a) A 2D metamaterial with microstructures in the form of elastically connected internal masses. (b) Representative mass-in-mass lattice model.

2. Elastic Solid with Anisotropic Effective Mass Density

2.1 Representative Elastic Solid

An effective elastic continuum model was derived by Huang and Sun [11] to represent the lattice model of Fig. 1.1b. Let x_1, x_2 be a set of rectangular Cartesian coordinates of reference in principal directions and x, y be an arbitrary Cartesian coordinate system (see Fig. 1.1). The equations of motion for the continuum with anisotropic mass density ρ_{ij} in the principal coordinate system can be directly obtained from the classical theory of elastodynamics. We have

$$\rho_{ij}\frac{\partial^2 u_j}{\partial t^2}=\frac{\partial \sigma_{ij}}{\partial x_j},\ (\mathrm{i, j} = 1, 2) \tag{2.1.1}$$

where

$$\rho_{11}=\rho_{eff,1},\ \rho_{22}=\rho_{eff,2},\ \rho_{12}=\rho_{21}=0 \tag{2.1.2}$$

in which

$$\rho_{eff,\gamma}=\frac{m_{eff,\gamma}}{V}=\frac{1}{L^2}\left(m_1+\frac{m_2k_{2\gamma}}{k_{2\gamma}-m_2\omega^2}\right),\ (\gamma = 1, 2) \tag{2.1.3}$$

In the equations (2.1.3), $V = L^2$ is the volume of the unit cell including the cavity, and $m_{eff,1}$ and $m_{eff,2}$ are the effective masses for plane harmonic waves propagating in the x_1 and x_2 directions, respectively. These expressions were derived for a one-dimension one-resonator lattice system [9]. Since the four components of the effective mass density given in (2.1.2) form a second order tensor, for an arbitrary coordinate system, the components in the new coordinate system are obtained from the usual coordinate transformation law:

$$\begin{bmatrix}\bar{\rho}_{11} & \bar{\rho}_{12}\\ \bar{\rho}_{21} & \bar{\rho}_{22}\end{bmatrix}=\begin{bmatrix}C^2\rho_{eff,1}+S^2\rho_{eff,2} & CS(\rho_{eff,2}-\rho_{eff,1})\\ CS(\rho_{eff,2}-\rho_{eff,1}) & S^2\rho_{eff,1}+C^2\rho_{eff,2}\end{bmatrix} \tag{2.1.4}$$

where $C\equiv\cos\delta$ and $S\equiv\sin\delta$ and δ is the angle between the x-y axes and the principal axes.

The equivalent 2D elastic solid for the lattice system (Fig. 1.1b) is orthotropic with the constitutive equations:

$$\begin{Bmatrix} \sigma_{11} \\ \sigma_{22} \\ \sigma_{12} \end{Bmatrix} = \begin{bmatrix} Q_{11} & 0 & 0 \\ 0 & Q_{11} & 0 \\ 0 & 0 & Q_{33} \end{bmatrix} \begin{Bmatrix} e_{11} \\ e_{22} \\ 2e_{12} \end{Bmatrix} \tag{2.1.5}$$

in which $Q_{11} = K_1$ and $Q_{33} = G$. In the *x-y* coordinate, the constitutive equations can be derived by using coordinate transformation and written as

$$\begin{Bmatrix} \sigma_{xx} \\ \sigma_{yy} \\ \sigma_{xy} \end{Bmatrix} = \left[\bar{Q}\right] \begin{Bmatrix} e_{xx} \\ e_{yy} \\ 2e_{xy} \end{Bmatrix} \tag{2.1.6}$$

where [$\bar{Q}$] is obtained from coordinate transformation of [Q].

2.2 Dispersion Curves

In the present study, we adopt the following material properties of metamaterials for numerical results.

Masses: $m_1 = 0.263\ Kg/m$, $m_2 = 0.527\ Kg/m$ (2.2.1)
External spring constants: $K_1 = 210\ GN/m$, $G = 82\ GN/m$ (2.2.2)
Internal spring constants: $k_{21} = 20\ GN/m$, $k_{22} = 100\ GN/m$ (2.2.3)
Lattice spacing: $L = 0.01\ m$ (2.2.4)

The components of the anisotropic effective mass density in the principal directions are obtained from equations (2.1.3), and plotted in Fig. 2.1. They are obviously frequency dependent.

To verify the validity of the elastic continuum model, we compare the dispersion curves of harmonic waves propagating in the principal directions (x_1 and x_2), respectively, with those of the original lattice model. The results are shown in Fig. 2.2. For each direction, there are two wave modes, namely, the longitudinal mode (L) and transverse shear mode (S). For each mode there are two branches, namely, the acoustic mode and the optic mode (e.g., AC-L and OP-L). It is evident that they match perfectly for the acoustic mode and fairly well for the optical mode.

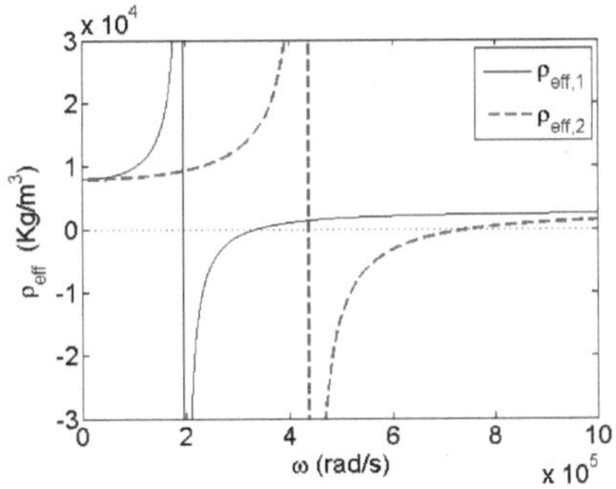

Fig. 2.1 Components of the effective mass density in principal directions with respect to wave frequency.

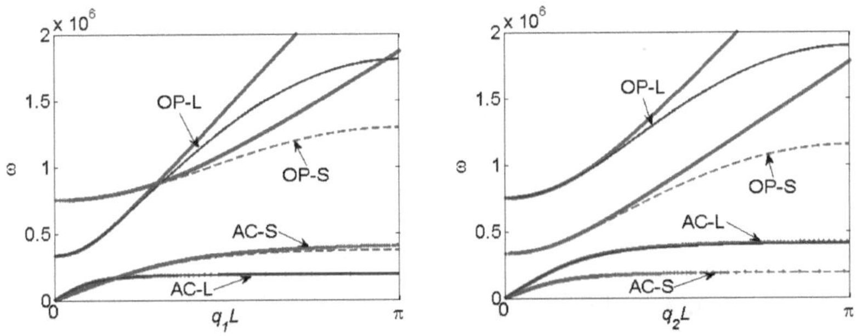

Fig. 2.2 The dispersion curves obtained from the mass-in-mass lattice model shown in Fig.1.1b (blue lines: Longitudinal mode; green dots: Shear mode) and the anisotropic continuum model (red lines) for the x_1 direction (left figure) and the x_2 direction (right figure).

3. Wave Reflection and Transmission

Of particular interest is the wave reflection and transmission between the metamaterial and other "ordinary" media. We consider a pressure wave in a fluid-like material (FL-mat) impinging on the acoustic metamaterial that is oriented in an arbitrary direction. Because of the anisotropy of the mass density tensor, both longitudinal wave and transverse shear waves are induced in the metamaterial. These two modes of wave have different stop bands and may be stopped separately by selecting the local resonance frequencies of the resonator (the local spring-mass in the cavity). It is conceivable that a metamaterial slab may be used to stop the longitudinal wave and only allow the shear wave to pass through it. Since the transverse shear wave cannot be transmitted into a FL-mat, thus, no disturbance would be able to propagate into the FL-mat behind the slab of the metamaterial. Thus, it (the FL-mat) would remain undisturbed.

3.1 Wave propagation from a Fluid-Like Material to the Metamaterial

Consider a FL-mat interfaced with the metamaterial as shown in Fig. 3.1. The equations of motion for plane waves propagating in the x direction in the equivalent elastic medium that represents the 2D metamaterial can be obtained from equation (2.1.1) after performing coordinate transformation. We have

$$\bar{\rho}_{11}\frac{\partial^2 u_x}{\partial t^2}+\bar{\rho}_{12}\frac{\partial^2 u_y}{\partial t^2}=\bar{Q}_{11}\frac{\partial^2 u_x}{\partial x^2}+\bar{Q}_{13}\frac{\partial^2 u_y}{\partial x^2} \tag{3.1.1}$$

$$\bar{\rho}_{12}\frac{\partial^2 u_x}{\partial t^2}+\bar{\rho}_{22}\frac{\partial^2 u_y}{\partial t^2}=\bar{Q}_{13}\frac{\partial^2 u_x}{\partial x^2}+\bar{Q}_{33}\frac{\partial^2 u_y}{\partial x^2} \tag{3.1.2}$$

The FL-mat is assumed to have mass density ρ_f and bulk modulus λ_f with the normal stress given by [14]

$$\sigma_{xx}=\lambda_f\frac{\partial u_x}{\partial x} \tag{3.1.3}$$

and the equation of motion by

$$\frac{\partial \sigma_{xx}}{\partial x}=\rho_f\frac{\partial^2 u_x}{\partial t^2} \tag{3.1.4}$$

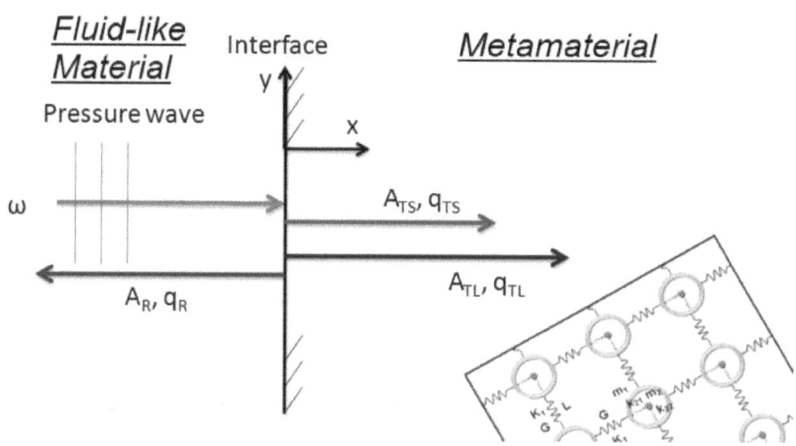

Fig. 3.1 An illustration showing wave propagation from a FL-mat into a metamaterial. "A" represents the amplitude of the reflected and transmitted waves, "q" is the wavenumber. Subscript "R" means the reflected wave, "T" the transmitted wave, "L" for "extension-dominated longitudinal mode" and "S" for "shear-dominated mode".

Consider a plane harmonic incident wave of unity amplitude of the form in the FL-mat:

$$u_{x,I} = \exp[i(q_I x - \omega t)] \tag{3.1.5}$$

in which q and ω denote wavenumber and angular frequency, respectively. For brevity, we shall henceforth omit the time harmonic term, exp[-$i\omega t$]. The reflected wave is then given as

$$u_{x,R} = A_{Rx} \exp[-iq_R x] \tag{3.1.6}$$

Two modes of the transmitted plane wave in the metamaterial are expected. One mode is dominated by longitudinal (extensional) deformation and the other is dominated by shearing deformation. The displacement fields of the transmitted wave are expressed in the form

$$u_{j,T} = A_{TLj} \exp[iq_{TL} x] + A_{TSj} \exp[iq_{TS} x]\,,\ (j = x,\, y) \tag{3.1.7}$$

in which q_{TL} and q_{TS} are wavenumbers and the subscripts are defined in the caption of Fig. 3.1.

The wave number for incident and reflected waves is related to the angular frequency as

$$q_I = q_R = \omega\sqrt{\rho_f / \lambda_f} \tag{3.1.8}$$

Substitution of the transmitted wave solution (3.1.7) in equations (3.1.1) and (3.1.2) yields, respectively,

$$\begin{aligned} &-\omega^2 \bar{\rho}_{11}(A_{TLx} \exp[iq_{TL} x] + A_{TSx} \exp[iq_{TS} x]) \\ &-\omega^2 \bar{\rho}_{12}(A_{TLy} \exp[iq_{TL} x] + A_{TSy} \exp[iq_{TS} x]) \\ &\quad = \bar{Q}_{11}(-q_{TL}^2 A_{TLx} \exp[iq_{TL} x] - q_{TS}^2 A_{TSx} \exp[iq_{TS} x]) \\ &\qquad + \bar{Q}_{13}(-q_{TL}^2 A_{TLy} \exp[iq_{TL} x] - q_{TS}^2 A_{TSy} \exp[iq_{TS} x]) \end{aligned} \tag{3.1.9}$$

and

$$\begin{aligned} &-\omega^2 \bar{\rho}_{12}(A_{TLx} \exp[iq_{TL} x] + A_{TSx} \exp[iq_{TS} x]) \\ &-\omega^2 \bar{\rho}_{22}(A_{TLy} \exp[iq_{TL} x] + A_{TSy} \exp[iq_{TS} x]) \\ &\quad = \bar{Q}_{13}(-q_{TL}^2 A_{TLx} \exp[iq_{TL} x] - q_{TS}^2 A_{TSx} \exp[iq_{TS} x]) \\ &\qquad + \bar{Q}_{33}(-q_{TL}^2 A_{TLy} \exp[iq_{TL} x] - q_{TS}^2 A_{TSy} \exp[iq_{TS} x]) \end{aligned} \tag{3.1.10}$$

The two equations above have to hold for any x. Thus, we obtain

$$\omega^2(\bar{\rho}_{11}A_{TLx}+\bar{\rho}_{12}A_{TLy})=q_{TL}^2(\bar{Q}_{11}A_{TLx}+\bar{Q}_{13}A_{TLy}) \tag{3.1.11}$$
$$\omega^2(\bar{\rho}_{11}A_{TSx}+\bar{\rho}_{12}A_{TSy})=q_{TS}^2(\bar{Q}_{11}A_{TSx}+\bar{Q}_{13}A_{TSy}) \tag{3.1.12}$$

from equation (3.1.9), and

$$\omega^2(\bar{\rho}_{12}A_{TLx}+\bar{\rho}_{22}A_{TLy})=q_{TL}^2(\bar{Q}_{13}A_{TLx}+\bar{Q}_{33}A_{TLy}) \tag{3.1.13}$$
$$\omega^2(\bar{\rho}_{12}A_{TSx}+\bar{\rho}_{22}A_{TSy})=q_{TS}^2(\bar{Q}_{13}A_{TSx}+\bar{Q}_{33}A_{TSy}) \tag{3.1.14}$$

from equation (3.1.10).

At the interface, $x = 0$, continuity conditions on displacement and stress must be satisfied. We have

(1) Displacement continuity: $u_{x,FL-mat}=u_{x,Metamat}$ at $x = 0$. This condition yields

$$1+A_R=A_{TLx}+A_{TSx} \tag{3.1.15}$$

(2) Stress continuity: $\sigma_{xx,FL-mat}=\sigma_{xx,Metamat}$ at $x = 0$.

In the FL-mat, the normal stress is given by

$$\sigma_{xx,FL-mat}=\lambda_f\left(iq_I\exp[iq_Ix]-iq_RA_R\exp[-iq_Rx]\right) \tag{3.1.14}$$

In the metamaterial, we obtain

$$\begin{aligned}\sigma_{xx,Metamat}&=\bar{Q}_{11}e_{xx}+\bar{Q}_{12}e_{yy}+\bar{Q}_{13}(2e_{xy})\\&=\bar{Q}_{11}(iq_{TL}A_{TLx}\exp[iq_{TL}x]+iq_{TS}A_{TSx}\exp[iq_{TS}x])\\&\quad+\bar{Q}_{13}(iq_{TL}A_{TLy}\exp[iq_{TL}x]+iq_{TS}A_{TSy}\exp[iq_{TS}x])\end{aligned} \tag{3.1.17}$$

Equating (3.1.16) and (3.1.17) at $x = 0$ yields

$$\begin{aligned}&\lambda_fq_RA_R+\bar{Q}_{11}q_{TL}A_{TLx}+\bar{Q}_{11}q_{TS}A_{TSx}\\&\qquad+\bar{Q}_{13}q_{TL}A_{TLy}+\bar{Q}_{13}q_{TS}A_{TSy}=\lambda_fq_I\end{aligned} \tag{3.1.18}$$

(3) Shear traction free boundary: $\sigma_{xy,Metamat}=0$ at $x = 0$.

This boundary condition is necessary since the FL-mat does not resist shear stresses. The shear stress in the metamaterial is

$$\begin{aligned}\sigma_{xy,Metamat} &= \bar{Q}_{13}e_{xx} + \bar{Q}_{23}e_{yy} + \bar{Q}_{33}(2e_{xy}) \\ &= \bar{Q}_{13}(iq_{TL}A_{TLx}\exp[iq_{TL}x] + iq_{TS}A_{TSx}\exp[iq_{TS}x]) \\ &\quad + \bar{Q}_{33}(iq_{TL}A_{TLy}\exp[iq_{TL}x] + iq_{TS}A_{TSy}\exp[iq_{TS}x])\end{aligned} \tag{3.1.19}$$

The shear traction free condition leads to

$$\bar{Q}_{13}q_{TL}A_{TLx} + \bar{Q}_{13}q_{TS}A_{TSx} + \bar{Q}_{33}q_{TL}A_{TLy} + \bar{Q}_{33}q_{TS}A_{TSy} = 0 \tag{3.1.20}$$

First, we solve for the unknown wave numbers q_{TL} and q_{TS} using equations (3.1.11) through (3.1.14). From equations (3.1.11) and (3.1.13) we have

$$q_{TL}^2 = \frac{\omega^2(\bar{\rho}_{11}A_{TLx} + \bar{\rho}_{12}A_{TLy})}{(\bar{Q}_{11}A_{TLx} + \bar{Q}_{13}A_{TLy})} = \frac{\omega^2(\bar{\rho}_{12}A_{TLx} + \bar{\rho}_{22}A_{TLy})}{(\bar{Q}_{13}A_{TLx} + \bar{Q}_{33}A_{TLy})} \tag{3.1.21}$$

By denoting $\alpha = A_{TLx}/A_{TLy}$, equation (3.1.21) can be written as

$$q_{TL} = \left[\frac{\omega^2(\bar{\rho}_{11}\alpha + \bar{\rho}_{12})}{\bar{Q}_{11}\alpha + \bar{Q}_{13}}\right]^{\frac{1}{2}} = \left[\frac{\omega^2(\bar{\rho}_{12}\alpha + \bar{\rho}_{22})}{\bar{Q}_{13}\alpha + \bar{Q}_{33}}\right]^{\frac{1}{2}} \tag{3.1.22}$$

Equation (3.1.22) leads to the following equation

$$B_1^*\alpha^2 - B_2^*\alpha + B_3^* = 0 \tag{3.1.23}$$

where

$$B_1^* = \bar{\rho}_{11}\bar{Q}_{13} - \bar{\rho}_{12}\bar{Q}_{11} \tag{3.1.24}$$
$$B_2^* = (\bar{\rho}_{12}\bar{Q}_{13} + \bar{\rho}_{22}\bar{Q}_{11}) - (\bar{\rho}_{11}\bar{Q}_{33} + \bar{\rho}_{12}\bar{Q}_{13}) \tag{3.1.25}$$
$$B_3^* = \bar{\rho}_{12}\bar{Q}_{33} - \bar{\rho}_{22}\bar{Q}_{13} \tag{3.1.26}$$

Thus, the wave number q_{TL} for the transmitted extension-dominated mode is given by equation (3.1.22) with α determined by equation (3.1.23). Similarly, from equations (3.1.12) and (3.1.14), we obtain the wave number for the shear-dominated mode as

$$q_{TS} = \left[\frac{\omega^2(\bar{\rho}_{11}\beta + \bar{\rho}_{12})}{\bar{Q}_{11}\beta + \bar{Q}_{13}}\right]^{\frac{1}{2}} \tag{3.1.27}$$

in which $\beta = A_{TSx}/A_{TSy}$ and, like α, it satisfies equation (3.1.23).

It is now possible to solve the amplitudes of the reflection and transmitted waves based on equations (3.1.15), (3.1.18), and (3.1.20) which are the continuity conditions at the interface. For numerical studies, we adopt the following material properties for the FL-mat:

$$\rho_f = 7900\ Kg/m^3,\ \lambda_f = 210\ GN/m^2 \tag{3.1.28}$$

Two cases with different wave frequencies are investigated:

Case (1):

$$\omega = 2.1\times10^5\ rad/s,\ \rho_{eff,1} = -3.0\times10^4\ Kg/m^3,\ \rho_{eff,2} = 9.5\times10^3\ Kg/m^3$$

Case (2):

$$\omega = 3.6\times10^5\ rad/s,\ \rho_{eff,1} = 4.5\times10^2\ Kg/m^3,\ \rho_{eff,2} = 1.9\times10^4\ Kg/m^3$$

For Case (1), the wave frequency is selected so that the effective mass in the x_1 direction is negative and that in the x_2 direction is positive. The amplitudes of the reflected and transmitted waves when the incident wave strikes the metamaterial from the FL-mat are obtained and shown in Fig. 3.2a. It is known that a displacement field in general has the form $u = A\exp[i(qx-\omega t)]$. In this equation, the wavenumber q is generally complex and responsible for wave attenuation. In addition, the amplitude A can also be complex so that

$$A = |A|\exp[i\psi] \tag{3.1.29}$$

where ψ represents the phase angle. It is therefore conceivable that the steady-state reflected and transmitted waves could be in a different phase relative to that of the incident wave. In this study, it is the absolute value of the wave amplitude that is of interest to us.

The symbols in the legend are defined based on the deformation mode at $\delta = 0$. For example, $|A_{TLx}|$ denotes the amplitude of the displacement component in the x direction of the transmitted extension-dominated wave. The ratio $|A_{TLx}|/|A_{TLy}|$ is always larger than unity at $\delta = 0$ because this is how the symbol is defined. As the angle δ increases, the content of shear motion also increases, and ratio $|A_{TLx}|/|A_{TLy}|$ may become less than unity, and the wave becomes shear-dominated. It is expected that the pressure wave is totally reflected in this case when $\delta = 0$ because of the effect of the band-gap associated with the negative effective mass density in the x direction. More details were discussed in Refs. 9 and 10 for one-dimensional acoustic metamaterials.

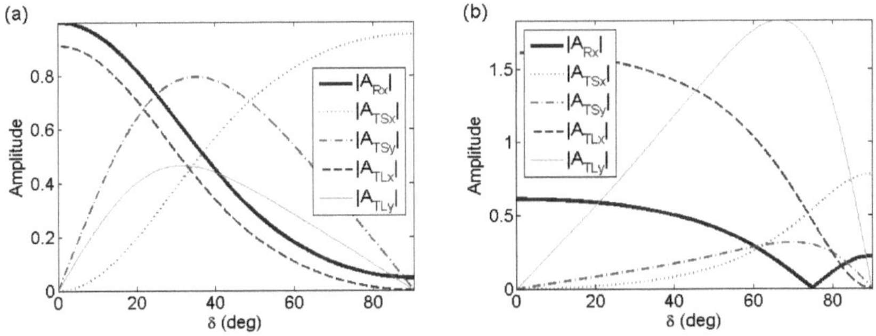

Fig. 3.2 Amplitudes of the reflected and transmitted waves when an incident wave of unity amplitude strikes the metamaterial from a FL-mat. (a) case 1; (b) case 2.

An interesting phenomenon is observed in Case (2) presented in Fig. 3.2b. Note that when δ is in between 60° and 70°, the pressure wave from the FL-mat is mostly transmitted and converted to a strong shear-dominated mode (i.e., $|A_{TLy}|/|A_{TLx}|$ is larger than unity). Potential applications of this behavior will be discussed in the next subsection.

3.2 A Metamaterial Sandwiched between two FL Media

When the acoustic metamaterial is subjected to a pressure wave, two modes of wave motion are generated. They propagate at different wave numbers and speeds. If the metamaterial is sandwiched between two FL-mat media, then these two waves would both experience multiple reflections and transmissions as depicted in Fig. 3.3. Numerical solutions for this multiple reflection/transmission problem can be readily obtained. Details of the idea are discussed as follows.

For waves propagating from the metamaterial to the FL-mat, the general form of the incident wave in the metamaterial is a combination of two propagating modes for which the displacement fields can be expressed in the form

$$u_{j,I} = A_{ILj} \exp[iq_{IL}x] + A_{ISj} \exp[iq_{IS}x] \qquad (j = x, y) \tag{3.2.1}$$

The reflected wave in the metamaterial and the transmitted wave in the FL-mat can be obtained in a similar manner as described in Section 3.1. We express the displacement fields of the reflected and transmitted waves as

$$u_{j,R} = A_{RLj} \exp[-iq_{RL}x] + A_{RSj} \exp[-iq_{RS}x] \qquad (j = x, y) \tag{3.2.2}$$

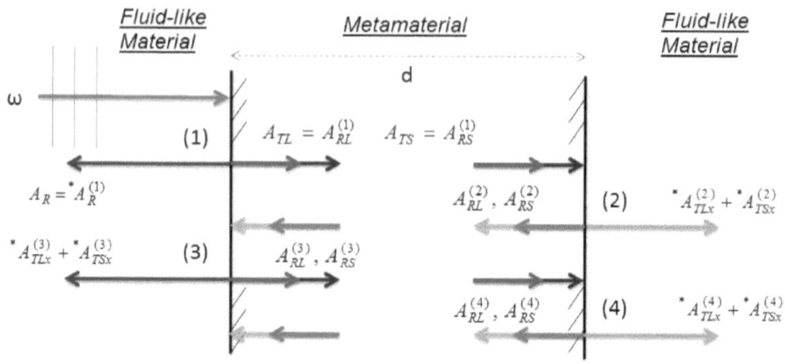

Fig. 3.3 A schematic showing wave propagation through a metamaterial. "A" denotes the amplitude of the reflected or transmission wave.

and

$$u_{x,T} = \left({}^*A_{TLx} + {}^*A_{TSx}\right)\exp[iq_{TL}x] \quad , \tag{3.2.3}$$

respectively. Here the star sign (*) denotes the quantity associated with the FL-mat. As depicted in Fig. 3.3, wave reflections and transmissions continue to take place after the wave enters the metamaterial. It is noted that a reflected wave at the boundary subsequently becomes the incident wave to the other interface which produces the next reflected wave. Based on continuity conditions, the relation between the (n+1)th and the (n)th reflected waves in the metamaterial is obtained for the L-mode as

$$\begin{bmatrix} 1 & 0 & -1 \\ \overline{Q}_{11}q_{RL} & \overline{Q}_{13}q_{RL} & \lambda_f q_{RL} \\ \overline{Q}3q_{RL} & \overline{Q}_{33}q_{RL} & 0 \end{bmatrix} \begin{Bmatrix} A_{RLx}^{(n+1)} \\ A_{RLy}^{(n+1)} \\ {}^*A_{TLx}^{(n+1)} \end{Bmatrix} = \begin{bmatrix} -1 & 0 & 0 \\ \overline{Q}_{11}q_{IL} & \overline{Q}_{13}q_{IL} & 0 \\ \overline{Q}_{13}q_{IL} & \overline{Q}_{13}q_{IL} & 0 \end{bmatrix} \begin{Bmatrix} A_{RLx}^{(n)} \\ A_{RLy}^{(n)} \\ {}^*A_{TLx}^{(n)} \end{Bmatrix} \tag{3.2.4}$$

The wavenumbers of the reflected and incident waves in the metamaterial are identical, i.e., $q_{RL} = q_{IL} = q_L$. Equation (3.2.4) can be used to obtain the amplitudes of the waves when the waves propagate from the metamaterial to the FL-mat. We obtain

$$\begin{Bmatrix} A_{RLx}^{(n+1)} \\ A_{RLy}^{(n+1)} \\ {}^*A_{TLx}^{(n+1)} \end{Bmatrix} = \left[T_{ij,L}\right] \begin{Bmatrix} A_{RLx}^{(n)} \\ A_{RLy}^{(n)} \\ {}^*A_{TLx}^{(n)} \end{Bmatrix} = \ldots = \left[T_{ij,L}\right]^n \begin{Bmatrix} A_{RLx}^{(1)} \\ A_{RLy}^{(1)} \\ {}^*A_{TLx}^{(1)} \end{Bmatrix} \tag{3.2.5}$$

where $A_{RLx}^{(1)}$ and $A_{RLy}^{(1)}$ are the amplitudes of the initial transmitted wave as indicated in Fig. 3.3, and

$$\left[T_{ij,L}\right]=\begin{bmatrix}1 & 0 & -1\\ \bar{Q}_{11}q_L & \bar{Q}_{13}q_L & \lambda_f q_{TL}\\ \bar{Q}_{13}q_L & \bar{Q}_{33}q_L & 0\end{bmatrix}^{-1}\begin{bmatrix}-1 & 0 & 0\\ \bar{Q}_{11}q_L & \bar{Q}_{13}q_L & 0\\ \bar{Q}_{13}q_L & \bar{Q}_{33}q_L & 0\end{bmatrix} \tag{3.2.6}$$

Similarly, we obtain the amplitudes for the S-mode wave by replacing subscript L with subscript S in equations (3.2.5) and (3.2.6).

The recurrence relations given by equations (3.2.5) and (3.2.6) provides the amount of reflection and transmission of the wave when it hits the FL-mat/metamaterial each time and can be used to estimate the total amounts of waves that are reflected/transmitted back to the first and second FL-mats, respectively. The total amplitude of the waves that are reflected or transmitted back to the first FL-mat for an incident wave of unity amplitude is given by

$$\begin{aligned}&{}^*A_R^{(1)}+{}^*A_{TLx}^{(3)}\exp[2iq_Ld]+{}^*A_{TSx}^{(3)}\exp[2iq_Sd]\\&\quad+{}^*A_{TLx}^{(5)}\exp[4iq_Ld]+{}^*A_{TSx}^{(5)}\exp[4iq_Sd]+\ldots\\&\quad+{}^*A_{TLx}^{(2n+1)}\exp[2n\cdot iq_Ld]+{}^*A_{TSx}^{(2n+1)}\exp[2n\cdot iq_Sd]+\ldots\end{aligned} \tag{3.2.7}$$

The total amplitude of the waves that pass through the metamaterial is

$$\begin{aligned}&{}^*A_{TLx}^{(2)}\exp[iq_Ld]+{}^*A_{TSx}^{(2)}\exp[iq_Sd]\\&\quad+{}^*A_{TLx}^{(4)}\exp[3iq_Ld]+{}^*A_{TSx}^{(4)}\exp[3iq_Sd]+\ldots\\&\quad+{}^*A_{TLx}^{(2n)}\exp[(2n-1)\cdot iq_Ld]+{}^*A_{TSx}^{(2n)}\exp[(2n-1)\cdot iq_Sd]+\ldots\end{aligned} \tag{3.2.8}$$

where d is the thickness of the metamaterial.

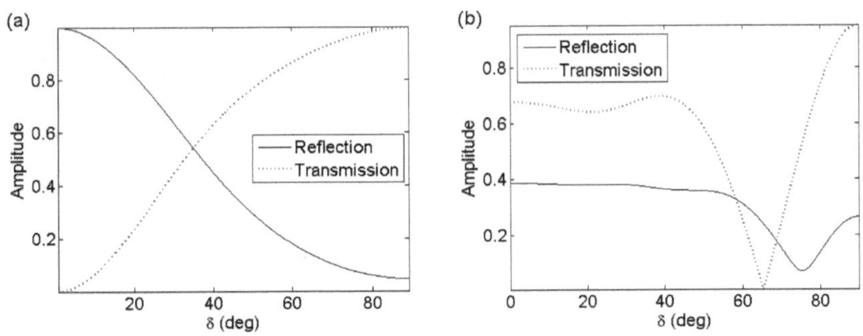

Fig. 3.4 Amplitudes of the total reflected and transmitted waves when a metamaterial is sandwiched between two FL-mats. (a) case 1; (b) case 2. In both cases n = 4.

Fig. 3.4 shows the numerical results of the total reflection and transmission obtained from equations (3.2.7) and (3.2.8) with n = 4. It was found that the solutions given by these two equations sufficiently converge for $n \geq 4$. It is seen that the steady-state pressure wave impinging the metamaterial can be almost totally "reflected" as demonstrated by Case (1) (see Fig.3.4a) if the metamaterial is oriented with $\delta = 0^{\circ}$. From the results of Case (2) presented in Fig. 3.4b, a pressure wave from a FL-mat striking the metamaterial can be nearly totally suppressed by orienting the metamaterial at an angle around $\delta = 65^{\circ}$. This phenomenon is not the result of a total reflection at the first fluid/solid interface. As indicated by the result of Fig. 3.2b, the incident longitudinal wave from the FL-mat is transmitted into the metamaterial in two wave modes; one of them is a shear-dominated wave whose amplitude of the y component is A_{TLy}, and the amplitude achieves the greatest amplitude at $\delta = 65^{\circ}$. The shear-dominated wave in the metamaterial cannot be easily transmitted back to the FL-mat and may be trapped. It is conceivable that this wave mode conversion capability provided by the metamaterial may be used to partially isolate the FL-mat behind it from acoustic disturbances.

4. Conclusion

Harmonic wave propagation in a 2D acoustic metamaterial is studied using an equivalent elastic solid with a frequency-dependent anisotropic mass density. In general, two modes of wave motion are excited in the metamaterial; one is dominated by extensional deformation and the other by in-plane shear deformation. For each mode, there are two branches representing the acoustic mode and optical mode, respectively. Since there are two direction-dependent stopping bands which may be selected to cover different frequency ranges, impinging waves can be selectively stopped from propagating into the metamaterial. For example, a pressure wave impinging on an acoustic metamaterial may be stopped directly by designing the gap frequency. Moreover, the impinging pressure wave can be converted into a strong shear-dominated wave mode accompanied by a weak extension-dominated wave. The shear-dominated wave motion can hardly be transmitted into a fluid-like medium and, thus, the FL-mat behind the metamaterial can remain mostly undisturbed.

Acknowledgments: This work was supported by AFOSR through a grant to Purdue University. Dr. Les Lee was the program manager.

References

[1] Veselago, V. G.: The electrodynamics of substances with simultaneously negative values of ε and μ. *Sov. Phys. Usp.* **10** 509-514 (1968).

[2] Pendry, J. B.: Negative refraction makes a perfect lens. *Phys. Rev. Lett.* **85** 3966-3969 (2000).

[3] Smith, D. R., Pendry, J. B., Wiltshire, M. C. K.: Metamaterials and negative refractive index, *Science* **305** 788-92 (2004).

[4] Liu, Z., Zhang, X., Mao, Y., Zhu, Y.Y., Yang, Z., Chan, C.T, Sheng, P.: Locally resonant sonic materials. *Science* **289** 1734-1736 (2000).

[5] Li, J., Chan, C.T.: Double-negative acoustic metamaterial. *Phys. Rev. E* **70** 055602 (2004).

[6] Fang, N., Xi, D., Xu, J., Ambati M., Srituravanich, W., Sun, C., Zhang, X.: Ultrasonic metamaterials with negative modulus. *Nat. Mater.* **5** 452-456 (2006).

[7] Torrent, D., Sánchez-Dehesa, J.: Acoustic metamaterials for new two-dimensional sonic devices. *New J. Phys.* **9** 323 (2007).

[8] Milton, G. W., Willis, J. R.: On modifications of Newton's second law and linear continuum elastodynamics. *Proc. R. Soc. A* **463** 855-880 (2007).

[9] Huang, H. H., Sun, C. T., Huang, G. L.: On the negative effective mass density in acoustic metamaterials. *Int. J. Eng. Sci.* **47** 610-617 (2009).

[10] Huang, H. H., Sun, C. T.: Wave attenuation mechanism in an acoustic metamaterial with negative effective mass density. *New J. Phys.* **11** 013003 (2009).

[11] Huang, H. H., Sun, C. T.: Locally resonant acoustic metamaterials with 2D anisotropic effective mass density. (In preparation)

[12] Milton, G. W., et al.: On cloaking for elasticity and physical equations with a transformation invariant form. *New J. Phys.* **8** 248 (2006).

[13] Torrent, D., Sánchez-Dehesa, J.: Anisotropic mass density by two-dimensional acoustic metamaterials. *New J. Phys.* **10** 023004 (2008).

[14] Brekhovskikh, L. M.: *Waves in layered media.* Academic Press, NY (1960).

Complex Band Structure of Phononic Crystals and the Diffraction Problem

Vincent Laude[1], Younes Achaoui[1], Sarah Benchabane[1] and Abdelkrim Khelif[2]

[1]Institut FEMTO-ST, Université de Franche-Comté, CNRS, ENSMM, UTBM; 32 avenue de l'Observatoire, F-25044 Besançon, France

[2]GeorgiaTech-CNRS, Atlanta, GA, USA

vincent.laude@femto-st.fr

Abstract. The diffraction of elastic waves on a two-dimensional finite phononic crystal is investigated by a plane wave technique. It is first remarked that a full solution to the phononic crystal problem requires that all modes of the periodic structure (Bloch waves) are indentified and incorporated in the solution, including evanescent Bloch waves. An extended plane-wave expansion (PWE) method is used to obtain the complex band structure of the phononic crystal, but also the band structure of diffracted waves in the incident and exit media. Complex isofrequency curves are presented and show sharp variations of the Bloch wave vector with the angle of propagation. Finally, the complex band structures are used to formulate a reflection/transmission problem similar to the one leading to Fresnel formulas for homogeneous media. Some examples of diffracted field computations are given.

1. Introduction

Phononic crystals [1,2] are two- or three-dimensional periodic structures that are made of two materials with different elastic constants. They possibly give rise to absolute stop bands under specific geometrical conditions, depending on the choice of materials. In addition, their unique dispersion properties can be used to design efficient waveguides or to obtain unusual refraction properties. Band structures are usually employed to describe infinite phononic crystals, as they provide information regarding any wave propagating in the periodic medium. Such solutions of the periodic and monochromatic wave equation are referred to as Bloch waves. Band structures for Bloch waves record the

T.-T. Wu and C.-C. Ma (eds.), *IUTAM Symposium on Recent Advances of Acoustic Waves in Solids*, IUTAM Bookseries 26, DOI 10.1007/978-90-481-9893-1_15,

occurrence of band gaps and measure dispersion. However, it is well-known that evanescent waves must be considered in propagation problems whenever scattering, diffusion, or diffraction by a finite object is investigated [3,4]. In experiments, of course, actual samples are always of a finite dimension. Even in the context of infinite phononic crystals, evanescent waves appear very naturally within frequency band gaps: since no waves can propagate within a band gap, only evanescent waves are left to explain the exponentially-decreasing transmission of acoustic waves. The purpose of this paper is to discuss the role of evanescent Bloch waves in the diffraction by phononic crystals. Our analysis is based on the recognition that as finite periodic structures, phononic crystal samples can be viewed as two- or three-dimensional diffraction gratings.

2. Diffraction in Phononic Crystals

The problem of diffraction by a square-lattice silicon-void phononic crystal is considered for definiteness and depicted in Figure 1(a). A plane and monochromatic wave is incident on the phononic crystal. As it impinges on the phononic crystal, it is diffracted in the incident medium (here homogeneous silicon) in either propagative or evanescent plane waves. Inside the phononic crystal, it is addi tionally converted to Bloch waves, again either propagative or evanescent. This conversion is dictated by the boundary conditions separating the incident medium from the phononic crystal itself, where the fields have to be matched. This boundary is obviously not uniquely defined. By convention, we decide to tile space using the unit cell of the phononic crystal, as depicted in Figure 1(b). The unit cell is here a square with lateral dimension a. Outside the phononic crystal the unit cell is the plain square, while inside the phononic crystal it has an additional hole in the center, with diameter d=0.85a. At the exit of the phononic crystal, conversion again occurs to propagative and evanescent plane waves. For simplicity, the exit medium is taken to be identical to the incident medium. As a consequence, the plane wave bases in both media are the same.

The mathematical treatment of the diffraction problem then proceeds as follows. The displacement and the stress fields inside the phononic crystal are expanded over the complete basis of the evanescent Bloch waves which will be described in the next section. Since the surrounding medium is homogeneous silicon, grating plane wave solutions are used to represent the incident plus the reflected wave fields, and the transmitted wave field. Specifically, these grating solutions are given by the reciprocal lattice wave vectors combined with the dispersion relation for propagation in homogeneous silicon. Boundary conditions at the interface between silicon and the phononic crystal then allow us to obtain the reflection and transmission coefficients for all diffracted waves, by solving a Fresnel-type matrix problem.

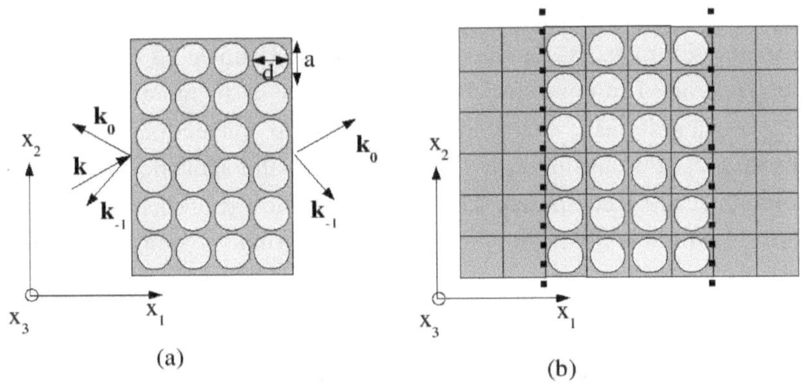

Fig. 1 (a) Diffraction on a finite size square-lattice phononic crystal made of holes in silicon. The pitch of the structure is a and the diameter of the holes is d=0.85a. A plane wave is incident on the phononic crystal and is diffracted in either propagative or evanescent plane waves. (b) Tiling of space used to solve the diffraction problem. The dotted line figures the matching boundaries considered in the computation of reflection and transmission coefficients.

3. Complex Band Structures

Though it has long been known that the Bloch-Floquet wave vector can assume both real values, for propagating Bloch waves, and complex values, for evanescent Bloch waves, the problem of obtaining all complex solutions for phononic crystals has not been treated in depth so far, to the best of our knowledge. It is certain that the layer multiple scattering (LMS) method naturally allows to retrieve such complex wave vector solutions [5,6], since this on-shell method allows one to assume a given monochromatic frequency and solve for the wave vector. However, this method lacks the generality (in terms of arbitrary materials and scatterer shapes) of the plane wave expansion (PWE) method [7,8], for instance. Conversely, whereas the computation of band structures by the LMS method is naturally conducted at a fixed frequency, in the PWE method one usually obtains eigenfrequencies for a fixed (real) Bloch-Floquet wave vector within the first Brillouin zone. The same remark applies as well to mesh-based methods such as the finite element method (FEM) [9], for instance. We have specifically chosen to extend the classical plane wave expansion (PWE) method so that it includes complex wave vectors in the direction of propagation [10]. The new complex PWE method has been used to generate band structures for two-dimensional silicon-void phononic crystals. Both propagative and evanescent solutions are found at once. The decay constants within band gaps are thus found and shown to depend on the wave polarization. The complex graphs also allow us to identify clearly the different branch systems in the band structure and to connect bands below and above band gaps. Further-more, the distribution of the acoustic fields of evanescent modes can be computed. Their transformation from below to above a band gap and within is observed to be perfectly continuous along any complex branch of the band structure.

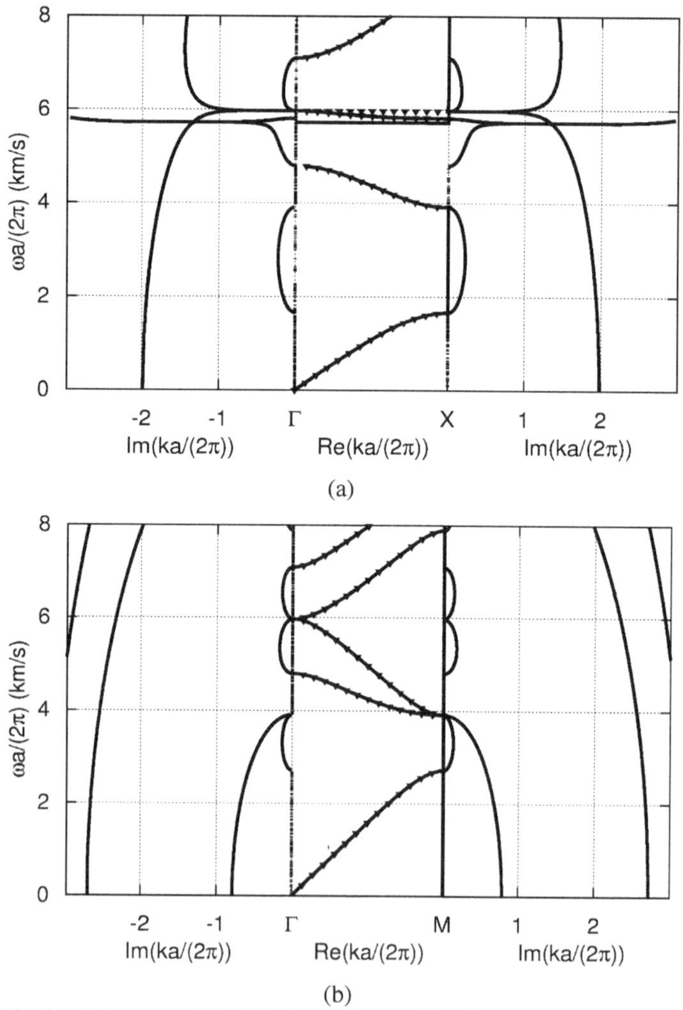

Fig. 2 Complex band structure of the 2D phononic crystal in (a) the ΓX and (b) the ΓM direction of the first Brillouin zone.

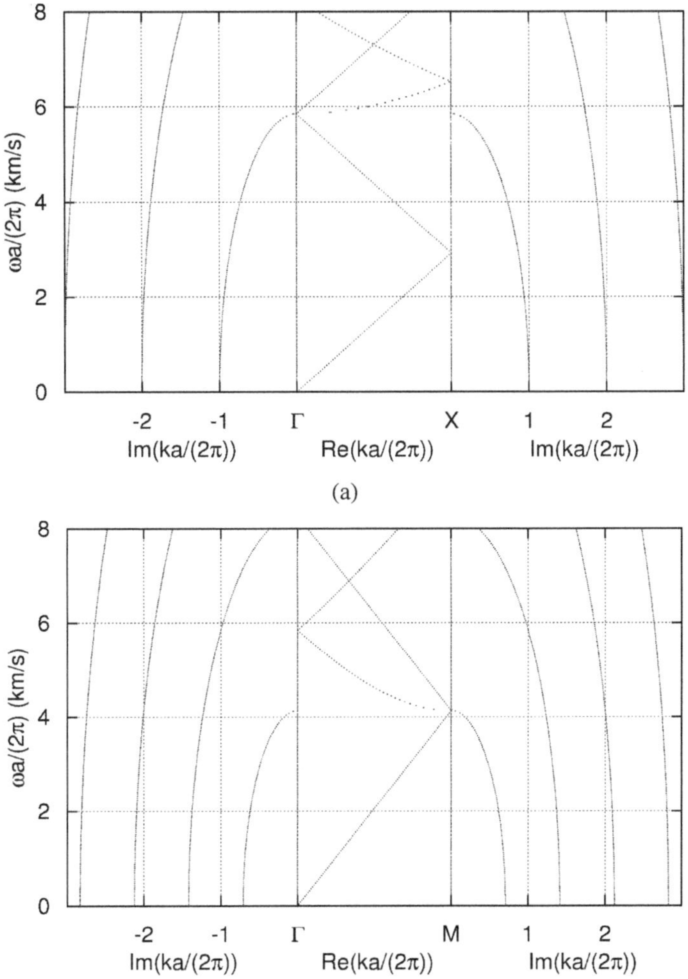

(b)

Fig. 3 Complex band structure of homogeneous silicon crystal in (a) the ΓX and (b) the ΓM direction of the first Brillouin zone.

Figure 2 displays the band structure of the phononic crystal of Figure 1 for waves polarized out-of-plane, i.e. for which the only non vanishing displacement is u_3. In this case, there is only one independent stress component which we choose to be T_3=T_{3n} where n is a unit vector in the direction of observation. The real part of the Bloch wave vector is shown in the center of the diagrams and is restricted to the first Brillouin zone. The imaginary part is shown on the left-hand side (for negative values) and on the right-hand side (for positive values). The points marked by triangles are the purely-real $\omega(k)$ solutions found with the usual

PWE method. It can be observed that many complex bands are not revealed by the traditional analysis; such bands are the seat of evanescent Bloch waves.

Figure 3 displays the complex band structure of homogeneous silicon, again restricted to the out-of-plane polarization only. A two-dimensional periodicity identical to the one of the phononic crystal has been considered. Of course, this periodicity is only artificial, but through this trick, the propagative and evanescent diffracted solutions attached to a surface enclosing this medium are revealed in a form that is directly compatible with the Bloch waves of the phononic crystal in Figure 2. The straight line in Figure 3 – folded at the Brillouin zone boundaries – is the usual pure shear wave of silicon, while the curved lines arise from adding reciprocal lattice vectors to the wave vector **k**.

4. Complex Isofrequency Plots

We can further use the extended PWE method to obtain information of wave propagation at a given frequency, for any angle of propagation. In this view, we have solved the $k(\omega)$ problem while simultaneously varying the angle of propaga tion, to obtain the complex isofrequency plots displayed in Figures 4 (for fa=4500 m/s) and 5 (for fa=6000 m/s). These plots can be related to Figure 2; for instance, for fa=4500 m/s there one purely real closed curve and one purely imaginary solution is visible (others appear outside the range of the right-hand-side plot).

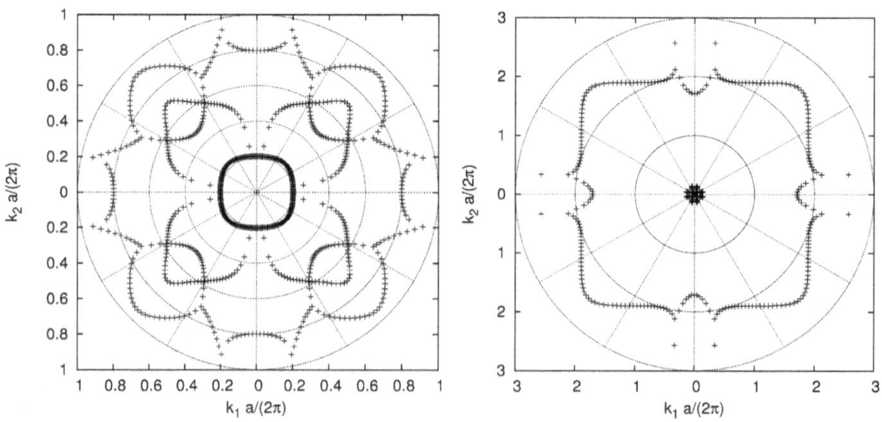

(a) (b)

Fig. 4 Complex isofrequency plots for the phononic crystal at the frequency given by fa=4500 m/s. The left panel (a) shows the real part of the complex Bloch wave vector, while the right panel (b) shows the imaginary part.

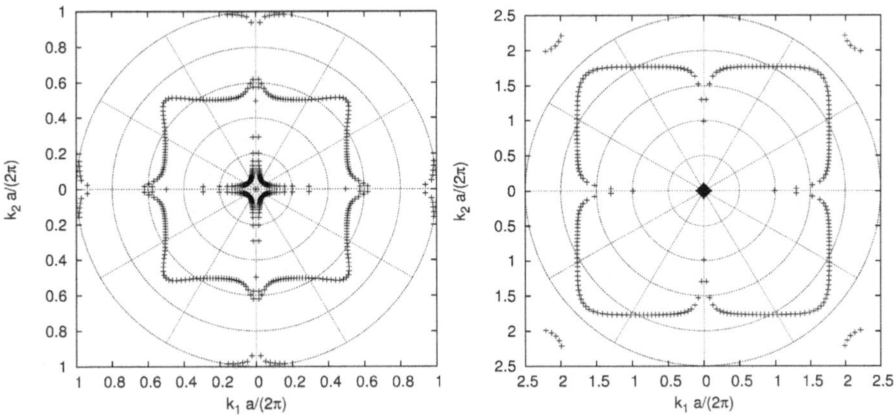

Fig. 5 Complex isofrequency plots for the phononic crystal at the frequency given by *fa*=6000 m/s. The left panel (a) shows the real part of the complex Bloch wave vector, while the right panel (b) shows the imaginary part.

It can be observed that the various complex bands in the isofrequency plots are strongly dependent on the angle of propagation but also on the selected frequency. The square symmetry of the periodic structuration clearly impacts the anisotropy of these bands.

5. A Diffraction Example

Figure 6 displays some examples of computations of diffraction on the phononic crystal following the method described in Section 2. We have specifically chosen the operating frequency given by *fa*=4500 m/s as in Figure 4 and a phononic crystal oriented for ΓX propagation. The angle of incidence is changed from normal incidence in order to reveal the +1 and -1 orders of diffraction. The generation of these orders of diffraction can especially be appreciated in Figure 6(e), where they are seen to cause a complex standing wave pattern on the left-hand side of the phononic crystal. Of course, this configuration is not intended to have any practical value and is only selected here for illustration purposes. Of particular interest is the possibility to combine band gap effects, in order to obtain high reflectivity for instance, with diffraction effects. It can be thought that two-dimensional or three-dimensional gratings possess more degrees of freedom than the traditional one-dimensional gratings, so that design possibilities are richer. Of course, this added capability as to be traded-off against increased complexity of modeling and numerical simulation.

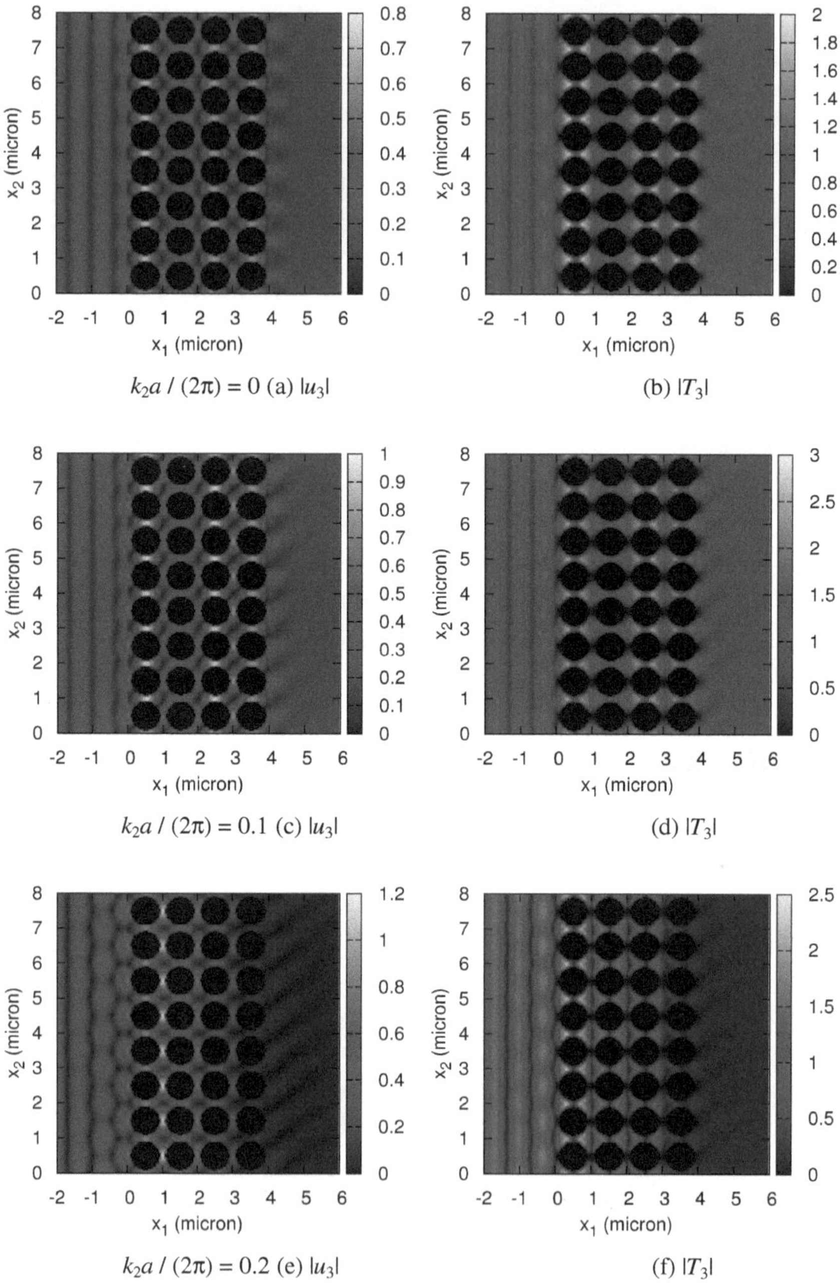

Fig. 6 Examples of diffraction on the square-lattice phononic crystal. The frequency is chosen to be fa=4500 m/s as in Figure 4 and the wave vector component along the phononic crystal boundary is varied. The field maps are chosen in arbitrary units, with the out-of-plane displacement on the left-hand side and the vertical stress on the right-hand side.

6. Conclusion

The diffraction of elastic waves on a two-dimensional finite phononic crystal made of hollow holes in silicon has been investigated by a plane wave technique. It has first been remarked that a full solution to the phononic crystal problem requires that all modes of the periodic structure (Bloch waves) are indentified, including both propagating and evanescent Bloch waves. An extended plane-wave expansion (PWE) method has been used to obtain the complex band structure of the phononic crystal, but also the band structure of diffracted waves in the incidence and exit media. Complex isofrequency curves have been presented and we have found sharp variations of the Bloch wave vector with the angle of propagation. Finally, the complex band structures have been used to formulate a reflection/transmission problem similar to the one leading to Fresnel formulas for homogeneous media. Some examples of diffracted field computations have been given at a frequency lying above the first band gap of the phononic crystal. It has especially been observed that several diffraction orders can be excited as the angle of incidence on the phononic crystal is varied. These results suggest that phononic crystals can be regarded as two- and three-dimensional gratings where diffraction can be combined with band gap effects. We finally note that appropriate designs could lead to very efficient phononic diffraction gratings.

References

[1] Kushwaha, M. S., Halevi, P., Dobrzynski, L., and Djafari-Rouhani, B.: Acoustic band structure of periodic elastic composites. *Phys. Rev. Lett.* **71** 2022-2025 (1993).

[2] Sigalas, M. M. and Economou, E. N.: Elastic and acoustic wave band structure. *J. Sound Vib.* **158** 377-382 (1992).

[3] Hsue, Y.-C., Freeman, A. J., Gu, B. Y.: Extended plane-wave expansion method in three-dimensional anisotropic photonic crystals. *Phys. Rev. B* **72** 195118 (2005).

[4] de Fornel, F.: Evanescent Waves. *From Newtonian Optics to Atomic Optics.* Springer (Berlin, 2001).

[5] Sainidou, R., Stefanou, N., Modinos, A.: Formation of absolute frequency gaps in three-dimensional solid phononic crystals. *Phys. Rev. B* **66** 212301 (2002).

[6] Sainidou, R., Stefanou, N., Psarobas, I. E. and Modinos, A.: A layer-multiple-scattering method for phononic crystals and heterostructures of such. *Computer Physics Communications* **166** 197-240 (2005).

[7] Wilm, M., Ballandras, S., Laude, V., Pastureaud, T.: A full 3D plane-wave-expansion model for 1-3 piezoelectric composite structures. *J. Acoust. Soc. Am.* **112** 943-952 (2002).

[8] Wu, T.-T., Huang, Z.-G., Lin, S.: Surface and bulk acoustic waves in two-dimensional phononic crystal consisting of materials with general anisotropy. *Phys. Rev. B* **69** 094301 (2004).

[9] Khelif, A., Aoubiza, B., Mohammadi, S., Adibi, A., Laude, V.: Complete band gaps in two-dimensional phononic crystal slabs. *Phys. Rev. E* **74** 046610 (2006).

[10] Laude, V., Achaoui, Y., Benchabane, S., Khelif, A.: Evanescent Bloch waves and the complex band structure of phononic crystals. Submitted (2009).

Versatile Phononic Slabs

I. E. Psarobas

Section of Solid State Physics, The University of Athens, Panepistimioupolis, 157 84 Athens, Greece

ipsarob@phys.uoa.gr

Abstract. Phononic slabs of an *fcc* phononic crystal consisting of close-packed (glued) rubber spheres in air, under the influence of mild dissipation in rubber, exhibit large absolute transmission gaps. Proper size variation of the spheres in a sequence of crystal slabs can shift and enlarge the frequency gap readily to comply with a variety of filtering needs in a phononic application. The aspects of such a versatile phononic slab are presented in a realistic theoretical approach, by means of the layer multiple-scattering method developed for phononic crystals.

1. Introduction

The study of the propagation of waves in inhomogeneous media is a problem of wide interest because of the implications in technology and the broad view that can provide in understanding a large area of physical problems [1]. In particular, classical wave transport in periodic media can provide the means to control light (electromagnetic waves), sound (vibrations) or both with the development of novel materials, also known as classical spectral gap materials. Such a periodic arrangement of scatterers can obviously open up several directional spectral gaps. When for all directions the spectral gaps overlap so that there is a forbidden range of frequencies in which the waves cannot propagate in any direction, we have a special type of material that exhibits an absolute frequency gap response. In this paper, we deal with composite materials whose elastic properties vary periodically in space, and are also known as phononic crystals [2].

Absolute frequency gaps in 3D phononic crystals can be more easily obtained in systems where solid high-density scatterers are placed in a low-density fluid host rather than solid. The problem is then how one can practically construct such a composition, given the fact that the scatterers are expected to float on the fluid host breaking up the periodicity. Self-assembled colloidal suspensions in a liquid

T.-T. Wu and C.-C. Ma (eds.), *IUTAM Symposium on Recent Advances of Acoustic Waves in Solids*, IUTAM Bookseries 26, DOI 10.1007/978-90-481-9893-1_16,

solution are definitely a good candidate, but nevertheless their use in broad applications is somewhat restricted. A close-packed periodic arrangement of glued scatterers in air is definitely an easy case that can be fabricated in reality. Therefore, a system to be examined here is an *fcc* crystal of closed-packed rubber spheres in air.

There are various methods available for the calculation of the elastic properties of phononic crystals [2], such as the traditional band-structure methods, which mainly deal with periodic, infinite, and nondissipative structures. However, in an experiment, one deals with finite-size slabs and the measured quantities are, usually, the transmission and reflection coefficients. Apart from that, realistic structures are dispersive and have losses. We remember that the usual band-structure calculation proceeds with a given wave vector **k** and compute the eigenfrequencies within a wide frequency range together with the corresponding eigenmodes. On the contrary, on-shell methods proceed differently: the frequency is fixed and one obtains the eigenmodes of the crystal for this frequency. These methods are ideal when one deals with dispersive materials (with or without losses). Moreover, on-shell methods are computationally more efficient than traditional band-structure methods [3].

The layer multiple-scattering method (LMS) for phononic crystals [4], is an on-shell method that uses multiple scattering techniques analogous to those applied to the treatment of electron scattering in solids and subsequently to the treatment of light scattering in photonic crystals. Whether one is interested in the transmission/reflection or absorption properties of a slab of a crystal, or the frequency band structure of an (infinite) phononic crystal – the angular frequency ω of the elastic field is a given conserved quantity. Additionally, when one deals with a slab of the material parallel to a given crystallographic plane, the reduced wave vector $\mathbf{k}_{\parallel}$, parallel to this plane, is also a given conserved quantity. In relation to the frequency band structure, the question we ask is: for the given ω, $\mathbf{k}_{\parallel}$, are there propagating Bloch waves in the infinite crystal? These represent of course the eigenmodes of the elastic field in the crystal. The method provides us with these propagating Bloch waves and at the same time with a number of evanescent waves which play an indirect role. By repeating the calculation for different values of ω, we obtain the frequency band structure over a selected region of frequency as required. From the above follows a number of advantages over the more conventional methods of calculation, such as the plane-wave expansion method. (a) The elastic coefficients of one or more of the constituents of the phononic crystal may depend on the frequency, and they can be complex (incorporating absorption). (b) The computation time can be reduced considerably by parallel processing at the same time of a number of frequencies. (c) One can calculate, with the same ease, the eigenmodes of the elastic field in the crystal and the transmittance/reflectance of a slab, for given ω, $\mathbf{k}\|$. (d) One can easily obtain the density of states and the local density of states of the elastic field. A multiple-scattering method is effective when the system under consideration consists of non-overlapping scatterers in a homogeneous medium. The shape of the scatterers can be arbitrary, but the calculations are simpler if one assumes that they are spherical, and in our work we assume that this is indeed the case. All of the above will provide the means to investigate the interesting properties of an fcc crystal of closed-packed rubber spheres in air, a slab of which, under the influence of mild dissipation, exhibits an appreciable omnidirectional phononic transmission gap. In what follows, the advantage of the scaling properties of the crystal is exploited by putting together heterostructures consisting of close-packed rubber spheres of different size intended to serve as a gap optimization mechanism. An arrangement which proves in a versatile manner its merits to competently reposition and enlarge the gap to almost any phononic application specs anticipated.

2. On the Properties of Viscoelastic Phononic Crystals

It is true that the crystal structure can be very important in the formation of absolute phononic gaps [5]. Nevertheless, the physical origin of the widest of these gaps lies beyond the Bragg gaps at the Brillouin zone boundaries. In reality, there are also bands originating from resonant elastic modes of the individual scatterers: resonant states on neighbor scatterers that couple weakly with each other, resulting in corresponding relatively narrow bands. These bands originating from the “rigid-body” resonance modes of the individual scatterers hybridize with the continuum bands corresponding to an almost free propagation in an effective homogeneous medium [2, 4]. The opening of the hybridization gap is favored by an increased volume filling fraction, but at the same time it is compromised by the widening of the resonance bands which is also favored by an increased value of the same property of the crystal. As we will show later, this compromising effect will be taken care of by introducing dissipation on the scatterers of the crystal [6].

The viscoelastic response of the system, we are going to study here, is accounted for by means of the Kelvin- Voigt model, which is well-suited for materials and ultrasonic frequencies of major interest. The problem of acoustic-wave scattering by a single viscoelastic sphere of radius S has been adequately addressed in the past [7] according to the Kelvin-Voigt viscoelastic model. In such a case a case the sphere is considered to be elastic with modified shear and compressional complex wavenumbers, the imaginary parts of which represent a measure of the loss. In particular, for an absorbing sphere in an inviscid fluid background, the complex compressional and shear wavenumbers are conveniently defined as

$$q_l = \frac{c_0}{c_l} = \frac{q_0}{\sqrt{1 - i\left[(\alpha+\beta)/\rho c_l^2\right]}}, \quad q_t = \frac{c_0}{c_t} = \frac{q_0}{\sqrt{1 - i(\beta/\rho c_l^2)}}, \tag{2.1}$$

where $q_0 = \omega/c_0$ refers to the fluid environment with ω being the angular frequency and c_0 the respective speed of sound. The real parts of the complex Lamé parameters of the sphere, $\lambda = \lambda_e - i\lambda_v$ and $\mu = \mu_e - i\mu_v$ combined with the sphere's density ρ yield the compressional and shear wave speeds respectively

$$c_l = \sqrt{(\lambda_e + 2\mu_e)/\rho}, \quad c_t = \sqrt{\mu_e/\rho}. \tag{2.2}$$

The imaginary parts of the Lamé parameters are connected to the viscous losses $\alpha+2\beta$, and β of the sphere as follows: $\alpha = \omega\lambda_v$, $\beta = \omega\mu_v$.

Multiple-scattering effects within planes of spheres, sonic crystals, and slabs of the same are taken into account by the LMS method described in Ref. [4]. This method views the crystal as a sequence of planes of spheres parallel to a given surface: a crystallographic plane described by a two-dimensional (2D) lattice $\{\mathbf{R}_n\}$. The corresponding 2D reciprocal lattice is denote by $\{\mathbf{g}\}$. In the host region between the nth and the (n+1)th planes, a Bloch-wave solution for the displacement field (harmonic time dependence is assumed), corresponding to a given frequency ω and a given reduced wave vector k‖ within the surface Brillouin zone (SBZ) of the given surface, can be expanded into plane waves propagating (or decaying) to the left and to the right, as follows:

$$\mathbf{u}(\omega;\mathbf{k}_\parallel) = \sum_{\mathbf{g}} \left\{ \mathbf{u}_{\mathbf{g}n}^{+} \exp\left[i\mathbf{K}_{\mathbf{g}0}^{+} \cdot (\mathbf{r} - \mathbf{A}_n)\right] + \mathbf{u}_{\mathbf{g}n}^{-} \exp\left[i\mathbf{K}_{\mathbf{g}0}^{-} \cdot (\mathbf{r} - \mathbf{A}_n)\right] \right\}, \tag{2.3}$$

where

$$\mathbf{K}_{\mathbf{g}0}^{\pm} = (\mathbf{k}_\parallel + \mathbf{g}, \pm[(\omega/c_0)^2 - (\mathbf{k}_\parallel + \mathbf{g})^2]^{1/2}) \tag{2.4}$$

and $\mathbf{A}_n$ is a point between the nth and the $(n+1)$th planes. It should be noted that, although both shear and compressional modes are considered within the spheres and enter the calculation through the scattering matrix, only longitudinal waves exist in the fluid, which in our case it will be air.

A generalized propagating or evanescent, Bloch wave satisfies the equation

$$\mathbf{u}^{\pm}_{gn+1} = \exp(i\mathbf{k}\cdot\mathbf{a}_3)\mathbf{u}^{\pm}_{gn} , \tag{2.5}$$

where $\mathbf{a}_3 = \mathbf{A}_{n-1} - \mathbf{A}_n$ and $\mathbf{k} = (\mathbf{k}_{\|}, k_z(\omega; \mathbf{k}_{\|}))$ is the Bloch wave vector. There are infinitely many such solutions for given $\mathbf{k}_{\|}$ and ω, corresponding to different values of the z component, $k_z(\omega; \mathbf{k}_{\|})$, of the reduced wave vector $\mathbf{k}$, but in practice one needs to calculate only a finite number (a few tens at most) of these generalized Bloch waves. We have propagating waves [for these $k_z(\omega; \mathbf{k}_{\|})$ is real] which constitute the normal modes of the infinite crystal, and evanescent waves [for these $k_z(\omega; \mathbf{k}_{\|})$ is imaginary] which do not represent real waves, but they are useful mathematical entities which enter into the evaluation of the reflection and transmission coefficients of a wave, with the same ω and $\mathbf{k}_{\|}$, incident on a slab of the crystal parallel to the given surface. The transmission and reflection matrices for a slab which consists of a stack of layers of spheres with the same 2D periodicity parallel to a given plane of the crystal are obtained from the transmission and reflection matrices of the individual layers in the manner described in Ref. [4].

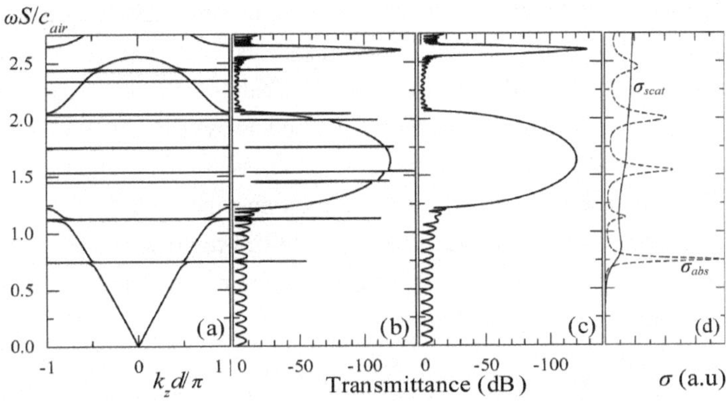

Fig. 2.1 The phononic band structure at the center of the SBZ of a (111) surface of an *fcc* crystal (of lattice constant a) of close-packed lossless rubber spheres in air (a). The corresponding transmittance curve of a slab of 16 layers parallel to the same surface is given in (b). In (c) the same transmittance curve is presented but with spheres of a low viscous level. d is the distance between successive (111) planes of the *fcc* crystal under consideration. Finally, (d) depicts the absorption (dashed line) and the scattering cross section of a single sphere with dissipation.

Knowing the transmission and reflection matrices for the slab we can readily obtain the transmission, reflection, and absorption coefficients of a plane acoustic wave incident on the slab.

The system examined here is a crystal of rubber spheres in air. The physical parameters entering the calculations are the mass density of air is ρ_{air}= 1.2 kg/m^3 and its respective speed of sound c_{air}=334 m/s. The rubber spheres have a mass density ρ=1130 kg/m^3, and c_l=1400 m/s and c_t=94 m/s are the compressional and shear speeds of sound, respectively. In addition, a mild viscoelastic level is considered for the rubber spheres with parameters α=0.5 MPa and β=0.01, which are typical values for commercial rubbers. We view the crystal as a succession of planes of spheres parallel to the (111) *fcc* surface. The radius of the spheres S=0.5 a0, where a0 (the distance between two touching spheres) is related to the lattice constant as $a = a0\sqrt{2}$. Figure 2.1 shows the frequency band structure normal to the *fcc* (111) plane ($\mathbf{k}_{\|}=\mathbf{0}$) and the corresponding transmission spectrum for waves incident normally on a slab of the crystal consisting of 16 layers of lossless spheres. The results are obtained with an angular momentum cutoff $l$max=7 (as a result of writing a plane wave as linear sum of spherical waves [4]) and 55 **g** vectors. The established convergence is within an accuracy of better than a tenth of a percent. One observes, besides a large Bragg gap extending from $\omega S/cair$=1.223 to 2.065, a number of flat bands which derive from the interacting sharp resonant modes localized on the individual rubber spheres. Because these bands are so narrow in the present case, they are hardly observable, except that they introduce small gaps, above and below the main gap, which result from the hybridization of these flat bands with the broadbands corresponding to nearly free propagating waves. These narrow gaps are seen more clearly in the transmission spectrum (part b of Fig. 2.1). Within the main gap these flat bands manifest themselves as sharp peaks in the transmission spectrum. The oscillations in the transmission coefficient, over the allowed regions of frequency, are due to interference effects resulting from multiple reflection at the surfaces of the slab of the crystal (Fabry-Perot-type oscillations).

When losses are present in the system, there are no true propagating waves and the band structure of the infinite lossless crystal is not of any help; therefore, the effect of dissipation is shown in the transmission spectrum, part c of Fig. 2.1. As expected from the results of the single sphere [6], the sharp peaks and dips of the resonant states disappear and a "clean" sonic gap without any resonant mode spikes is obtained. The existence of the frequency gap means that sound does not propagate through the crystal when its frequency lies within the gap (the intensity of the wave decays exponentially into the crystal for these frequencies), and if it cannot enter into the crystal, it cannot be absorbed either.

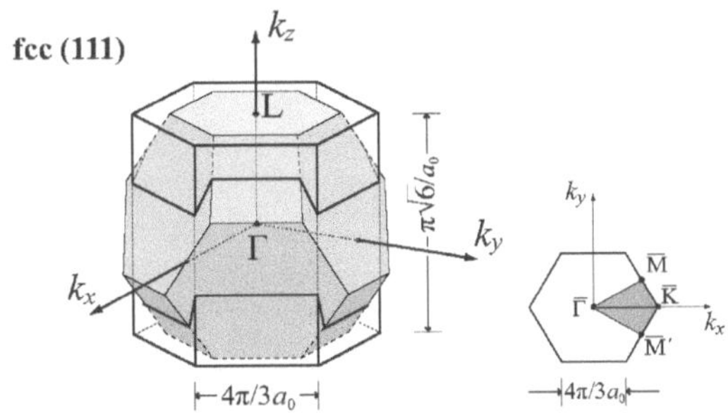

Fig. 2.2 The reduced **k** zone associated with the *fcc* (111) crystallographic plane (left) and the corresponding SBZ (right). The (bulk) *fcc* Brillouin zone (shaded decatetrahedron) is also shown for comparison.

The reduced zone of the 3D **k** (since the crystal is viewed as a sequence of identical crystallographic planes) is chosen as follows: ($\mathbf{k}_{\parallel}$, k_z) where $\mathbf{k}_{\parallel}=(k_x , k_y)$ extends over the SBZ of the given crystallographic plane. This reduced k zone is by all means equivalent to the commonly used, more symmetrical bulk Brillouin zone, in a sense that a point in one of them lies also in the other or differs by a vector of the 3D reciprocal lattice. This scheme is presented in Fig. 2.2, for a better understanding on how all the possible symmetry lines are taken into account in order to plot the band structure of the crystal.

In Fig. 2.3, the projection of the frequency band structure on the SBZ of the (111) plane of the *fcc* crystal along its symmetry lines is shown. This is obtained, for a given $\mathbf{k}_{\parallel}$, as follows: the regions of frequency over which there are no propagating states in the infinite crystal (the corresponding values of all k_z are complex) are shown in white, against the shaded areas which correspond to regions over which propagating states do exist. One clearly sees here how the sphere's resonances lead to narrow hybridization gaps above and below the main gap, and flat bands in the gap regions. When losses are present, the crystal under investigation exhibits an appreciable omnidirectional transmission gap extending from $\omega S/c_{air}$=1.595 up to 1.946.

Touching rubber spheres, obviously, can trigger highly non-linear effects (clapping). In our model, under the assumption of glued spheres, we predict that a linear interaction will more or less suffice to describe the dominant physical phenomena regarding the transport properties of such a phononic arrangement. In any case, the present study does not include any nonlinearities. Finally, although one may consider this specific phononic crystal as a porous insulator [8], no additional degrees of freedom (slow compressional waves) predicted by Biot's theory [9] are considered in our model and most certainly might have little (if they have) impact on the acoustic transmission gap properties of a slab of the crystal.

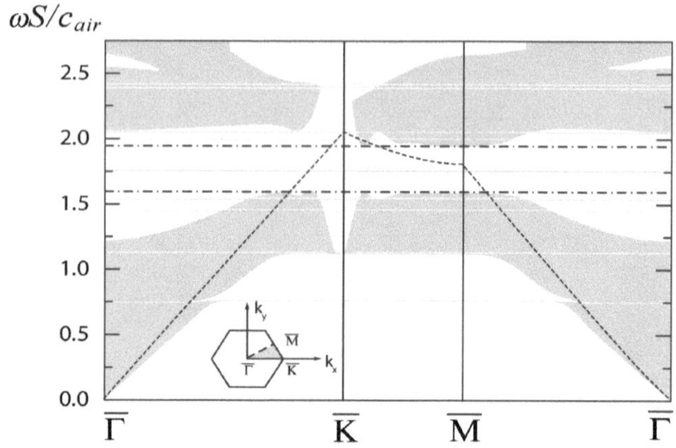

Fig. 2.3 Projection of the frequency band structure on the SBZ of the (111) surface of the *fcc* phononic crystal described in the caption of Fig.2.1, along the symmetry lines (inset). Propagating waves in the air about a slab of the crystal exist for frequencies above a threshold value (a function of $\mathbf{k}_{\parallel}$) $\omega_{\text{inf}} = c_{air}|\mathbf{k}_{\parallel}|$ denoted by the dashed line.

3. Slabs of Heterostructures and Gap Optimization

Recently, rubber/air viscoelastic phononic crystals have drawn serious attention [10], because they exhibit very wide band gaps in their transmission spectra that extend to frequencies in the audible range of the spectrum and they are a practical design of acoustic band gap sound barriers with small dimensions. In the past, certain full-band-gap enhancement techniques have been developed [11,12] to optimize the filtering spectrum of certain phononic crystals. The idea is to fabricate slabs of the crystal containing a minimal number of planes in order to preserve accurately the transport properties of the full 3D crystal. It has been shown that the number of 8 planes per slab safely preserves the transport properties and almost all underlying physics [13]. One then stacks a number of the slabs with spheres of different size each in such a way that the lower frequency edge of the absolute gap of one stack is at least equal or lower than the upper gap edge of the stack, or else the upper edge of one stack is at least equal or greater than the lower edge of the other. The former arrangement enhances the absolute gap to lower frequencies and the latter enhances it towards higher frequencies. The information to do so is coming from the projection of the frequency band structure in Fig. 2.3, which scales with S and provides the gap edges for different sizes of spheres, which stack in the close-packed (111) fcc arrangement. This is presented in Fig. 3.1. The actual omniderectional gap of a single viscoelastic stack has a gap width over the midgap frequency, $\Delta\omega/\omega MG$=19.8%.

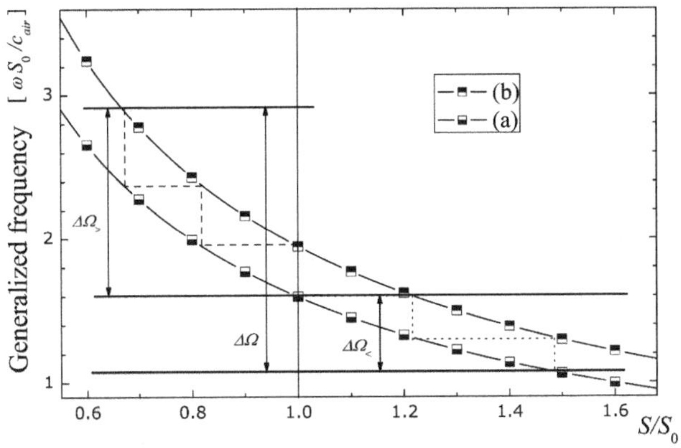

Fig. 3.1 Upper and lower edges of the absolute frequency gap of a close-packed *fcc* phononic crystal consisting of viscoelastic rubber spheres as a function of the radius of the spheres. (a) are the lower edges and (b) the upper, whereas 1 denotes the reference sphere size (this where lies the gap of 1 stack) and the gray dotted steps between the edges signify the scheme of overlapping gaps. $\Delta\Omega_<$ is the gap enhancement result towards the lower frequencies, and $\Delta\Omega>$ is the corresponding result towards higher frequencies. They both require 4 stacks of different size of spheres. $\Delta\Omega$ is a combination requiring 7 stacks.

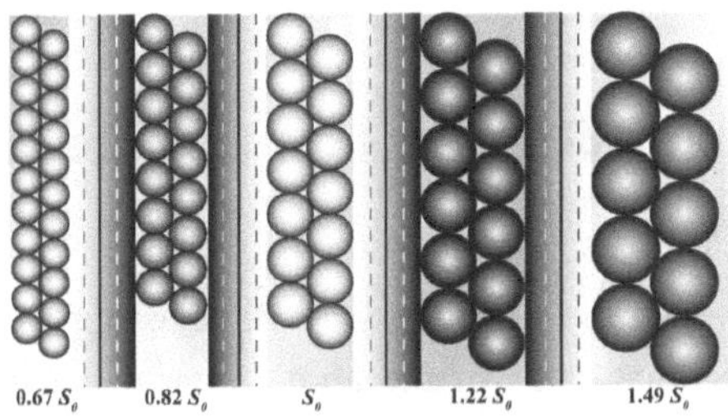

Fig. 3.2 Schematic representation of a combination of 5 stacks to form an optimized phononic slab of close-packed rubber spheres of different size.

According to the above if one creates a sequence of 5 stacks as shown in Fig. 3.2 will result to an absolute gap of 92% which is more than 4 times wider than the one of a stack. In Fig. 3.3, we present the basic combinations which result in widening the gap of the system and moving it up or down in the frequency range or both, with appropriate minimum number of stacks for each case, which can lead up to gaps of 160%. In theory one can go asymptotically up to 200%, which is the maximum. It is recommended that one should proceed with caution in pathological cases of having too many stacks, or pushing the scattering size parameter out of range, not to mention the probable change of viscoelastic properties as a function of size and frequency range. The cases considered here are safe and computationally accurate. One may note that local porosity effects are expected to be appreciably dissipated by rubber.

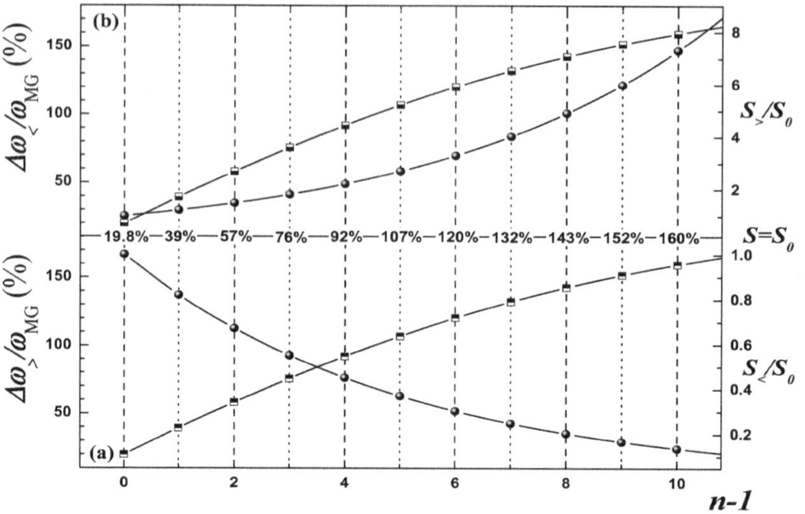

Fig. 3.3 Enhanced gap features for shifting up and down the gap in frequency or both. The number ***n*** (***n*** -1 plus the reference stack with spheres of radius S_0) of minimum stacks needed for each case as well as the size of the sphere for each stack (gray spheres) is represented in the above graph. The achieved gapwidth over midgap frequency is also given.

4. Conclusions and Remarks on the Versatility

On investigating the versatility of rubber/air viscoelastic phononic crystal slabs, it has been demonstrated that slabs of the above crystal (each of which is a finite part of a close-packed *fcc* crystal) with proper size variation of the spheres in a sequence of crystal slabs one can manipulate quite competently (this from where the term versatile comes) the absolute frequency gap of the crystal both in size and frequency. This has been only a small part of the capabilities of what such systems can do. One can also manipulate directional gaps by stacking slabs of different crystallographic orientation in order to make directional filters or sonic polarizers. And to say the least this is a good example of strong classical Anderson localization due to size disorder. Physically this is the effect that is responsible for widening the phononic transmission gap [14]. Therefore a systematic introduction of disorder, namely size disorder, can trigger Anderson localization of acoustic waves over a region of frequency much wider than the absolute gap of the ordered phononic crystal. Finally, for practical purposes, this study may be considered a guide for proper size variation of a slab of a disordered crystal of reasonable thickness that will produce the desired effect.

References

[1] Modinos, A., Stefanou, N., Psarobas, I.E., Yannopapas, V.: On wave propagation in inhomogeneous systems. *Physica B* **296** 167-173 (2001).

[2] Sigalas, M., Kushwaha, M.S., Economou, E., Kafesaki, M., Psarobas, I.E., Steurer, W.: Classical vibrational modes in phononic lattices: theory and experiment. *Z. Kristallogr.* **220** 765-809 (2005).

[3] Qiu, Y., Leung, K.M., Carin, L., Kralj, D.: Dispersion-curves and transmission spectra of a 2-dimensional photonic band-gap crystal theory and experiment. *J. Appl. Phys.* **77** 3631-3636 (1995).

[4] Sainidou, R., Stefanou, N., Psarobas, I.E., Modinos, A.: A layer-multiple-scattering method for phononic crystals and heterostructures of such. *Comput. Phys. Commun.* **166** 197-240 (2005).

[5] Sainidou, R., Stefanou, N., Modinos, A.: Formation of absolute frequency gaps in three-dimensional solid phononic crystals. *Phys. Rev. B* **66** 212301 (2002).

[6] Psarobas, I.E.: Viscoelastic response of sonic band-gap materials. *Phys. Rev. B* **64** 012303 (2001).

[7] Ayres, V.M., Gaunaurd, G.C.: Acoustic-resonance scattering by viscoelastic objects. *J. Acoust. Soc. Am.* **81** 301 (1987).

[8] Zimmerman, C., Stern, M.: Scattering of plane compressional waves by spherical inclusions in a poroelastic medium. *J. Acoust. Soc. Am.* **94** 527 (1993).

[9] Berryman, J.C.: Effective conductivity by fluid analogy for a porous insulator filled with a conductor. *Phys. Rev. B* **27** 7789 (1983).

[10] Merheb, B., Deymier, P.A., Jain, M., Aloshyna-Lesuffleur, M., Mohanty, S., Berker, A., Greger, R.W.: Elastic and viscoelastic effects in rubber/air acoustic band gap structures: A theoretical and experimental study. *J. Appl. Phys.* **104** 064913 (2008).

[11] Kushwaha, M.S., Djafari-Rouhani, D., Dobrzynski, L., Vasseur, J.O.: Sonic stop-bands for cubic arrays of rigid inclusions in air. *Eur. Phys. J. B* **3** 155 (1998).

[12] Psarobas, I.E., Sigalas, M.M.: Elastic band gaps in a fcc lattice of mercury spheres in aluminum. *Phys. Rev. B* **66** 052302 (2002).

[13] Psarobas, I.E., Stefanou, N., Modinos, A.: Phononic crystals with planar defects. *Phys. Rev. B* **62** 5536-5540 (2000).

[14] Sainidou, R., Stefanou, N., Modinos, A.: Widening of Phononic Transmission Gaps via Anderson Localization. *Phys. Rev. Lett.* **94** 205503 (2005).

Omnidirectional Band Gap Mirror for Surface Acoustic Wave

Abdelkrim Khelif [1,2] , Abdelkrim Choujaa[1], Sarah Benchabane[1] and Vincent Laude[1]

[1]Institut FEMTO-ST, Université de Franche-Comté, CNRS, ENSMM, UTBM; 32 avenue de l'Observatoire, F-25044 Besançon, France

[2] CNRS-Georgia Tech joint Laboratory, UMI 2958, Georgia Institute of Technology, Atlanta, GA 30332

abdelkrim.khelif@femto-st.fr

Abstract. We report on the theoretical and experimental evidence for the occurrence of an omnidirectional elastic band gaps in one-dimensional phononic crystal structures. The structure is constituted by a periodically layered mirror deposited on a specific substrate that exhibits total reflection of waves for all incident angles and polarizations in a given frequency range. We study in this paper the influence of the nature and the filling fraction of the layered structures as well as of the substrate nature on the omnidirectional band gap. By introducing a defect piezoelectric layer in the finite size layered structure, selective resonance modes can occur within the omnidirectional bang gap under certain conditions. In this case, the elastic energy is confined to the defect layer. The frequencies of the defect modes are sensitive to the nature of the material and to the layer thicknesses. We excite the localized mode in the defect layer via interdigital transducers deposited on the substrate surface. We show a high frequency one port resonator at 4.3 GHz with high phase velocity of 43 km/s. The omnidirectional character of the band gap is crucial to confine the elastic energy in the defect layer.

1. Introduction

Recently, phononic band gap materials, the so-called phononic crystals, have been made possible by using periodic structures, in analogy with electrons in semiconductor crystals [1, 2]. These materials allow the propagation of elastic waves to be regulated. In other words, they play the role of perfect mirrors for elastic waves in

T.-T. Wu and C.-C. Ma (eds.), *IUTAM Symposium on Recent Advances of Acoustic Waves in Solids*, IUTAM Bookseries 26, DOI 10.1007/978-90-481-9893-1_17,

the frequency window of the band gap, i.e., forbidding the propagation of elastic waves for all polarizations and directions. This area of research has received much attention because of the fundamental interest in localization of elastic energy and of the potential applications of phononic band gaps. The omnisaw device is the first illustration of the capabilities of these new materials to confine the elastic energy and to control the dispersion of waves through the geometry. The paper is organized as follows: first, we demonstrate the existence of a complete elastic band gap in Aluminium/Tungsten periodically layered structures deposited on a silicon substrate. The influence on the band gap of the filling fraction of the layers is presented. The results for an AlN defect layer added on the structure are presented and discussed in section 3. In section 4, we give a first theoretical and experimental result for a one port resonator based on the OmniSAW concept.

2. Omnidirectional Elastic Band Gap

In this section, we show the possibility of realizing one-dimensional structures that exhibit the property of omnidirectional 'complete' reflection for elastic waves. In the frequency range of forbidden propagation [3], the structure behaves analogously to 2D and 3D phononic crystals in which elastic waves are reflected for any incidence angle (equivalently, any incident wave vector) and any polarization. We consider the superlattice structure described in Fig. 1, in which N periods of the couple of layers (Al/W) are deposited on a semi-infinite silicon substrate. For this kind of materials, it possible to decouple the sagittal (x,z) and transverse (y) polarizations. We have focused, in this case, on the sagittal polarisation of elastic waves propagating parallel to the interfaces.

First, we examine the so-called projected band structure of the superlattice. This refers to the diagram obtained by plotting the frequency versus the parallel wave vector, $k_{//}$. Figure 2 displays the phononic band structure of six pairs of layers composed of Al and W with thicknesses d1 and d2, such that d1=d2=0.5D, D being the period of the superlattice (D=d1+d2). We display the frequency*period (f*D) as a function of the reduced parallel wave vector ($k_{//}$*D). The grayed and black areas in the projected band structure, respectively, correspond to the pass bands and gaps of the structure where propagation of elastic waves is allowed or forbidden. It can be easily noticed that the band structure shown in Fig. 2 exhibits an absolute band gap for every value of the reduced parallel wave vector in the frequency range (1400 m/s<f*D<2300 m/s). It means that no acoustic wave with sagittal polarization is allowed to propagate in this region. The two straight lines correspond to the shear and longitudinal velocities of the substrate (here silicon).

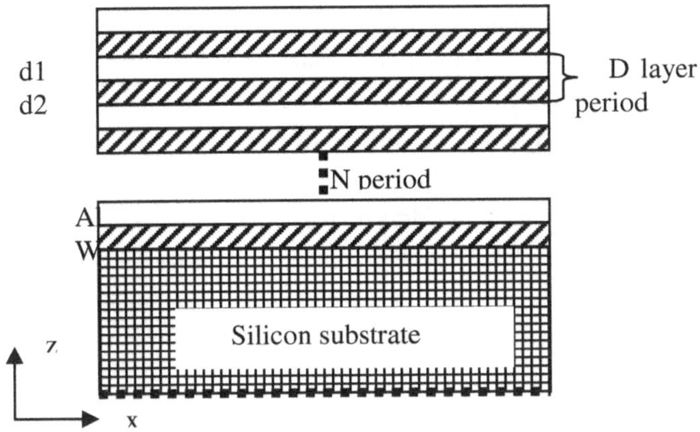

Fig. 1 Sketch of the superlattice structure.

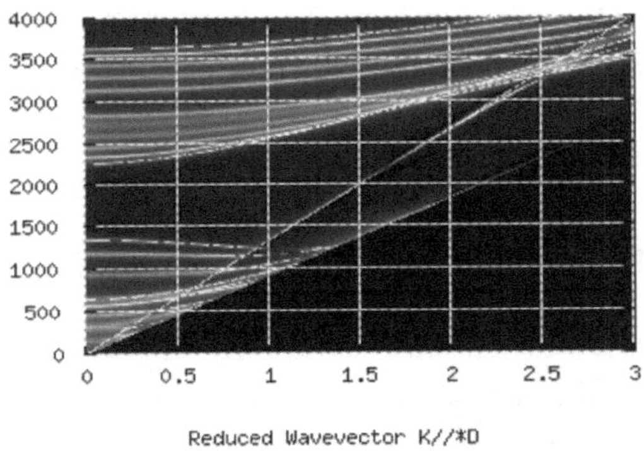

Fig. 2 Projected phononic band structure of a six-pair superlattice composed of Al and W with thicknesses d1 and d2, such as d1=d2=0.5D. The two non-dispersive modes (straight lines) show the signature of the longitudinal and transverse velocities of the silicon substrate.

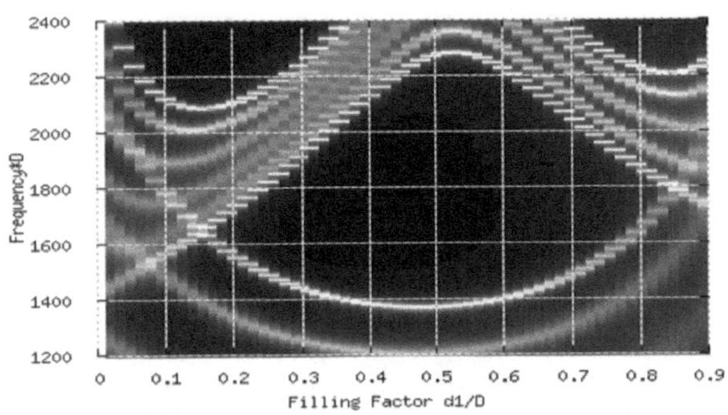

Fig. 3 Evolution of the band gap for an Al/W structure.

Three factors can influence the width of the Omnidirectional band gap: the filling fraction, the superlattice composition and the substrate nature. The first parameter that influences the width of the omnidirectional band gap is the filling fraction which is equal, in this case, to the thickness of one layer divided by the period, for instance, $F=d_l/D$. Figure 3 gives the evolution of the gap versus the ratio d_l/D, showing that, in this case the largest gap is obtained when the ratio is around 0.5. One can notice that for $0.2<F<0.85$ we still obtain an omnidirectional band gap in this material composition. The existence and the width of the band gap are also influenced the elastic parameters or the nature of the materials constituting the superlattice. As a general trend, a large band gap requires a huge contrast between the elastic constants and densities of the constituent materials.

3. Defect Mode Structures

In order to achieve mode confinement in the elastic band gap, we introduce a defect in the structure. This can be achieved by changing the thickness or the nature of a single layer. In the OmniSAW device we use a piezoelectric material such as AlN or ZnO to excite localized modes through the interdigital transducer (IDT) technology.

In Figure 4 we plot the dispersion of a defect elastic mode calculated from the local density of state technique for an Aluminium Nitride (AlN) layer added on top of a W/Al omnidirectional mirror structure with filling fraction $F = 0.5$.

The thickness of the AlN layer is equal to 1.1*D By comparing the dispersion curves to the band structure in Figure 2 for the complete band gap superlattice structure, we note that we have one defect mode occuring in the band gap with a frequency cut-off equal to 1800 m/s. This mode is the fundamental mode, which can give the maximum electro-mechanical coupling coefficient. By increasing the thickness of the defect layer we can introduce more and more localized modes in the band gap. We note that the defect mode at zero wave vector corresponds to the situation of "Bragg mirror" for a solidly mounted resonator.

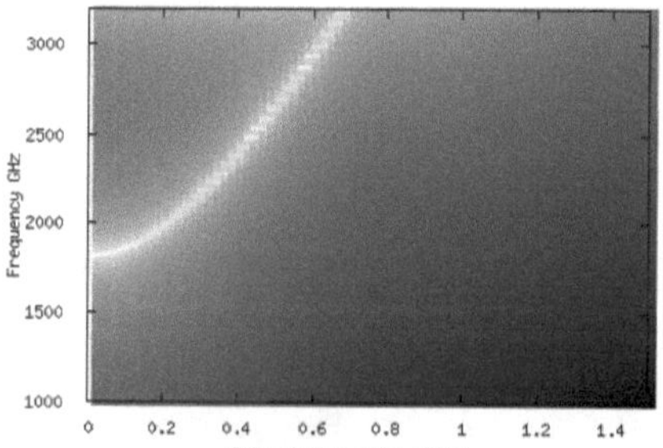

Fig. 4 Dispersion of a localized defect mode in the AlN layer with thickness equal to 1.1*D deposited on the omnidirectional mirror structure.

4. Omnisaw Devise Results

Before starting to give the parameters and the numerical results of a one port resonator based on the OmniSAW concept [4], we start to design the band gap and give the value of the period to fix the frequency range in order to achieve a one port resonator operating around 4.5 GHz. From Figure 4 giving the defect dispersion diagram, we know that the frequency*period product should be around 1800 m/s. The exact resonance frequency of the surface also depends on the value of the IDT period or on the wavelength obtained from the value of the parallel wave vector. For the central frequency of 4.5 GHz, the period of the structure is equal to 0.4 μm and d_1=200 nm and d_2=200 nm (F=0.5). The complete band gap then extends from 3 to 6.5 GHz. The relative bandwidth equals 77%, which gives very large latitude and reduces the impact of any error on the geometrical parameters of the structured layers.

After having designed the band gap and the thickness of the defect AlN layer, we have simulated the harmonic admittance of an infinite periodic electrode array deposited on top of the AlN surface. First, we assume that the thickness of the electrodes is negligible compared to the pitch of the structure. We assume the filling fraction of the IDT is equal to 50%. In figure 5, we plot the real and the imaginary parts of the harmonic admittance of the Omnisaw structure with an electrode period 5 μm. The thickness of the defect layer is equal to 0.45 μm. The figure clearly shows the occurrence of an acoustic resonance at 4.3 GHz, which is confirmed in figure 6 by the experimental result where the measurement of S_{11} exhibits the signature of the resonance at the same frequency estimated theoretically. The effective phase velocity of the acoustic mode in this case is 43000 m/s, which opens up novel possibilities for the design of surface acoustic wave devices operating at high frequency but long wavelength. The effective phase velocity is more than ten times larger that the one achievable with piezoelectric substrates such as quartz or lithium tantalate. Whereas traditional photolithography with UV light limits usual SAW devices to around operating frequencies 2GHz, OmniSAW devices have the potential to operate above 10 Ghz with the same technological constraints for the electrode mask.

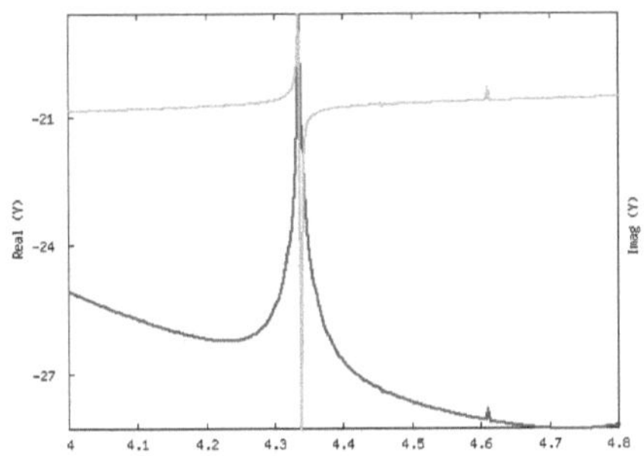

Fig. 5 Real and imaginary parts of the harmonic admittance of the structure.

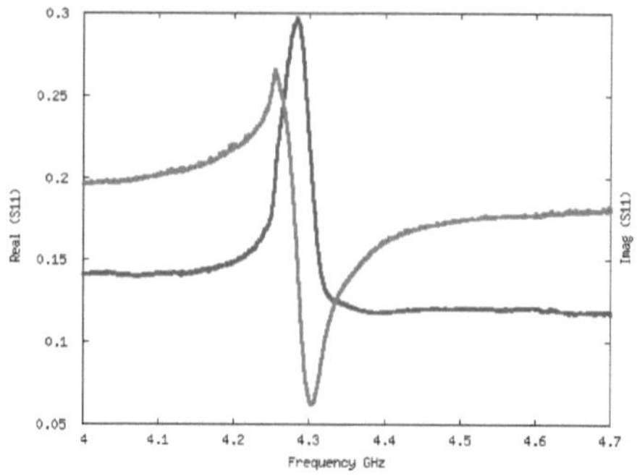

Fig. 6 Real and imaginary parts of S11 parameters.

5. Conclusion

In this paper we have theoretically and experimentally demonstrated the Omni-SAW device concept for an operating frequency of 4.3 GHz. We have shown the occurrence of omnidirectional elastic band gaps in one-dimensional phononic crystal structures. The structure is constituted by a periodic arrangement of layer pairs deposited on a specific substrate that exhibits total reflection for waves of any incident angle and polarization in a given frequency range called the omnidirectional frequency band gap. We have presented the influence of the filling fraction of the layer pairs on the complete band gap. By introducing a defect piezoelectric layer (AlN), selective resonance modes occur within the omnidirectional bang gap. In this case, the elastic energy is confined inside the defect layer. We have shown clearly the occurrence of an acoustic resonance at 4.3 GHz, which is confirmed by the measurement of the S_{11} electrical parameter. The effective phase velocity of the acoustic mode in this case is 43000 m/s, which opens new propects for surface acoustic wave devices operating at high frequency.

References

[1] Kushwaha, M. S., Halevi, P., Dobrzynski, L., and Djafari-Rouhani, B.: Acoustic band structure of periodic elastic composites. *Phys. Rev. Lett.* **71** 2022-2025 (1993).

[2] Sigalas, M. M. and Economou, E. N.: Elastic and acoustic wave band structure. *J. Sound Vib.* **158** 377-382 (1992).

[3] Bria, D. and Djafari-Rouhani, B.: Omnidirectional elastic band gap. *Phys. Rev. E.* **66**(13) (2002).

[4] Khelif, A., Choujaa, A., Laude, V.: Patent WO 2006/087496.

Band Structure of Three-Dimensional Phononic Crystals with an Opal Structure

Yukihiro Tanaka, Shin-ichiro Tamura and Takuro Okada

Department of Applied Physics, Hokkaido University, Sapporo 060-8628, Japan

yuki@eng.hokudai.ac.jp

Abstract. We made a systematic investigation on the band structure of sound waves in three-dimensional phononic crystals consisting of periodic arrays of spheres with an opal structure. The opal structure is made with close-packed solid spheres surrounded by air (or vacuum). We introduce a sintering parameter which characterizes the contact between adjacent spheres. For a proper range of parameters we find the existence of a complete band gap in the structure. We also find that several flat bands appear below the lowest eigenfrequency of vibration of the isolated sphere. To understand the vibrational motion of the modes in these flat bands, their intensity distributions are also shown.

1. Introduction

Recently, a great deal of attention has been attracted to synthetic, three-dimensional (3D) composite structures consisting of periodic arrays of materials. Such meta-materials are called phononic crystals when their elastic properties are emphasized. Coherent multiple scatterings arising from artificially designed periodicities allow phononic crystals to reveal interesting physical properties which do not exist in nature. One of the important properties is the existence of complete frequency gaps in which sound waves with wave vectors in any direction are prohibited to propagate. The perfect confinement of sound in these gaps suggests that phononic crystals can serve as cavities for sound and also leads to the potential applications to the design of new opto-acoustical devices which manipulate optical gaps.

T.-T. Wu and C.-C. Ma (eds.), *IUTAM Symposium on Recent Advances of Acoustic Waves in Solids*, IUTAM Bookseries 26, DOI 10.1007/978-90-481-9893-1_18,

A simple and hence widely studied 3D phononic crystal is a synthetic opal composed of self-assembled glass spheres with a sub-micron diameter that is expected to operate as a band gap material for both the visible lights and hypersonic waves.[1] The effects of hypersound of several GHz frequencies in opal films (consisting of silica nano-spheres with diameters of about 100 nm) have recently been observed experimentally but rather indirectly with a light scattering technique through the modulation of optical reflectivity.[2] Also Brillouin light scattering technique is applied to find phononic gaps in soft opals, i. e., the self-assembled FCC films of polystyrene spheres.[3] More direct experiments with ultrasound, such as transmission measurements, for example, should reveal the detailed acoustical properties of 3D phononic crystals. Such an experiment has recently been made with similar opal structures but of larger dimensions (~a few millimeters in diameter), i. e., mesoscopic opals.

In the present paper we will theoretically study the properties of sound waves in 3D phononic crystals with an opal structure. More precisely, the phononic crystals we consider are composed of periodic arrays of aluminum spheres in air (or in vacuum), which also form a FCC lattice structure.

In analyzing theoretically the vibrational properties of phononic crystals, the plane-wave expansion (PWE) and multiple-scattering (MS) methods have proved to be powerful tools. However, the MS method is applicable only to isotropic materials and the structures where no overlapping exist between scatterers.[4-7] The PWE method can be used to the systems where no large acoustic mismatch exists between constituent materials.[8] Synthetic opals which have recently been studied experimentally consists, however, of spheres connected each other by sintering. Also, as phononic crystals, they are composed of solid spheres surrounded by air, possessing large acoustic contrast. Therefore, neither the PWE nor the MS methods can be applied to the opals we study. Only possible numerical scheme employed is the finite-difference time-domain (FDTD) method.[9]

In our analysis, we first introduce a so-called sintering parameter which characterizes the contact between adjacent spheres, and calculate the dispersion relations of sound waves in the perfect, periodic opals with the FCC structure. Next, we clarify the dependence of the band structure on the sintering parameter in the $\Gamma-L$ direction which is important in experiments. Finally, we show the vibrational intensities distributions at several points in the band structure.

2. Model and Formulation

For a lossless inhomogeneous elastic material, the equations of motion and constitutive law of elastic waves are given by

$$\rho(\mathbf{r})\ddot{u}_i(\mathbf{r},t) = \frac{\partial T_{ij}(\mathbf{r},t)}{\partial x_j} + F_i, \tag{1}$$

$$T_{ij}(\mathbf{r},t) = c_{ijmn}(\mathbf{r})\frac{\partial u_m(\mathbf{r},t)}{\partial x_n}, \tag{2}$$

where $\rho(\mathbf{r})$ is the mass density, $u_i(\mathbf{r},t)$ ($i=1,2,3$) the lattice displacement, $T_{ij}(\mathbf{r},t)$ the stress tensor, F_i the body force, and the summation convention over repeated indices is assumed.

It should be noted that the mass density $\rho(\mathbf{r})$ and elastic constants $c_{ijmn}(\mathbf{r})$ depend on the position $\mathbf{r}$. In the FDTD scheme which we will employ to solve Eqs. (1) and (2), they should be transformed into the difference equations both in time and space and are solved recursively under appropriate initial and boundary conditions (BCs).

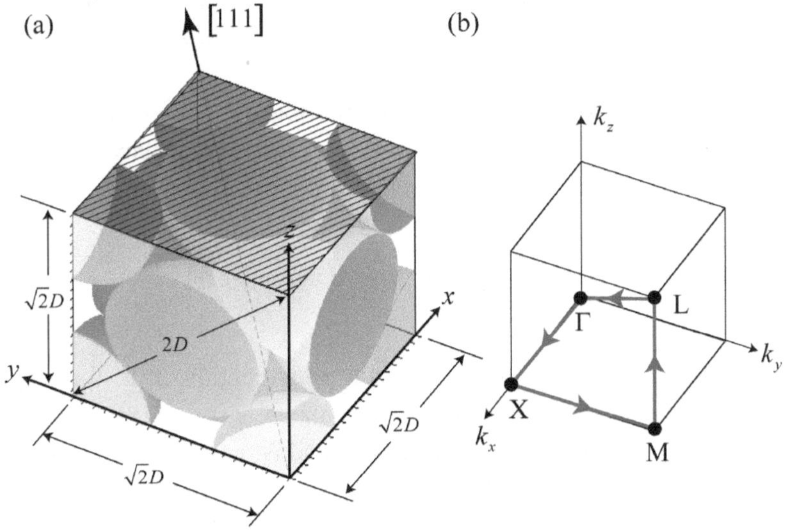

Fig. 1 (color online) (a) A conventional unit cell of face-centered cubic structure. All the faces are imposed on the periodic boundary conditions. The distance between adjacent spheres is D The distributions of vibrational intensities are shown on the face shaded in context. (b) the irreducible Brillouin zone corresponding to the conventional unit cell (a). The dispersion relations of elastic waves in Fig. 2 are drawn along the Brillouin zone boundaries (red line).

First we calculate the dispersion relation of the sound waves in opals with the FCC structure. For this purpose we consider a conventional unit cell shown in Fig. 1(a). At the boundary surfaces we impose the periodic BCs for both the displacements and stresses, which are expressed as

$$u_i(\mathbf{r}+\mathbf{a},t)=e^{i\mathbf{k}\cdot\mathbf{a}}u_i(\mathbf{r},t), \tag{3}$$

$$T_{ij}(\mathbf{r}+\mathbf{a},t)=e^{i\mathbf{k}\cdot\mathbf{a}}T_{ij}(\mathbf{r},t), \tag{4}$$

where $\mathbf{k}$ is a Bloch wave vector and $\mathbf{a}$ is one of the following lattice vectors $\mathbf{a}_1=(\sqrt{2}D,0,0)$, $\mathbf{a}_2=(0,\sqrt{2}D,0)$, $\mathbf{a}_3=(0,0,\sqrt{2}D)$, with D the distance between the centers of the adjacent spheres.

Briefly, the procedure of the calculation is as follows: First we specify a wave vector $\mathbf{k}$ in the first Brilloin zone and this determines the periodic BCs through Eqs. (3) and (4). Next, in order to excite initial lattice displacement an impulsive force $\mathbf{F}$ with frequency spectrum covering the range to be probed is given at a point in the unit cell. Then Eqs. (1) and (2) are solved numerically using finite difference formulas and the time evolution of the displacements $\mathbf{u}(\mathbf{r}_m,t_n)$ at several points $\mathbf{r}_m$ chosen at random in the unit cell are recorded at $t=t_n$. A large number of $\mathbf{u}$ data on the discretized time axis $t_1,t_2,\cdots,t_N$, ($N\gg 1$), are then Fourier transformed into the frequency space. The positions of the peaks in the spectrum are identified to the eigenfrequencies in the system.

Next we calculate vibrational intensity distribution. This is made by giving an external force $\mathbf{F}$ of a Gaussian form with a central frequency ω_0 and a narrow width $\Delta\omega$ in the frequency space. Explicitly,

$$\mathbf{F}(t)=\hat{\mathbf{F}}\int\exp\left[-\frac{(\omega-\omega_0)^2}{(\Delta\omega)^2}\right]\exp(-i\omega t)\,d\omega=\sqrt{\pi}(\Delta\omega)\hat{\mathbf{F}}\exp\left[-\frac{(\Delta\omega)^2t^2}{4}-i\omega_0t\right] \tag{5}$$

where $\hat{\mathbf{F}}$ is a constant vector which indicates the direction of the force applied. Then the vibrational intensity has been calculated according to the following formula after the vibration is well developed and is regarded as stationary.

$$I=\sum_{i=1}^{3}|u_i|^2. \tag{6}$$

3. Numerical Results

Numerical calculations have been developed for 3D arrays of aluminum spheres of a mm scale. The mass density ρ and the elastic constants C_{11} and C_{44} in the isotropic approximation are 2.7 g/cm^3, 109.9×10^{10} dyn/cm^2, and 26.3×10^{10} dyn/cm^2, respectively, for aluminum. The air surrounding the spheres is approximated as vacuum. An important disposal parameter is the sintering parameter χ expressing the degree of the overlap between adjacent spheres,

$$\chi = \frac{2r}{D} - 1, \tag{7}$$

where r is the radius of an aluminum sphere. A positive χ means that adjacent spheres are overlapped each other.

Fig. 2(a) shows for $\chi = 0.01$ the band structure of bulk sound waves along the irreducible Brillouin-zone boundaries shown in Fig. 1(b). In this calculation, both the longitudinal and transverse vibrations are simultaneously excited and detected. We find the existence of a complete band gap at around $\omega D / v_t = 4.0\text{-}4.8$ where the propagation of sound is prohibited. In the experiments conducted recently, the propagation characteristics of sound waves in the [111] direction have been investigated [2]. So we concentrate our attention to the band structure in this direction (the $\Gamma - \mathrm{L}$ direction in Fig. 2). The modes on the branches overlapped with solid lines (in the low frequency region) have longitudinal polarization. For these longitudinal sound the band gap extends from 3.0 to 5.0 (in the normalized frequency), as determined by the width of gap at the Γ point.

Fig. 2(b) shows the similar band structure for $\chi = 0.04$. In this case the complete band gap is not seen. This is because larger χ means larger channels through which sound waves are transmitted effectively. Hence, coherent backscatterings of sound which induce band gaps become less probable.

Fig. 3 shows the χ dependence of the band structures along the $\Gamma - \mathrm{L}$ direction (the [111] direction). For these results, the vibrations with longitudinal polarization are exclusively selected. Therefore, the transverse branches existing in Fig. 2, the lowest linear branch, for instance, are not present. Now we see the existence of two band gaps, i. e., a wider gap around $\omega D / v_t = 4.0$ and a narrower gap near $\omega D / v_t = 8.0$. The increase of χ results in the reduction of the width of the phononic band gap as clearly seen in this figure. For $\chi \geq 0.06$ these band gaps disappear completely (not shown here). An interesting observation is the fact that the upper edges of the lowest gap are almost independent of χ but the lower edges are drastically shifted upward with increasing χ. That is because the branches below the 1st band gap are due to collective motion of the spheres, while the ones above this gap stem from the vibrations of isolated spheres. The collective motion is more sensitive to the interaction between adjacent spheres.

We also find that there exist a few flat bands below the lowest vibrational frequency of an isolated sphere, which is marked by an arrow in Fig. 2 (a). To make the vibrational motion of spheres at these flat bands clear, we have illustrated in Fig. 4 the vibrational intensities $I = \sum_i |u_i|^2$ at four frequencies shown by red dots in Fig. 2 (a).

At point A ($\omega D / v_t = 2.082$) only the sphere at the center vibrates and almost no oscillation is excited at the rest of four spheres surrounding it. Thus, the vibrational energy is strongly localized. In contrast, at point B ($\omega D / v_t = 2.722$) the spheres at the center hardly vibrate, though the rest of spheres strongly oscillate. At point C ($\omega D / v_t = 3.094$) the vibrational amplitude extends over all of the spheres and hence the branch is sloped, or depends on the wave number. On the other hand, at point D ($\omega D / v_t = 5.230$) just above the lowest band gap and close to the lowest frequency of an isolated sphere, the spheres execute vibrations in a complicated manner.

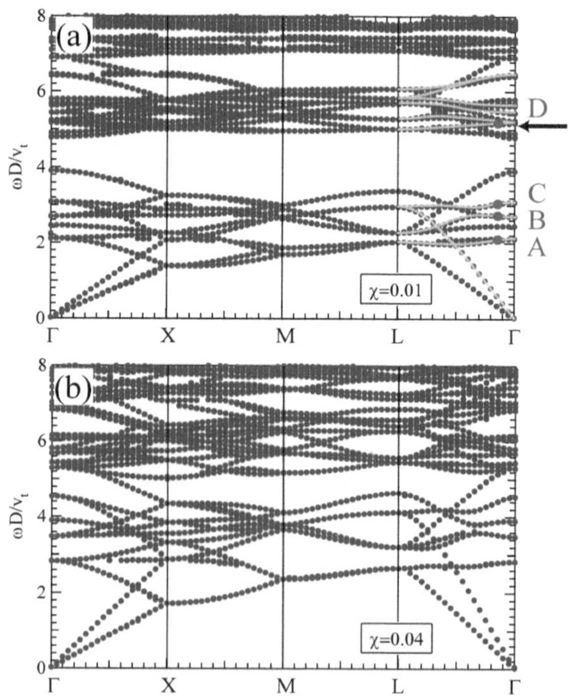

Fig. 2 (color online) Calculated band structures of opals consisting of aluminum spheres with FCC structure along the irreducible Brillouin zone boundaries shown in Fig. 1(b). (a) $\chi = 0.01$ and (b) $\chi = 0.04$. $v_t = 3.12\times10^5$ cm/s is the velocity of transverse sound of aluminum. In Fig. 2(a), the solid lines in the $\Gamma - \mathrm{L}$ region represent the branches with longitudinal polarization and the arrow indicates the position of the lowest eigenfrequency of an isolated sphere.

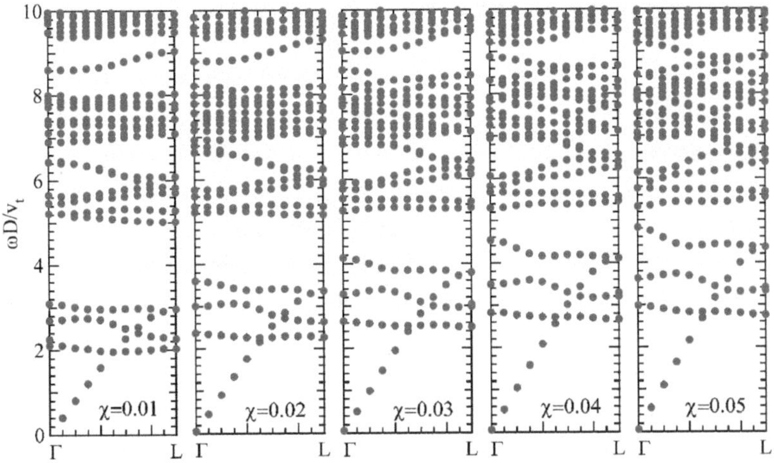

Fig. 3 (color online) Calculated band structures of opals consisting of aluminum spheres with FCC structure along the $\Gamma - \mathrm{L}$ direction for several sintering parameters: $\chi = 0.01$, 0.02, 0.03, 0.04, and 0.05. Hatched regions show the band gaps for longitudinal sound.

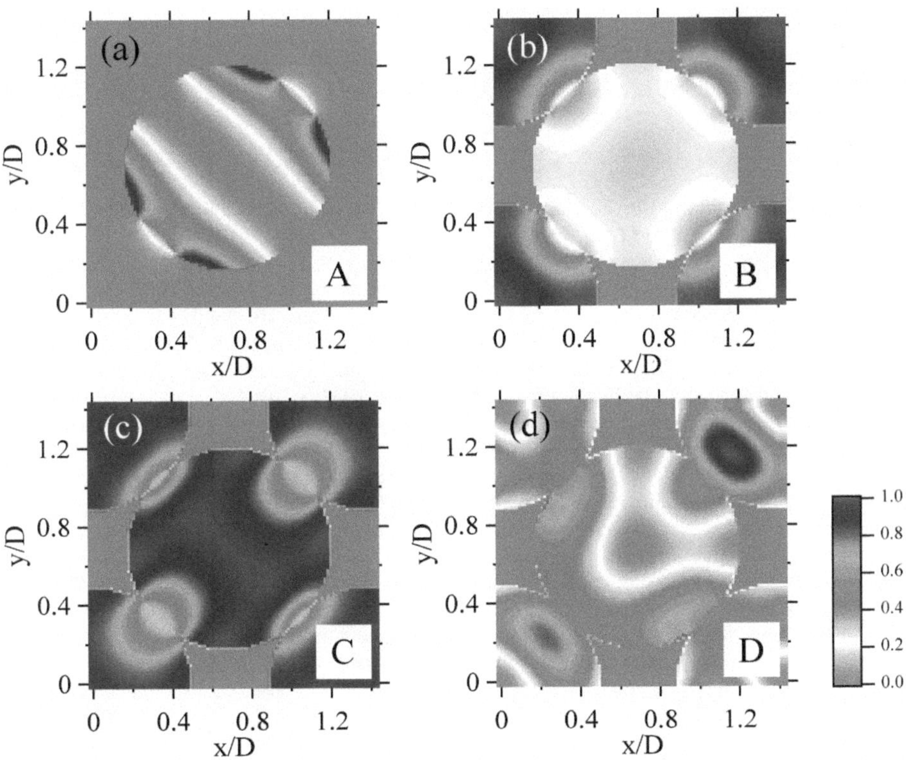

Fig. 4 (color online) Calculated vibrational intensity ($I = \sum_i |u_i|^2$) patterns in the hatched face of Fig. 1(a). Frequencies chosen are those labeled *A* to *D* in Fig. 2(a).

4. Concluding Remarks

We have made a systematic investigation on the band structure of sound waves in three-dimensional phononic crystals consisting of periodic arrays of spheres with an opal structure. The opal structure is constructed by close-packed solid spheres (FCC structure) surrounded by air (or vacuum). We introduced a sintering parameter which characterizes the connection between adjacent spheres. For a sintering parameter below 0.06 we find a complete band gap for only longitudinal sound in the opal structure consisting of aluminum spheres and for both longitudinal and transverse sound waves below $\chi = 0.03$. We also find that several flat bands appear below the lowest vibrational eigenfrequency of an isolated sphere. The vibrations of spheres in those flat bands proved to be strongly localized in a sphere without extending over neighboring spheres.

Aknowledgements: The work at Hokkaido University was supported in part by the CASIO Science Promotion Foundation, the Murata Science Foundation and the Grant-in-Aid for Scientific Research from the Ministry of Education, Science, Sports and Culture of Japan (Grant No.17510106).

References

[1] Gorishnyy, T., Maldovan, M., Ullal, C., and Thomas, E. L.: Phononic crystals go hypersonic. *Phys. World* **18** 24 (2005).

[2] Akimov, A. V., Tanaka, Y., Pevtsov, A. B., Kaplan, S. F., Golubev, V. G., Tamura, S., Yakovlev, D. R., and Bayer, M.:Hypersonic Modulation of Light in Three-Dimensional Photonic and Phononic Band-Gap Materials. *Phys. Rev. Lett.* **101** 33902 (2008).

[3] Still, T., Cheng, W., Retsch, M., Sainidou, R., Wang, J., Jonas, U., Stefanou, N., and Fytas, G.: Simultaneous Occurrence of Structure-Directed and Particle-Resonance-Induced Phononic Gaps in Colloidal Films. *Phys. Rev. Lett.* **100** 194301 (2008).

[4] Kafesaki, M. and Economou, E. N.: Multiple-scattering theory for three-dimensional periodic acoustic composites. *Phys. Rev. B* **60** 11993 (1999).

[5] Psarobas, I. E., Stefanou, N., and Modinos, A.: Scattering of elastic waves by periodic arrays of spherical bodies. *Phys. Rev. B* **62** 278 (2000).

[6] Liu, Z., Chan, C. T., Sheng, P., Goertzen, A. L., and Page, J. H.: Elastic wave scattering by periodic structures of spherical objects: Theory and experiment. *Phys. Rev. B* **62** 2446 (2000).

[7] Psarobas, I. E., Modinos, A., Sainidou, R., and Stefanou, N.: Acoustic properties of colloidal crystals. *Phys. Rev. B* **65** 064307 (2002).

[8] Kushwaha, M. S., Halevi, P., Dobrzynski, L., and Djafari-Rouhani, B.: Acoustic band structure of periodic elastic composites. *Phys. Rev. Lett.* **71** 2022 (1993).

[9] Tanaka, Y., Tomoyasu, Y., and Tamura, S.: Band structure of acoustic waves in phononic lattices: Two-dimensional composites with large acoustic mismatch. *Phys. Rev. B* **62** 7387 (2000).

Bandgap Characteristics of a 3D Phononic Meta Material Composed of Ordered Quantum Dots

Yu-Chieh Wen[1], Tzu-Ming Liu[1], Christian Dais[3], Detlev Grützmacher[3,4], Tzung-Te Chen[5], Yang-Fang Chen[5], Jin-Wei Shi[6] and Chi-Kuang Sun[1,2]

[1]Department of Electrical Engineering and Graduate Institute of Photonics and Optoelectronics, National Taiwan University, Taipei 10617, Taiwan

[2]Research Center for Applied Sciences, Academia Sinica, Taipei 115, Taiwan

[3]Laboratory for Micro- and Nanotechnology, Paul Scherrer Institut, 5232 Villigen, Switzerland

[4]Institute of Bio- and Nanosystems, Forschungszentrum Jülich, 52425 Jülich, Germany

[5]Department of Physics, National Taiwan University, Taipei 10617, Taiwan

[6]Department of Electrical Engineering, National Central University, Jhongli 32054, Taiwan

sun@ee.cc.ntu.edu.tw

Abstract. The authors demonstrated an artificial material composed of ordered semiconductor quantum dots (QDs) with functional acoustic properties. The phononic bandgap of this three-dimensional (3D) artificial crystal was characterized by Raman and transient reflectivity spectroscopies. The Raman spectra exhibited interferometric patterns caused by the coherent interference of acoustic-phonon-related Raman signals, revealing the structural information of the crystal. In addition, the acoustic tunneling in the QD crystal was studied by the transient spectroscopy with the picosecond ultrasonic technique. The measured signals indicated that the QDs are responsible for the scatterings of the acoustic phonons at the bandgap frequency, while the QD crystal serves as a phononic meta material for low-frequency phonons.

1. Introduction

Semiconductor quantum dot (QDs) is sometime called artificial atom due to its unique characteristics for zero-dimensional charge confinement. With a controlled distribution of QDs, those "atoms" have capability to create new artificial materials with designed functionalities for electronic and biological applications [1-2]. In addition, the behaviors of phonons in a QD system are also dominated by the ordering of

T.-T. Wu and C.-C. Ma (eds.), *IUTAM Symposium on Recent Advances of Acoustic Waves in Solids*, IUTAM Bookseries 26, DOI 10.1007/978-90-481-9893-1_19,

QDs [1]. To understand and to manipulate the 3D QD ordering for phononic meta materials can open a lot of new possibilities for the controlling of the phonon-phonon, the electron-phonon, and the phonon-polariton interactions [3].

In the past decade, the dynamics of coherent acoustic phonons in QDs has attracted a lot of attention. Several literatures reported the observations of confined acoustic mode in QDs and also studied the exciton-coherent phonon interactions by femtosecond pump-probe technique [4,5]. A recent report by Devos et. al. demonstrated an efficient generation of coherent acoustic phonons in self-assembled QD layers by laser excitation [6]. Those contributors revealed not only the fundamental interests in this issue but also its potential in phonon engineering [7]. However, the present studies on coherent phonons are restricted to either confined mode in QDs or collected propagating mode from a single QD layer. The structural effect on the coherent phonon dynamics in 3D regularized QDs has not been investigated so far. The purpose of this paper is to demonstrate and discuss the phononic bandgap induced by the ordering of semiconductor QDs using Raman and transient reflection spectroscopies. The measurement results showed a great contribution of ordered QDs to acoustic scatterings at the frequency of the mini-gap, while those ordered QDs also serve together as a phononic meta material for low-frequency acoustic phonons.

2. Sample Structure

The studied sample consists of 3D ordered SiGe QDs (QD crystal [3]) grown by templated self-organization. The templates were fabricated by extreme ultraviolet interference lithography using a wavelength of 13.5 nm [8]. Reactive ion etching was employed to transfer the two-dimensional hole arrays with a periodicity of 90nm x 100nm into the Si(001) substrate. This prepatterned substrate was then deposited by a sequence of Ge and Si layers for fabricating a stack of 11 QD layers. It is noted that the growth condition of the first QD layer (7ML) on the substrate is different from the other layers, resulting in a relatively large dot size in the first layer. Each subsequent island layer was grown by depositing 5 ML of Ge on top of the 10 nm Si spacer layer. The lateral and vertical periodicities of QDs were confirmed by TEM and X-ray analyses. In addition, AFM investigations indicated that the height and diameter of QDs are 2.96 ± 0.34 nm and 34.21 ± 2.98 nm, respectively. Furthermore, analysis of the x-ray diffraction measurements shows that the SiGe dots have a Ge content of 60% [9]. The sample has a 100 nm-thick Si cap layer serving as a propagation layer for separating detected acoustic signals temporally. The details of fabrication and characterization of the studied QD crystal has been reported elsewhere [3,10].

3. Experimental Results and Discussions

3.1 Raman Measurement

We first performed low-frequency resonant Raman scattering measurement on the QD crystal using an Argon laser ($\lambda = 512$ nm) and a Jobin Yvon T64000 spectrometer. The measured Raman spectrum shown in Fig. 1 exhibits a clear interferometric pattern due to the ordering of QD crystal. It is caused by the coherent interference of acoustic-phonon-related Raman signals originated from the electrons in different QDs. These interferences depend on the electronic confinement within QD (form factor) and on the relative QD positions (structure factor) [1]. The form factor determines the spectral envelope. In the studied case, the weaker spatial confinement of electrons on the lateral dimensions results in a narrower momentum distribution and thus a restricted Raman spectral envelope. With a limitation on low-frequency resolution, the Raman measurement reveals the information of ordered vertical arrangement of the QDs. The structure factor determines the oscillation period and contrast, which is proportional to [1]

$$H^2(q_z + \Delta k_z) + H^2(q_z - \Delta k_z) - 2H(q_z + \Delta k_z)H(q_z - \Delta k_z)\cos[2q_z(t + (N-1)d/2)] \tag{3.1.1}$$

where q_z is the wavevector of the phonons, $\Delta k_z = 4n\pi/\lambda$ (n is the refractive index of silicon) is the momentum change of Raman scattering, t is the thickness of the cap layer, d is the spacing between QD layers, and N is the number of layers. H is the usual interference function,

$$H(Q) = \frac{\sin(\frac{Nd}{2}Q)}{\sin(\frac{d}{2}Q)}. \tag{3.1.2}$$

Taking $t = 100$ nm, $N = 11$, and $n = 4.23$ [11], the structure factor can be calculated with a fitting parameter $d = 10.97 \pm 0.16$ nm, as also shown in Fig. 1. Here the average sound velocity used in the calculation is determined by the literature values of sound velocities [12] and the relative thickness of the Si spacer and the $Si_{0.4}Ge_{0.6}$ island layer. The vertical periodicity estimated by Raman measurement agrees well with the observations in TEM imaging and is crucial structural information for calibrating contributions of SiGe alloys to phonon scatterings, which will be discussed in details as follows.

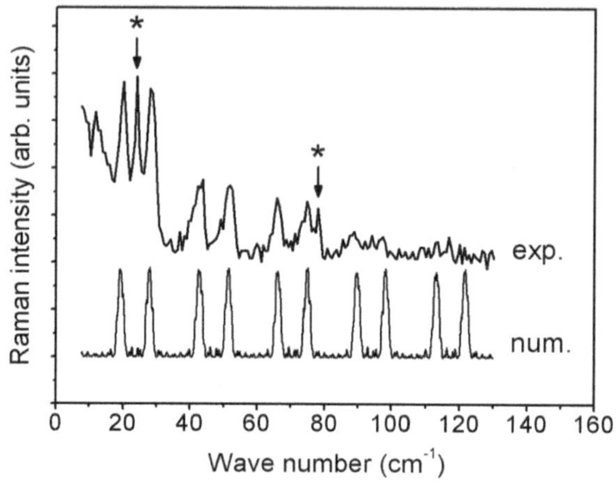

Fig. 1 Low-frequency Raman spectrum of the QD sample. The bottom line is the calculated structure factor with a fitting parameter of d = 10.97 ± 0.16 nm, which was vertically shifted for clarity. The starred peaks are artifacts induced by the plasma absorption of the Argon laser.

3.2 Transient Reflectivity Measurement

To study the propagation properties of coherent acoustic phonons in the QD crystal, the picosecond ultrasonic technique [13] was then adopted with a femtosecond pump-probe system. The light source was a mode-locked Ti:sapphire laser with ~150 fs pulsewidth. Both optical pump and probe beams had the same wavelength of 780 nm and were focused onto the sample with a spot diameter of ~25 μm. The pumping fluence was ~200 μJ/cm^2, while the fluence of the probe pulse was one order lower than the optical pump. This measurement was also done at room temperature.

To generate and detect coherent acoustic phonons, an additional ~17 nm-thick Al film was coated on the sample surface. As we shined the sample with optical pump pulses, the transient temperature rise in the Al film induced thermal stress, which launches coherent acoustic phonons toward the QD crystal. The resultant coherent phonons have a broad-bandwidth with a central frequency determined by the thickness and sound velocity of the Al film [14]. When they propagates in the QD crystal, parts of phonons were scattered by the ordered QDs. The reflected acoustic phonons were detected by the optical probe pulse through the Al film again, while the transmitted ones can also be found in the measured transient signals due to backward Brillouin scattering [13].

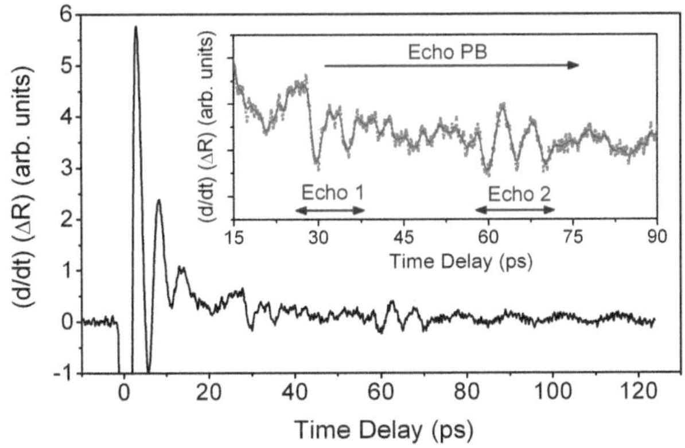

Fig. 2 Differential reflection changes of the QD crystal as a function of time delay. Differentiation of the data is helpful to eliminate slowly varied background signal and thus clarify the acoustic dynamics. The inset is the magnified data for observing the features of acoustic echos clearly, where the red line is a typical fitting curve.

Figure 2 shows the differential reflection changes as a function of time delay. To analyze the oscillation components, we performed Fourier transform for the differential data after 20 ps, as shown in Fig. 3. The spectrum obviously indicates three oscillation features (79, ~200, 344 GHz), which can be further divided into two parts according to their origins. The oscillation component with the lowest frequency (79.35 GHz) is contributed from the acoustic propagation in silicon. This frequency is determined by the momentum conservation between photons and phonons, and can be formulated as $f = 2nV\cos\theta / \lambda$, where n (λ= 780 nm) = 3.69 [11], V = 8440 m/s is the sound velocity of Si [12], and $\theta \sim 0^{\circ}$ is the incident angle of the optical probe inside the material. With the reported material parameters, f was calculated to be 79.85 GHz, agreeing well with our observations. The backward Brillouin oscillations can also be seen in the time-domain data clearly after a time delay of ~90 ps.

Besides the Brillouin signals, the other oscillation features are attributed to the detection of coherent acoustic phonons by the Al film, which mainly occurs before ~80 ps. From Fig. 2, we can find that the Al film vibrated at its natural frequency after optical excitation. It results in initial oscillations of ~200 GHz appearing at 0-20 ps. The consequent acoustic phonons then propagated toward the QD crystal, and the "acoustic echo" can be observed at a delayed time after ~25 ps, corresponding to the round trip time of the Si cap layer. The echo signals are emphasized in the inset of Fig. 2. From this figure, two high-amplitude acoustic echos following with a train of high-frequency oscillations can be found. They are denoted by Echos 1, 2, and PB individually.

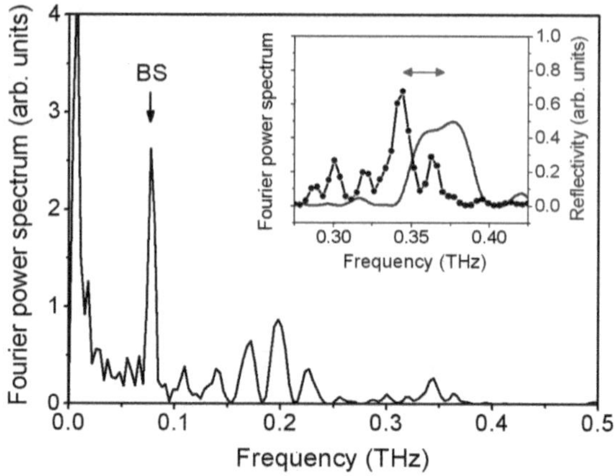

Fig. 3 Fourier power spectrum for the transient data (Fig. 2) after 20 ps. The initial data was neglected in order to avoid the influence of huge oscillations near time zero on the data analysis. The splitting of spectral peaks near 200 GHz is due to multiple echos which were separated temporally. The inset shows the magnified power spectrum near 350 GHz and the calculated acoustic reflectivity with the reduced model.

Echos 1 and 2 have an oscillation frequency of ~200 GHz. According to their arrival times at which the acoustic echos are sensed by the Al film, Echos 1 and 2 originate from the reflection at the top and the bottom QD layers in the QD crystal, respectively. The wavelength of 200 GHz acoustic phonons is 42 nm, which is ~4 time longer than the vertical periodicity of the QD layers. It means that 200 GHz is well below the first phononic bandgap of the QD crystal along the direction of phonon propagation. In addition, if we used a simplified 1D model (Rytov model [15]) to investigate phonon dispersion along the growth direction by considering uniform SiGe alloy layers, the calculated curve (not shown here) exhibited a dispersion-free feature with an average sound velocity at ~200 GHz. Consequently, the QD crystal can be accounted a homogeneously effective medium (phononic meta material) for 200 GHz acoustic phonons propagating along the direction of the sample growth. From this viewpoint, the excited 200 GHz phonons would experience only two interfaces: top and bottom boundaries of the QD crystal, so that two echos can be seen.

Echo PB seems to be followed by Echo 1 so that it can be explained by the multiple scatterings of coherent phonons in the QD crystal. With the calculated dispersion relation for the QD crystal, Echo PB was attributed to the acoustic reflection at the first zone-edge mini-gap of the reduced Brillouin zone. Besides QDs, the wetting layers also scatter acoustic phonons and could thus open phononic bandgaps in an actual QD crystal. To characterize the contributions of different scattering elements, we then compared the measured frequency of Echo PB with the calculated reflection spectrum, as shown in the inset of Fig. 3. The reflectivity was calculated by using transfer matrix method [7] with considerations of the multiple wetting layers, but without the existence of the QDs. This reduced model was used to examine the effect of the wetting layers on acoustic scatterings. The vertical periodicity of QD layers used in the calculation was obtained from the Raman measurement. From the inset of Fig. 3, the frequency of the spectral peak of the calculated reflectivity was found to be different from the experimental observations (~70 % of the calculated spectral width). This notable discrepancy thus shows that the wetting layers are not the only scattering elements for 344 GHz acoustic phonons, i.e. other scattering origins should be considered. To include the effect of layered QDs, we applied an additional $Si_{0.4}Ge_{0.6}$ layer to the top of each wetting layer. It was regarded as an effective medium instead of laterally distributed dots. By fitting the spectrum of Echo PB, the thickness of the effective layers is 1.7-2.1 nm, which is twice thicker than the wetting layer and is ~60 % of the height of QDs. Our 1D calculation, therefore, indicated that besides the wetting layers, the QDs could play crucial role in the scatterings of coherent acoustic phonons at the mini-gap frequencies.

4. Conclusions

Raman and transient reflectivity spectroscopies were performed for studying the propagation properties of coherent acoustic phonons in a QD crystal. The ordering of QDs resulted in not only the interferometric patterns in the Raman spectrum but also acoustic echos at the frequency of phononic mini-gaps. By comparing those measurements, the structure information was extracted, and the contribution of QDs to the phonon scatterings can thus be characterized. Our demonstration thus shows future possibility for semiconductor QDs to create artificial nanocrystal with functional acoustic and thermal properties.

Acknowledgements: The authors wish to acknowledge J. Groenen at CNRS-Université Paul Sabatier for data analysis, and also thank J.-H. Sun, J.-C. Hsu, and Prof. T.-T. Wu at National Taiwan University for stimulating scientific discussion. This work was sponsored by the National Science Council of Taiwan under Grant No. NSC-97-2120-M-002-014. One of the authors (Y.-C. Wen) would like to thank the MediaTek Fellowship for financial support.

References

[1] Cazayous, M., Groenen, J., Zwick, A., Mlayah, A., Carles, R., Bischoff, J. L., and Dentel D.: Resonant Raman scattering by acoustic phonons in self-assembled quantum-dot multilayers: From a few layers to superlattices. *Phys. Rev. B* **66** 195320 (2002).

[2] Friesen, M., Rugheimer, P., Savage, D. E., Lagally, M. G., van der Weide, D. W., Joynt, R., and Eriksson, M. A.: Practical design and simulation of silicon-based quantum-dot qubits. *Phys. Rev. B* **67** 121301 (2003).

[3] Grützmacher, D., Fromherz, T., Dais, C., Stangl, J., Müller, E., Ekinci, Y., Solak, H. H., Sigg, H., Lechner, R. T., Wintersberger, E., Birner, S., Holý, V., and Bauer, G.: Three-dimensional Si/Ge quantum dot crystals. *NanoLett.* **7** 3150 (2007).

[4] Krauss, T. D. and Wise, F. W.: Coherent acoustic phonons in a semiconductor quantum dot. *Phys. Rev. Lett.* **79** 5102 (1997).

[5] Cerullo, G., Silvestri, S. D., and Banin, U.: Size-dependent dynamics of coherent acoustic phonons in nanocrystal quantum dots. *Phys. Rev. B* **60** 1928 (1999).

[6] Devos, A., Poinsotte, F., Groenen, J., Dehaese, O., Bertru, N., and Ponchet, A.: Strong generation of coherent acoustic phonons in semiconductor quantum dots. *Phys. Rev. Lett.* **98** 207402 (2007).

[7] Lanzillotti-Kimura, N. D., Fainstein, A., Balseiro, C. A., and Jusserand, B.: Phonon engineering with acoustic nanocavities: Theoretical considerations on phonon molecules, band structures, and acoustic Bloch oscillations. *Phys. Rev. B* **75** 024301 (2007).

[8] Solak, H. H.: Nanolithography with coherent extreme ultraviolet light. *J. Phys. D* **39** R171 (2006).

[9] Holy, V., Stangl, J., Fromherz, T., Lechner, R. T., Wintersberger, E., Bauer, G., Dais, C., Müller, E., Grützmacher, D.: X-ray diffraction investigation of a three-dimensional Si/SiGe quantum dot crystal. *Phys. Rev. B* **79** 035324 (2009).

[10] Dais, C., Solak, H. H., Ekinci, Y., Müller, E., Sigg, H., Grützmacher D.: Ge quantum dot molecules and crystals: Preparation and properties. *Surf. Sci.* **601** 2787 (2007).

[11] Philipp, H. R. and Taft, E. A.: Optical constants of silicon in the region 1 to 10 eV. *Phys. Rev.* **120** 37 (1960).

[12] Mechanical properties of Si and Ge were taken from Lockwood, D. J., Dharma-wardana, M. W. C., Baribeau, J.-M., and Houghton, D. C.: Folded acoustic phonons in Si/GexSi1-x strained-layer superlattices. *Phys. Rev. B* **35** 2243 (1987); Properties of aluminum were taken from Ref. 11, which was used for estimating the effect of the cap layer on the acoustic reflectivity.

[13] Devos, A. and Louarn, A. L.: Strong effect of interband transitions in the picosecond ultrasonics response of metallic thin films. *Phys. Rev. B* **68** 045405 (2003).

[14] Lin, H.-N., Stoner, R. J., Maris, H. J., and Tauc, J.: Phonon attenuation and velocity measurements in transparent materials by picosecond acoustic interferometry. *J. Appl. Phys.* **69** 3816 (1991).

[15] Rytov, S. M., Akust, Zh., *Sov. Phys. Acoust.* **2** 68 (1956).

Analytical and Experimental Analysis of Bandgaps in Nonlinear one Dimensional Periodic Structures

Nicholas Boechler[1], Chiara Daraio[1], Raj K Narisetti[2], M. Ruzzene[2] and M. J. Leamy[3]

[1]Aeronautics and Applied Physics, California Institute of Technology, Pasadena, CA, USA

[2]Department of Aerospace Engineering, Georgia Institute of Technology, Atlanta, GA, USA

[3]Department of Mechanical Engineering, Georgia Institute of Technology, Atlanta, GA, USA

massimo.ruzzene@aerospace.gatech.edu

Abstract. Wave propagation characteristics of nonlinear one-dimensional periodic structures are investigated analytically, numerically and experimentally. A novel perturbation analysis is first applied to predict the band gap location and extent in terms of linear and nonlinear system parameters. Approximate closed-form expressions capture the effect of nonlinearities on dispersion and depict amplitude dependent cut-off frequencies. The predictions from the perturbation analysis are verified through numerical simulations of harmonic wave motion. Results indicate the possibility of input amplitude as a tuning parameter through which cut-off frequencies can be adjusted to achieve filtering properties over selected frequency ranges. A periodic diatomic chain of stainless steel spheres alternating with aluminium spheres is experimentally investigated. The dynamic behavior of the chain is governed by Hertzian interaction of spheres and by a compressive pre-load which can be adjusted to obtain linear, weakly nonlinear and highly nonlinear behavior. For a weakly nonlinear case, preliminary results in experiments show the tendency for a shift in the band gap edges by varying input amplitude. The paper is a work in progress, for which the experimental results for a weakly nonlinear system are interpreted by the perturbation analysis developed for a specific case of linear and nonlinear power law interaction of exponent 3/2.

T.-T. Wu and C.-C. Ma (eds.), *IUTAM Symposium on Recent Advances of Acoustic Waves in Solids*, IUTAM Bookseries 26, DOI 10.1007/978-90-481-9893-1_20,

1. Introduction

The study of wave propagation in linear periodic structures of infinite extent has shown that for harmonic oscillation, these structures admit propagating and attenuating wave solutions depending on the frequency of the motion. This includes the property of "Stop bands" where the waves attenuate allow periodic structures to behave as acoustic filters, wave-guides, resonators, etc. A detailed study on techniques to analyze continuous linear periodic structures is given by Mead [1]. Under the assumption of linearity, periodic structures demonstrate interesting propagation properties; however, nonlinear periodic structures may exhibit enhanced wave transmission behavior. Localization and wave characteristics in nonlinear one dimensional elastic media have been studied by invoking continuum approximation [2-5]. The concept of nonlinear mode was introduced in [6] and was characterized by presenting periodic motions with all coordinates passing equilibrium points as well as extremum points simultaneously. Apart from continuous nonlinear systems, discrete nonlinear periodic systems, for example composed by granular chains of spheres [7-10], have also been subjects of great interest. Significant wave speed tunability has been demonstrated experimentally by varying the induced pre-load on one-dimensional nonlinear phononic crystals formed by chains of viscoelastic PTFE (polytetrafluoroethylene) and elastic (stainless-steel) beads [11].

There exist a number of analytical methods to study wave propagation in nonlinear periodic structures such as nonlinear mapping [12] and harmonic balance [13]. A number of perturbation techniques have been documented as well, such as the Modified Lindstedt Poincaré method [14], homotopy perturbation [15] and several other analysis methods are described in detail by Nayfeh and Mook [16]. In this paper, a straight-forward perturbation analysis [17] similar in spirit to the Lindstedt Poincaré technique is applied to a discrete one-dimensional nonlinear periodic chain with dynamics governed by a Hertzian contact model to obtain closed-form amplitude dependent dispersion relations. Next, the focus is subsequently placed on experimental validation of this amplitude dependent dispersion behavior. A diatomic chain consisting of aluminum spheres alternating with stainless steel spheres with a lever actuated preload system and piezoelectric exciter is setup and tested. The paper then concludes summarizing the main findings of this work and presenting future directions of research.

2. Analytical Analysis of Dispersion

2.1 Equations of Motion

Consider a one-dimensional chain of identical masses subject to initial pre-load, which is a constant compressive force F_0 applied at the boundaries [8]. This reduces the distance between the mass centers by a magnitude δ_0, which is equal to the initial displacement of the mass centers. Figure 2.1 describes the schematic of the chain. The displacement of the ith mass from its equilibrium position is defined by coordinate u_i and positive direction is in the direction of wave travel. Hertz's interaction law models the force between two masses by $F \propto (\delta_0 + \delta_d)^{3/2}$, where δ_d is relative displacement between the masses. Assuming the Hertz contact force law, the equation of motion governing the ith mass is given by [8],

$$\ddot{u}_i = A(\delta_0 + u_{i-1} - u_i)^{3/2} - A(\delta_0 + u_i - u_{i+1})^{3/2}, \quad N-1 \geq i \geq 2 \tag{2.1}$$

$$A = \frac{E(2R)^{1/2}}{3(1-\upsilon^2)m} \tag{2.2}$$

where m is the mass of the sphere, R is the radius of the sphere, υ is the Poisson's ratio and E is the elastic modulus.

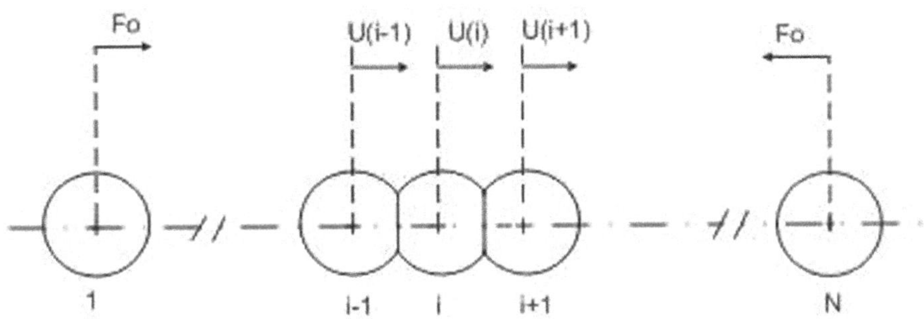

Fig. 2.1 A schematic of pre-loaded one-dimensional chain with Hertz force interaction model [8].

2.2 Perturbation Analysis

Described next is a perturbation analysis applied to the system governed by equations of motion given by Eq. (2.1). A more general approach of this novel perturbation procedure applied to discrete systems is documented by Narisetti, Leamy and Ruzzene [16] wherein various one-dimensional unit cell configurations are examined and closed-form dispersion relations are obtained. Eq. (2.1) is transformed by Taylor series expansion more suitable for perturbation analysis. It is assumed that [10]

$$\frac{|u_{i-1} - u_i|}{\delta_0} << 1. \tag{2.3}$$

With the Taylor series expansion, under the assumption of Eq. (2.3), Eq. (2.1) transforms into the following form,

$$\begin{aligned} \ddot{u}_i &= \alpha\left(u_{i-1} + u_{i+1} - 2u_i\right) + \beta\left(\left(u_{i-1} - u_i\right)^2 - \left(u_i - u_{i+1}\right)^2 \right) \\ &\quad + \gamma\left(\left(u_{i-1} - u_i\right)^3 - \left(u_i - u_{i+1}\right)^3 \right) \\ \alpha &= \frac{3A\delta_0^{1/2}}{2m},\ \beta = \frac{3A\delta_0^{-1/2}}{8m},\ \gamma = \frac{-A\delta_0^{-3/2}}{16m} \end{aligned} \tag{2.4}$$

Introducing non-dimensional time $\tau = \omega t$ and substituting the following displacement and frequency expansions about linear solution into Eq. (2.4),

$$\begin{aligned} \omega &= \omega_0 + \varepsilon\omega_1 + \varepsilon^2\omega_2 + O(\varepsilon^3) \\ u_i &= \varepsilon u_i^{(0)} + \varepsilon^2 u_i^{(1)} + \varepsilon^3 u_i^{(2)} + O(\varepsilon^4) \end{aligned} \tag{2.5}$$

The following ordered equations are obtained,

$$\varepsilon^0 : \omega_0^2 \frac{d^2 u_i^{(0)}}{d\tau^2} + \alpha\left(2u_i^{(0)} - u_{i-1}^{(0)} - u_{i+1}^{(0)}\right) = 0 \tag{2.6}$$

$$\begin{aligned} \varepsilon^1 : \omega_0^2 \frac{d^2 u_i^{(1)}}{d\tau^2} + \alpha\left(2u_i^{(1)} - u_{i-1}^{(1)} - u_{i+1}^{(1)}\right) &= -2\omega_0\omega_1 \frac{d^2 u_i^{(0)}}{d\tau^2} + \\ \beta\left(u_{i+1}^{(0)} + u_{i-1}^{(0)} - 2u_i^{(0)}\right)\left(u_{i-1}^{(0)} - u_{i+1}^{(0)}\right) & \end{aligned} \tag{2.7}$$

$$\varepsilon^2 : \omega_0^2 \frac{d^2 u_i^{(2)}}{d\tau^2} + \alpha\left(2u_i^{(2)} - u_{i-1}^{(2)} - u_{i+1}^{(2)}\right) = -2\omega_0\omega_1 \frac{d^2 u_i^{(1)}}{d\tau^2} - \left(\omega_1^2 + 2\omega_0\omega_2\right)\frac{d^2 u_i^{(0)}}{d\tau^2}$$
$$+\beta\left(u_{i-1}^{(0)} + u_{i+1}^{(0)} - 2u_i^{(0)}\right)\left(u_{i-1}^{(1)} - u_{i+1}^{(1)}\right) + \beta\left(u_{i-1}^{(1)} + u_{i+1}^{(1)} - 2u_i^{(1)}\right)\left(u_{i-1}^{(0)} - u_{i+1}^{(0)}\right)$$
$$+\gamma\left(\left(u_{i-1}^{(0)} - u_i^{(0)}\right)^3 - \left(u_i^{(0)} - u_{i+1}^{(0)}\right)^3\right) \tag{2.8}$$

Next, a plane wave solution is imposed to yield the following generator solutions and the distance between two adjacent cells is $a = 2R - \delta_0$,

$$u_i^{(0)} = \frac{A_0}{2} e^{j(\kappa a i)} e^{j\tau} + \frac{\bar{A}_0}{2} e^{-j(\kappa a i)} e^{-j\tau}$$
$$u_{i\pm1}^{(0)} = \frac{A_0}{2} e^{j(\kappa a (i\pm1))} e^{j\tau} + \frac{\bar{A}_0}{2} e^{-j(\kappa a (i\pm1))} e^{-j\tau} \tag{2.9}$$

The linear dispersion relation is obtained by substituting Eq. (2.9) into Eq. (2.6) [10],

$$\omega_0 = \sqrt{2\alpha\left(1 - \cos\left(\kappa a\right)\right)} \tag{2.10}$$

The updated ε^1 order equation after substitution of Eq. (2.9) into Eq. (2.7) is given by,

$$\varepsilon^1 : \omega_0^2 \frac{d^2 u_i^{(1)}}{d\tau^2} + \alpha\left(2u_i^{(1)} - u_{i-1}^{(1)} - u_{i+1}^{(1)}\right) = c_1 e^{j(\kappa a i)} e^{j(\tau)} + c_2 e^{j(2\kappa a i)} e^{j(2\tau)} + c.c. \tag{2.11}$$

where $c.c.$ denotes complex conjugate of all the preceding terms and the coefficients c_1 and c_2 are given by,

$$c_1 = 2\omega_0\omega_1, \quad c_2 = j\left(\frac{\beta A_0^2}{2}(\sin(2\kappa) - 2\sin(\kappa))\right) \tag{2.12}$$

The homogenous part of Eq. (2.11) is similar to Eq. (2.6) and it can be shown that all the terms in the forcing function (i.e. RHS of Eq. (2.11)) with spatial-temporal form of $e^{j(\kappa a i)} e^{j\tau}$ yield an unbounded solution for $u_i^{(1)}$ and therefore need to be eliminated (secular) [17]. By setting $c_1 = 0$, the first order correction of frequency $\omega_1 = 0$. But the remaining forcing in Eq. (2.11) always occurs at $e^{j(2\kappa a i)} e^{j2\tau}$ and hence a particular solution for $u_i^{(1)}$ is found at this order for a subsequent substitution in the next ordered equation to find second order correction to the frequency.

$$u_i^{(i)} = A_1 e^{j(\kappa a i)} e^{j(\tau)} + B_1 e^{j(2\kappa a i)} e^{j(2\tau)} + c.c.$$
$$B_1 = \frac{-c_2}{4\omega_0^2 + \omega_s^2}, \ \omega_s = \sqrt{2\alpha(\cos(2\kappa a) - 1)} \tag{2.13}$$

By substituting Eq. (2.12) and Eq. (2.9) into Eq. (2.8), the following equation results,

$$\varepsilon^2 : \omega_0^2 \frac{d^2 u_i^{(2)}}{d\tau^2} + \alpha\left(2u_i^{(2)} - u_{i-1}^{(2)} - u_{i+1}^{(2)}\right) = d_1 e^{j(\kappa a i)} e^{j(\tau)} + d_3 e^{j(3\kappa a i)} e^{j(3\tau)} + c.c. \tag{2.14}$$

Observing that the linear kernel of Eq. (2.14) is similar to Eq. (2.6) and that forcing term $e^{j(\kappa a i)} e^{j\tau}$ yields an unbounded solution for $u_i^{(2)}$, d_1 is set to zero which yields an expression for frequency correction at second order.

2.3 Dispersion Relations

The closed form expression for frequency correction at second order in terms of linear and nonlinear parameters is given by,

$$\omega_2 = \frac{|A_0|^2 ((3\gamma\alpha - 2\beta^2)\cos(\kappa a)^2 - 6\gamma\alpha\cos(\kappa a) + 2\beta^2 + 3\gamma\alpha)}{2\alpha\omega_0} \tag{2.15}$$

And $\omega = \omega_0 + \varepsilon^2 \omega_2$ where ω_0 and ω_2 are given by Eq. (2.10) and Eq. (2.15) respectively. From Eq. (2.4), it is observed that the nonlinearity is present in quadratic as well as in the cubic form. The quadratic nonlinearity is hardening while the cubic is softening. It is seen that hardening stiffness increases the cutoff frequencies while the softening decreases the cutoff frequencies with increasing wave amplitude [16]. Figure (2.1a) depicts the dispersion trend predicted by perturbation analysis with change in amplitude for a generic Hertzian-modeled periodic chain.

Figure (2.1b) depicts dispersion trends of a nonlinear Hertzian diatomic chain and the closed from expressions were generated by applying similar perturbation approach as explained above. Figure (2.2) shows the correction to lower and upper band gap limits with varying amplitude of the wave (assuming weakly nonlinear behavior throughout).

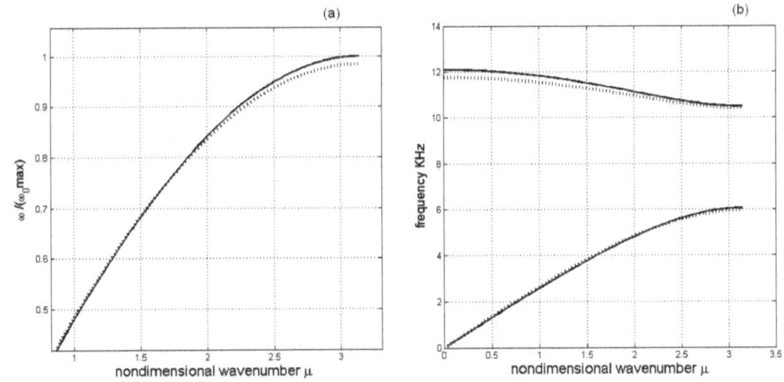

Fig. 2.1 (a) Monoatomic Chain (Hertzian) (b) Diatomic Chain (Hertzian) Dispersion vs. amplitude. ——— Low amplitude — — — · High Amplitude.

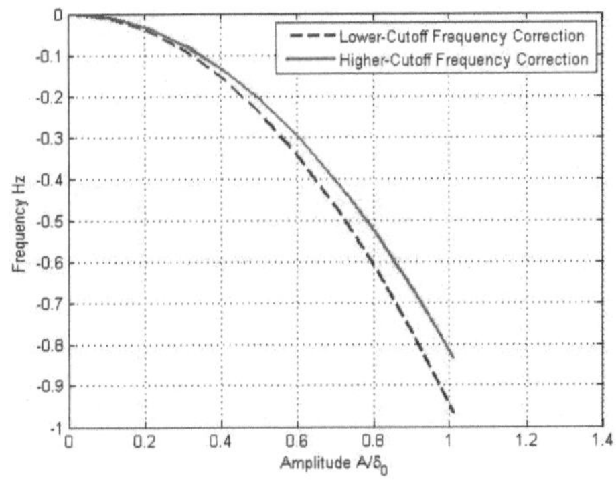

Fig. 2.2 Upper and lower band-gap corrections with changing amplitude of the wave in a diatomic chain consisting of Aluminium and Stainless steel spheres.

3 Experimental Results

3.1 Experiment Setup

We constructed an experimental setup to test the presence and tunability of acoustic band gaps in heterogeneous periodic (dimer) granular systems, and the effects of the nonlinearity in the overall acoustic response [15]. We assembled a one dimensional chain of 81 spherical beads composed of alternating aluminum (6061-T6, McMaster-Carr) and stainless steel (316-type, McMaster-Carr) in a horizontally mounted holder. We applied a fixed static precompression (F_0=20 N) to the chain to linearize the otherwise fully nonlinear dynamic response of the system. The aluminum beads were 19.05 mm in diameter, with an elastic modulus E_{Al}=68.9 GPa, poisson ratio ν_{Al}=0.35 (Matweb, Aluminum 6061-T6; 6061-T651), and mass m=9.73 g. The steel beads were similarly 19.05 mm in diameter, E_s=193 GPa, ν_s=0.3 (Matweb, 316 Stainless Steel, annealed bar), and m=29.1 g. The chain of beads was kept in a one-dimensional configuration by four polycarbonate restraining bars (12.7 mm in diameter) that were held in place by a 12.7 mm thick polycarbonate guide plates placed approximately at 168 mm intervals down the length of the chain.

The chain was pre-compressed between two "walls" placed at the two ends of the system. Precompression was added by an actuated lever arm placed at one end of the chain that transfers force from a weight hanging on a horizontal bar mounted in the horizontal direction (Fig. 1(a,b)). The chain was excited at the desired frequencies by a P-820.10 (Physik Instrumente) preloaded piezo-stack actuator placed at the other end of the chain. The chosen actuator has a resonant frequency of 22 kHz, maximum displacement to 45 m, and maximum force up to 50 N. We drove the actuator with a sine function voltage signal from an Agilent 33220A signal generator passed through a Piezo Systems Inc. EPA-104 linear amplifier. The AC voltage post-amplifier was read by a standard voltmeter.

The evolution of the force-time history of the excitations propagating in the dimer system was visualized by means of periodically placed piezo-sensors, sandwiched inside selected particles. We used 15 mm diameter, 0.9 mm thick discs (StemInc) with a resonant frequency of 2.2 MHz. These piezo-sensors were glued between two hemispheres with carved lodging for the wires preserving the diameter (and mass) of the original bead [9]. Output was read by two synched Tektronix TDS 2024B oscilloscopes. The sensors were placed approximately 16 particles apart from each other.

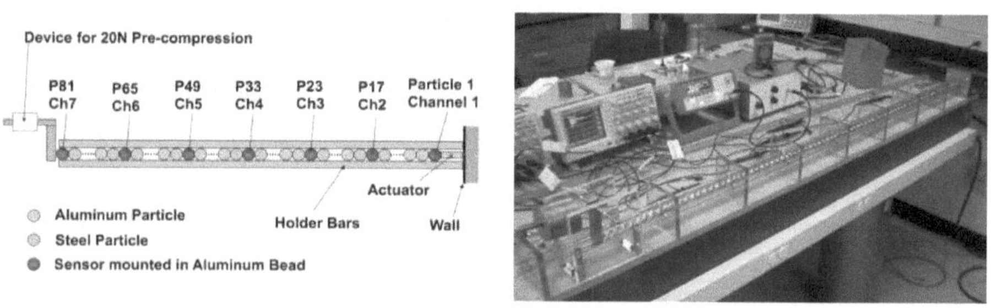

(a) (a)

Fig. 3.1 (a) Schematic diagram and (b) Digital image of experimental setup.

3.2 Results

To observe the change of the band edges in a dimer system as a function of input dynamic force amplitude, we fixed the input voltage of the actuator (keeping the system under constant static precompression) and burst a sine signal at a fixed number of cycles for a set frequency. We recorded the output at each of the sensors for the different cycles. This process was iterated through the frequency range of the predicted band gap with increasing resolution near the predicted band edges, and repeated multiple times for different input voltages (and therefore different driving forces).

Tests were run with 10-cycle and 5000-cycle bursts, and the maximum force was recorded for each sensor. Running the short 10-cycle burst tests allowed the wave packets at selected frequencies to travel across the entire length of the chain without interacting with the walls (and being affected by reflections). The 5000-cycle test was intended to study the effects of a quasi-continuous excitation. Data was recorded 20 ms after turning on the actuator. Accordingly, the measured signal involves reflections from both boundaries, and the excitation of the system as a whole. Figure 3.2 shows the force ratio transmitted between the particle #34 and the first bead (next to the actuator) during a 5000-cycle test as a function of frequency and input voltage. Figure 3.3 shows a result of force ratio transmitted through sensors at Particles 17 / Particle 1 for a 10 cycle test. A small shift of the band edges frequency is reported and a sharp transition can be seen as the input amplitude reaches 50 V. This value corresponds to a driving force amplitude of ~ 23.5 N, comparable to the level of the precompressive force. In this regime, the dynamic response of the system is expected to be in the strongly nonlinear regime, which is not covered by the assumptions made in the perturbation analysis followed for this study.

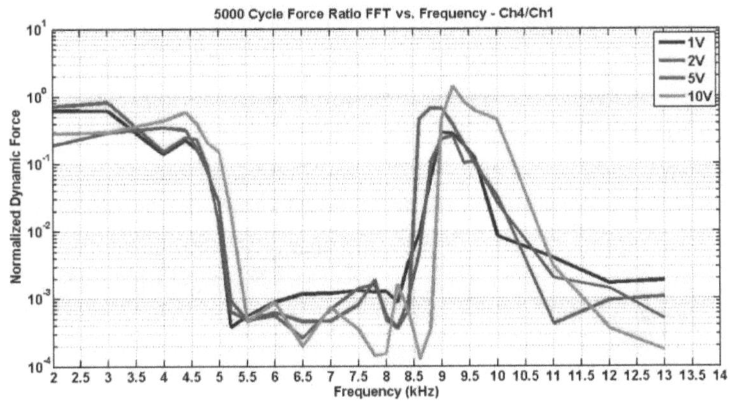

Fig. 3.2 Transmission diagram for a 5000-cycle burst test showing the output force ratio for bead #33 (counting from the actuator) normalized over the bead nearest to the actuator. The transmission and stop bands (acoustic and optical bands and the band gap in between) are plotted for different driving force amplitudes.

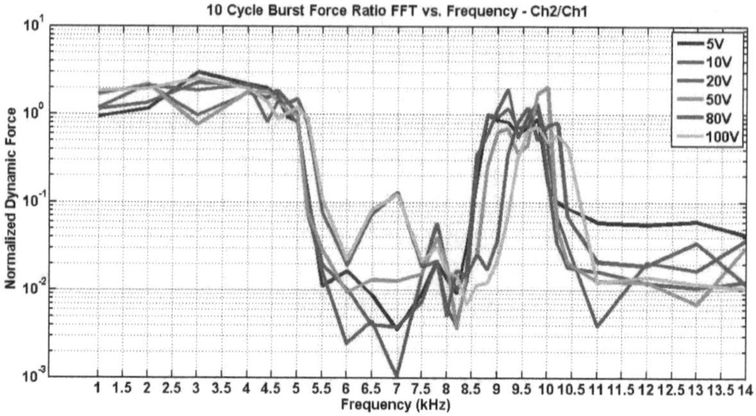

Fig. 3.3 Transmission diagram for a 10-cycle burst test showing the output force ratio for the bead #17 (counting from the actuator) normalized over the bead nearest to the actuator. The transmission and stop bands are plotted for different driving force amplitudes. A distinct transition can be seen around a driving voltage of 50V, which corresponds to an input amplitude being approximately equal to the precompression force (~20 N).

4. Conclusions

A novel perturbation procedure suitable for nonlinear discrete periodic systems has been applied to obtain approximate dispersion relation. Amplitude dependent dispersion (in the higher order estimation) for a discrete system whose dynamics are governed by Hertz interaction force is derived. It is observed that the wave speed changes with increase in amplitude at certain frequency leading to small changes in the band edges. Additional work is needed in order to correlate the analytical predictions with the experimental observations. Specifically, near term studies will verify that the applied preload actually leads to a weak nonlinearity within the theoretical assumption at the basis of the perturbation analysis. On the analytical point of view, the presented approach will be extended to investigate the strongly nonlinear behavior, so that the theory can predict the effect of a large range of amplitudes on the wave properties of the chain.

References

[1] Mead, D.J.: Wave propagation in continuous periodic structures: Research contributions from Southampton 1964-1995. *J. Sound Vibrat.* **190**(3) 495-524 (1996).

[2] Vakakis, A.F, King, M.E, Pearlstein, A.J.: Forced Localization in a periodic chain of nonlinear oscillators. *Int. J. Nonlinear Mech.* **29**(3) 429-447 (1994).

[3] Vakakis, A.F., King, M.E.: Non-linear wave transmission in a monocoupled elastic periodic system. *J. Acoust. Soc. Am.* **98** 1534-1546 (1995).

[4] Kosevich, A.M., Kovalev, A.S.: Self localization of vibrations in a one-dimensional anharmonic chain. *Sov. Phy. JETP* **40** 891-896 (1975).

[5] Rosenau, P., Quasicontinuous spatial motion of a mass-spring chain. *Physica D.* **27** 224-234, (1987).

[6] Rosenberg, R.M.: On non-linear vibration of systems with many degrees of freedom. *Adv. Appl. Mech.* 9155-243 (1966).

[7] Nesterenko, V.F.: Propagation of nonlinear compression pulses in granular media. *J. Appl. Mech. Tech. Phys.* **24**(5) 733-43 (1983).

[8] Nesterenko, V.F.: Dynamics of Heterogenous Materials. NY Springer, Verlag (2001).

[9] Porter, M.A., Daraio, C., Herbold, E.B., Szelengowicz, I., Kevrekidis, P.G.: Highly nonlinear solitary waves in phononic crystal dimmers. *Phys. Rev. E.* **77** 015601(R) (2008).

[10] Herbold, E.B., Kim, J., Nesterenko, V.F., Wang, S., Daraio, C.: Tunable frequency band-gap and pulse propagation in a strongly nonlinear diatomic chain, *Acta Mech.* DOI 10.1007/s00707-009-0163-6 (2009).

[11] Daraio, C., Nesterenko, V.F., Herbold, E.B, Jin, S.: Tunability of solitary eave properties in one-dimensional strongly nonlinear phononic crystals. *Phys. Rev. E.* **73** 026610 (2006).

[12] Romeo, F., Rega, G.: Wave propagation properties in oscillatory chains with cubic nonlinearities via nonlinear map approach. *Chaos, Solitons and Fractals.* **27** 606–617 (2006).

[13] Lazarov, B.S, Jensen, J.S.: Low-frequency band gaps in chains with attached non-linear oscillators. *Int. J. Nonlinear Mech.* **42** 1186-1193 (2007).

[14] He, Ji-Huan: Modified Lindstedt-Poincare methods for some strongly non-linear oscillations Part II: a new transformation. *Int. J. Nonlinear Mech.* **37** 315-320 (2002).

[15] He, Ji-Huan: The homotopy perturbation method for nonlinear oscillators with discontinuities, *Appl. Math. Comput.* **151** 287–292 (2004).

[16] Nayfeh, A. and Mook, D.: *Nonlinear Oscillations*, Wiley, NewYork (1985).

[17] Narisetti, R.K, Leamy, M.J., Ruzzene, M.: A perturbation approach for predicting wave propagation in one dimensional nonlinear periodic structures. *J. Vib. Acoust.* (accepted) (2009).

Lamb Waves in Phononic-Crystal Plates: Numerical Studies and Experiments

Tsung-Tsong Wu[1], Jin-Chen Hsu[2] and Jia-Hong Sun[1]

[1]Institute of Applied Mechanics, National Taiwan University, Taipei 106, Taiwan

[2]Department of Mechanical Engineering, National Yunlin University of Science & Technology, Yunlin 640, Taiwan

wutt@ndt.iam.ntu.edu.tw

Abstract. In this paper, the Lamb waves inside phononic-crystal (PC) plates are investigated numerically and experimentally. Two types of PC plates are studied: flat PC plates that composed of binary constituents and surface-stubbed PC plates with single constituting material but periodic cylindrical stubs protruded on one of the plate surfaces. The dispersion and characteristic displacement of Lamb waves are analyzed. Further, resonant cavities and waveguides are defined and the related acoustic modes were investigated. The analyses help understand the fundamental of Lamb waves inside PC plates, cavities and waveguides. Thus new acoustic devices based on PC could be designed accordingly.

1. Introduction

Phononic crystal (PC) is a periodic composite structure comprised of two or more elastic materials which differ in mechanical properties. The existence of complete phononic band gaps, frequency ranges in which acoustic waves are forbidden to propagate, suggests many possible applications of phononic structures, such as acoustic lens, acoustic filters, and efficient waveguides. In the past decade, the band gaps for bulk waves in unbounded PC and surface waves in semi-infinite PC have been considerably investigated [1-3].

Similar to the surface waves that can be guided by the surface of a half space, Lamb waves are guided to propagate by a plate of finite thickness. Therefore, they are especially favorable in many applications for the guiding characteristics. Guided acoustic-wave energies are much more effective to be used than bulk acoustic

T.-T. Wu and C.-C. Ma (eds.), *IUTAM Symposium on Recent Advances of Acoustic Waves in Solids*, IUTAM Bookseries 26, DOI 10.1007/978-90-481-9893-1_21,

waves. For example, surface waves and Lamb waves are controlled generated and received by the patterning the source and detection to efficiently interact with environmental physical quantities around their guiding surfaces, like surface-acoustic-wave and Lamb-wave sensors. Recently, Lamb waves in two-dimensional PC plates have received increasing attention in the community of PC research because the released researches show that Lamb waves in two-dimensional PC plates also exhibit complete band gaps. [4, 5] As a result, many efforts has devoted to the studies of band gaps in PC plate and their applications, such as Lamb-wave filters, Lamb-wave resonant cavities for amplification of acoustic energy.

In this paper, we report our investigation on PC plates and their applications utilizing the band gaps. Two categories of PC plates are studied: flat PC plates that composed of binary constituents and surface-stubbed PC plates with single constituting material but periodic cylindrical stubs protruded on one of the plate surfaces. This paper is organized as follows. In Sec. 2, propagation of Lamb waves in flat PC plate and in that with a cavity is characterized. Section 3 demonstrates the propagation and waveguiding properties of Lamb waves in the surface-stubbed PC plates and the usage for frequency selection from broadband acoustic energy. Finally, Sec. 4 briefly summarizes and concludes this work.

2. Flat Phononic-Crystal Plates

The basic type of PC plate is a plate composed of binary materials. The plate studied in this section has inclusions arranged periodically and two flat free surfaces. The dispersion and band gaps of a square lattice flat PC plate consisting of tungsten cylinders embedded in a silicon matrix were analyzed firstly and then resonant cavities of acoustic waves inside the PC plate were mentioned.

2.1 Dispersion and Complete Band Gaps

A schema of the square lattice W/Si PC plate is shown in Fig. 1(a). The cylindrical inclusions are arranged to form a square lattice on the x_1-x_2 plane. The normal of the free surfaces of the plate is in the x_3-axis. The [100] direction of the (001) silicon plate coincides with the x_1-axis of the PC plates. The lattice constant a is defined as 10 μm, and the radius r of tungsten cylinder is 2.5 μm. Thus, the filling fraction of square lattice crystal equals 0.196. We chose the thickness h as 10 μm for the obvious complete band gap.

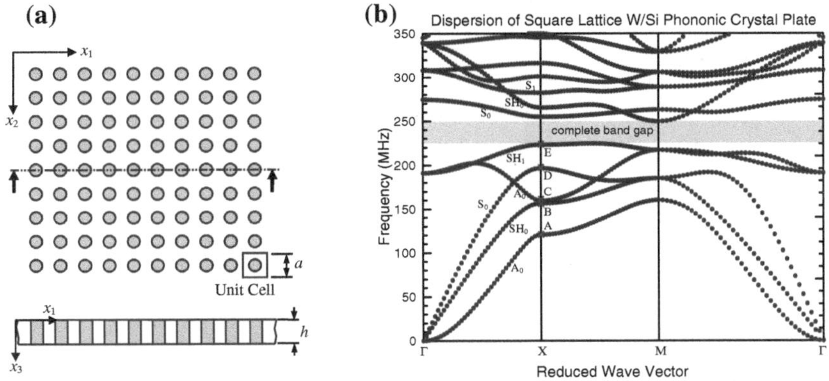

Fig. 1 (a) The schema of a square lattice PC plate with circular cylinders embedded in a background material. (b) The dispersion of the W/Si PC plate.

The band structures of PC plates are decided by the filling fraction, the materials, geometry conditions (i.e., lattice symmetry), and also thickness of plates [5, 6]. We adopted the finite-difference time-domain (FDTD) method [7] to analyze the dispersion of acoustic waves inside the W/Si PC plate. On the curves, every point represents one eigenmode of the phononic-crystal plate. Obviously, there is a range without any eigenmode, and thus the complete band gap is identified as the range from 223.08 to 249.56 MHz. Besides, a partial band gap is observed between 249.56 and 254.76 MHz at the ΓX-direction.

In the PC, the finite size of thickness and inclusions support mode conversion, and thus the eigenmodes can't be simply decoupled into the in-plane modes or the anti-plane modes. However, they still can be classified as flexural (anti-symmetric), longitudinal (symmetric) and transverse (shear horizontal) waves, similar to the Lamb waves in the classical plates [6].

The modes of the bands were investigated by calculating their displacement distribution. A segment consisting of four unit cells, as shown in Fig. 2(a), is defined to demonstrate the polarization. Two-dimensional vector plots were plotted to show the characteristic displacement distribution. Four eigenmodes were demonstrated in Figs. 2(b)-(e) to present the first four bands at the ΓX - direction. They are points A, B, C and D in Fig. 1 with wave vector **k** of $(\pi/a,0)$. The directions of cones indicate the direction of displacement vector, and the sizes reflect the magnitudes of vectors which were normalized to their maximum value.

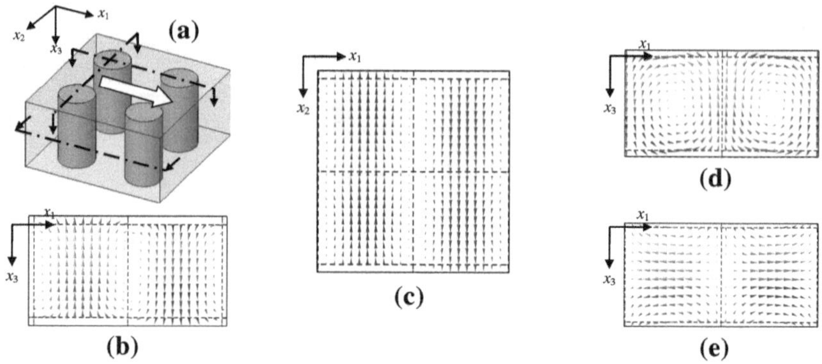

Fig. 2 (a) A $2a \times 2a$ segment of the PC plate. The dashed-dot lines indicate the planes which the following vector plots lie on. (b) the displacement filed of A_0 mode. Dashed lines indicate the boundaries of unit cells. (c) SH_0 mode; (d) A_0 mode of the folded band; (e) S_0 mode.

With the displacement distributions, these modes were classified. Fig. 2(b) shows the displacement distribution of point A in Fig. 1(b). The displacement components u_1 and u_3 dominate the behavior of the mode. Thus the first band basically corresponds to the lowest flexural (A_0) mode in a plate. Fig. 2(c) presents the mode of point B. The eigenmode has a primary polarization in the x_2-direction and this is the zero-th order mode of shear horizontal wave (SH_0). The displacement of point C is shown in Fig. 2(d). The main components consisting of u_1 and u_3 indicate that the third mode also belongs to the A_0 mode, a folded band of the first band. The fourth mode of point D has a higher phase velocity than the previous two bands. Fig. 2(e) shows displacement mainly along x_1-direction, and the field is symmetric with respect to the plane at $x_3=h/2$. This is the lowest longitudinal mode (S_0). In summary, in this PC plate, the acoustic waves present similar modes as the Lamb modes in a classical plate.

2.2 Resonant Cavities

In the flat W/Si PC plate, the periodic tungsten cylinders scatter Lamb waves and result in band gaps. Within the range of band gaps, propagation is forbidden and thus Lamb waves can not penetrate the crystal. Based on the property, resonant cavities of Lamb waves were constructed. A cavity is a defect in the PC plate, and the defect allows some modes at the range of band gaps. We defined a dislocation by arranging extra space in the PC plate and formed a resonant cavity of Lamb waves. A typical resonant cavity is shown in Fig. 3(a). An extra space of ΔL is inserted and then the cavity length L is defined as the distance between two neighboring cylinders of the defect.

To investigate the property of the cavity, the supercell technique is used to analyze the dispersion. First, an extra dislocation of ΔL=7.5 μm is defined and thus the cavity has a length L=12.5 μm (i.e. 1.25a). A supercell of 10.75a×1a×1a is adopted to calculate the wave modes inside the range of band gaps. The dispersion is plotted in Fig. 3(b). We focused on the frequency range of 190-280 MHz to observe cavity modes inside the band gap. The boundaries of complete band gaps are marked with lines at 223.08 and 249.56 MHz. Besides, a line at 254.76 indicates the upper limit of the partial band gap at the ΓX -direction. Numerous extended modes outside the band gap are not concerned, and the regions are presented in gray. Two noticeable flat bands appear at 232.40 and 247.66 MHz, inside the complete band gap. Since the bands are horizontal, they have zero group velocity and present the resonant modes.

In Fig. 3(b), the bands of Lamb waves in the perfect PC plate were plotted again as hollow circles. The X point presents wave vectors **k**=(π/a,0) for the perfect PC plate and **k**=($\pi/10.75a$,0) for the cavity. Although the values of the boundaries of the Brillouin zone are not the same, the figure still is useful to observe the gap and resonant modes.

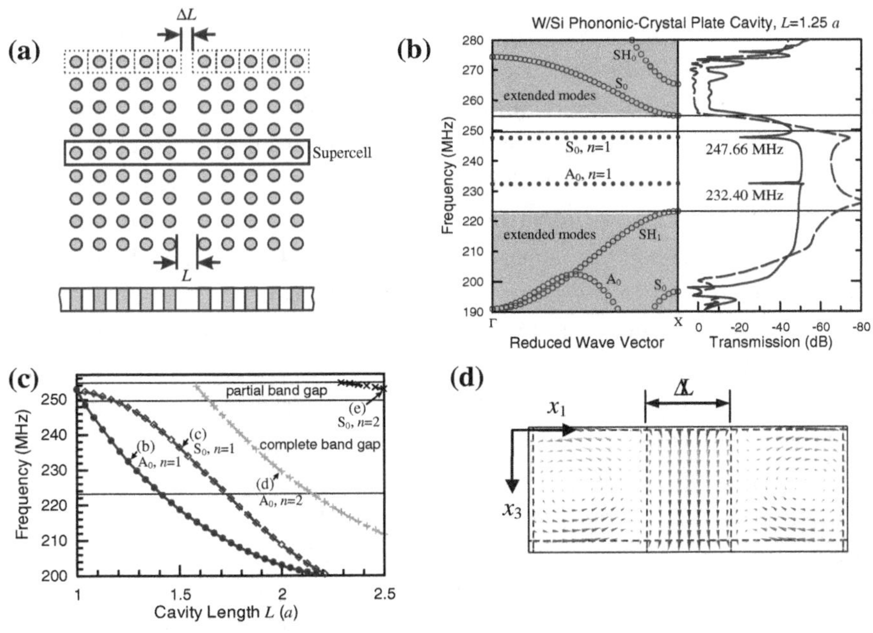

Fig. 3 (a) A resonant cavity based on a PC plate. (b) The dispersion shows resonant modes of a cavity of L=1.25a. The transmission is on the right. The solid and dotted lines present results of Lamb waves propagating through the cavity and a ten-layer PC plate, respectively. (c) The resonant modes excited by the source polarized at x_3-direction on the top surfaces of plates. (d) The displacement field of the 232.40-MHz mode inside a cavity of L=1.25a.

The transmission of Lamb waves passing the resonant cavity is also shown in Fig. 3(b). A line source is defined along the x_2-direction to launch an x_3-polarization wide-band wave packet. The displacement is recorded on the other side of the PC cavity to calculate the spectrum and the transmission. The transmission for the cavity is plotted as the solid line; besides, the transmission of a ten-layer PC plate is also shown as the dashed line. In the transmission, two peaks of f=232.40 and 247.66 MHz are obtained in the solid line which is not obtained in the dashed line. This consists with the resonant modes shown in the dispersion. In addition, comparing the transmission with the dispersion of the PC plate (the dashed line and the hollow circles), two bands are not excite by the x_3 polarization line source. Thus the ranges 202.40-223.08 and 274.40-282.04 MHz become non-excitation gaps, and low transmission coefficient is obtained.

The Fabry-Perot-type standing wave resonant condition, $n\lambda_n/2=L_n$, is useful to understand the resonant waves inside the PC cavities, [8, 9] where n is the order of the resonant modes, λ_n is the wavelength, and L_n is the equivalent cavity length for the nth order resonant mode. However, the resonant frequency cannot be calculated simply according to the condition, because the waves can penetrate the geometric boundary of cavities and Lamb waves are dispersive. To figure out the relationship between the resonant frequency and the cavity length, we directly calculate the transmission of cavities with different length. The cavity length changes from $1a$ to $2.5a$ and the resonant frequencies of all cavities are identified and marked in Fig. 3(c). The resonant modes are identifiable at the range 202.40 MHz to 254.76 MHz and they result in four curves. In these curves, the lower resonant frequency is obtained while the cavity length is increased. The variation consists with the trend expected by the Fabry-Perot condition.

Further, for distinguishing the difference of these four curves, the displacement distributions of these curves are calculated. One example is shown as Fig. 3(d) for the mode of 232.40-MHz inside a cavity of $L=1.25a$, the point (a). The displacement field shows obvious u_3 components which are anti-symmetric with respect to the plane at $x_3=h/2$. This is the lowest order flexural mode (A_0, n=1) with only a half of a complete wavelength inside the cavity. Similarly, we identified point (c) belongs to the second-order resonant flexural mode (A_0, n=2), point (b) the first longitudinal resonant mode (S_0, n=1), and point (d) the second longitudinal resonant mode (S_0, n=2). The result could be used to design proper cavities for the specific resonant frequency, and further a magnified Lamb wave source was designed accordingly [9].

3. Stubbed Phononic-Crystal Plates

This section investigates the PC plate structure with a stubbed surface [10, 11] instead of the binary flat plate. The idea is to produce a periodic variation in the geo metries rather than the material properties, which generates similar influence on the acoustic waves as well. It can be shown that complete band gaps and related characteristics exist in the proposed periodic stubbed plates.

3.1 Dispersion Relations of the Plate Modes

Consider a thin plate (plate thickness of h_1=1mm) with periodic cylindrical stubs on one of the plate surfaces [a unit cell of the structure is shown in Fig. 4(a)]. The material chosen to constitute the whole structure is aluminum 6061. The cylindrical stubs are arranged in square lattice with a lattice constant a=10 mm. The radius of the cylindrical stubs is r=3.5 mm, and the filling factor defined as $F=\pi r^2/a^2$ is 0.385.

For waves propagating with a Bloch wave vector along the boundaries of the irreducible part of the first Brillouin zone, the calculated band structure by using finite-element method (with Comsol Multiphysics software) is shown in Fig. 4(b). The band structure exhibits a complete band gap ranging from 114 to 143 kHz and three partial band gaps in the Γ-X direction. Furthermore, the band structure displays many flat bands which reveal the resonant characteristics of the PC plate.

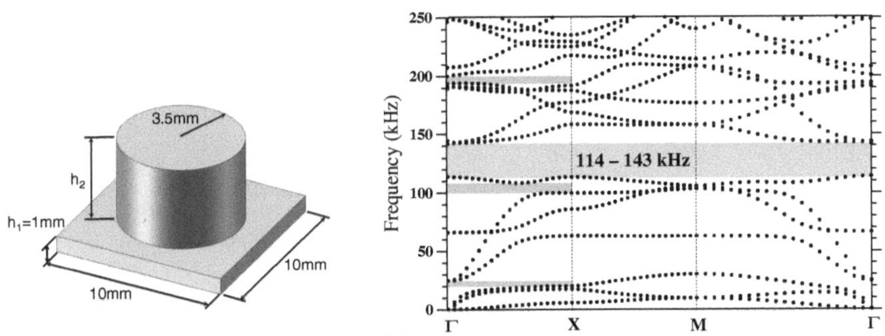

(a) (b)

Fig. 4 (a) Schematic of a unit cell of the PC plate with a periodic stubbed surface. (b) Band structure of Lamb waves in the stubbed PC plate.

3.2 Complete Band Gaps and Resonant Modes

To demonstrate the evidence of the complete band gap and resonance in the stubbed PC plate, a series of laser ultrasonic experiments are conducted. The fabricated structure contains 16×10 stubs on one of the plate surfaces. In the experiments, Nd:YAG pulsed laser is applied on the PC structure to generate broadband wave energy, and a He-Ne laser interferometer is used to receive the signal of acoustic waves propagating in the structure. The source and receiver are spaced by four unit cells in the structure. Both the Γ-X and Γ-M directions are considered. The digitized sampling rate for the received wave signal is 50 MHz so good resolution of the spectrum is obtained.

Shown in Fig. 5 are the measured reference spectra in the Γ-X and Γ-M directions, respectively. The reference spectrum is defined as the ratio of the spectrum measured on the stubbed PC plate to that measured on a uniform thin plate of thickness 1 mm. The spectra show very low intensity between 114 and 143 kHz which coincides with the predicted range of the complete band gap in Sec. 3.1. The spectrum measured along the Γ-X direction also significantly drops off in the ranges that exactly conform to the partial band gaps. Moreover, in Fig. 5(a), there are peaks in the spectrum at the band-edge frequencies of 20, 100, 110, and 205 kHz, which demonstrate the evidence of resonances of the structure.

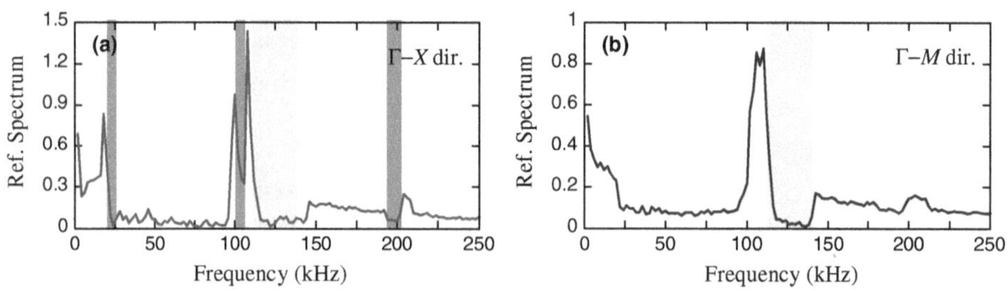

Fig. 5 Experimental measurement of reference spectra for the waves propagating in the stubbed PC plate structure along the (a) Γ-X and (b) Γ-M directions.

3.3 Waveguiding and Frequency Selection

Since the Lamb waves with frequencies in the band gaps are expected to be only able to propagate along the line defect, one can make a channel based on the stubbed PC plate to guide the wave energy flow two-dimensionally if the band gap is complete. Accordingly, a design of a bent waveguide created in the stubbed PC plate used is proposed, and results of laser ultrasonic measurement are demonstrated below. The bent waveguide is formed with two-graduation turns produced by removing some of the stubs from the stubbed PC plate, and the schematic of the designed and fabricated structure is shown in Fig. 6(a). When broadband elastic waves are generated in the thin plate, all the waves with frequencies either outside or inside the complete band gap are incident. It may be judged that most of the energy of the waves with frequencies outside the complete band gap may propagate straightforwardly through the stubbed PC plate; however, for those inside the complete band gap, they can not keep going straight. The waves are, therefore, compelled to make a turn and propagate along the bent waveguide. For comparison, measurements of the ultrasonic signals are carried out at points **a, b, c**, and **d**, respectively, shown in Fig. 6(a). Note that the spectrum measured at point **a** (55 mm apart from the source) is regarded as the reference corresponding to the measurements at other points.

Figure 6(b) shows the reference spectra of the measurements. For the measurement at point **b**, which is 60 mm apart from point **a**, the reference spectrum exhibits several peaks which associate with the resonance of the flat bands in the band structure [Fig. 4(b)]. Also, very low intensities are displayed in the ranges corresponding to the complete and partial band gaps. Since the periodic stubs are ordered between the points **a** and **b**, the measurement at point **b** is, therefore, intrinsically associated with the characteristics of waves in the perfect periodic structure. Observe the spectra measured at point **c** and point **d** located inside the waveguide. An interesting phenomenon is found that, in the complete band gap, the intensities of the spectra are obviously outstanding rather than sunken, with values about 0.3 and 0.2. That is to say, the original broadband acoustic signals at point **a** are separated into the resonant energy with frequencies outside the complete band gap and guided energy within the complete band gap propagating along the bent waveguide. Based on the experimental results, we may conclude that the measured results conform to the prediction about the band gaps schemed in Fig. 6(a), and the proposed phononic waveguiding structure may be characterized by the selection of acoustic frequency.

Furthermore, it should be noticed that, in Fig. 6(b) where the frequency range from 66 to 100 kHz, the measured spectra in the waveguide exhibit an obviously higher intensity in that range which corresponds to no band gaps. However, an outstanding signal is still found. This phenomenon is discussed in the following. On observing the band structure in Fig. 4(b), there are two frequency bands extended in the range from 66 kHz to 100 kHz. By calculating the displacement fields, the eigenmode shapes of the bands show the displacement fields are mainly governed by the vibrations associated with the symmetric (S0) and transverse (T0) plate modes in the frequency range. However, the broadband elastic waves generated by Nd:YAG pulsed laser are major in anti-symmetric modes, minor in symmetric modes, and lacked in transverse plate modes. Therefore, in Fig. 6(b), as considering the ranging from 66 to 100 kHz, the modes are not suitably coupling to and converted from the anti-symmetric modes that can be effectively generated by the pulsed laser to propagate in the stubbed PC plate; therefore, most of the energy excited with frequencies in this range can only stay in the non-stubbed regions of the plate, i.e., the waveguides. As a result, for the laser ultrasonic excitation, this frequency range behaves as a complete band gap. It is called the deaf bands. Based on the existence of the deaf band with selectivity due to the applied source, it is reasonable to claim that waves with frequency ranging from 66 to 100 kHz should appear and be measured at points **c** and **d** in the experiment. Obviously, broadband signals can be separated into a lot of energy flows according to the wave properties in different frequency bands and can be measured at different positions through a proper design of the phononic structure.

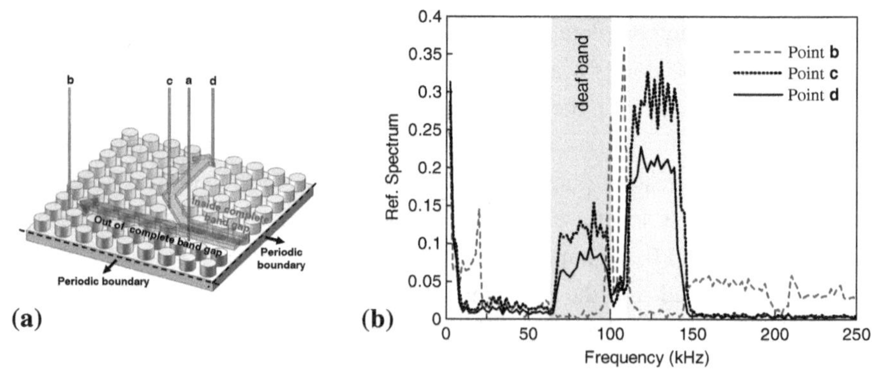

Fig. 6 (a) Paths of the waves with frequency respectively outside and inside the complete band gap propagating in the structure with a bent waveguide. (b) Comparison among the spectra measured at point **b**, **c**, and **d**.

4. Conclusion

In summary, for the flat PC plates, the dispersion of the square lattice W/Si PC plate is calculated and a complete band gap is observed. By investigating the displacement distribution of eigenmodes, the flexural, longitudinal, and shear horizontal waves are classified and help understand wave propagation in the PC plate. The related resonant cavities were also investigated to show the relation between the cavity length and resonant frequencies. For the surface-stubbed PC plate, the evidence of the complete band gap and resonances in the plate is demonstrated. Based on the complete band gap, a stubbed PC plate with a bent waveguide is designed and applied to use for frequency selection of Lamb waves. The experimental results by using the laser ultrasonic technique show that wave signals with the frequencies inside the complete band gap can be guided to follow the designed route. Finally, we note that the results of this paper could be further applied to innovated design of acoustic-wave devices.

References

[1] Kushwaha, M. S., Halevi, P., Dobrzynski, L. and Djafari-Rouhani, B.: Acoustic band structure of periodic elastic composite. *Phys. Rev. Lett.* **71** 2022-2025 (1993).

[2] Tanaka, Y. and Tamura, S.: Surface acoustic waves in two-dimensional periodic elastic structures. *Phys. Rev. B* **58** 7958-7965 (1998).

[3] Sun, J.-H. and Wu, T.-T.: Propagation of surface acoustic waves through sharply bent two-dimensional phononic crystal waveguides using a finite-difference time-domain method. *Phys. Rev. B* **74** 174305 (2006).

[4] Zhang, X., Jackson, T., Lafond, E., Deymier, P. and Vasseur, J.: Evidence of surface acoustic wave band gaps in the phononic crystals created on thin plates. *Appl. Phys. Lett.* **88** 041911 (2004).

[5] Khelif, A., Aoubiza, B., Mohammadi, S., Adibi, A. and Laude, V.: Complete band gaps in two-dimensional phononic crystal slabs. *Phys. Rev. E* **74** 046610 (2006).

[6] Sun, J.-H. and Wu, T.-T.: Propagation of acoustic waves in phononic-crystal plates and waveguides using a finite-difference time-domain method. *Phys. Rev. B.* **76** 104304 (2007).

[7] Hsieh, P.-F., Wu, T.-T. and Sun, J.-H.: Three-Dimensional Phononic Band Gap Calculations Using the FDTD Method and a PC Cluster System. *IEEE Trans. Ultrason., Ferroelect., Freq. Contr.* **53**(1) 148-158 (2006).

[8] Wu, T.-T., Hsu, C.-H. and Sun, J.-H.: Design of a highly magnified directional acoustic source based on the resonant cavity of two-dimensional phononic crystals. *Appl. Phys. Lett.* **89** 171912 (2006).

[9] Sun, J.-H. and Wu, T.-T.: A Lamb Wave Source Based on the Resonant Cavity of Phononic-Crystal Plates. *IEEE Trans. Ultrason., Ferroelect., Freq. Contr.* **56**(1) 121-128 (2009).

[10] Wu, T.-T., Huang, Z.-G., Tsai, T.-C. and Wu, T.-C.: Evidence of complete band gap and resonance in a plate with periodic stubbed surface. *Appl. Phys. Lett.* **93** 111902 (2008).

[11] Wu, T.-C., Wu, T.-T. and Hsu, J.-C.: Waveguiding and frequency selection of Lamb waves in a plate with a periodic stubbed surface. *Phys. Rev. B* **79** 104306 (2009).

Part IV
Wave Electronics

Phase-Sensitive and Fast-Scanning Laser Probe System for RF SAW/BAW Devices

Ken-ya Hashimoto[1], Nan Wu[1], Keisuke Kashiwa[1], Tatsuya Omori[1], Masatsune Yamaguchi[1], Osamu Takano[2], Sakae Meguro[2], Naoki Kasai[2] and Koichi Akahane[2]

[1]Graduate School of Engineering, Chiba University, Chiba 263-8522, Japan

[2]Neoark Co. Ltd, Hachi-Ohji, Tokyo 192-0015, Japan

k.hashimoto@ieee.org

Abstract. This paper reviews a phase-sensitive and fast-scanning laser probing system developed by the authors' group for the diagnosis of radio frequency (RF) surface and bulk acoustic wave (SAW/BAW) devices. The system is based on the Sagnac interferometer, which is insensitive to low frequency vibration. From this feature, we can maximize the scanning speed without sacrificing the signal-to-noise ratio of the measurement. It is demonstrated that high quality two-dimensional (2D) image of SAW/BAW field patterns can be captured in minutes order. Currently the maximum applicable frequency is about 2.5 GHz. Because of the phase sensitivity, the measured field in the space domain is readily converted into the wavenumber domain by the 2D Fourier Transform. It is also demonstrated how effective the wavenumber domain analysis is for the purpose.

1. Introduction

Currently, use of laser probing systems[1-3] is essential for the research and development of sophisticated radio frequency (RF) surface and bulk acoustic wave (SAW/BAW) devices, which are used widely in various communication systems[4].

For detecting acoustic vibration, the Michelson and Mach-Zender interferometers has widely been used[1,3]. The method detects surface vertical motion through the interference between two beams reflected by the vibrating surface and static mirror. The SAW/BAW field can be visualized by either scanning a laser beam-spot on the vibrating surface or by mechanically translating the device under test (DUT).

T.-T. Wu and C.-C. Ma (eds.), *IUTAM Symposium on Recent Advances of Acoustic Waves in Solids*, IUTAM Bookseries 26, DOI 10.1007/978-90-481-9893-1_22,

One of the drawbacks of this method is that the sensitivity is independent of the frequency of the vibration, and that the whole system is occasionally affected by a low-frequency vibration caused by the scan. Therefore, the scanning speed is limited so that the mechanical disturbance is minimized[1,3].

The knife-edge method has been also used for the purpose[4-7]. Since the method optically detects a surface micro-bending caused by the wave propagation, the system is inherently insensitive to a low-frequency vibration. Thus together with a fast scanning technique, the system is made capable to capture high quality two-dimensional (2D) images of SAW field patterns in minutes order[6,7].

Although this method is effective in the characterization of SAW devices, it is not suitable for BAW device diagnosis, in which the vertical displacement component is dominant. This is because the knife-edge method is, in principle, only sensitive to a surface bending.

This paper reviews a phase-sensitive and fast-scanning laser probe system developed by the authors' group, which is applicable to the diagnosis of RF SAW/BAW devices[8].

For the optical sensing, the Sagnac interferometer[9] composed of micro-optic elements is employed. Although the Sagnac system is able to detect vertical motion of vibration similar to the Michelson/Mach-Zender interferometer, its significant difference is that owing to its intrinsic frequency dependence, RF surface vertical motion can selectively be detected free from the mechanical disturbance. From this feature, we can maximize the scanning speed without sacrificing the signal-to-noise ratio (SNR) and sensitivity of the measurement.

As a demonstration, the system is applied to the characterization of RF SAW/BAW devices operating in 2 GHz range. It is shown that high quality two-dimensional (2D) image of SAW/BAW field patterns can be captured in minutes order.

Because of the phase sensitivity, the measured field in the real space (x-y) domain is readily converted into the wavenumber (β_x-β_y) domain by the 2D Fast Fourier Transform (FFT).

The wavenumber-domain analysis offers various information; the types and characteristics of spurious signals, and where and how the spurious signals are generated and propagated[10,11]. Here we apply this technique to the diagnosis of an RF BAW resonator, and its effectiveness is demonstrated.

2. System Setup

Fig. 1 shows the basic setup of the Sagnac interferometer for detecting vertical motion [8,9].

A linearly-polarized laser beam launched from a laser diode (LD, λ_0=660 nm, P_{max}=120 mW) is beam-shaped and applied to a $\lambda/2$ plate to adjust polarization of the incident laser beam 45o off from the base plane.

The laser beam transmitted through the non-polarizing beam splitter (NPBS) is incident to the Sagnac loop composed two polarizing beam splitters (PBSs), two static mirrors and the $\lambda/4$ plate.

First, the incident laser beam is divided into two beams at the PBS1 and illuminates the DUT after re-combination at the PBS2: s- beam with polarization normal to the base plane experiences reflections by two PBSs and mirrors while p-beam with polarization parallel to the base plane passes through two PBSs. Here we designate the difference in the arrival time to the surface as Δ_τ.

When the reflected beam arrives at the PBS2, the p-beam is converted to normal polarization and the s-beam is converted to parallel polarization due to two way transmission through the $\lambda/4$ plate. The incident beam is divided again into two beams at the PBS2, and they are recombined at the PBS1.

Finally two beams are guided to high-speed photo detectors (PD, Newport D-200, f_T=2.5 GHz) after transmitting the $\lambda/4$ plate and the PBS3. The arrangement makes these two beams interfere after giving 90° optical phase difference between them[8,9]. The phase difference is important to detect not only amplitude but also phase information from the interference output.

Since initially s- and p-beams propagate the identical Sagnac loop clockwise and counterclockwise, respectively, these two beams arrive at the detector at the same time. However, this is only true when the DUT surface is stationary.

Let us assume that the surface of the DUT vibrates as a form of $A_{RF}\sin(2\pi f_{RF}t)$, where A_{RF} is the vibration amplitude, and f_{RF} is the frequency. Due to asymmetry of the Sagnac loop, the surface vibration causes optical phase difference given by $4\pi\lambda_o^{-1}A_{RF}\sin(\pi f_{RF}\Delta_\tau)\cos(2\pi f_{RF}t)$, where λ_o is the optical wavelength. Since A_{RF} is much smaller than λ_o, the PD output is proportional to the phase difference. Thus when Δ_τ is set at close to $(2f_{RF})^{-1}$, we can maximize the sensitivity and make the system insensitive to low frequency vibration.

In the developed system, Δ_τ is fixed at 1/6 ns for all the measurement targets since the frequency dependence is not so obvious.

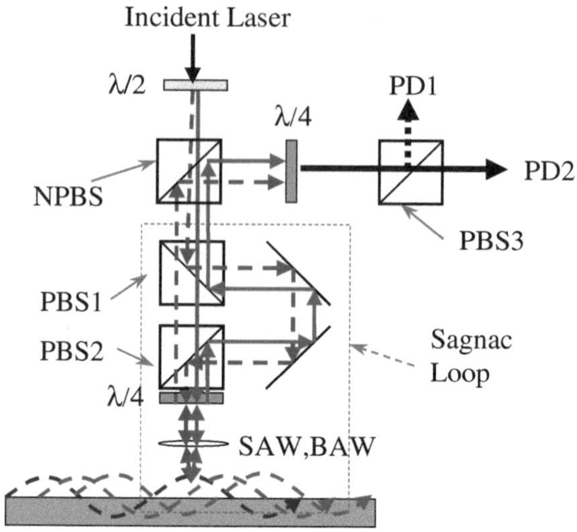

Fig. 1 Sagnac interferometer for detecting vertical vibration.

Fig. 2 shows the system setup of the fast-scanning laser probe system[7]. Measurement is carried out by the following procedure. The translation stage moves continuously along the longitudinal (x) direction between the specified starting and ending points. The high-precision linear scale attached to the stage outputs two-phase pulse trains every 40 nm movement. Then analog output signals of the detection circuit are acquired by the high-speed data-logger synchronously with the pulse trains. After one x scan is completed, the stage returns to the original position, moving simultaneously along the lateral (y) direction at a given step. This process is repeated until the two-dimensional scan is completed.

In the present system, the translation stage moves at its maximum speed of about 1.0 mm/s, and the sampling rate of 25 kS/s is theoretically achievable.

Since the spatial resolution is limited by the employed objective lens, the sampling interval of 40 nm step is generally too dense. So the N-divider circuit was prepared and inserted between the linear-scale output and the data-logger (see Fig. 2). Here, N can be set arbitrarily by a program; when N = 25, for example, the data interval becomes 1 μm (=40 nm × 25) step.

For the fast-scanning, data acquisition must be also performed in high speed. Fig. 3 shows present setup of the detection circuit[12]. An RF signal is applied to the DUT, and the PD output is fed to the mixer after RF amplification. Then the output signal is down-converted to the IF frequency f_{IF}=10.7 MHz through the mixing with a local signal with a frequency of f_{LO} and detected synchronously by an RF lock-in amplifier (Stanford Research SR844, f_{max}=200 MHz). The RF signal is also directly down converted to the IF frequency and fed to the lock-in amplifier as a reference signal. This setup allows us to detect both amplitude and phase information of the output signal in high sensitivity upto 2.5 GHz.

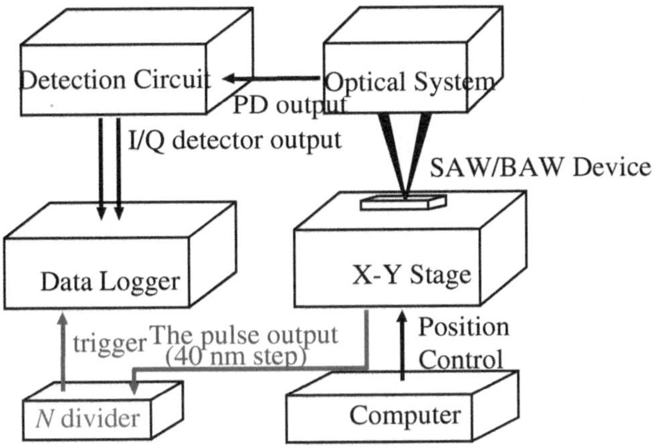

Fig. 2 System setup.

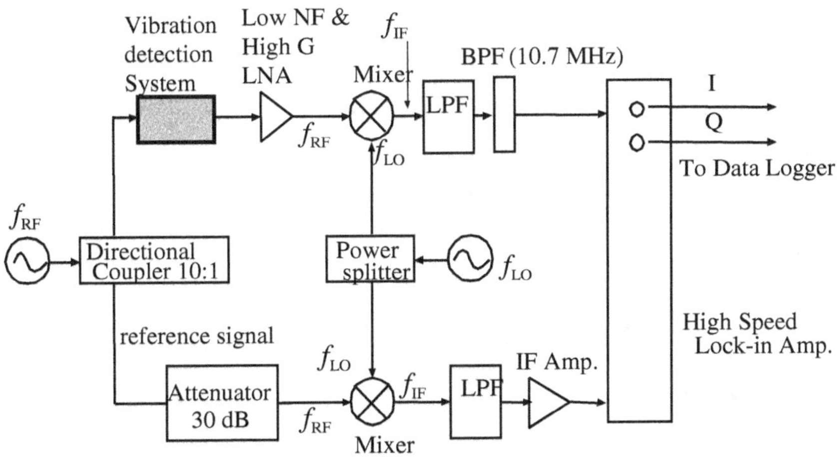

Fig. 3 Detection circuit.

By the way, when high spatial resolution is necessary, we must employ high magnification objective lens with extremely shallow focal depth to reduce the laser spot size. In the case, tiny inclination of the measurement sample will cause severe defocus resulting in blurred images. Because of the fast mechanical scanning, in situ auto-focusing is not applicable in our system.

The authors developed a focus adjustment technique particularly for our laser probe[13]. The focus adjustment is carried out by the following procedure. First, the objective lens is manually adjusted into focus at three different points on a surface of a device, and the lens heights and their corresponding device positions are recorded. If the surface is flat enough, the relation between the lens height and surface position (inclination) can be modeled mathematically by the recorded data. Then, in accordance with the measuring position during the fast mechanical scanning, the height of the objective lens is controlled continuously by using the mathematical model. In practice, the lens height should be controlled monotonically to avoid problems occurring with the mechanical backlash.

Fig. 4 shows the whole system developed. Owing to the Sagnac interferometer, which is insensitive to low frequency vibration, no tight anti-vibration systems are needed, and the optical system occupies only a small area of 66 × 56 cm^2.

3. Measurement Examples

First, effectiveness of this laser probe was examined by using an RF BAW resonator[14] operating in a 2 GHz range.

Fig. 5(a) shows a surface image of the RF BAW resonator used as the DUT. In the oval region, a piezoelectric AlN thin film is sandwiched in two Ru electrodes, and the structure is floating on the Si substrate through the air gap. The major axis length of the resonance area is 115 μm, and the minor axis length is 95 μm. Fig. 5(b) shows the return loss characteristic of the DUT. The main resonance frequency can be seen around 1,820 MHz, and nearby there are several spurious transverse resonances.

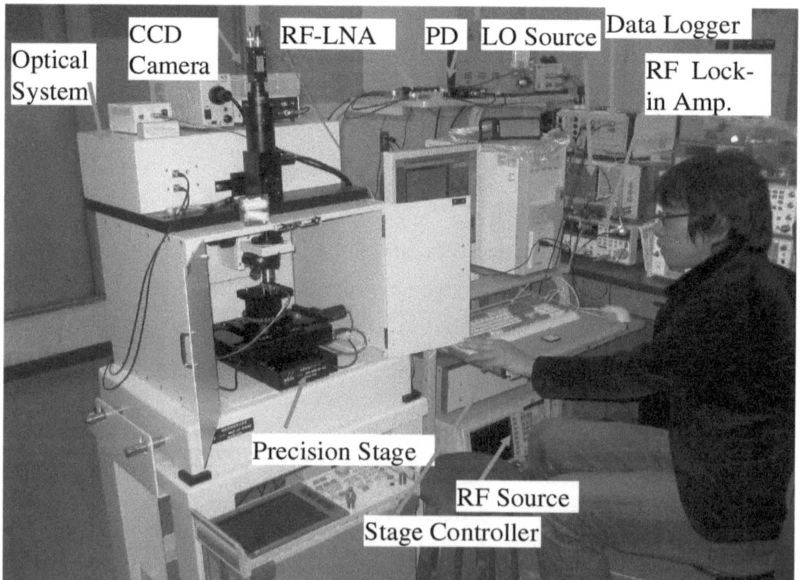

Fig. 4 Appearance of the developed laser probe.

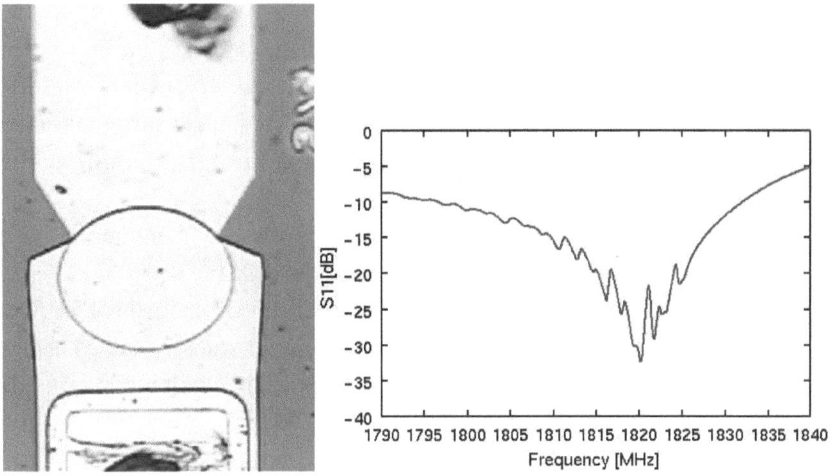

(a) Chip photo (b) S11 characteristic

Fig. 5 DUT used for the experiment.

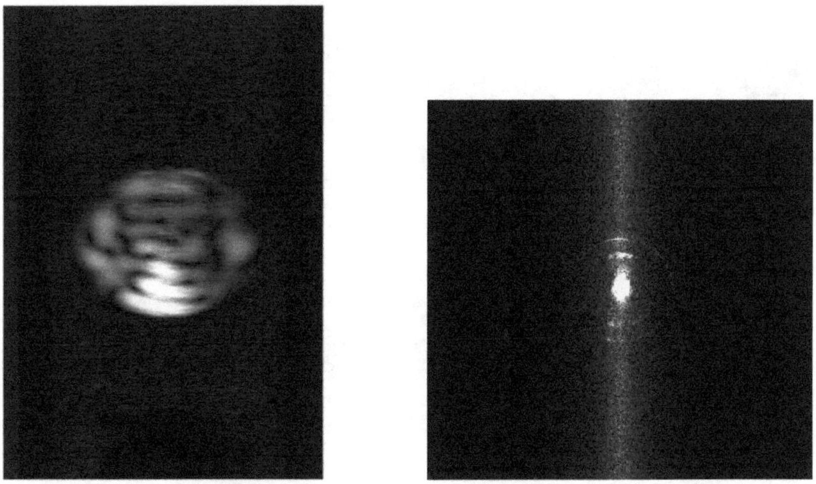

(a) space (x-y) domain (b) wavenumber (x-y) domain

Fig. 6 Measured field distribution (1.816 GHz).

Fig. 6(a) shows 2D image acquired by the laser-probe at 1,860 MHz. It took about 17 minutes to scan 500×750 ($x \times y$) points with 0.4 μm step. The transversal resonance pattern is clearly observed in the oval region where the resonator is placed. For this measurement, the RF input power was set at 13 dBm, and the maximum amplitude is roughly estimated to 50 pm.

Next, the wavenumber domain analysis is performed by using the integrated software developed by the authors[15].

Fig. 6(b) shows the result in wavenumber (β*x*-β*y*) domain obtained by the FFT conversion of the data shown in Fig. 6(a). Horizontal and vertical directions represent β*x*, and β*y* components, respectively, where the origin (β*x*, β*y*)=(0,0) is located in the middle. Several concentric circles are seen. Lack of the angular dependence indicates isotropic propagation of acoustic waves.

We extracted spectral data corresponding to each circle numerically, and the inverse FFT was applied to reconvert to real (*x*-*y*) space.

Fig. 7(d) shows the result when only the inner portion in the FFT image was extracted as shown in Fig.7(a) and inverse Fourier transformed. The result is almost identical with Fig. 6(a). Since |**β**| is small, this portion corresponds to the thickness vibration of the longitudinal wave.

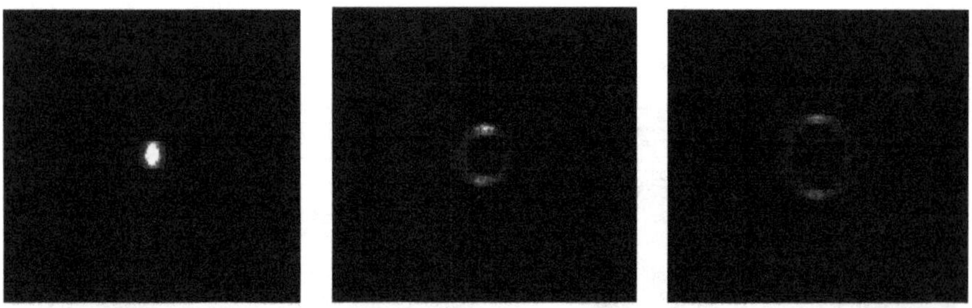

(a)Inner circles (b) second largest circle (c) outermost circle
Extracted data in wavenumber domain.

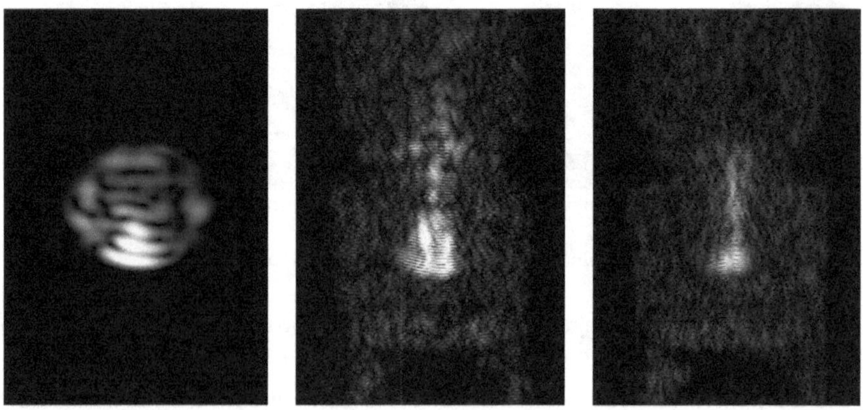

(b)Inner circles (e) second largest circle (d) outermost circle
Field distribution reconverted from extracted data.

Fig. 7 Selective observation of Lamb waves generated by the scattering.

Fig. 7(e) shows the result when the image processing was applied to the second largest circle in the FFT image (see Fig.7(b)). It is seen that the energy of the mode leaks through the upper electrode of the resonator. Since $|\beta|$ is relatively large, this circle corresponds the Lamb mode generated at the resonator edge. It should be noted that the AlN layer extends under the upper interconnecting electrode, which is not mechanically isolated from the AlN layer in the resonator area.

Fig. 7(f) shows the result when this image processing was applied to the outer-most circle in the FFT image (see Fig.7(c)). Lamb wave generation at the resonator edge is clearly visible also in this case.

Fig. 8(a)-(f) show the results when the same procedure with Fig. 7(e) was applied to the measured data at different frequencies. From comparison of these six figures, it is clearly seen that the energy leakage is very frequency-dependent. At frequency of 1,790MHz, 1,810.5MHz and 1,816MHz, the energy leakage through the upper electrode can be seen very clearly.

Next, the laser probe is applied to the characterization of a ZnO/diamond SAW filter[16] operating at 2.441 GHz. In the measurement, an objective lens with magnifying power of 100× and long working distance was employed.

Figs. 9(a) and (b) compare the results without and with the focus adjustment described in Section 3. Effectiveness of the focus adjustment is clearly visible; it compensated the surface inclination successfully, and SAW field distribution is only visible when the focus adjustment was applied.

4. Conclusions

The paper reviewed a phase-sensitive and fast-scanning laser probe system developed by the authors' group, and demonstrated how effective the system is for the diagnosis of RF SAW/BAW devices.

As a next step, we are attempting to speed up the measurement, to increase the maximum frequency, and to enhance the sensitivity. Calibration of the measured data, namely estimation of the absolute vibration amplitude, should be realized for further enhancement of the usefulness of this system.

Acknowledgments: The authors thank Mr. Ueda of Fujitsu Laboratories, Ltd. and Mr. Fujii of Epson Corp. for supplying the SAW/BAW devices used in this work. This work was partially supported by the Mitsubishi Foundation, a Project to Develop "Innovative Seeds" from the Japan Science and Technology Agency, and a Grant-in-Aid for Scientific Research from Japan Society for the Promotion of Science.

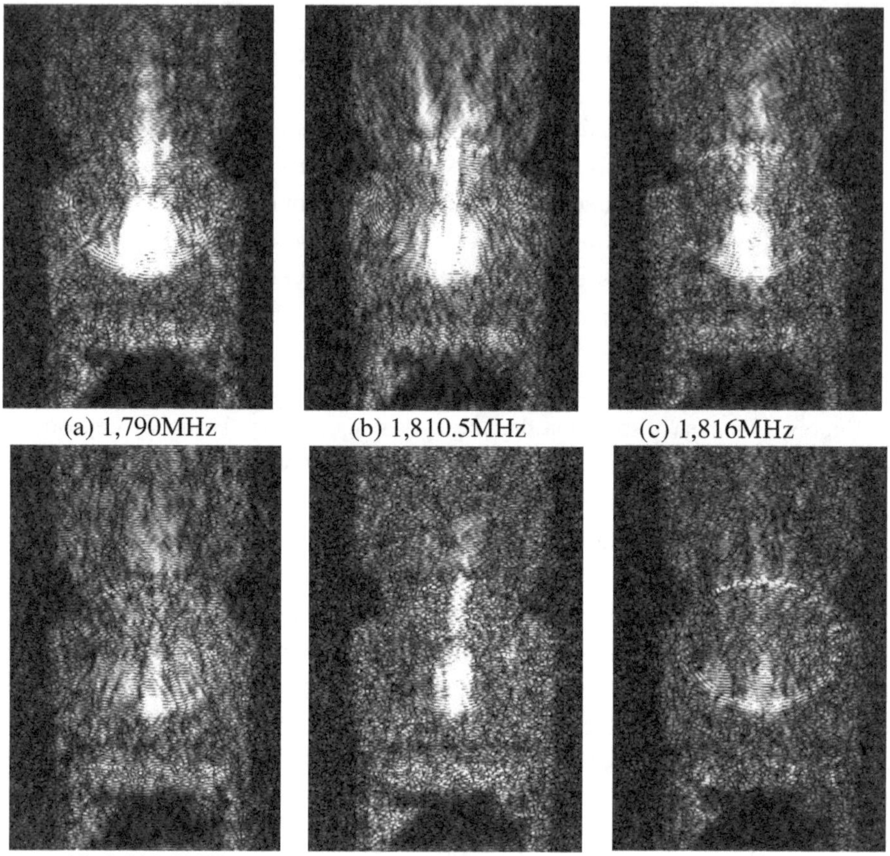

(a) 1,790MHz (b) 1,810.5MHz (c) 1,816MHz

(d) 1,817.5MHz (e) 1,820MHz (f) 1,840MHz

Fig. 8 Scattered field distribution.

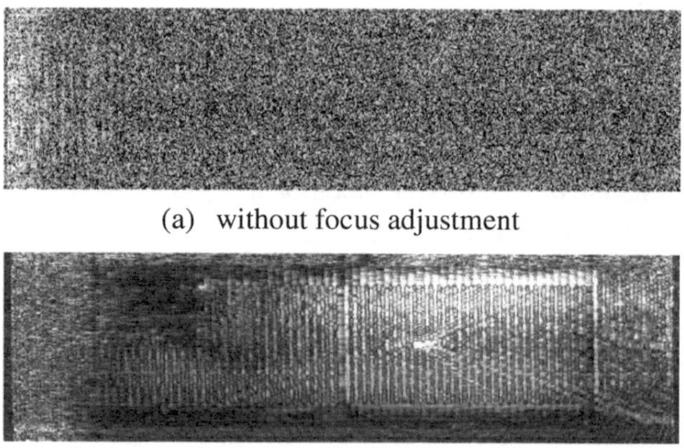

(a) without focus adjustment

(b) with focus adjustment

Fig. 9 Measurement result for the ZnO/diamond SAW filter at 2.441 GHz.

References

[1] Knuuttila, J.V., Tikka, P.T., Salomaa, M.M.: Scanning michelson inteferometer for imaging surface acouctic wave fields. *Optics Lett.* **25**(9) 613-615 (2000).

[2] Chiba, T.: Optical measurement and numerical analysis of SAW propagation at dispersive delay line on Y-Z LiNbO3 substrate. *Proc. IEEE Ultrason. Symp.* 1718-1721 (2003).

[3] Miyamoto, M., Wakana, S, Ito, A.: Novel optical observation technique for shear horizontal wave in SAW resonators on 42°YX-cut Lithium Niobate. *Proc. IEEE Ultrason. Symp.* 89-92 (2002).

[4] Weigel, R., Morgan, D.P., Owens, J.M., Ballato, A., Lakin, K.M., Hashimoto, K., Ruppel, C.C.W.: Microwave acoustic materials, devices and applications. *IEEE Trans. Microwave Theory and Tech.* **50**(3) 738-749 (2002).

[5] Engan, H.: Phase sensitive laser probe for high frequency surface acoustic wave measurements. *IEEE Trans. Sonics and Ultrason.* SU-**25** 372-377 (1978).

[6] Jen, S., Hartmann, C.S.: Recent advances in SAW laser probe. *Proc. IEEE Ultrason. Symp.* 33-36 (1996).

[7] Kamizuma, H., Hashimoto, K., Omori, T., Yamaguchi, M.: Development of fast-scanning laser probe system based on knife-edge method for diagnosis of RF surface acoustic wave devices. *IEEE Ultrason. Symp.* 1604-1609 (2005).

[8] Kamizuma, H., Hashimoto, K., Omori, T., Yamaguchi, M.: Development of fast-scanning laser probe system based on knife-edge method for diagnosis of surface acoustic wave Devices. *IEEE Trans. Ultrason., Ferroelec. Freq. Contr.* **52**(6) 1186-1191 (2006).

[9] Hashimoto, K., Kashiwa, K., Omori, T., Yamaguchi, M., Takano, O., Meguro, S., Akahane, K: A fast scanning laser probe based on sagnac interferometer for RF surface and bulk acoustic wave devices: technical digest. *IEEE Microwave Symp* 851-854 (2008).

[10] Tachizaki, T., Murota, T., Matsuda, O., Sugawara, Y., Hurley, D.H., Wright, O.B.: Scanning ultrafast sagnac interferometry for imaging surface wave propagation. *Review of Scientific Instruments* **77** 043713 (2006).

[11] Hashimoto, K., Kamizuma, H., Watanabe, M., Omori, T., Yamaguchi, M.: Wavenumber domain analysis of two-dimensional SAW images captured by phase-sensitive laser probe system: *IEEE Trans. Ultrason.*, Ferroelec. and Freq. Contr. **53**(5) 1072-1075 (2007).

[12] Wu, N., Hashimoto, K., Kashiwa, K., Omori, T., Yamaguchi, M.: Study on the frequency dependence of lateral energy leakage in RF BAW device by fast-scanning laser probe system. *Proc. IEEE Ultrason. Symp.* 94-97 (2008).

[13] Kashiwa, K., Omori, T., Hashimoto, K., Yamaguchi, M.: Improvement of detection system for a fast-scanning and phase-sensitive laser probe for surface acoustic wave devices in 2 GHz range. *Jpn. J. Appl. Phys.* **47**(5B) 4108-4110 (2008).

[14] Wu, N., Hashimoto, K., Omori, T., Yamaguchi, M., Kasai, N.: Focus adjustment system for a fast-scanning and phase-sensitive laser probe for radio frequency surface and bulk acoustic wave devices. *IEEE Freq. Contr. Symp.* to be published (2009).

[15] Nishihara, T., Yokoyama, T., Miyashita, T., Satoh, Y.: High performance and miniature thin film bulk acoustic wave filters for 5 GHz. *Proc. IEEE Ultrason. Symp.* 969-972 (2002).

[16] Wu, N., Hashimoto, K., Kashiwa, K., Omori, T., Yamaguchi, M.: Integrated software for image processing in RF SAW/BAW laser probe system. *Proc. IEEE Freq. Contr. Symp.* 773-778 (2008).

[17] Nakahata, H., Kitabayashi, H., Hachigo, A., Higaki, K., Fujii, S., Uemura, T., Shikata, S.: Diamond SAW Filter for 2.488 Gbps Retimig. *Proc. IEEE Ultrason. Symp.* 319-322 (1998).

A Two-Dimensional Analysis of Surface Acoustic Waves in Finite Piezoelectric Plates

Ji Wang, Rongxing Wu and Jianke Du

Piezoelectric Device Laboratory, Department of Mechanics and Engineering Science, School of Engineering, Ningbo University, 818 Fenghua Road, Ningbo, Zhejiang 315211, China

wangji@nbu.edu.cn

Abstract. The analysis of surface acoustic waves in finite elastic solids is of fundamental and practical importance, giving the fact that those typical problems concerning solutions of the well-known wave propagation equations are hard to obtain. In searching of an accurate analytical method, we compare this problem with the known bulk acoustic wave problems in finite plates and corresponding Mindlin and Lee plate theories which have been instrumental in solving these problems with accuracy and simplicity. Analogous to power and trigonometric series expansions of displacements, we use exponential functions obtained from semi-infinite solutions of surface acoustic waves to represent the decaying displacements along thickness direction. We present a two-dimensional theory specifically for surface acoustic waves in finite solids with goals to use it for surface acoustic wave resonator analysis and design. The two-dimensional theory for piezoelectric plates is presented and considered through the effective elastic constants for simplification.

1. Introduction

Surface acoustic waves propagating in the vicinity of surface of semi-infinite elastic solids have been well presented in many popular textbooks [1-4] and essential results and properties from these simple and elegant solutions have been widely used as foundations for further research on surface acoustic waves (SAW) [5-7]. Since analytical solutions of practical problems concerning finite elastic solids are not available, approximations have been made for simplified equations, which are essentially one-dimensional and solvable, but less rigorous in comparison with the usual three-dimensional equations that can only be treated numerically [8-9].

T.-T. Wu and C.-C. Ma (eds.), *IUTAM Symposium on Recent Advances of Acoustic Waves in Solids*, IUTAM Bookseries 26, DOI 10.1007/978-90-481-9893-1_23,

Consequently, practical equations and methods concerning SAW in finite solids are also oversimplified, leaving accurate solutions based on a systematic approach similar to the sophisticated plate theories for bulk acoustic waves (BAW) in finite solids yet to be established.

Noting the fact that accurate solutions from an infinite plate have been used for the development of two-dimensional theories, or plate theories, for BAW in finite plates by Mindlin [10], Lee [11], and Peach [12], we start with accurate solutions of surface acoustic waves in semi-infinite solids for a similar two-dimensional analysis. Naturally, for isotropic materials, we expect the dominant SAW modes in a finite elastic solid will closely resemble to the two modes appearing in semi-infinite solids, thus enabling the expansion of displacements in the known exponential functions from semi-infinite solutions. Then a two-dimensional theory for the analysis of SAW in finite elastic solids can be established based on the known three-dimensional solutions. This implies that we are intended to use the eigensolutions of Rayleigh waves for the construction of a two-dimensional theory by following earlier successes. For isotropic materials, as we stated before, there will be only two dominant modes in a semi-infinite solid. However, there will be three dominant modes in case of anisotropic materials. All these will be reflected in the two-dimensional theory, which will be material-dependent in the order of the expansion and equations. In case of piezoelectric materials, there will be four eigenmodes which will pose further challenges in the formulation and solution.

2. Two-Dimensional Equations

For SAW in a semi-infinite isotropic solid, solutions include the wave velocity and two parameters related to the decaying of displacements along the depth of solids for the calculation of vibration modes. Since we assume the surface acoustic waves in finite anisotropic solids are dominated by modes appearing in the semi-infinite solid of the same material, following Mindlin [10] and Lee [11], we write displacements with three components as

$$u_j(x_1, x_2, x_3, t) = \sum_{n=1}^{3} u_j^{(n)}(x_1, x_3, t)\varphi_n(x_2),\ j = 1,2,3, n = 1,2,3, \tag{1}$$

with x_j, t and $u_j^{(n)}$ as the coordinates, time, and the nth-order displacements and expansion functions defined as

$$\varphi_n(x_2) = \left\{\exp(k\beta_1 x_2), \exp(k\beta_2 x_2), \exp(k\beta_3 x_2)\right\}, n = 1,2,3, \tag{2}$$

where k is wavenumber and the decaying parameters $\beta_n (n=1,2,3)$ are obtained from SAW solutions of semi-infinite solids. In case of isotropic materials, there are only two decaying parameters and the displacement expansion is simplified accordingly. The results have been presented before in our previous paper with complete analysis of mode shapes and frequency spectra [13]. For anisotropic materials, there will be three decaying parameters in general with complicated displacement expansion [14]. In this paper, we focus on rotated Y-cut quartz crystals which are commonly used as substrates in SAW resonators.

With displacements in (1) and basis functions for expansion in (2), corresponding strains in abbreviated notation [10, 11, 15, 16] are defined as

$$\begin{aligned} &S_p = \sum_{n=1}^{3} S_p^{(n)} \varphi_n, S_1^{(n)} = u_{1,1}^{(n)}, S_2^{(n)} = k\beta_n u_2^{(n)}, S_3^{(n)} = u_{3,3}^{(n)}, \\ &S_4^{(n)} = u_{2,3}^{(n)} + k\beta_n u_3^{(n)}, S_5^{(n)} = u_{3,1}^{(n)} + u_{1,3}^{(n)}, S_6^{(n)} = u_{2,1}^{(n)} + k\beta_n u_1^{(n)}, \\ &p = 1,2,3,4,5,6, n = 1,2,3. \end{aligned} \tag{3}$$

Consequently, for anisotropic materials the stresses are

$$T_p = \sum_{n=1}^{3} \hat{T}_p^{(n)} \varphi_n, \hat{T}_p^{(n)} = c_{pq} S_q^{(n)}, p, q = 1,2,3,4,5,6; n = 1,2,3, \tag{4}$$

where c_{pq} are elastic constants.

For an elastic plate with thickness h shown in Fig. 1, stress equations of motion in variational form are

$$\int_A \int_{-h}^{0} \left(T_{ij,i} - \rho \ddot{u}_j\right) \sum_{m=1}^{3} \delta u_j^{(m)} \varphi_m dx_2 dA = 0, \tag{5}$$

and through integration over the thickness coordinate we have the corresponding two-dimensional equations as

$$T_{ij,i}^{(n)} - \overline{T}_{2j}^{(n)} + F_{2j}^{(n)} = \rho \sum_{m=1}^{3} A_{mn} \ddot{u}_j^{(m)}, i, j = 1,2,3, n = 1,2,3, \tag{6}$$

where the two-dimensional quantities are defined as

$$\begin{aligned} &T_{ij}^{(n)} = \int_{-h}^{0} T_{ij} \varphi_n dx_2, \overline{T}_{2j}^{(n)} = k\beta_n T_{2j}^{(n)}, F_{2j}^{(n)} = T_{2j}(0)\varphi_n(0) - T_{2j}(-h)\varphi_n(-h), \\ &A_{mn} = \int_{-h}^{0} \varphi_m \varphi_n dx_2, i, j = 1,2,3; n = 1,2,3. \end{aligned} \tag{7}$$

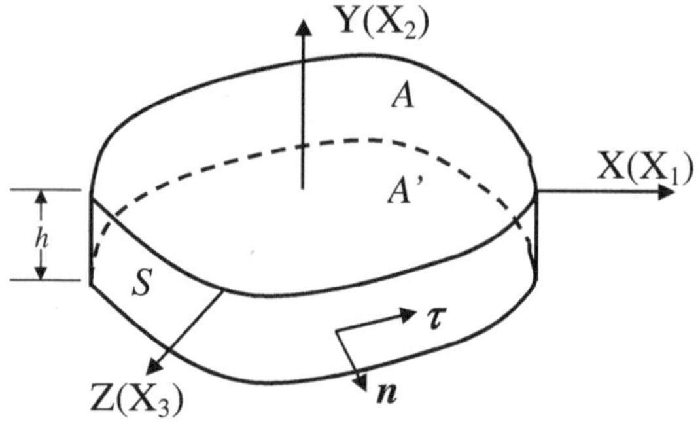

Fig. 1 A finite elastic solid with coordinate system.

With two-dimensional variables, we have the nth-order stresses in (7) as

$$T_p^{(n)} = \sum_{m=1}^{3} A_{mn} \Big[c_{p1} u_{1,1}^{(n)} + c_{p2} k\beta_n u_2^{(n)} + c_{p3} u_{3,3}^{(n)} + c_{p4} \left(u_{2,3}^{(n)} + k\beta_n u_3^{(n)} \right) + c_{p5} \left(u_{3,1}^{(n)} + u_{1,3}^{(n)} \right) + c_{p6} \left(u_{2,1}^{(n)} + k\beta_n u_1^{(n)} \right) \Big], \tag{8}$$

and equations of motion in (6) can be expanded to

$$\begin{aligned} T_{1,1}^{(n)} + T_{5,3}^{(n)} - k\beta_n T_6^{(n)} + F_6^{(n)} &= \rho \sum_{m=1}^{3} A_{mn} \ddot{u}_1^{(m)}, \\ T_{6,1}^{(n)} + T_{4,3}^{(n)} - k\beta_n T_2^{(n)} + F_2^{(n)} &= \rho \sum_{m=1}^{3} A_{mn} \ddot{u}_2^{(m)}, \\ T_{5,1}^{(n)} + T_{3,3}^{(n)} - k\beta_n T_4^{(n)} + F_4^{(n)} &= \rho \sum_{m=1}^{3} A_{mn} \ddot{u}_3^{(m)}, n = 1,2,3. \end{aligned} \tag{9}$$

The integral (7) with known expansion functions from (2) are

$$A_{mn} = \int_{-h}^{0} \varphi_m \varphi_n dx_2 = \frac{1}{k(\beta_m + \beta_n)} \left[1 - \frac{1}{\exp 2\pi(\beta_m + \beta_n) H} \right], H = \frac{h}{\lambda}, m, n = 1,2,3, \tag{10}$$

where λ is wavelength. For a finite plate-like solid, we can always use the wavelength as a measure of plate thickness. Since the value of $\beta_m \, (m = 1,2,3)$ is close to unity (1.0) for materials we have studied, larger H will make the integral in (10) to a simple expression

$$A_{mn} = \frac{1}{k(\beta_m + \beta_n)}, m, n = 1,2,3. \tag{11}$$

Referring to Fig. 1, boundary conditions are

$$\oint_C \int_{-h}^{0} n_i T_{ij} \delta u_j ds dx_2 = \oint_C \int_{-h}^{0} n_i T_{ij} \sum_{n=1}^{3} \delta u_j^{(n)} \varphi_n ds dx_2 = \sum_{n=1}^{3} \oint_C n_i T_{ij}^{(n)} \delta u_j^{(n)} ds = 0, \tag{12}$$

where C is the cylindrical surface. Basically it states that on the cylindrical surface we can specify either higher-order displacements or stress components.

We now examine SAW travelling along the x_1 direction. For simplicity, and the solutions are assumed as

$$u_1^{(m)} = A_1^{(m)} \sin k\bar{k}x_1 e^{ikct}, u_2^{(m)} = A_2^{(m)} \cos k\bar{k}x_1 e^{ikct}, u_3^{(m)} = A_3^{(m)} \cos k\bar{k}x_1 e^{ikct}, \tag{13}$$

where $A_i^{(m)} (i = 1,2,3), k, \bar{k}, c$, and t are amplitudes, wavenumber, associated wavenumber, velocity, and time, respectively. It should be emphasized that the associated wavenumbers $\bar{k}$ only appear in the trigonometric function arguments.

By substituting (13) into (9), we have the equations in expansion as

$$\begin{aligned}
&T_{1,1}^{(1)} - k\beta_1 T_6^{(1)} + F_6^{(1)} = \rho\left(A_{11}\ddot{u}_1^{(1)} + A_{21}\ddot{u}_1^{(2)} + A_{31}\ddot{u}_1^{(3)}\right),\\
&T_{6,1}^{(1)} - k\beta_1 T_2^{(1)} + F_2^{(1)} = \rho\left(A_{11}\ddot{u}_2^{(1)} + A_{21}\ddot{u}_2^{(2)} + A_{31}\ddot{u}_2^{(3)}\right),\\
&T_{5,1}^{(1)} - k\beta_1 T_4^{(1)} + F_4^{(1)} = \rho\left(A_{11}\ddot{u}_3^{(1)} + A_{21}\ddot{u}_3^{(2)} + A_{31}\ddot{u}_3^{(3)}\right),\\
&T_{1,1}^{(2)} - k\beta_2 T_6^{(2)} + F_6^{(2)} = \rho\left(A_{12}\ddot{u}_1^{(1)} + A_{22}\ddot{u}_1^{(2)} + A_{32}\ddot{u}_1^{(3)}\right),\\
&T_{6,1}^{(2)} - k\beta_2 T_2^{(2)} + F_2^{(2)} = \rho\left(A_{12}\ddot{u}_2^{(1)} + A_{22}\ddot{u}_2^{(2)} + A_{32}\ddot{u}_2^{(3)}\right),\\
&T_{5,1}^{(2)} - k\beta_2 T_4^{(2)} + F_4^{(2)} = \rho\left(A_{12}\ddot{u}_3^{(1)} + A_{22}\ddot{u}_3^{(2)} + A_{32}\ddot{u}_3^{(3)}\right),\\
&T_{1,1}^{(3)} - k\beta_3 T_6^{(3)} + F_6^{(3)} = \rho\left(A_{13}\ddot{u}_1^{(1)} + A_{23}\ddot{u}_1^{(2)} + A_{33}\ddot{u}_1^{(3)}\right),\\
&T_{6,1}^{(3)} - k\beta_3 T_2^{(3)} + F_2^{(3)} = \rho\left(A_{13}\ddot{u}_2^{(1)} + A_{23}\ddot{u}_2^{(2)} + A_{33}\ddot{u}_2^{(3)}\right),\\
&T_{5,1}^{(3)} - k\beta_3 T_4^{(3)} + F_4^{(3)} = \rho\left(A_{13}\ddot{u}_3^{(1)} + A_{23}\ddot{u}_3^{(2)} + A_{33}\ddot{u}_3^{(3)}\right).
\end{aligned} \tag{14}$$

With solutions of $\bar{k}$ and the amplitude ratios from above

$$\alpha_{mi} = \frac{A_k^{(n)}\left(\bar{k}_i\right)}{A_3^{(3)}\left(\bar{k}_i\right)}, n, k = 1,2,3, i = 1,2,3,\cdots,9, m = 3(n-1)+k, \tag{15}$$

we can rewrite stresses as

$$
\begin{aligned}
T_1^{(n)} &= \sum_{i=1}^{9}\Big\{A_{1n}\left[c_{11}\alpha_{1i}+c_{12}\beta_1\alpha_{2i}+c_{14}\beta_1\alpha_{3i}\right]+A_{2n}\left[c_{11}\alpha_{4i}\bar{k}_i+c_{12}\beta_2\alpha_{5i}+c_{14}\beta_2\alpha_{6i}\right] \\
&\quad + A_{3n}\left[c_{11}\alpha_{7i}\bar{k}_i+c_{12}\beta_3\alpha_{8i}+c_{14}\beta_3\alpha_{9i}\right]\Big\}A_3^{(3)}\left(\bar{k}_i\right)k\cos\bar{k}_i k x_1 e^{ikct}, \\
T_6^{(n)} &= \sum_{i=1}^{9}\Big\{A_{1n}\left[c_{66}\left(\beta_1\alpha_{1i}-\alpha_{2i}\bar{k}_i\right)-c_{65}\bar{k}_i\alpha_{2i}\right]+A_{2n}\left[c_{66}\left(\beta_2\alpha_{4i}-\alpha_{5i}\bar{k}_i\right)-c_{65}\bar{k}_i\alpha_{6i}\right] \\
&\quad + A_{3n}\left[c_{66}\left(\beta_3\alpha_{7i}-\alpha_{8i}\bar{k}_i\right)-c_{65}\bar{k}_i\alpha_{9i}\right]\Big\}A_3^{(3)}\left(\bar{k}_i\right)k\sin\bar{k}_i k x_1 e^{ikct}, \\
T_5^{(n)} &= \sum_{i=1}^{9}\Big\{A_{1n}\left[c_{56}\left(\beta_1\alpha_{1i}-\alpha_{2i}\bar{k}_i\right)-c_{55}\bar{k}_i\alpha_{2i}\right]+A_{2n}\left[c_{56}\left(\beta_2\alpha_{4i}-\alpha_{5i}\bar{k}_i\right)-c_{55}\bar{k}_i\alpha_{6i}\right] \\
&\quad + A_{3n}\left[c_{56}\left(\beta_3\alpha_{7i}-\alpha_{8i}\bar{k}_i\right)-c_{55}\bar{k}_i\alpha_{9i}\right]\Big\}A_3^{(3)}\left(\bar{k}_i\right)k\sin\bar{k}_i k x_1 e^{ikct}.
\end{aligned}
\tag{16}
$$

The traction-free boundary conditions are

$$
T_1^{(n)}\left(x_1=\pm a\right)=T_6^{(n)}\left(x_1=\pm a\right)=T_5^{(n)}\left(x_1=\pm a\right)=0, n=1,2,3. \tag{17}
$$

By substituting (16) into (17), we have frequency spectra which is the most important solutions for finite solids we are interested.

3. Surface Acoustic Waves in Piezoelectric Plates

To develop a two-dimensional theory for the SAW in finite piezoelectric solids, we expand the displacements and electrical potential in known basis functions of the thickness coordinate as

$$
\begin{aligned}
u_j\left(x_1,x_2,x_3,t\right) &= \sum_{n=1}^{4} u_j^{(n)}\left(x_1,x_3,t\right)\varphi_n\left(x_2\right), \\
\varphi\left(x_1,x_2,x_3,t\right) &= \sum_{n=1}^{4} \varphi^{(n)}\left(x_1,x_3,t\right)\varphi_n\left(x_2\right), j=1,2,3,
\end{aligned}
\tag{18}
$$

where $u_j^{(n)}\left(j=1,2,3\right),\varphi^{(n)}$ and $\varphi_n\left(n=1,2,3,4\right)$ are higher-order displacements, electrical potential components, and basis functions, respectively. The basis function will be from SAW solutions of semi-infinite piezoelectric solids in the form

$$
\varphi_n\left(x_2\right)=\left\{\exp\left(k\beta_1 x_2\right),\exp\left(k\beta_2 x_2\right),\exp\left(k\beta_3 x_2\right),\exp\left(k\beta_4 x_2\right)\right\}, n=1,2,3,4, \tag{19}
$$

where decaying parameters $\beta_n\,(n=1,2,3,4)$ are obtained from a semi-infinite piezoelectric solid with propagating surface acoustic waves.

Based on the piezoelectricity theory and displacements and electrical potential in (18) and (19), we have the corresponding strain and electrical field as

$$S_p=\sum_{n=1}^{4}S_p^{(n)}\varphi_n, E_i=\sum_{n=1}^{4}E_i^{(n)}\varphi_n, p=1,2,3,4,5,6, i=1,2,3, \tag{20}$$

where $S_p^{(n)}$ and $E_i^{(n)}\,(n=1,2,3,4)$ are higher-order two-dimensional strain and electrical filed components, respectively. With the given displacements and electrical potential in (18), it is straight-forwarded to have the following components

$$\begin{aligned}&S_1^{(n)}=u_{1,1}^{(n)}, S_2^{(n)}=k\beta_n u_2^{(n)}, S_3^{(n)}=u_{3,3}^{(n)}, S_4^{(n)}=u_{2,3}^{(n)}+k\beta_n u_3^{(n)}, S_5^{(n)}=u_{3,1}^{(n)}+u_{1,3}^{(n)},\\&S_6^{(n)}=u_{2,1}^{(n)}+k\beta_n u_1^{(n)}; E_1^{(n)}=-\varphi_{,1}^{(n)}, E_2^{(n)}=-k\beta_n\varphi^{(n)}, E_3^{(n)}=-\varphi_{,3}^{(n)}, n=1,2,3,4.\end{aligned} \tag{21}$$

Consequently, we have the stresses and electrical displacements for an anisotropic piezoelectric material as

$$T_p=\sum_{n=1}^{4}\hat{T}_p^{(n)}\varphi_n, D_i=\sum_{n=1}^{4}\hat{D}_i^{(n)}\varphi_n, p=1,2,3,4,5,6; i=1,2,3, \tag{22}$$

with

$$\begin{aligned}&\hat{T}_p^{(n)}=c_{pq}S_q^{(n)}-e_{pk}E_k^{(n)}, \hat{D}_i^{(n)}=e_{ip}S_p^{(n)}+\varepsilon_{ik}E_k^{(n)},\\&p,q=1,2,3,4,5,6; i,k=1,2,3; n=1,2,3,4,\end{aligned} \tag{23}$$

where c_{pq}, e_{pk} and ε_{ik} are elastic, piezoelectric, and dielectric constants, respectively.

With displacements and electrical potential in (18) and strains and electrical fields in (21), we follow Mindlin [10] and Lee [11] with the variational forms of stress equations of motion and charge equation of linear piezoelectricity to a piezoelectric solid as illustrated in Fig. 1

$$\int_A\int_{-h}^{0}\left(T_{ij,i}-\rho\ddot{u}_j\right)\sum_{m=1}^{4}\delta u_j^{(m)}\varphi_m dx_2 dA=0, \int_A\int_{-h}^{0}D_{i,i}\sum_{m=1}^{4}\delta\varphi_j^{(m)}\varphi_m dx_2 dA=0, \tag{24}$$

where A is the faces, or the domain of the solids, h is the thickness, and ρ is the density of material. Through integration of (24), we have

$$\begin{aligned}&T_{ij,i}^{(n)}-\overline{T}_{2j}^{(n)}+F_{2j}^{(n)}=\rho\sum_{m=1}^{4}A_{mn}\ddot{u}_j^{(m)},\\&D_{i,i}^{(n)}-\overline{D}_2^{(n)}+D^{(n)}=0, i,j=1,2,3; n=1,2,3,4,\end{aligned} \tag{25}$$

with two-dimensional quantities defined as

$$T_{ij}^{(n)} = \int_{-h}^{0} T_{ij}\varphi_n dx_2, \bar{T}_{2j}^{(n)} = k\beta_n T_{2j}^{(n)}, F_{2j}^{(n)} = T_{2j}(0)\varphi_n(0) - T_{2j}(-h)\varphi_n(-h),$$
$$D_i^{(n)} = \int_{-h}^{0} D_i\varphi_n dx_2, \bar{D}_2^{(n)} = k\beta_n D_2^{(n)}, D^{(n)} = D_2(0)\varphi_n(0) - D_2(-h)\varphi_n(-h), \quad (26)$$
$$A_{mn} = \int_{-h}^{0} \varphi_m\varphi_n dx_2, i, j = 1,2,3; n = 1,2,3,4.$$

Now we have the stress and electrical displacement components as

$$T_p^{(n)} = \sum_{m=1}^{4} A_{mn}\Big[c_{p1}u_{1,1}^{(m)} + c_{p2}k\beta_m u_2^{(m)} + c_{p3}u_{3,3}^{(m)} + c_{p4}\left(u_{2,3}^{(m)} + k\beta_m u_3^{(m)}\right) + c_{p5}\left(u_{3,1}^{(m)} + u_{1,3}^{(m)}\right)$$
$$+ c_{p6}\left(u_{2,1}^{(m)} + k\beta_m u_1^{(m)}\right) + e_{1p}\varphi_{,1}^{(m)} + e_{2p}k\beta_m\varphi^{(m)} + e_{3p}\varphi_{,3}^{(m)}\Big],$$
$$D_i^{(n)} = \sum_{m=1}^{4} A_{mn}\Big[e_{i1}u_{1,1}^{(m)} + e_{i2}k\beta_m u_{2,2}^{(m)} + e_{i3}u_{3,3}^{(m)} + e_{i4}\left(u_{2,3}^{(m)} + k\beta_m u_3^{(m)}\right) + e_{i5}\left(u_{3,1}^{(m)} + u_{1,3}^{(m)}\right) \quad (27)$$
$$+ e_{i6}\left(u_{2,1}^{(m)} + k\beta_m u_1^{(m)}\right) - \varepsilon_{i1}\varphi_{,1}^{(m)} - \varepsilon_{i2}k\beta_m\varphi^{(m)} - \varepsilon_{i3}\varphi_{,3}^{(m)}\Big].$$

As a result, the stress equations of motion and charge equations in terms of the higher-order stress and electrical displacement components are

$$T_{1,1}^{(n)} + T_{5,3}^{(n)} - k\beta_n T_6^{(n)} + F_6^{(n)} = \rho\sum_{m=1}^{4} A_{mn}\ddot{u}_1^{(m)},$$
$$T_{6,1}^{(n)} + T_{4,3}^{(n)} - k\beta_n T_2^{(n)} + F_2^{(n)} = \rho\sum_{m=1}^{4} A_{mn}\ddot{u}_2^{(m)},$$
$$T_{5,1}^{(n)} + T_{3,3}^{(n)} - k\beta_n T_4^{(n)} + F_4^{(n)} = \rho\sum_{m=1}^{4} A_{mn}\ddot{u}_3^{(m)}, \quad (28)$$
$$D_{1,1}^{(n)} + D_{3,3}^{(n)} - k\beta_n D_2^{(n)} + D^{(n)} = 0, n = 1,2,3,4.$$

We now examine SAW travelling along the x_1 direction in a straight-crested manner. For simplicity, solutions are assumed as

$$u_1^{(m)} = A_1^{(m)}\sin k\bar{k}x_1 e^{ikct}, u_2^{(m)} = A_2^{(m)}\cos k\bar{k}x_1 e^{ikct},$$
$$u_3^{(m)} = A_3^{(m)}\cos k\bar{k}x_1 e^{ikct}, \varphi^{(m)} = A_4^{(m)}\sin k\bar{k}x_1 e^{ikct}. \quad (29)$$

The usual solution procedure, which is to substitute (29) back into the 16 equations in (28), does not yield the results we are expected. Namely, the associated wavenumbers are not conjugated complex numbers which can be utilized. The solutions of (28) are to be examined for further interpretation and the construction of displacements and electrical potential.

The approximation, on the other hand, can be made based on the fact that SAW velocity of piezoelectric solids can be obtained with the piezoelectrically stiffened elastic constants in the form of [5]

$$\overline{c}_{pq} = c_{pq} + \frac{e_{1p} e_{1q}}{\varepsilon_{11}}, p, q = 1, 2, 3, 4, 5, 6, \tag{30}$$

where e_{1p}, e_{1q}, and ε_{11} are specific piezoelectric and dielectric constants, respectively. Accordingly, the two-dimensional equations can be simplified also.

This approximation can be traced to the three-dimensional equation of charge of piezoelectricity with the truncation

$$\varepsilon_{11} \varphi_{,11} = e_{11} u_{1,11} + e_{15} u_{3,11} + e_{16} u_{2,11}. \tag{31}$$

Through integration of this equation, we have

$$\varphi = \frac{e_{11} u_1 + e_{15} u_3 + e_{16} u_2}{\varepsilon_{11}} + f_1(x_2) x_1 + f_2(x_2), \tag{32}$$

where two arbitrary functions $f_i(x_2)(i = 1, 2)$ are to be determined. Naturally, we can approximate the two-dimensional electrical potential with (32) as

$$\varphi = \frac{e^{ikct}}{\varepsilon_{11}} \sum_{i=1}^{9} A_{3i}^{(3)} \left(e_{11} \sin k\overline{k}_i x_1 + e_{16} \cos k\overline{k}_i x_1 + e_{15} \cos k\overline{k}_i x_1 \right) + f_1(x_2) x_1 + f_2(x_2). \tag{33}$$

For typical piezoelectric devices with electrodes on the upper and lower faces shown in Fig. 1, we can conveniently treat the two arbitrary functions as the static electrical potentials and let

$$f_1(x_2) = 0, f_2(x_2) = \frac{\varphi(0)}{h}(h - x_2) + \frac{\varphi(h)}{h} x_2, \tag{34}$$

where $\varphi(0)$ and $\varphi(h)$ are the potentials on the lower and upper faces. The solution in (34) will enable the calculation of electrical parameters of a resonator with known electrical voltage on the electrodes, which can be discrete, or periodic, as in actual structures we are familiar with.

4. Numerical Examples

As a numerical example, we study an AT-cut quartz crystal strip of $H = 10$ with straight-crested surface acoustic waves. The SAW velocity normalized by $\sqrt{c_{66}/\rho}$ *vs.* length is given in Fig. 2. A similar study for different materials is done by Wang and Hashimoto [14]. The results show that velocities of SAW mode and its overtones are generally sensitive to the plate length, while mode conversion occurs at certain lengths which should be avoided in the selection of substrates for SAW resonators. This has been observed in our earlier studies with isotropic and anisotropic materials [13-14, 17]. Although there are complication factors such as the periodic electrodes affect the SAW propagation in finite anisotropic solids, proper selection of substrates is essential in resonator performance improvement. To this purpose, the velocity spectra based on our two-dimensional analysis is of practical importance, and further refinement with the consideration of electrodes will be more useful [18].

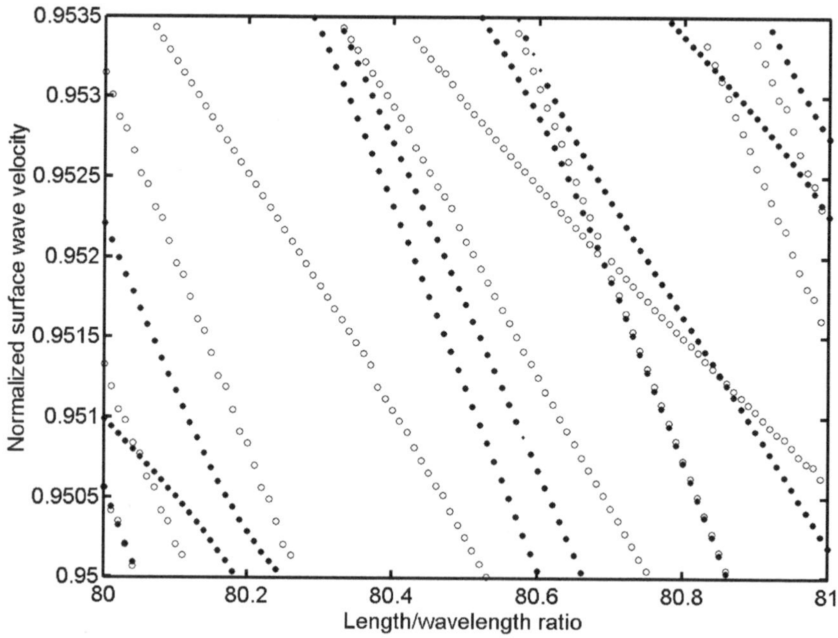

Fig. 2 Normalized surface acoustic wave velocity *vs.* length in an AT-cut quartz plate with thickness $H = 10$. The computed frequency spectra from the mechanical vibrations (circle ○) is compared the piezoelectrically stiffened constants (dot •).

It is also clearly shown that the piezoelectricity has noticeable effect on the SAW velocity in a finite structure based on the analysis with the two-dimensional theory and piezoelectrically stiffened elastic constants. Our approximation is based on the fact that the piezoelectric effect through three-dimensional equations is very small [5], and we are expecting the same from the two-dimensional equations with stiffened elastic constants. The results in Fig. 2 show that there are changes we cannot make adequate comparison at this moment. We expect the further calculation on the frequency spectra in a much larger frequency and parameter range will enable us to visualize the changes more clearly. The simplified consideration of piezoelectric effect will make the analysis of SAW resonators with larger number of pairs of periodic electrodes more convenient [19-21], particularly for the electrical circuit parameters which are more important in the design and applications.

5. Conclusions

A two-dimensional theory for the analysis of SAW in finite anisotropic solids is established based on the exponential expansion of displacements in the thickness direction with representative parameters from semi-infinite substrates. The two-dimensional theory is analogous to the well-known plate theories by Mindlin and Lee for the propagation of BAW in finite elastic solids. These two-dimensional equations can be used for analytical solutions in a manner similar to plate equations for BAW problems, or they can be implemented in finite element method to improve the efficiency of numerical techniques. The limited order of variables and the absence of correction procedure make the two-dimensional theory for SAW simple and effective in practical applications involving finite anisotropic piezoelectric solids. With the calculated velocity (frequency) spectra and identified wave modes, the selection of substrates of SAW resonators can be optimal, thus easing the burden of trial-and-error iteration in the design process. The extension of the two-dimensional theory to anisotropic materials and approximate consideration of piezoelectric effect through the stiffened elastic constants are the necessary steps for the refinement to meet the needs of SAW resonator design and optimization.

Acknowledgements: This research is supported by grants from the National Natural Science Foundation of China (Nos. 10572065 and 10772087). Additional support is from the K. C. Wong Magna Fund administered by Ningbo University.

References

[1] Achenbach, J. D.: *Wave Propagation in Elastic Solids*. North-Holland (1987).
[2] Graff, K. F.: *Wave Motion in Elastic Solids*. Dover (1991).
[3] Auld, B. A.: *Acoustic Fields and Waves in Solids (I & II)*. Krieger, Malabar, FL (1990).
[4] Royer, D., Dieulesaint, E.: *Elastic Waves in Solids (I & II)*. Springer, Berlin (2000).
[5] Farnell, G.W.: Properties of elastic surfaces waves. In: Mason, W.P., Thurston, R. N. (eds.): *Physical Acoustics*, VI, pp. 109-166. Academic Press, New York (1970).
[6] Wu, Y.: Waves in Elastic Solids. In: Zhang, F., Wang, L. (eds.): *Modern Piezoelectricity*, I, pp. 100-140. Science Press, Beijing (2001). (In Chinese)
[7] Hashimoto, K.-Y.: *Surface Acoustic Wave Devices in Telecommunications. Springer*, Berlin (2000).
[8] Yong, Y.-K.: Analysis of periodic structures for BAW and SAW resonators. *Proc. of 2001 IEEE Intl. Ultrasonics Symp.* 781-790 (2001).
[9] Yoon, S., Yu, J.-D., Kanna, S., Oshio, M., Tanaka, M.: Finite element analysis of the substrate thickness on traveling leaky surface acoustic waves. *Proc. of 2004 IEEE Intl. Ultrasonics Symp.* 1696-1699 (2004).
[10] Mindlin, R.D.: High frequency vibrations of piezoelectric crystal plates. *Intl. J. Solids Struct.* **8** 891-906 (1972).
[11] Lee, P.C.Y., Yu, J.-D., Lin, W.-S.: A new two-dimensional theory for vibrations of piezoelectric crystal plates with electroded faces. *J. Appl. Phys.* **83**(3) 1213-1223 (1998).
[12] Peach, R.C.: A normal mode expansion for the piezoelectric plates and certain of its applications. *IEEE Trans. UFFC.* **35**(5) 593-611 (1988).
[13] Wang, J., Hashimoto, K.-Y.: A two-dimensional theory for the analysis of surface acoustic waves in finite elastic solids. *J. Sound Vib.* **295** 838-855 (2006).
[14] Wang, J., Hashimoto, K.-Y.: A two-dimensional theory for the analysis of surface acoustic waves in anisotropic elastic solids. *Proc. of 2003 IEEE Intl. Ultrasonics Symp.* 637-640 (2003).
[15] Mindlin, R.D. (Yang, J.S., ed.): An Introduction to the Mathematical Theory of Vibrations of Elastic Plates. *World Scientific* Hackensack, NJ (2006).
[16] Tiersten, H. F.: *Linear Piezoelectric Plate Vibrations*. Plenum, New York (1969).
[17] Wang, J., Du, J., Pan, Q.: A two-dimensional analysis of surface acoustic waves in finite plates with eigensolutions. *Sci. China Ser. G.* **50**(5), 631-649 (2007).
[18] Wang, J., Li, Z., Du, J.: An analysis of the effect of periodic electrodes on surface acoustic wave resonators. *Proc. of IEEE 2006 Intl. Freq. Control Symp.* 161-164 (2006).
[19] Wang, J., Du, J., Lin, J., Li, Z.: Two-dimensional analysis of the effect of an electrode layer on surface acoustic waves in a finite anisotropic plate. *Ultrasonics* **44** (S1) 935-939 (2006).
[20] Wang, J., Lin, J., Wan, Y., Zhong, Z.: A two-dimensional analysis of surface acoustic waves in finite solids with considerations of electrodes. *Int. J. Appl. Electrom.* **22** 53-68 (2005).
[21] Rahman, S., Langtangen, H.P., Barnes, C. H. W.: A finite element method for modelling electromechanical wave propagation in anisotropic piezoelectric media. *J. Comput. Phys.* **2** (2), 271-292 (2007).

Wireless Drive of Piezoelectric Components

Junhui Hu and Satyanarayan Bhuyan

School of Electrical & Electronic Engineering, Division of Microelectronics, Nanayang Technological University, Singapore 639798, Singapore

ejhhu@ntu.edu.sg

Abstract. In this work, electromagnetic wave is used to drive piezoelectric components. We proposed and investigated three new techniques of transmitting electric energy wirelessly to piezoelectric components which include wireless drive of piezoelectric components by parallel plate capacitor structure, by focused electric field structure and by dipole antenna-like structure. The output power of the piezoelectric component wirelessly driven by focused electric field structure is significantly higher than that wirelessly driven by the other two structures. In the focused electric field structure, when the operating frequency is close to mechanical resonance frequency of the piezoelectric component operating in the thickness mode, a maximum real output power of 0.26 W and energy conversion efficiency of 1.02% have been achieved with an input power of 26 W and 1 cm electrode separation. Experimentally it has been found that the real output power achieved by the piezoelectric component depends on the frequency, vibration mode, and electrical load of the piezoelectric component, and electric field. We studied the electric field pattern by finite element method to assess the electric field on the surface of the piezoelectric plate. It is seen that the electric field on the surface of the wirelessly driven piezoelectric plate is non-uniform. We also derived and then proposed an equivalent circuit for the wirelessly driven piezoelectric component operating in the thickness mode, which can be used to analyze the vibration characteristics of the wirelessly driven piezoelectric plate.

1. Introduction

Piezoelectric devices which include transducers [1], motors and actuators [2- 4], transformers [5], oscillators, filters, generators, sensors, [6] etc. have the advantages such as compact size, high power density, high output force, high precision positioning, etc. They have applications in various areas such as precision

T.-T. Wu and C.-C. Ma (eds.), *IUTAM Symposium on Recent Advances of Acoustic Waves in Solids*, IUTAM Bookseries 26, DOI 10.1007/978-90-481-9893-1_24,

positioning, particle manipulations, switching power supplies, communication systems and others. In most applications, electric energy is applied to the devices via lead wires soldered on the electrodes of piezoelectric components. But the lead wires may fall off at large vibration, high input voltage or high temperature, and this causes the breakdown of piezoelectric devices. The lead wires also hinder the applications of piezoelectric devices in rotary mechanisms and micro systems. The fundamental limitation of applying electric energy to the devices via lead wires soldered on the electrodes necessitates the pursuit of wireless drive of piezoelectric components. To widen the application range of piezoelectric devices, wireless drives of piezoelectric components have been investigated by the author using a properly designed electric field [7-9]. It has been observed that the energy in an ac electric field can be transmitted to a piezoelectric plate. However, so far the output power of the driven piezoelectric component has not been high enough. Therefore, the methods of improving the output power of wirelessly driven piezoelectric components need to be investigated further.

2. Experimental Methods

We proposed and investigated three new techniques of transmitting electric energy wirelessly to piezoelectric components which include wireless drive of piezoelectric components by parallel plate capacitor structure, by focused electric field structure and by dipole antenna-like structure. Fig. 2.1(a) shows the experimental setup to measure the output power of the piezoelectric plate wirelessly driven by parallel plate capacitor structure. An ac input voltage is applied to two brass electrodes with dimensions of $5\times2\times0.3$ cm^3, mounted on a plastic plate with a separation distance of 4 mm. The piezoelectric plate is inserted parallel into the gap between the two brass electrodes and the plate is aligned along the central axis of the gap. To measure the output power, two lead wires are soldered onto the output electrodes and a load resistor is connected across the output electrodes of the piezoelectric plate. To transmit a large electric energy to the piezoelectric plate, a focused ac electric field structure is used as shown in Fig. 2.1(b). With a live electrode, a needle ground electrode is used to focus the ac electric field onto the piezoelectric plate. The needle ground electrode is placed below perpendicular to the live electrode which is suspended above the piezoelectric plate. In order to increase the driving power, the focused electric field structure is in series with an inductor. When the electric field generator and inductor are in electric resonance, the transmitted power is increased because of the large voltage across the focused electric field structure.

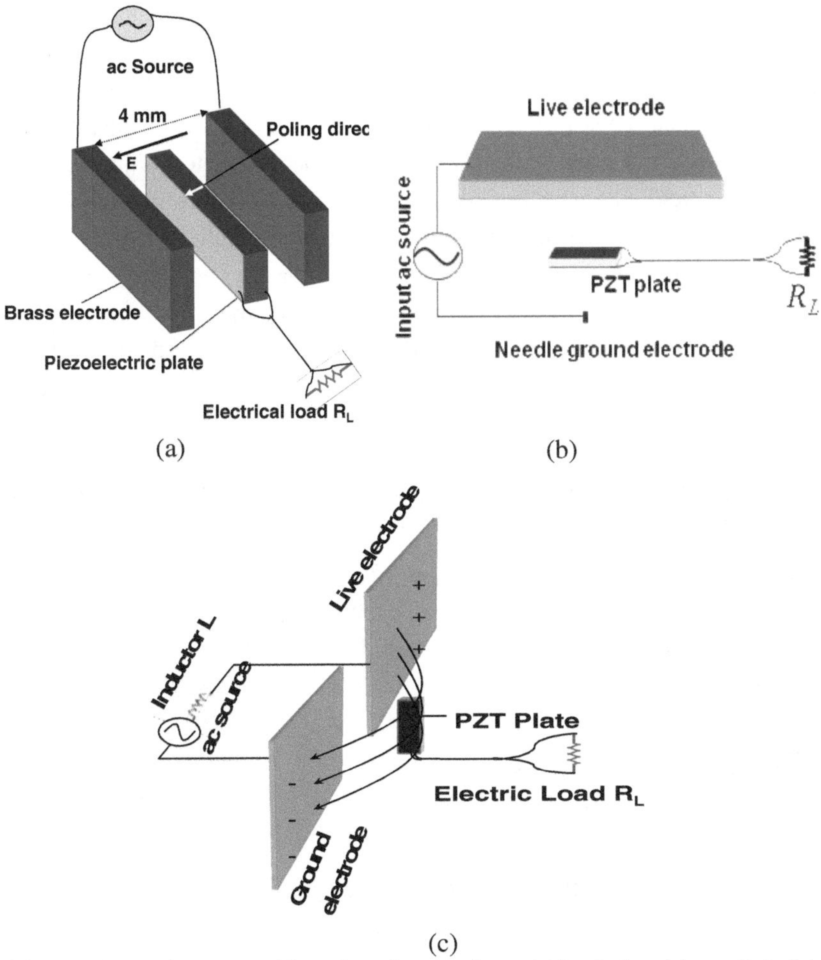

Fig. 2.1 Experimental setup to drive piezoelectric plate wirelessly by: (a) parallel plate capacitor structure, (b) focused electric field structure, (c) dipole antenna-like structure.

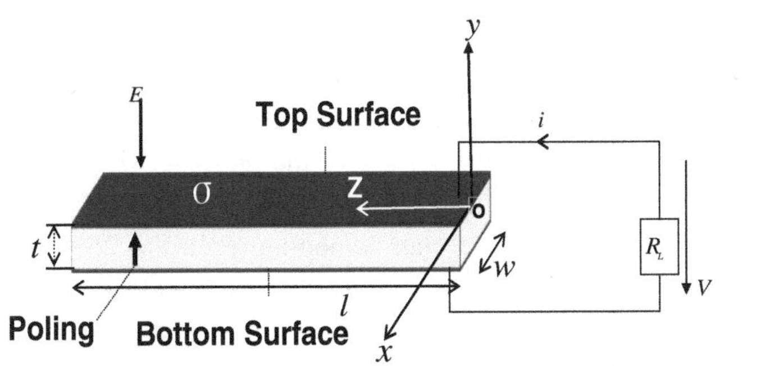

Fig. 2.2 Configuration of the piezoelectric plate wirelessly driven by E-field.

This experiment is performed under the following conditions. The live electrode area is 30×30 cm^2; the ground electrode is a metal needle whose tip is assumed to have zero area; the distance between the live and ground electrodes is 1 cm; the inductance (L) is 1.10 mH; input source voltage is 150 V rms; the medium between the live and ground electrodes is air. The earlier techniques such as focused electric field and capacitor-like structures are not robust for the application in rotary machines. Thus, the wireless drive of piezoelectric plate by dipole antenna-like structure has been investigated further for the application in rotary machines. In Fig. 2.1(c), an electric dipole antenna-like structure is used to transmit electric energy to PZT plat. This experimental study is performed under the following conditions. The live and electrode have the same area of 50×50 cm^2; the distance between the electrodes is 5 cm; input source voltage is 150 Vrms; piezoelectric plate is placed at the center 6 mm away from the antenna plane. Fig. 2.2 shows the configuration of the piezoelectric plate operating in the thickness mode wirelessly driven by electric field. The piezoelectric plate is made of lead zirconate titanate (PZT) ceramic material (Fuju C201), and poled along the thickness direction. Piezoelectric charge constant d_{33}, mechanical Q, dissipation factor $tan\delta$, and relative dielectric constant $\varepsilon_{33}{}^{T}/\varepsilon_0$ are 325×10^{-12} m/V, 2000, 0.3 and 1450, respectively.

3. Theoretical Analyses

3.1 Theoretical Calculation of Electric Field

The finite element method (COMSOL Multiphysics) simulation has been carried out in order to assess the electric field on the surface of the piezoelectric plate operating in the thickness mode wirelessly driven by electric field structure as shown in Fig.2.1(c).

The calculated 2-D electric field pattern around the wirelessly driven piezoelectric plate by an electric dipole antenna-like structure is shown in Fig. 3.1.1. The distribution of electric field on the surface of the piezoelectric plate along the x-direction is shown in Fig. 3.1.2. It is seen that the electric field on the surface of the wirelessly driven piezoelectric plate is non-uniform. For 2500 cm^2 electrode area of the dipole antenna-like structure and input source voltage of 150 Vrms, the average value of the electric field on the surface of the piezoelectric plate is 5.72×10^4 V/m, which is used in our calculation in Sec. 4.

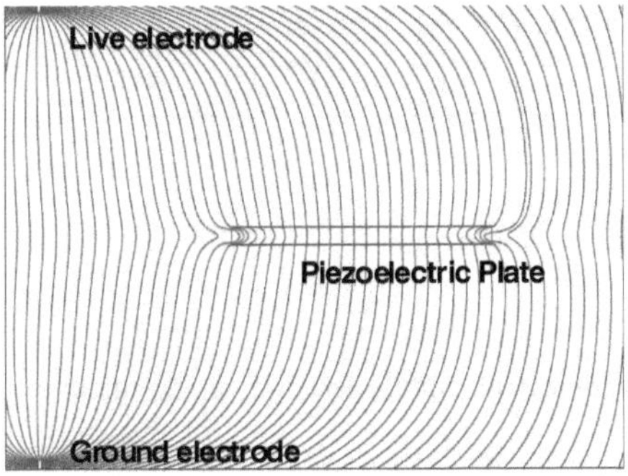

Fig. 3.1.1 Simulated 2-D electric field pattern around the piezoelectric plate wirelessly driven by dipole antenna-like structure.

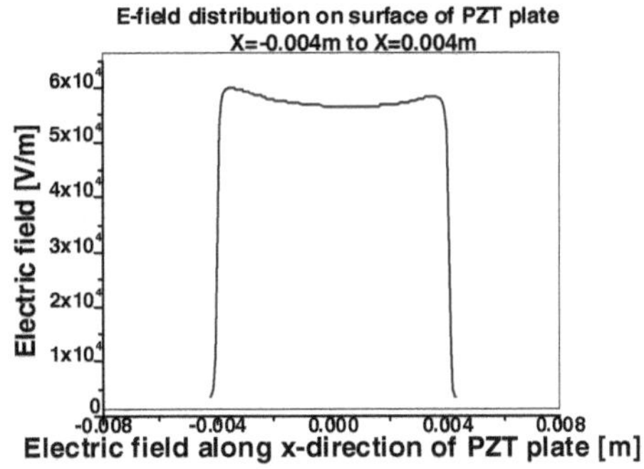

Fig. 3.1.2 Calculated electric field on the surface of the piezoelectric plate.

3.2 Equivalent Circuit Model

To understand the operation principle and output power characteristics, an equivalent circuit is derived for the wirelessly driven piezoelectric plate operating in the thickness vibration mode.

According to Gauss's law, the charge on the surface electrode of the piezoelectric plate is

$$\sigma(x) = D_3 - \varepsilon_0 E(x) \tag{1}$$

Where D_3 is the electric displacement vector, ε_0 is the permittivity of the free space and $E(x)$ is the external electric field on the surface of the piezoelectric plate. So, the current flowing to the piezoelectric plate with an electrode area A can be obtained as:

$$i = j\omega\left(VC_d + 2nu_0 - \varepsilon_0 \tilde{E}A\right) \tag{2}$$

where C_d is clamped capacitance, n is turn ratio, u_0 is the vibration displacement of the piezoelectric plate. From Eq. (2) the equivalent circuit of the wirelessly driven piezoelectric plate operating in the thickness vibration mode as shown in Fig.3.2.1 is derived.

When the piezoelectric plate operates near its resonance, the real power delivered to the electrical load resistor can be obtained from the derived equivalent circuit

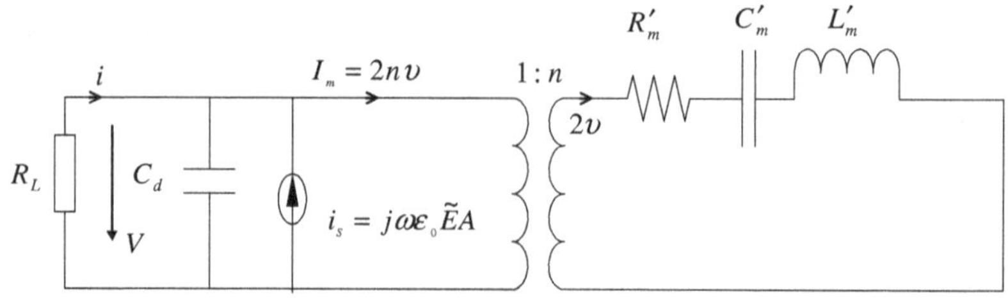

Fig. 3.2.1 Derived equivalent circuit of the piezoelectric plate operating in the thickness mode wirelessly driven by electric field.

$$P_L = \frac{\left(\omega\varepsilon_0\tilde{E}A\right)^2 R_L\left[R'^2_m + \left(\omega L'_m - \frac{1}{\omega C'_m}\right)^2\right]}{\left(R'_m + n^2R_L - \omega^2 C_d R_L L'_m + \frac{C_d R_L}{C'_m}\right)^2 + \left(\omega L'_m - \frac{1}{\omega C'_m} + \omega C_d R_L R'_m\right)^2}. \tag{3}$$

The vibration displacement of the wirelessly driven piezoelectric plate operating in the thickness mode can be obtained from the equivalent circuit as:

$$u_0 = \frac{\left(0.5\varepsilon_0\tilde{E}An^2R_L\right)}{\left[\left(R'_m + n^2R_L - \omega^2 C_d R_L L'_m + \frac{C_d R_L}{C'_m}\right)^2 + \left(\omega L'_m - \frac{1}{\omega C'_m} + \omega C_d R_L R'_m\right)^2\right]^{1/2}} \tag{4}$$

4. Results and Discussion

4.1 Parallel Plate Capacitor Structure

The frequency characteristic of output power of the piezoelectric plate which operates in thickness mode and has the optimum electrical load, is shown in Fig.4.1.1. The optimum load is the load resistance for which the output power at resonance is maximum for a given input source voltage. The size of the piezoelectric plate used in the experiment is 30×8×2 mm^3. It is observed that at the resonance frequency 772 kHz, a maximum output power is achieved. When the frequency of electric field in the gap is close to the mechanical resonance frequencies of piezoelectric plate, a relatively large vibration can be excited in the plate by the converse piezoelectric effect. This mechanical resonance generates a relatively large voltage by the piezoelectric effect.

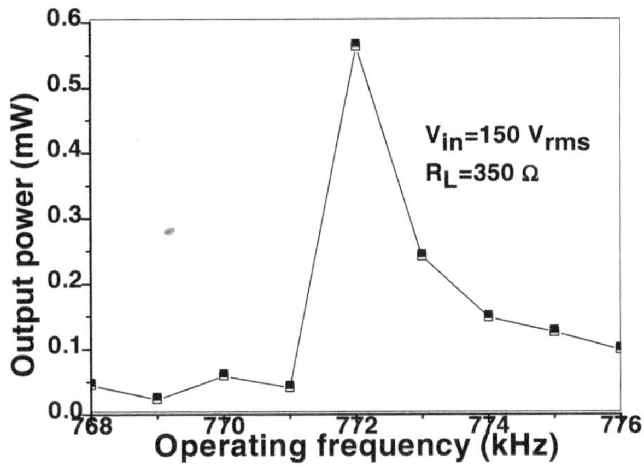

Fig. 4.1.1 Frequency characteristic of the output power of the piezoelectric plate wirelessly driven by parallel plate capacitor structure.

4.2 Focused Electric Field Structure

Fig. 4.2.1 shows the electrical load characteristics of the output power of the piezoelectric plate operating in the thickness vibration mode at resonance. The size of the piezoelectric plate used in this experiment is 8×3×2 mm^3. It is seen that the output power at resonance reaches the maximum at an optimum load resistance.

At resonance frequency 782 kHz and optimum load resistance 1365 Ω, a maximum output power of 0.26 W has been achieved by the piezoelectric plate operating in the thickness mode with a needle ground electrode, input voltage of 150 Vrms, 1 cm electrode separation, and a live electrode area of 30×30 cm^2.

Fig. 4.2.2 shows the calculated frequency characteristics of vibration displacement of the wirelessly driven piezoelectric plate when the electrical load resistances are $R_L = \infty$ (open circuit), $R_L = 0$ (short circuit), and R_L= 350 Ω (output power is maximum across the PZT plate). In the calculation, the dimension of the piezoelectric plate is 30×8×2 mm^3, and the electric field is 1.2×10^6 V/m. It is found that a relatively large vibration displacement of 4.44 nm is achieved when the load branch is open circuited.

The reason for this phenomenon can be explained by the equivalent circuit, depicted in Fig. 3.2.1. When the load branch is open circuited, the current i thorough the load resistance R_L is zero. So, the motional current $I_m = 2nv$ is relatively large. But, when the load branch is short circuited, all i_S passes thorough the load branch, and thus the motional current I_m is zero. Hence the vibration displacement is relatively large when the load branch is open circuited, and the vibration displacement is zero when the load branch is short circuited.

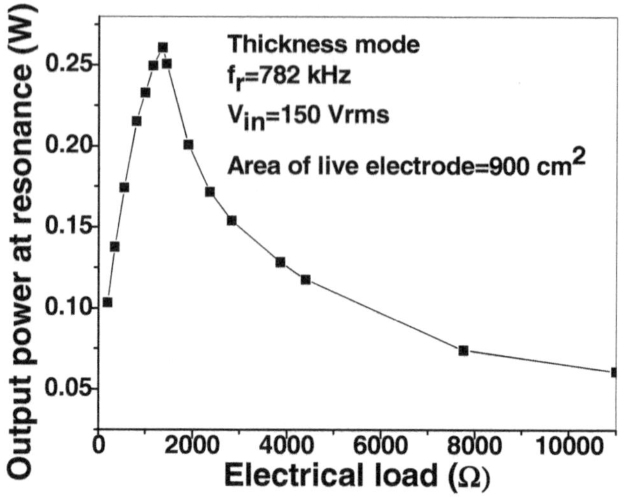

Fig. 4.2.1 Electric load characteristic of the output power of the piezoelectric plate wirelessly driven by focused electric field structure.

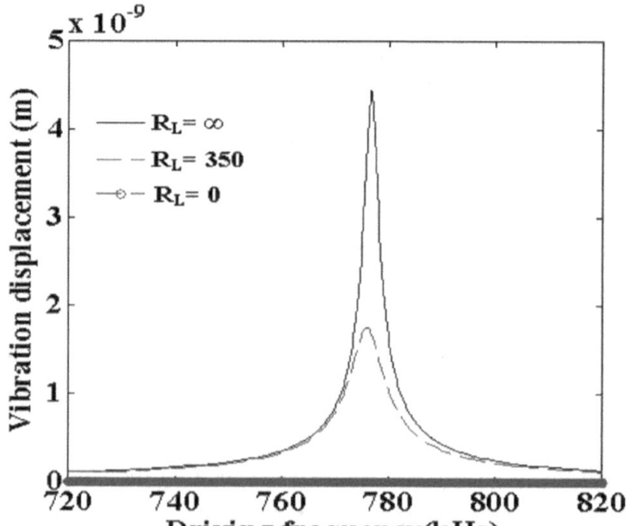

Fig. 4.2.2 Calculated vibration displacement characteristic of the piezoelectric plate wirelessly driven by focused electric field structure.

4.3 Dipole Antenna-Like Structure

Fig. 4.3.1 shows the theoretical and experimental dependence of the output power on the electrical load at resonance frequency of the piezoelectric plate. The dimension of the piezoelectric plate used here is $30\times8\times2$ mm^3, and the electric field is 5.72×10^4 V/m as shown in Fig.3.1.2, which is used in our calculation. In the experiment, the resonance frequency decreases slightly as the load resistance increases. It is around 772 kHz. It is seen that the theoretical result well agrees with the experimental one, and the output power at resonance reaches the maximum at an optimum load resistance of 350 Ω. The optimum load resistance can be explained by the equivalent circuit of piezoelectric plate operating in the thickness mode, as shown in Fig. 3.2.1. A maximum power can be delivered to the electric load resistor, when

$$\frac{dP_L}{dR_L} = 0\,. \tag{5}$$

Using eqs. (8) and (10), we get the following equation of the optimum load resistance for which a maximum power is delivered to the load.

$$\begin{aligned}&\left[\left(n^4 + \omega^2 C_d^2 R_m'^2 + \omega^4 C_d^2 L_m'^2 - 2n^2\omega^2 C_d L_m'\right) + \left(C_d^2/C_m'^2\right) + \left(2n^2 C_d/C_m'\right) - \left(2\omega^2 C_d^2 L_m'/C_m'\right)\right] R_L^2 \\ &- \left[R_m'^2 + \omega^2 L_m'^2 + \left(1/\omega^2 C_m'^2\right) - \left(2L_m'/C_m'\right)\right] = 0\end{aligned} \tag{6}$$

Using eq. (11), we found an optimum load resistance of 358 Ω for the wirelessly driven piezoelectric plate operating in the thickness mode, which well aggress with the measured value of 350 Ω.

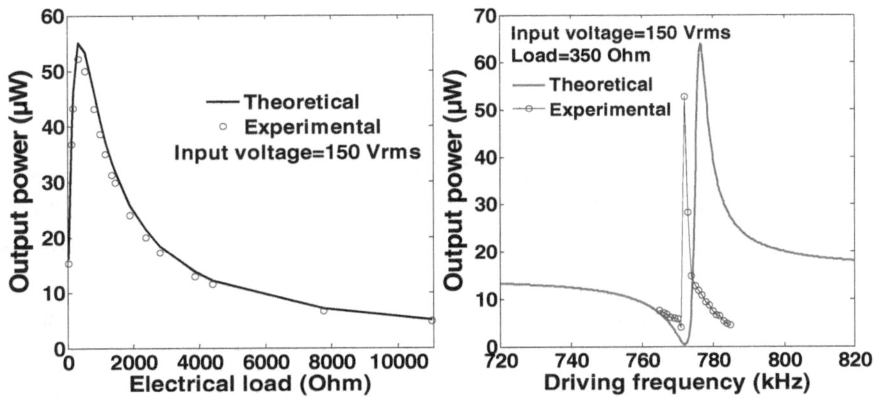

Fig. 4.3.1 shows the effect of area of the piezoelectric plate on the output power at resonance with the constant thickness and optimum electric load of the piezoelectric plate operating in thickness mode wirelessly driven by dipole antenna-like structure.

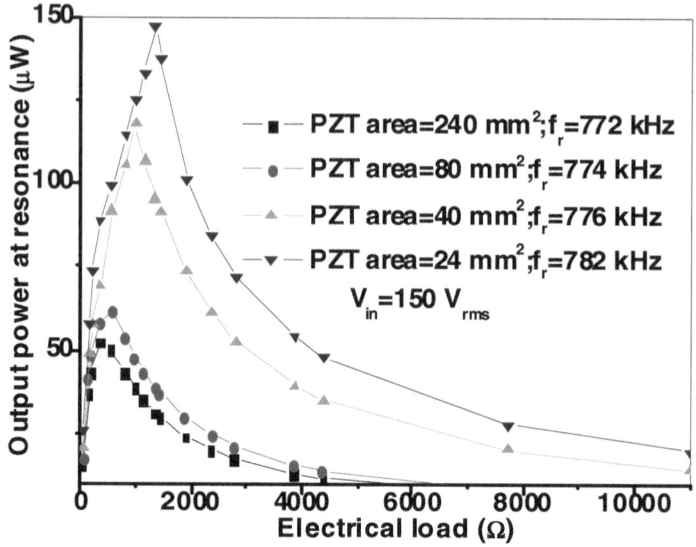

Fig. 4.3.2 Effect of the area of piezoelectric plate on the output power at resonance.

Fig. 4.3.2 shows the theoretical and experimental frequency characteristics of output power of the piezoelectric plate. It is seen that the theoretical results agree with the experimental results quite well, and the output power reaches the maximum at resonance frequency.

It is seen that the measured output power at resonance becomes larger when the size of the piezoelectric plate is reduced. The change in output power at resonance may be due to the change of electric field and equivalent resistance of the piezoelectric components. This experimental result shows the feasibility to drive micro-piezoelectric components wirelessly by using an electromagnetic wave.

5. Summary

In this work, new techniques of transmitting electric energy wirelessly to piezoelectric components have been investigated both theoretically and experimentally. An equivalent circuit is derived, and electric field pattern is also studied for the wirelessly driven piezoelectric component operating in the thickness mode. It has been observed that the output power of the piezoelectric component wirelessly driven by focused electric field structure is significantly higher than that wirelessly driven by the other two structures i.e. parallel plate capacitor structure and dipole antenna-like structure. In the focused electric field structure, at resonance frequency 782 kHz and optimum load 1365 Ω, a maximum real output power of 0.26 W and energy conversion efficiency of 1.02% have been achieved with an input voltage of 150 Vrms (to the series of the capacitor-like structure and resonance inductance) and 1 cm electrode separation. It is found that the output power and vibration displacement of the wirelessly driven piezoelectric component depends on the frequency, electric load, size of live electrodes of the electric field structures. The method of wireless drive of piezoelectric component by dipole antenna-like structure is robust for the application in rotary machines. The wireless drive of piezoelectric components technique provides a new method of driving micro-piezoelectric components and piezoelectric devices in rotary machines.

References

[1] Arnau A.: *Piezoelectric Transducers and Applications*, Berlin, Germany: Springer, 225-253 (2003).
[2] Kumada A.: A piezoelectric ultrasonic motor. *Jpn. J. Appl. Phys.* **24** Suppl.24-1,739-741 (1985).
[3] Ueha S.: *General outline of ultrasonic motors in Ultrasonic Motors- Theory and Applications*, Oxford, Oxford University Press, 1-7 (1993).
[4] Uchino K.: *Ferroelectric Devices*, New York: Marcel Dekker, Inc., 145-197 (2000).
[5] Rosen C. A.: Ceramic transformers and filters. *Proc. Electron. Component. Symp.* 205-211 (1956).
[6] Zelenka J.: *Piezoelectric Resonators and their Applications*. New York: Elsevier, 242-287 (1986).
[7] Bhuyan S. and Hu J. H.: Wireless drive of piezoelectric plate by focused electric field. *Appl. Phys. Lett.* **91** 264101-3 (2007).
[8] Bhuyan S., Hu J. H., and Sun C. Q.: Wireless drive of piezoelectric plate by capacitorlike strcuture in electric resonance with an inductor. *J. Appl. Phys.* **103**, 4915-3 (2008).
[9] Bhuyan S., Hu J. H. and Sun C. Q.: Wireless energy transmission to piezoelectric components. *Jpn. J. Appl. Phys.* **47** 4226-4230 (2008).

Novel High Frequency, Temperature Stable, Quartz Gyroscopes

Yook-Kong Yong[1] and Jianke Du[2]

[1]Department of Civil and Environmental Engineering, Rutgers University, 623 Bowser Road, Piscataway, NJ 08854, U.S.A

[2]Department of Engineering Mechanics, Ningbo University, Ningbo, China

yyong@rci.rutgers.edu

Abstract. An AT-cut quartz plate with lateral tines was proposed as an angular velocity sensor. The plate itself formed the driving component while the lateral tines formed the angular velocity sensors. The angular velocity effect on the plate-gyroscope was demonstrated via its effect on the electric potential at edge electrodes, and on the admittance at driving electrode. The change in electric potential at edge electrodes varied with the magnitude and sign of the angular velocity. An AT-cut plate-gyroscope may offer advantages in terms of (a) frequency stability (b) frequency-temperature stability, and (c) separation of the driving component from the sensing component. The separation of the driving component from the sensing component allowed for a wider variety of tine geometries and modes for detecting angular velocity. Since the Coriolis force was a function of the displacement velocities, the gyroscopic effect therefore a nonlinear problem, albeit weakly nonlinear. Furthermore the frequency responses of change in electric potential at the edge electrode, and change in admittance at the driving electrode were shown to be affected by the nonlinear elastic constants.

1. Introduction

Quartz resonators have been shown to have better aging, jitter and phase noise characteristics than the current MEMS resonators [1]. In addition, the AT-cut quartz resonator is well known for its short term and long term frequency stabilities as well as its good frequency-temperature characteristics. These are important characteristics in a good vibratory gyroscope if they could be realized. A vibratory gyroscope is a type of gyroscope that senses angular velocity by the momenta of its vibrating masses.

T.-T. Wu and C.-C. Ma (eds.), *IUTAM Symposium on Recent Advances of Acoustic Waves in Solids*, IUTAM Bookseries 26, DOI 10.1007/978-90-481-9893-1_25,

When the gyroscope undergoes a rotation the momenta generate Coriolis forces that in turn effect changes in the electric potential or charge at its sensing electrodes. We study and analyze in this paper the use of a quartz plate resonator as an angular velocity sensor. Our analyses are valid also for langasite plate resonators since the two crystals share similar characteristics. Langasite plate resonator-gyroscopes will have higher gyroscopic sensitivity than their quartz counterparts due to their higher electromechanical coupling factors [2]. However at present the material properties of quartz are better known, and high quality quartz plates are available commercially at low cost.

Currently most quartz gyroscopes are tuning fork types that use their vibrating tines as angular velocity sensors. These gyroscopes needed to overcome technical challenges from complex electrical leads, driving electrodes, sensing electrodes, mounting supports sensitivity, mechanical drift and low Q. The AT-cut quartz plate resonator-gyroscopes may on the other hand have simple driving and sensing electrodes, simple electrical leads, low mechanical drift and high Q. The high energy trapping in an AT-cut plate resonator generally isolates the active region of the resonator from the mounting supports, and hence a gyroscope derived from such a resonator may have much less mounting supports sensitivity. The idea of using energy trapped modes to reduce mounting supports sensitivity was also proposed by Nakamura, et. al. [3], Abe, et. al. [4], and Nakamura, et. al. [5].

2. Principles of the Quartz AT-Cut Vibratory Gyroscope

2.1 AT-Cut Quartz Plate-Gyroscope

In order to illustrate the operational principles of the quartz AT-cut vibratory gyroscope we consider one resonator structure in Fig. 2.1 below consisting of three drawings. The structural components of this resonator may be varied in geometry and dimensions to yield optimal gyroscopic responses. The separation of the driving component from the sensing component allows for a wide variety of gyroscopic sensing designs. The sensing component was the tine that could vibrate in a variety of modes such as flexure, torsion, or extension, while the driving mode was the thickness shear mode of the plate. The top drawing shows the tines and plate resonator, the middle drawing shows the mounting support and bottom electrode details, and the bottom drawing shows the top electrode details. The bottom electrodes were grounded. The top electrode details consisted of the driving electrode at the center of plate, and sensing electrodes at the edges near the tines. The sensing electrodes could be designed to sense changes in admittance if shorted, or sense changes in voltage if open circuited.

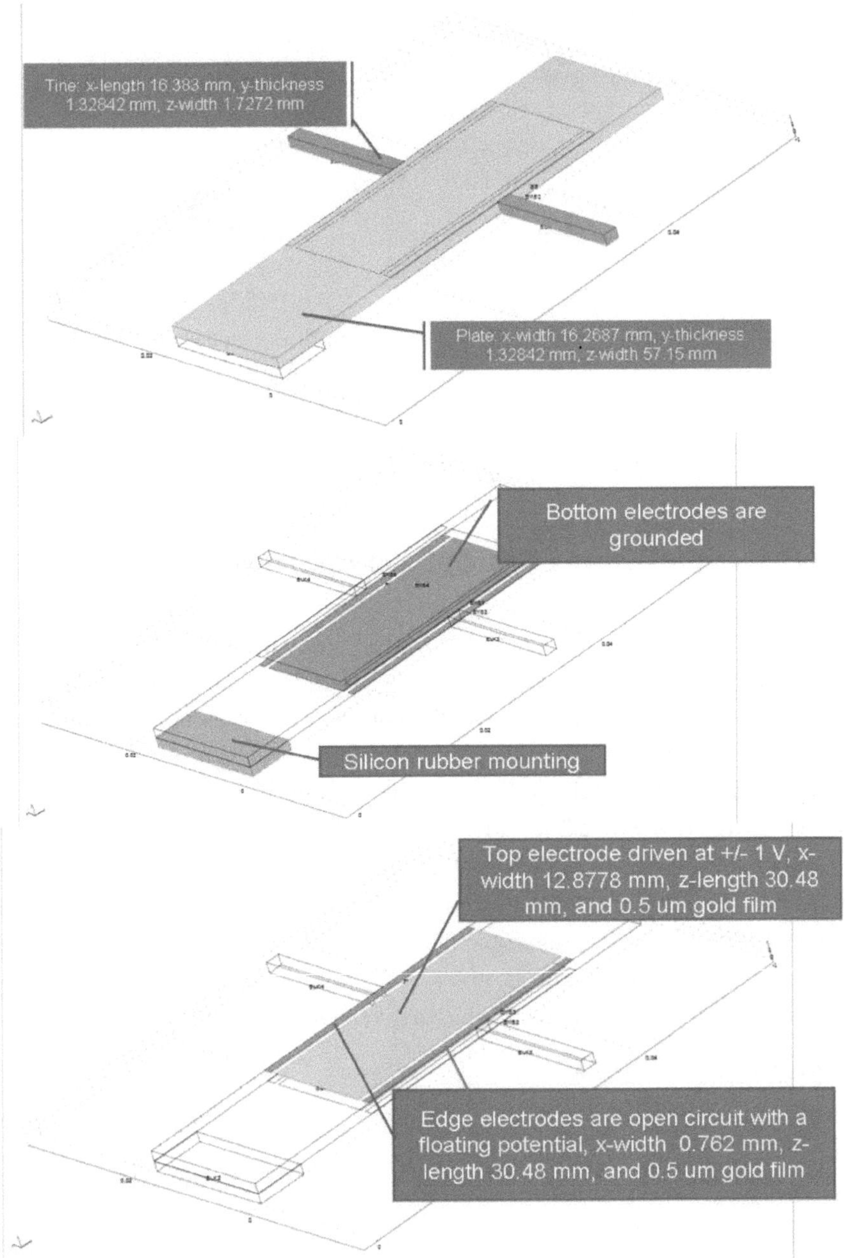

Fig. 2.1 AT-cut quartz plate-gyroscope structure: (top) plate and tine dimensions, middle) mounting support and bottom electrode details, and (bottom) top electrode details, including the dimensions.

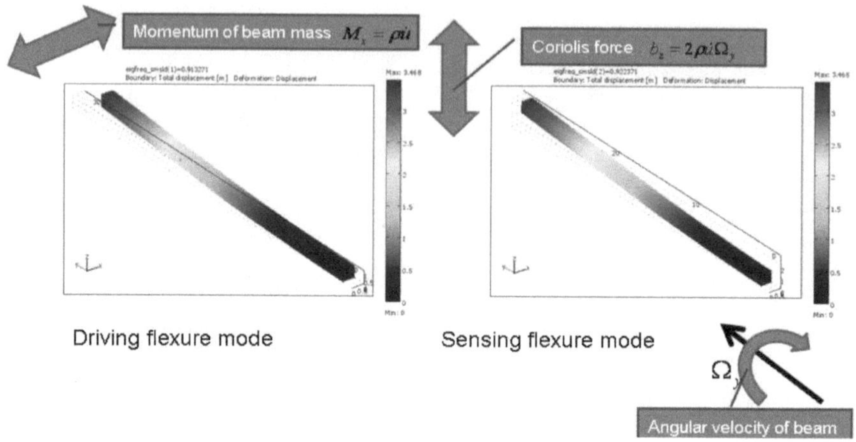

Fig. 2.2 Gyroscopic sensing principles of the tines.

2.2 Gyroscopic Sensing Principles of the Tines

The energy for driving the tines are derived from the thickness shear driving mode of the plate. In order to maximize gyroscopic sensing, the tines must be designed to satisfy two criteria simultaneously: (1) the tines must vibrate and resonate well with the thickness shear mode, and (2) the Coriolis force generated by the momenta and angular velocity of the tines must excite another tine mode. This is illustrated in Fig. 2.2 for a flexure mode sensing. When the tine flexes in a X-Z plane (left), the tine momenta will interact with the angular velocity about say the X-axis to produce a Coriolis force in the X-Y plane. If the tine was designed to resonate well with the Coriolis force then the tine will sense the angular velocity. In general the Coriolis force is given by cross product of the momentum of displacement velocities $\rho\underline{\dot{u}}$ and the angular velocity vector $\underline{\Omega}$

$$\underline{b} = 2\rho\underline{\dot{u}} \otimes \underline{\Omega} \tag{2.2.1}$$

3. Nonlinear Analysis of the Vibratory Gyroscope

The Coriolise force was observed from Eq. 2.2.1 to be nonlinear as it was dependent on the displacement velocity; hence the vibratory gyroscope problem is inherently nonlinear, albeit weakly nonlinear. Another source of nonlinearity was from the material constants when the gyroscopic effect was weak due to mechanical drifts and very small angular velocity.

3.1 Governing Equations for the AT-Cut Quartz Plate-Gyroscope

Although the quartz crystal is weakly piezoelectric, its effects must be considered in our analysis because the sensing of angular velocity is based on small changes in admittance or voltage at the edge electrodes. Also, although the viscosity coefficients of quartz are small, they must be considered to allow for a mechanism to calculate changes in the quality factor Q via significant energy losses at the mounting supports. The Q plays an important role in the interaction of the driving mode and the sensing mode: if the Q is very high for both the modes, the two modes will not "see" or "hear" each other even when their two frequencies are relatively near to each other.

3.1.1 Strain-displacement, and electric field-potential relations

$$\begin{aligned} s_{ij} &= \frac{1}{2}(u_{j,i} + u_{i,j} + u_{k,i}u_{k,j} + u_{k,j}u_{k,i}) \\ E_i &= -\varphi_{,i} \end{aligned} \tag{3.1.1}$$

s_{ij}, u_i, E_i, φ are respectively the mechanical strains, mechanical displacements, electric field and electric potential.

3.1.2 Constitutive relations

$$\begin{aligned} t_{ij} &= (C_{ijkl} + C_{ijklmn}s_{mn})s_{kl} + \eta_{ijkl}\dot{s}_{kl} - e_{kij}E_k \\ D_i &= e_{ijk}s_{jk} + \varepsilon_{ik}E_k \end{aligned} \tag{3.1.2}$$

t_{ij}, C_{ijkl}, C_{ijklmn}, η_{ijkl}, e_{kij}, D_i, ε_{ik} are respectively the mechanical stress, linear elastic constants, nonlinear elastic constants, viscosity constants, piezoelectric stress constants, electric displacements, and dielectric permittivity. The values of the material constants of quartz are well known and could be found in the literature [6, 7, 8, 9].

3.1.3 Equations of motion and electrostatics

$$\begin{aligned} (t_{ij} + t_{jk}u_{i,k})_{,j} + \rho b_i &= \rho\ddot{u}_i \\ D_{i,i} &= 0 \end{aligned} \tag{3.1.3}$$

4. Results of the AT-Cut Quartz Plate-Gyroscope

Results were obtained for the plate-gyroscope of Fig. 2.1. Since the top electrode was driven at +/- 1 V while the bottom electrode was grounded, the admittance of the quartz resonator has the same values as the current calculated at the top electrode. Fig.4.1 shows admittance frequency response curve of the plate-gyroscope (left) and the floating potential at the edge electrodes (right).

4.1 Frequency Response of the Plate-Gyroscope to Angular Velocity about the Y-Axis

Frequency response of the plate-gyroscope to angular velocity about the Y-axis was studied. While the plate-gyroscope was not optimized for sensitivity to angular velocity, we nevertheless found small changes in the electric potential at top edge electrodes and in admittance at the top driving electrode.

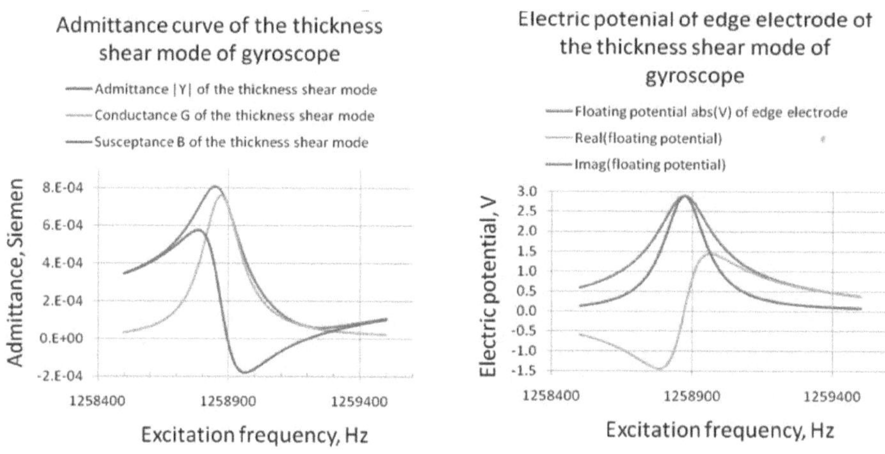

Fig. 4 Frequency response of the AT-cut quartz plate-gyroscope: (left) Admittance at the top driving electrode, and (right) floating potential at the edge electrode.

4.1.1 Changes in the electric potential at edge electrodes due to angular velocity about the Y-axis

Angular velocities of 1 rad/s, and 10 rad/s were successively applied to the plate-gyroscope, and changes in the electric potential at edge electrode were observed in Figs. 4.1.1, and 4.1.2 respectively.

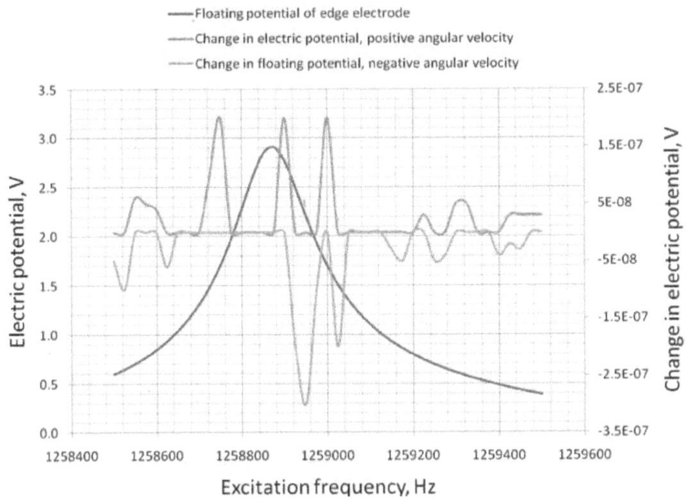

Fig. 4.1.1 Change in electric potential at the edge electrode as a function of excitation frequency for an angular velocity of 1 rad/s about the Y-axis.

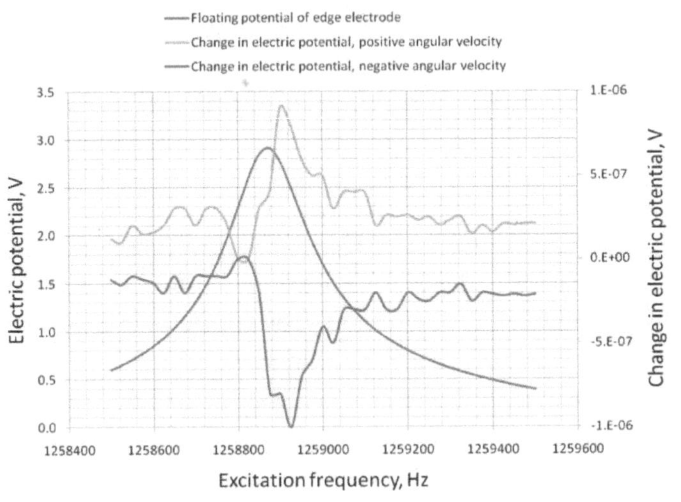

Fig. 4.1.2 Change in electric potential at the edge electrode as a function of excitation frequency for an angular velocity of 10 rad/s about the Y-axis.

4.1.2 Changes in the admittance at top driving electrode due to angular velocity about the Y-axis

Changes were also observed in the admittance at the top driving electrode due to the angular velocities. As an example, Fig. 4.1.3 shows the change in admittance as a function of excitation frequency.

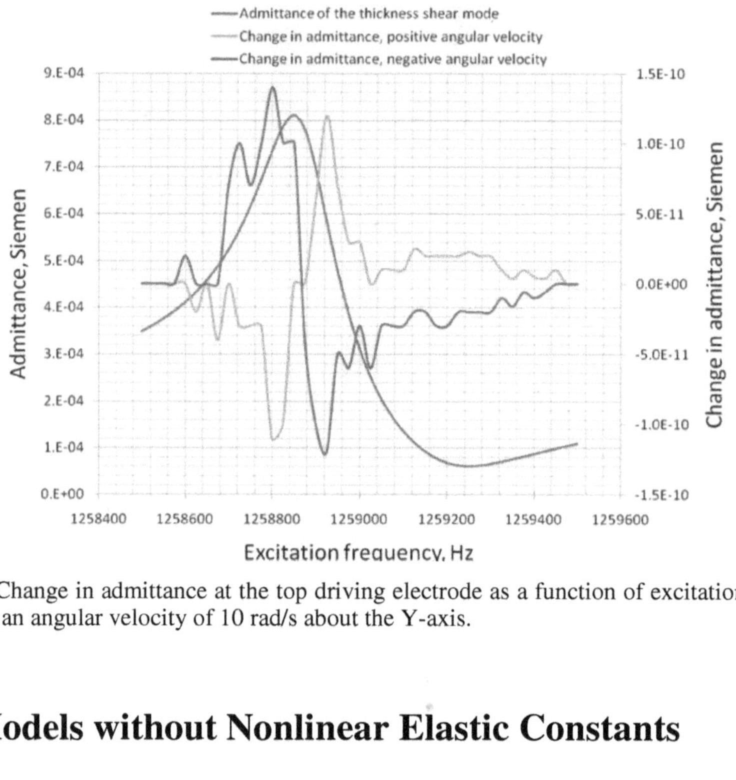

Fig. 4.1.3 Change in admittance at the top driving electrode as a function of excitation frequency for an angular velocity of 10 rad/s about the Y-axis.

4.2 Models without Nonlinear Elastic Constants

It was observed in section 3 that vibratory gyroscope is inherently a nonlinear problem since the Coriolis force is a function of the displacement velocity. The modeling efforts could be reduced if the material constants and strains were considered as linear, resulting in less iterations and computation time. The figure below (Fig. 4.2.1) shows the changes in admittance of the previous figure (Fig. 4.1.3) if the nonlinear elastic constants and nonlinear strains were neglected. A comparison of Figs. 4.1.3 and 4.2.1 (they are of the same scale and subjected to the same angular velocity of 10 rad/s) showed significant differences in the frequency response of changes in admittance. Hence, nonlinear elastic constants and nonlinear strains should be included in the models when possible

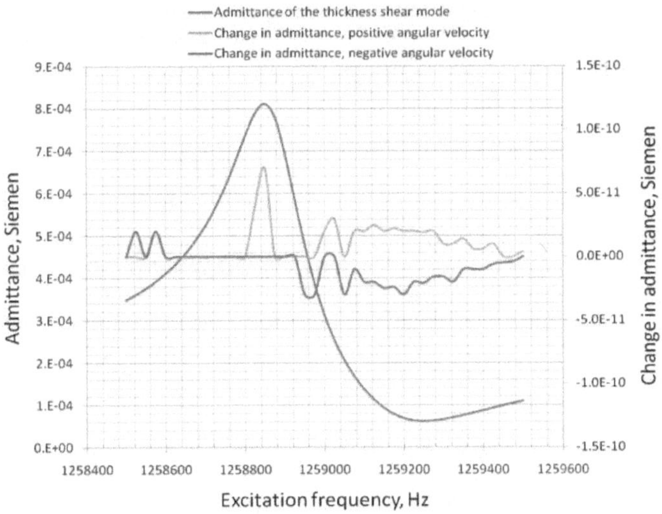

Fig. 4.2.1 Change in admittance at the top driving electrode as a function of excitation frequency for an angular velocity of 10 rad/s about the Y-axis. Nonlinear elastic constants were neglected.

5. Summary and Conclusion

A novel design for an AT-cut plate-gyroscope was proposed. It was shown analytically that the angular velocity sensing function could be realized using an essentially thickness shear mode and tines. Since the AT-cut plate resonator is temperature stable, and well isolated from its mounting supports, the proposed plate-gyroscope should have better temperature stability, and less dependency on the mounting conditions. Furthermore the electrode configurations will be as simple as those in an AT-cut plate resonator. The separation of the driving component from the sensing component allows for a wider variety of designs of the sensing component and function.

Analytical results on changes in electric potential at the edge electrodes with angular velocity demonstrated the theoretical feasibility of the plate-gyroscope. The change in electric potential was dependent on the magnitude of the angular velocity and its sign. When negative angular velocity was applied, the change in electric potential changed sign too. The inclusion of nonlinear elastic constants causes significant differences in the frequency responses of the change in electric potential.

Viscosity coefficients of quartz were included in the analysis to introduce a mechanism for modeling the changes in quality factor Q. The Q plays a role in the interaction of the driving mode and the sensing mode.

Acknowledgments: Jianke Du gratefully acknowledged the support by the National Natural Science Foundation of China (No.10772087). Yook-Kong Yong gratefully acknowledged the support of Seiko Epson, Inc. for research on quartz gyroscopes.

References

[1] Henry, R., and Kenny, D.: Comparative Analysis of MEMS, Programmable, and Synthesized Frequency Control Devices Versus Traditional Quartz Based Devices. *Proceedings of the 2008 IEEE International Frequency Control Symposium,* 396-401 (2008).

[2] Yang, J. S.: Some Analytical Results on Piezoelectric Gyroscopes. *Proceedings of the 1998 IEEE International Frequency Control Symposium,* 733-741 (1998).

[3] Nakamura, K., and Ohsaki, M.: Trapped-energy Vibratory Gyroscopes Using Rotated Y-Cut LiNbO3. *Proceedings of the 1999 Joint Meeting EFTF – IEEE IFCS,* 797-799 (1999).

[4] Abe, H., Yoshida, T., Ishikawa, T., Miyazaki, N., and Watanabe, H.: Trapped – Energy Vibratory Gyroscopes Using a Partially Polarized Piezoelectric Ceramic Plate. *Electronics and Communications in Japan*, Part 2, Vol. **84** No. 3, 44-52 (2001).

[5] Nakamura, K., and Abe, T.: Vibratory Gyroscopes Using Trapped-Energy Vibrators of Rotated Y-cut LiTaO3. *Proceedings of the 2001 IEEE Ultrasonics Symposium*, 611-614 (2001).

[6] Lee, P.C.Y., Liu, N.H. and Ballato, A.: Thickness Vibrations of Piezoelectric Plates with Dissipation. *IEEE Transactions on Ultrasonics, Ferroelectrics and Frequency Control,* **51** 52-62 (2004).

[7] Lamb, J., and Richter, J.: Anisotropic Acoustic Attenuation with New Measurement for Quartz at Room Temperatures. *Proceedings of the Royal Society* **293A** 479-492 (1966).

[8] Yong, Y.-K.,: Three-Dimensional Finite Element Solution of the Lagrangean Equations for the Frequency Temperature Behavior of Y-Cut and NT-Cut Bars. *IEEE Transactions on Ultrasonics, Ferroelectrics, and Frequency Control,* Vol. **UFFC-34** No.5, 491-499 (1987).

[9] Bechmann, R., Ballato, A.D., and Lukaszek, T.J.: Higher Order Temperature Coefficients of the Elastic Stiffnesses and Compliances of Alpha-Quartz. *Proceedings of the IRE*, Vol. **50** No. 8, 1812-1822 (1962).

COM Model Characterization for RF SAW Devices

Ben Abbott[1], Kevin Gamble[1], Natalya Naumenko[2], Svetlana Malocha[1] and Marc Solal[1]

[1]TriQuint Semiconductor, 1818 S. Hwy 441, Apopka, FL 32703, USA

[2]Moscow Steel and Alloys Institute, Moscow, Russia

bpabbott@tqs.com

Abstract. The *coupling-of-modes* (COM) model was applied to the analysis of surface acoustic wave (SAW) devices in the 1980's. At that time, the COM parameters were calculated using perturbation techniques or extracted from experimental data. With the improvements in computational resources it became possible to characterize the COM model's parameters using this robust harmonic admittance using FEM/BEM. The topic of this paper is on a method to accomplish the COM parameter characterization for RF type SAW, leaky SAW (LSAW), and buried interdigital transducer (BIDT) SAW devices.

1. Introduction

During the 1980's the coupling-of-modes (COM) model was applied to the analysis of Rayleigh type SAW devices [1, 2]. To characterize the COM model's parameters, perturbation, variational, and empirical methods were applied [3, 4, 5].

During the 1990's the robust FEM/BEM and FEM/SDA models were introduced [6, 7]. Due to the computational expense of analyzing finite devices, models such as the COM model and *p-matrix model* (PMA model) have remained popular [6, 8, 9, 10].

To improve the predictive reliability of the COM model, the FEM/BEM model has been applied as a standard for characterization of the model [11, 12, 13].

In the following sections, a method for the characterization of the COM model's parameters is described. Examples have been prepared for both conventional RF LSAW filters as well as the more recent devices, such as those buried/over-coated (BIDT) with SiO2 for the purpose of temperature compensation.

T.-T. Wu and C.-C. Ma (eds.), *IUTAM Symposium on Recent Advances of Acoustic Waves in Solids*, IUTAM Bookseries 26, DOI 10.1007/978-90-481-9893-1_26,

2. Harmonic Admittance Approximation

The harmonic admittance is the admittance of a single electrode in an infinitely periodic grating with a harmonic excitation applied [14]. For example, the unit cell for such a structure, with period *p*, is illustrated in Fig. 2.0.1. The voltage and current of each individual electrode, V_n and I_n are harmonically related, Eq. 2.0.1

$$V_n = V_0 \cdot e^{j2\pi s(x/p-n)}, \quad I_n = I_0 \cdot e^{j2\pi s(x/p-n)} \tag{2.0.1}$$

$$Y(s,f) = I_n / V_n \tag{2.0.2}$$

For the purpose of characterizing the COM model's parameters, three approximations to the harmonic admittance are employed.

The first is a *physical rational approximation* [15] represents the admittance of one or more surface modes, as well as a *single* neighboring bulk acoustic wave (BAW). This approximation includes the interaction between the SAW modes and the BAW.

The second approximation is a *spectral domain model,* which is physically representative of one or more surface modes in an infinitely periodic grating, but is not physically representative of BAW modes.

The third, and final, approximation is that of the *COM model.* Like the spectral model, the COM model is physically representative of surface mode propagation, and when included the effect of BAW modes is included as a parasitic admittance. The COM model differs from the spectral model in it includes the reflective coupling between counter propagating surface modes and may be applied to the analysis of non-periodic, and/or finite devices.

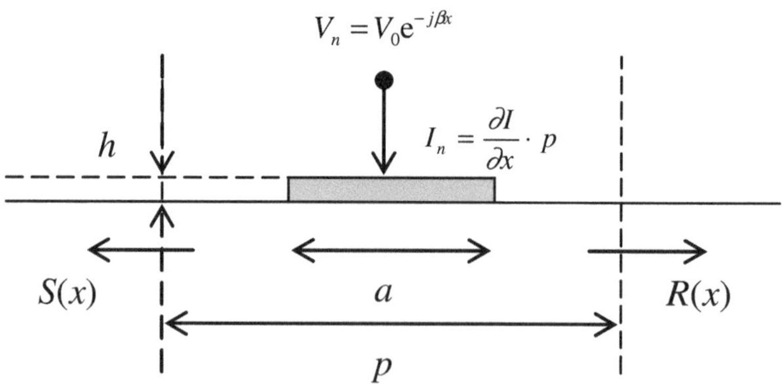

Fig. 2.0.1 Elementary cell represented by the harmonic admittance.

2.1 Physical Rational Approximation

The *physical rational approximation* [15] is an analytic representation of the admittance of one or more surface modes' and a BAW mode. This approximation, Eqs. 2.1.1 – 2.1.4, physically represents the branch of the BAW mode as well as its interaction between the BAW and the surface modes.

$$Y(s,f) = 2j \cdot \omega \cdot \sin(\pi s) \cdot \varepsilon_g(s,f) \tag{2.1.1}$$

$$\varepsilon_g(s,f) = \sum_{n=1}^{N} \frac{R_{\chi,n}}{\chi - \chi_n} + \sum_{k=0}^{K} c_k \cdot \chi^k \tag{2.1.2}$$

$$\chi = \sqrt{s - \bar{f}} + \sqrt{(1-s) - \bar{f}} \tag{2.1.3}$$

$$\bar{f} = f/2f_B, f_B = v_B/2p \tag{2.1.4}$$

The variable v_B is the cut off velocity for acoustic radiation into the bulk of the substrate by an infinitely long conventional IDT. The bulk mode whose velocity is represented by v_B should be selected to coincide with the bulk mode with the greatest influence on the device.

Because this approximation includes SAW and a BAW mode it is able to accurately represent the harmonic admittance function over a broad range of spectral frequencies. This makes this approximation well suited for the reliable extraction of the surface mode dispersive wavenumber and coupling coefficient.

2.2 Spectral Domain Model

The spectral domain model is representative of surface mode propagation in infinitely periodic gratings.

An equivalence exists between the spectral domain model and the physical rational approximation, Eqs. 2.2.1 and 2.1.1 & 2.1.2. While the spectral domain model, Eqn. 2.2.1, does not represent the physics for BAW radiation, the effect of the BAW may be included as a parasitic admittance Y_B.

$$\frac{Y(s,f)}{\omega} = \sum_{n=1}^{N} \frac{R_{s,n}}{\Delta s^2 - \Delta s_n^2} - j \cdot C_s(s) + Y_B \tag{2.2.1}$$

$$\Delta s_n = s_n - 0.5 \tag{2.2.2}$$

$$C_s(s) = \varepsilon_\infty \cdot \frac{P_{-s}(\cos(\pi a/p))}{P_s(-\cos(\pi a/p))} \cdot \sin(\pi s) \tag{2.2.3}$$

The variable, ε_∞, respects the *effective* static dielectric constant. The pole amplitudes for the rational approximation and spectral model, $R_{\chi,n}$ and $R_{s,n}$, in Eqs. 2.1.2 and 2.2.1 are related by Eq. 2.2.4.

$$R_{s,n} = 2j \cdot R_{\chi,n} \cdot \chi_n \cdot D_{B,n} \cdot \sin(\pi s_n) \tag{2.2.4}$$

The detuned spectral wavenumber, Δsn, is related to the poles locations of the physical rational approximation by Eq. 2.2.5.

$$\Delta s_n = \pm\sqrt{\Delta f^2 - (0.5\chi_n^2 + \Delta f)^2}, \quad \Delta f = \bar{f} - 0.5 \tag{2.2.5}$$

The variable, D_B, is representative of the BAW wavenumber, which is implicit in the rational approximation, but missing in the spectral domain model.

$$D_B = \pm\sqrt{\Delta f^2 - \Delta s^2} \tag{2.2.6}$$

2.3 COM Harmonic Admittance Model

The COM model has been applied previously to represent the harmonic admittance [12]. Eqs. 2.3.1-2.3.5 are representative of the electrical and acoustic wave amplitudes for a periodic grating of unit width, with a spectral excitation applied.

$$\frac{\partial \bar{R}}{\partial x} = -j\delta \cdot \bar{R} + j\kappa \cdot \bar{S} + j\alpha \cdot V_0 \cdot e^{-j\beta x} \tag{2.3.1}$$

$$\frac{\partial \bar{S}}{\partial x} = j\delta \cdot \bar{S} - j\kappa \cdot \bar{R} - j\alpha \cdot V_0 \cdot e^{-j\beta x} \tag{2.3.2}$$

$$\frac{\partial I}{\partial x} = -j2\alpha \cdot \bar{R} - j2\alpha \cdot \bar{S} - j\omega \cdot C_s(s) \cdot V_0 \tag{2.3.3}$$

$$\delta = \omega/v_u - \pi/p, \quad \beta = 2\pi s/p - \pi/p \tag{2.3.4}$$

$$R = \bar{R} \cdot e^{-j2\pi x/p}, \quad S = \bar{S} \cdot e^{-j2\pi x/p} \tag{2.3.5}$$

The right and left propagating waves are R and S, respectively. The COM model's parameters are the unperturbed detuning, δ, the distributed electrode reflectivity, κ, the transduction, α, and the spectral capacitance, $C(s)$. The detuning is expressed in relation to the unperturbed velocity, v_u.

Using the COM model, the harmonic solution to the admittance of a single electrode [12] is given by Eq. 2.3.6.

$$Y(\beta,\delta) = -j4\alpha^2 p\left(\frac{\delta+\kappa}{D^2-\beta^2}\right) - j\omega C_s(s) + Y_B \tag{2.3.6}$$

$$D = \pm\sqrt{\delta^2-\kappa^2} \approx \pm 2\pi\,\Delta s / p \tag{2.3.7}$$

The COM transduction coefficient, α, may be related to the pole amplitude, R_s, of the spectral model, by the comparison of Eqs. 2.3.6 and 2.2.1.

$$R_{s,n} = (\alpha_n \cdot p)^2 \cdot (\delta_n + \kappa_n) / \pi^2 \tag{2.3.8}$$

3. COM Model Characterization

The COM parameters, in Eqs. 2.3.1- 2.3.4, are the unperturbed surface mode velocity, v_u, the periodic reflection coefficient, κ, which *couples* the forward and reverse propagating modes, the electro-mechanical COM transduction coefficient, α, the electrostatic IDT capacitance, *C(s)*, and the BAW admittance, Y_B.

The COM parameters are characterized at discrete frequencies. Interpolation is used to obtain a continuous representation. The strategy to characterize the COM parameters consists of first obtaining a physical rational approximation to the FEM/BEM's harmonic admittance. Second, the desired mode's pole position and amplitude, χ_n and $R_{\chi,n}$, are isolated from the rational approximation. Next the COM detuned wavenumber, *D*, and transduction coefficient, α, are obtained by transformation of the rational approximation's pole position and amplitude. From the COM detuned wavenumber the COM reflection coefficient, κ, is extracted. Finally, the static capacitance, Eq. 3.0.1, and the remaining BAW admittance, Eq. 3.0.2 are evaluated.

$$C_s(s) = \lim_{f\to 0}\frac{Y(s,f)}{j\omega} \tag{3.0.1}$$

$$\begin{aligned} Y_B(s,f) &\approx Y_B(0.5,f) \\ &\approx Y(0.5,f) - j\omega C_s(s) - j4\alpha^2 p\left(\frac{\delta+\kappa}{D^2-\beta^2}\right) \end{aligned} \tag{3.0.2}$$

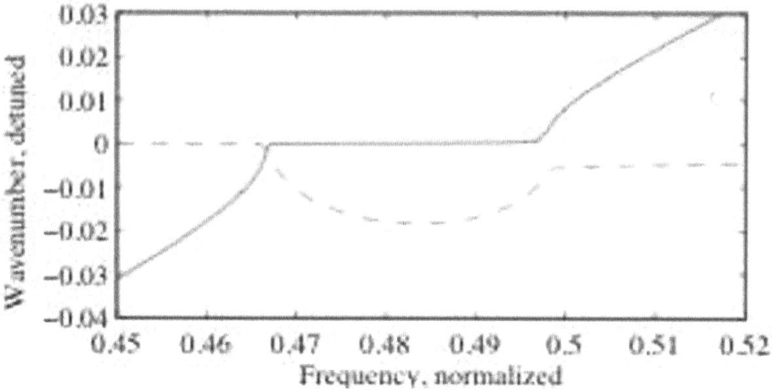

Fig. 3.0.1 The real (–) and imaginary (– –) parts of the detuned wavenumber, *D*

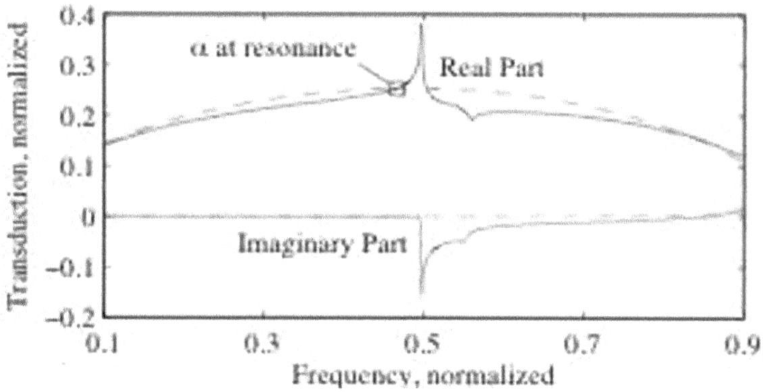

Fig. 3.0.2 The real (–) and imaginary (–) parts of α, and the Real (– –) and imaginary (– –) parts with the influence of the BAW suppressed.

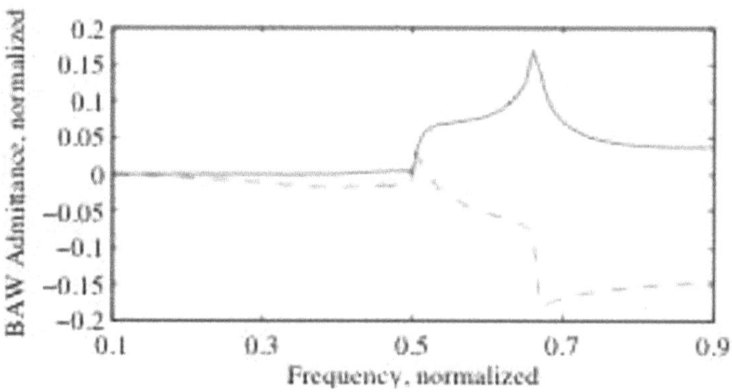

Fig. 3.0.3 The real (–) and imaginary (– –) parts of the remaining BAW harmonic admittance for $Y_B(s = 0.5, f) = Y(s = 0.5, f) - Y(\beta = 0, \delta)$.

3.1 Application of the Rational Approximation

For the purpose of fitting to the numerically obtained harmonic admittance the form of the rational approximation in Eq. 3.1.1 may be used.

$$\varepsilon_g(s,f) = \sum_{m=0}^{M} a_m \cdot \chi^m \Big/ \sum_{n=0}^{N} b_n \cdot \chi^n \tag{3.1.1}$$

The polynomial coefficients in Eq. 3.1.1 may be solved by a simultaneous, or linear least squares solution. Fig. 3.1.1 illustrates an example of the quality of fit provided by the physical rational approximation.

Eq. 3.1.1 may be manipulated into the form of Eq. 2.1.2. The subsequent evaluation of the spectral domain poles, Δs_n, and residues, $R_{s,n}$, as well as their COM model equivalents, D and α, are evaluated by applying Eqs. 2.2.4, 2.2.5, 2.3.7, and 2.3.8.

Fig. 3.0.1 illustrates an example of the detuned wavenumber, Fig. 3.0.2 illustrates the transduction coefficient, and Fig. 3.0.3 illustrates the BAW admittance. The extracted COM transduction coefficient is effected by the neighboring BAW mode at $\bar{f} = 0.5$. The effect of the BAW mode must be suppressed. The suppression may be accomplished by assuming the transduction coefficient's frequency dependence is consistent with the phenomenological representation in Eq. 3.1.2.

$$\alpha^2 \approx \alpha_r^2 \cdot \sin(\pi s_n) = \alpha_r^2 \cdot \cos(Dp) \tag{3.1.2}$$

where α_r is the transduction coefficient at resonance. In Fig. 3.0.2 the value of α_r is indicated with a box, □.

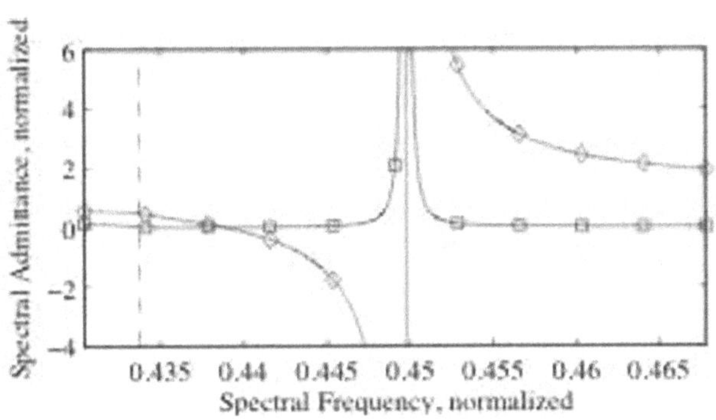

Fig. 3.1.1 The numerical results (symbols) and rational fit (lines) to the harmonic admittance at a frequency below resonance are illustrated. The □ and the ◇ symbols represent the real and imaginary parts of the numerical harmonic admittance. The spectral cut off frequency for the BAW is marked by the dashed line (- -).

3.2 COM Reflection Coefficient

Prior works have derived the parameters for the PMA and COM models by making reasonable approximations regarding to the frequency dependence of the reflectivity [8], or the unperturbed velocity [13].

When the reflection coefficient, κ in Eq. 2.3.7, is assumed to be a constant the derivative of the COM detuned wavenumber is unrelated to the reflection coefficient, Eq. 3.2.1.

$$\frac{\partial \delta^2}{\partial f} = \frac{\partial D^2}{\partial f} \tag{3.2.1}$$

In the context of the COM formalism, the unperturbed detuning, δ, and the unperturbed velocity, v_u, in Eq. 2.3.4, are continuous with frequency. In the vicinity of the stop band center there is a frequency, f_v, where the detuning is zero, Eq. 3.2.2.

$$\delta(f_v) = 0 \tag{3.2.2}$$

Examination of Eqs. 2.3.7, 3.2.1, and 3.2.2 leads to a solution to the reflection coefficient, κ, which is consistent with the COM formalism, Eq. 3.2.3.

$$\kappa^2 = -D(f_v)^2 \tag{3.2.3}$$

$$\left.\frac{\partial D^2}{\partial f}\right|_{f=f_v} = 0 \tag{3.2.4}$$

To evaluate Eq. 3.2.3 a numerical method must be applied to solve Eq. 3.2.4 for the complex valued f_v and to evaluate $D(f_v)$.

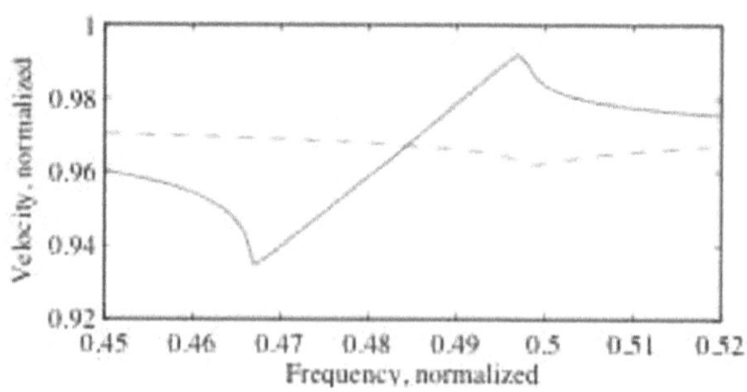

Fig. 3.2.1 The normalized perturbed (–) and unperturbed (– –) COM velocity.

In order for the resonance of a SAW device to manifests at the lower stop band, the sign of the real part of κ must be negative.

The sign ambiguity for δ in Eq. 2.3.4, may be resolved by ensuring the energy carried by the SAW dissipates as it propagates. As an illustration of a proper result, the normalized perturbed velocity, $\overline{v_p} = \overline{f}/s_1$, and the normalized unperturbed velocity, $\overline{v_u} = v_u/v_B$, are illustrated in Fig. 3.2.1.

4. Results

As a demonstration of the COM model, three devices have been constructed and modeled. These devices are (1) a one-port resonator on 46°YX $LiTaO_3$, (2) a one-port BIDT type resonator on 128°YX $LiNbO_3$, and a coupled-resonator filter (CRF) on 42°YX $LiTaO_3$.

4.1 46°YX $LiTaO_3$ One-Port Resonator

The first example is a one-port resonator whose metallization is proportionately composed of Aluminum, and whose physical geometries of the one-port resonator on 46°YX $LiTaO_3$ are given below.

Table 4.1.1 46° YX $LiTaO_3$ One Port Resonator

Length	Aperture	Period	*h/p*	*a/p*
400*p*	100*p*	1.08 um	0.197	0.4

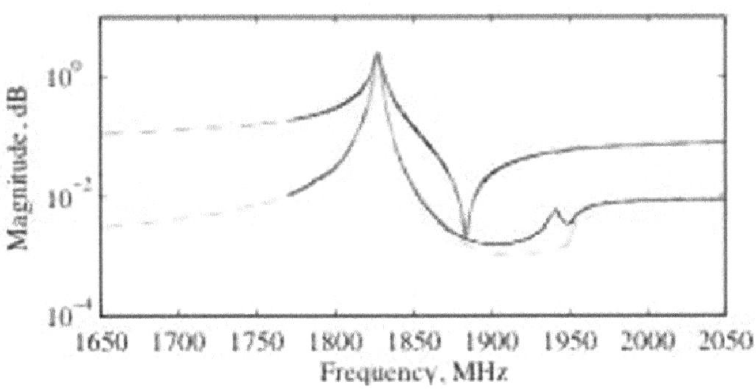

Fig. 4.1.1 Magnitude (–) and real (–) part of the measured admittance for a one-port resonator on 46°YX LT; accompanied by the magnitude (– –) and real (– –) part of the modeled admittance.

The small resonance seen in the experiment's conductance in the vicinity of 1940 MHz is that of the SAW-BAW hybrid (SBH) mode [16]. As the COM model represents a single acoustic mode, the SBH resonance is not represented by the theoretical result.

The electrode thickness and width are approximate values. The excellent agreement between results and calculations in Fig. 4.1.1 is, in part, the result of adjusting the model's parameters to account for the specifics of the fabrication processes used.

4.2 128°YX $LiNbO_3$ BIDT One-Port Resonator

The second example is a one-port resonator whose electrodes are proportionately composed of Copper, and which is buried under a planar overcoat of SiO_2, Fig. 4.2.1. The physical geometries of the one-port resonator on 128°YX $LiNbO_3$ are presented in Table 4.2.1.

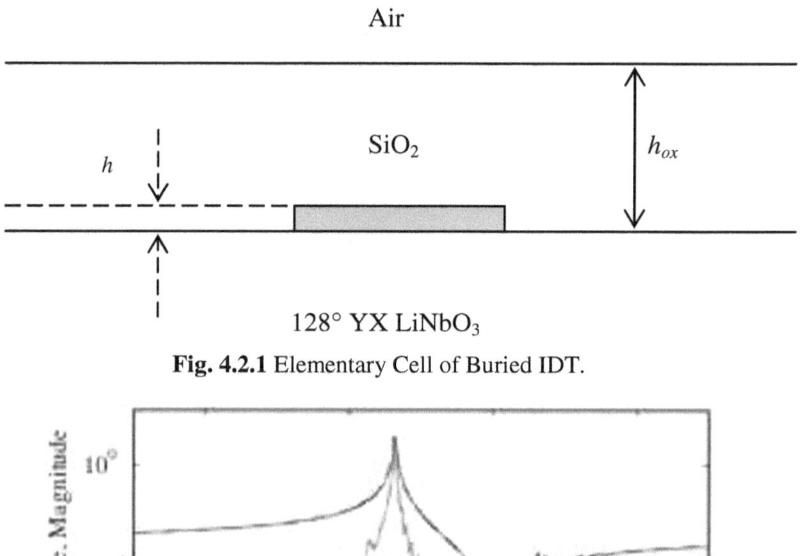

Fig. 4.2.1 Elementary Cell of Buried IDT.

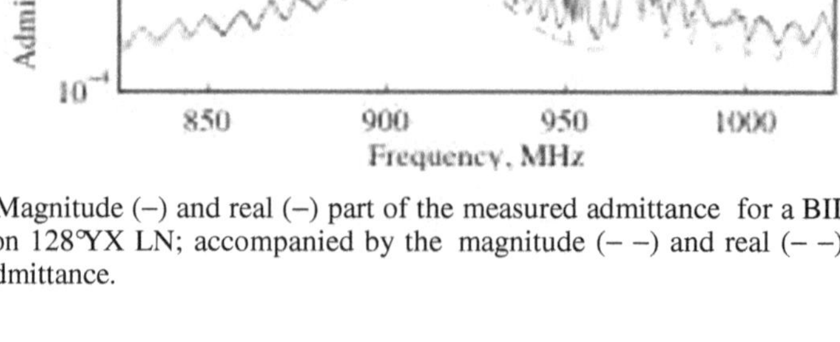

Fig. 4.2.2 Magnitude (–) and real (–) part of the measured admittance for a BIDT one-port resonator on 128°YX LN; accompanied by the magnitude (– –) and real (– –) part of the modeled admittance.

Table 4.2.1. 128° YX $LiNbO_3$ BIDT One Port Resonator

Length	Aperture	Period	h/p	a/p	h_{ox}/p
$200p$	$80p$	2.0 um	0.086	0.5	0.335

In Fig. 4.2.2, the resonances in the vicinity of the stop band, are the result of transverse modes.

The electrode thickness and width are approximate values. The excellent agreement between results and calculations in Fig. 4.2.2 is, in part, the result of adjusting the model's parameters to account for the specifics of the fabrication processes used.

4.3 48°YX $LiTaO_3$ BIDT Coupled Resonator

The third example is a two pole coupled resonator whose electrodes are proportionately composed of Copper, and buried beneath a planar SiO_2 overcoat.

Due to the more complex structure of this device the physical geometries are not itemized. The two port coupled resonator is specifically designed for on wafer measurement using RF probes. Fig. 4.3.1 illustrates a comparison between the COM model responses and measured responses for s_{11} and s_{12}.

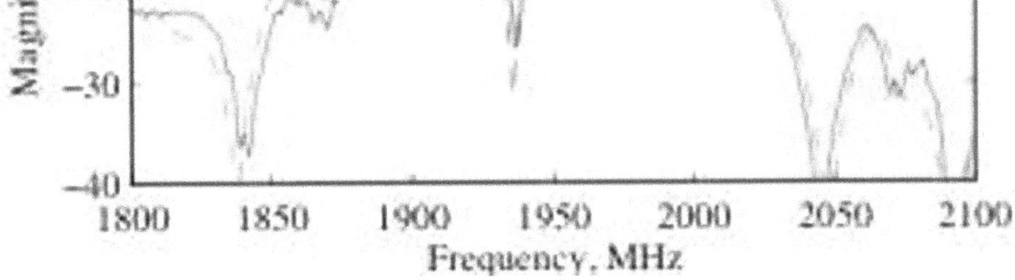

Fig. 4.3.1 Response for the measured (– –) and modeled (–) s_{11}, and the measured (–) and modeled (– –) s_{12}.

5. Conclusion

A method for charactering the COM model has been documented. The method relies upon a physical rational approximation of the harmonic.

Using these characterized parameters, the COM model has been applied to the analysis of RF type SAW devices. These SAW devices include both conventional and BIDT SAW one-port and two-port resonator. Good agreement between the responses obtained using the COM model and the experimental measurements has been demonstrated.

References

[1] Hartmann, C. S., Wright, P. V., Kansy, R. J., Earger, E. M.: An Analysis of SAW Interdigital Transducers with Internal Reflections and the Application to the Design of Single-Phase Unidirectional Transducers, *IEEE Ultras. Symp. Proc.* 40-45 (1982).

[2] Wright, P. V.: A New Generalized Modeling of SAW Transducers and Gratings, *43rd Symp. Freq. Control Proc.* 596-605 (1989).

[3] Chen, D.-P., Haus, H.A.: Analysis of Metal-Strip SAW Gratings and Transducers, *Sonics and Ultrasonics, IEEE Transactions* **32** 395-408 (1985).

[4] Thorvaldsson, T.: Analysis of the Natural Single Phase Unidirectional SAW Transducer, *IEEE Ultras. Symp. Proc.* 91-96 (1989).

[5] Hartmann, C. S., Abbott, B. P.: Experimentally Determining the Transduction Magnitude and Phase and Reflection Magnitude and Phase of SPUDT Structures, *IEEE Ultras. Symp. Proc.,* 37-42 (1990).

[6] Ventura, P., Hodé, J. M., Desbois, J, Solal, M.: Combined FEM and Green's Function Analysis of Periodic SAW Structure, Application to the Calculation of Reflection and Scattering Parameters, *IEEE Trans. Ultra. Ferro. and Freq. Control* **48** 1259-1274 (2001).

[7] Hashimoto, K.: *Surface Acoustic Wave Devices in Telecommunications*, Springer-Verlag, 293-318 (2000).

[8] Pastureaud, T.: Evaluation of the P-Matrix Parameters Frequency Variation Using Periodic FEM/BEM Analysis, *IEEE Ultras. Symp. Proc.* 80-84 (2004).

[9] Plessky, V., Koskela, J.: *Coupling-of-Modes Analysis of SAW Devices; Advances in Surface Acoustic Wave Technology, Systems & Applications*, World Scientific Publishing Co. Pte. Ltd., 1-82 (2000).

[10] Malocha, S., Abbott, B: LSAW COM Analysis, *IEEE Ultras. Symp. Proc.* (2008).

[11] Royer, D, Dieulesaint, E: *Elastic Waves in Solids II: Generation, Acousto-optic Interaction, Applications*. Springer-Verlag, 123-129 (1999).

[12] Koskela, J, Plessky, V, Salomaa, M: COM Parameter Extraction from Computer Experiments with Harmonic Admittance of a Periodic Array of Electrodes, *IEEE Ultras. Symp. Proc.,* 119-122 (1997).

[13] Hashimoto, K, Omori, T., Yamaguchi, M.: Recent Progress in Modelling and Simulation Technologies of Shear Horizontal Type Surface Acoustic Wave Devices. *Chiba University Symposium,* 169-178 (2001).

[14] Zhang, Y., Desbois, J., Boyer, L.: Characteristic Parameters of Surface Acoustic Waves in a Periodic Metal Grating on a Piezoelectric Substrate, *IEEE Trans. on Sonics and Ultra.* **40** 183-192 (1993).

[15] Naumenko, N., Abbott, B.: Fast Numerical Technique for Simulation of SAW Dispersion in Periodic Gratings and its Application to Some SAW Materials, *IEEE Ultras. Symp. Proc.* 166-170 (2007).

[16] Naumenko, N, Abbott, B.: Hybrid Surface-Bulk mode in Periodic Gratings. *IEEE Ultras. Symp. Proc.,* 234-248 (2001).

Propagation Characteristics of Surface Acoustic Wave and Properties of Gas Sensors on Quartz

Chi-Yen Shen[1], Cheng-Liang Hsu[1], Ming-Yau Su[1], Rume-Tze Tsai[1] and Shih-Han Wang[2]

[1]Department of Electrical Engineering, I- Shou University, Kaohsiung County 84001, Taiwan

[2]Department of Chemical Engineering, I- Shou University, Kaohsiung County 84001, Taiwan

cyshen@isu.edu.tw

Abstract. The simulation of wave propagation in periodic piezoelectric surface acoustic wave (SAW) structures was reported in this study. Firstly, space harmonics method (SHM) was used to analyze Rayleigh SAW propagation under a periodic *Al* grating. The results of SAW propagation analysis were applied to design a two-port resonator with *Al* grating on ST-cut quartz. The measured frequency response of the two-port resonator was approximately similar to the simulation one. Then, the chemical interface of polyaniline/WO_3 composites was coated on the SAW resonator for ammonia detection. The SAW sensor responded to the ammonia gas and could be recovered using dry nitrogen. Detecting to 9.3ppm ammonia, the frequency shift was 5.9ppm, the noise level was 0.18ppm, and a signal-to-noise ratio was 32.8.

1. Introduction

Ammonia is a toxic gas. The release of ammonia into the atmosphere has been causing global environmental issues [1]; therefore detection of ammonia gas is an important task. Among all kinds of gas sensors, acoustic wave-based sensors have been widely used for detection of the hazardous compounds in environments. With the advanced microfabrication techniques, SAW devices offer significant advantages of real-time and sensitive responses. Therefore, it has greatly received attention on development of the SAW gas sensors [2, 3].

T.-T. Wu and C.-C. Ma (eds.), *IUTAM Symposium on Recent Advances of Acoustic Waves in Solids*, IUTAM Bookseries 26, DOI 10.1007/978-90-481-9893-1_27,

The basic structures of the SAW devices consist of a piezoelectric substrate and periodic metal gratings. Due to the complexity of boundary condition, it is not easy to study SAW propagation with a consideration of mass loading effect of the periodic metal grating [4, 5]. Several analysis techniques have been developed for this purpose, including finite element method (FEM) [6, 7], Green's function [8, 9], and space harmonic method (SHM) [10, 11].

The propagation analysis of SAW by FEM must reduce the models in size and is calculated in the limit of infinitely thin electrodes. Green function method is also limited by its complexity and computational cost. In SHM, the boundary integral equations are derived from method of weighted residuals for a period of each region, such as substrate, metal, and free space, and are solved to satisfy the periodic boundary conditions. Therefore, SHM can integrate along boundary so that it is easy to apply to arbitrarily periodic shaped structures.

In this work, we report a sensor for measuring ammonia at room temperature near intensive farming. Ammonia detectors with a detection limit of 25ppm below 40°C are required for this application [1]. However, the commercially available sensors, which are generally based on metal oxide semiconductor materials operating at 250-500°C, can not reliably detect an allowable concentration of ammonia [12, 13]. We used SHM to investigate Rayleigh SAW propagating based on ST-cut quartz under the periodic *Al* grating. ST-cut quartz has been widely utilized to design SAW devices due to high temperature stability around room temperature. *Al* is a common material of the grating because of its well developed manufacture technique. We applied the results of propagation analysis to design a two-port resonator to detect ammonia at room temperature. The chemical interface was polyaniline/WO3 nanocomposite. Polyaniline film is easily prepared from aqueous solutions and is an effective interface for detecting ammonia [14, 15]. The material of WO3 has also shown good responses to ammonia [16]. The combination of polyaniline and WO3 in the nanostructure forms has the potential to increase sensitivity to ammonia. In this work, responses of an SAW sensor to ammonia were measured by the frequency shifts of SAW.

2. Theoretical Analysis

The SHM method only requires analysis of one period and expresses the displacements and the acoustic potential as the sum of space harmonics when it is used to analyze propagation under periodic gratings. The SAW propagates in the x_1 direction with period p in this work. h is thickness of the metal grating.

In each region, the displacements U_i and the electric potential φ are written as:
substrate :

$$U_i^s = \sum_{m=-\infty}^{\infty} \sum_{n=1}^{4} A^{(m,n)} \beta_i^{(m,n)} e^{j\frac{\pi}{p}\left\{\gamma\alpha^{(m,n)} x_3 + (\gamma+2m)x_1 - f(2p)t\right\}}, \quad (i = 1,2,3,4) \tag{2.1}$$

metal :

$$U_i^m = \sum_{m=-\infty}^{\infty} \sum_{n=5}^{10} A^{(m,n)} \beta_i^{(m,n)} e^{j\frac{\pi}{p}\left\{\gamma\alpha^{(m,n)} x_3 + (\gamma+2m)x_1 - f(2p)t\right\}}, \quad (i = 1,2,3), \tag{2.2}$$

free space :

$$U_4^f = \varphi^f = \sum_{m=-\infty}^{\infty} A^{(m,0)} \beta^{(m,0)} e^{j\frac{\pi}{p}\left\{\gamma\alpha^{(m,0)} x_3 + (\gamma+2m)x_1 - f(2p)t\right\}}, \tag{2.3}$$

where $U_4^s = \varphi^s$. m denotes the space harmonic and n is the index of the partial waves inside the layer. A(m,n) is the weighting factor, βi(m,n) is the component of the normalized eigenvector corresponding to the m-th space harmonic, α(m,n) is the decay factor in the x3 direction, and γ is the normalized wave number.

The wave equations under conditions of a quasi static approximation are as follows:
substrate :

$$T_{ij,j}^s = c_{ijkl}^s U_{k,lj}^s + e_{kij}^s U_{4,kj}^s = \rho^s \ddot{U}_i^s, \qquad (i, j, k, l=1, 2, 3) \tag{2.4}$$

$$D_{i,i}^s = e_{ikl}^s U_{k,lj}^s - \varepsilon_{ik}^s \varphi_{,ik}^s = 0 \tag{2.5}$$

metal :

$$T_{ij,j}^m = c_{ijrs}^m U_{r,sj}^m = \rho^m \ddot{U}_i^m \tag{2.6}$$

free space :

$$D_{i,i}^f = -\varepsilon_0 \varphi_{,ii}^f = 0 \tag{2.7}$$

where T_{ij} is the stress, D_i is the electric displacement, c is the elastic tensor, e is piezoelectric tensor, ε is the dielectric tensor, and ρ is the mass density. In equations (2.4)-(2.7), a dot denotes differentiation with respect to time and a comma denotes differentiation with respect to a space coordinate.

Substituting the equations of the displacements U_i and the electric potential φ into the wave equations, the $\alpha(^{m,n})$ and $\beta_i(^{m,n})$ can be obtained. In general, four $\alpha(^{m,n})$ are selected according to physical considerations in the substrate region, six $\alpha(^{m,n})$ are applied in the metal region, and one $\alpha(^{m,n})$ is selected according to physical considerations in the free space.

For the electrical boundary conditions, the potential and the normal component of electric displacement must be continuous across the boundaries, and the potential must be a constant value on the metal region. Moreover, the electric potential on the metal strips is zero in the shorted grating, or the total charge on the metal region is zero in the open case. For the mechanical boundary conditions, the displacements and the stress across the boundary between the substrate and the metal region must be continuous, and the stress across the boundary between the metal region and the free space must be zero.

Consequently, a matrix equation in the real and imaginary parts of the voltage V0 and the weighting factors A(m,n) can be obtained, it is as follows:

$$\mathbf{C}\cdot\mathbf{Y}=0 \tag{2.8}$$

where

$$\begin{aligned}\mathbf{Y}=[&\mathrm{Re}(V_0),\mathrm{Im}(V_0),\mathrm{Re}(A^{(-\infty,0)}),\mathrm{Im}(A^{(-\infty,0)}),\ldots,\mathrm{Re}(A^{(m,n)}),\mathrm{Im}(A^{(m,n)}),\\&,\ldots,\mathrm{Re}(A^{(\infty,10)}),\mathrm{Im}(A^{(\infty,10)})]^T\end{aligned} \tag{2.9}$$

The exact solution can be obtained by evaluating the normalized wave number and the normalized frequency, which satisfy the condition $det|\mathrm{C}| = 0$. The stopband frequencies converge in a similar manner as described in [16] when m increases. Hence, the number of space harmonics m was truncated to 4 in this work.

The dispersion curves of the Rayleigh wave under shorted and open *Al* grating on ST-cut quartz as a function of the normalized *Al* thickness, *h/2p,* are illustrated in Fig. 2.1. The stopband width for the shorted grating is narrower than that for the open grating. It indicates the open grating shows the enhanced reflectivity. The sensitivity of the upper stopband edge to *Al* strip thickness is less than that of the lower stopband edge. The imaginary part of the normalized wave number also increases with the *Al* strip thickness. It indicates the thick *Al* strip thickness induces the large reflection.

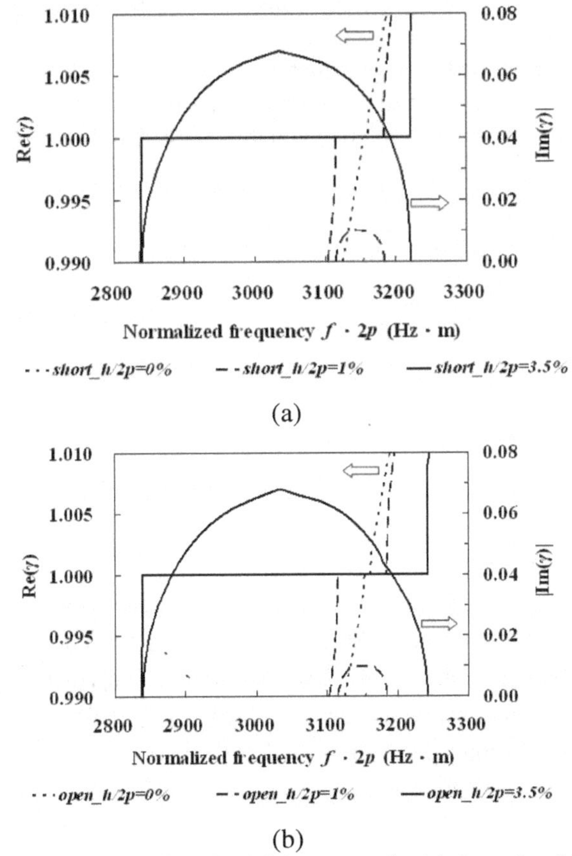

Fig. 2.1 The dispersion curves of the Rayleigh wave under (a) shorted and (b) open *Al* grating on ST-cut quartz as a function of the normalized *Al* thickness.

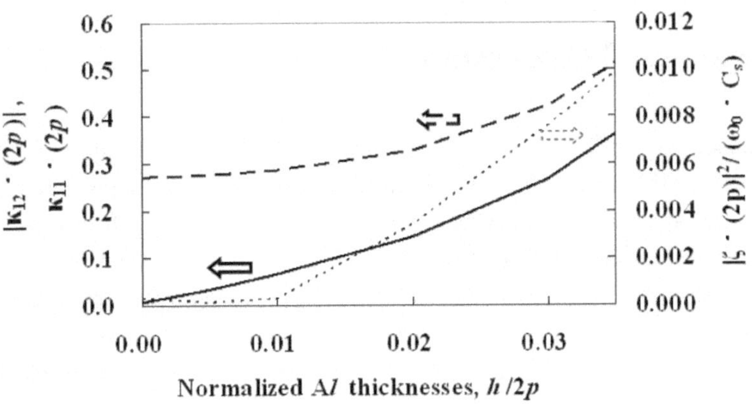

Fig. 2.2 COM parameters under the *Al* periodic grating on ST-cut quartz. : $|\kappa_{12}{\cdot}2p|$ (solid line), $\kappa_{11}{\cdot}2p$(dashed line), and $\left|\xi(2p)\right|^2/(\omega_0 C_s)$ (dotted line).

The COM theory is based on the concept that the progressing wave and counter-progressing wave couple with each other in periodic structures. The parameters using in the COM theory include κ_{12} (mutual coupling coefficient), κ_{11} (self-coupling coefficient), and ζ (transduction coefficient) that can be determined from the dispersion curves. The relative equations are as follows,

$$\kappa_{11} \cdot 2p = 2m\pi\left(1 - \frac{f_{ls} + f_{us}}{2mf_o}\right), \tag{2.10}$$

$$\left|\kappa_{12} \cdot 2p\right| = \pi\frac{f_{us} - f_{ls}}{f_0}, \tag{2.11}$$

$$\frac{\left|\zeta(2p)\right|^2}{\omega_0 C_s} = \frac{\pi}{2}\frac{1}{f_0^2}\left\{\left(f_{u0}^2 + f_{l0}^2 + f_{u0}f_{l0} + f_{us}f_{ls}\right) - \left(f_{us} + f_{ls}\right)\left(f_{u0} + f_{l0}\right)\right\}, \tag{2.12}$$

where f_0 is the Bragg frequency and C_s is the static capacitance per grating pair. f_{us}, f_{ls}, f_{uo}, and f_{lo} are four stopband edge frequencies of the dispersion curves for shorted and open gratings, respectively.

Fig. 2.2 shows the relationships between the COM parameters and grating thickness. $|\kappa_{12} \cdot 2p|$ is determined by stopband width and corresponds to the reflection coefficient per period. The increasing thickness of the gratings broadens stopband width. Hence, the thick grating with large effective reflection is suitable for application of the resonator with a small size. $\kappa_{11} \cdot 2p$ corresponds to phase shift per periodic. The parameter of $\kappa_{11} \cdot 2p$ evidently varies with grating thickness above $h/2p$ of 0.01. It is because the energy stored near the thick strip edges is not weak, which leads to large phase shift. The parameter of $\left|\xi(2p)\right|^2/(\omega_0 C_s)$ is also sensitive to grating thickness. Above $h/2p$ of 0.01, $\left|\xi(2p)\right|^2/(\omega_0 C_s)$ apparently increases with increasing metal grating thickness.

3. Results and Discussion

We designed and realized a two-port resonator with a shorted grating to investigate experimentally the accuracy of the theoretical results. Each of the input and output interdigital transducers (IDT) of *Al* has 104.5 finger pairs, each of reflector has 150 *Al* strip gratings, aperture of IDT is 960μm, the thickness of *Al* is 300nm, and the center-to-center distance between the IDTs is 605μm. The operating frequency of the resonator is 98.47MHz. Fig. 3.1 illustrates the frequency response of the two-port resonator. It shows that measurement data is similar to the simulation one. The oscillation frequency measured by the spectrum analyzer (4395A, Agilent, USA) and shown in Fig. 3.2 was stably at 98.47MHz after SAW resonator being connected with the oscillator.

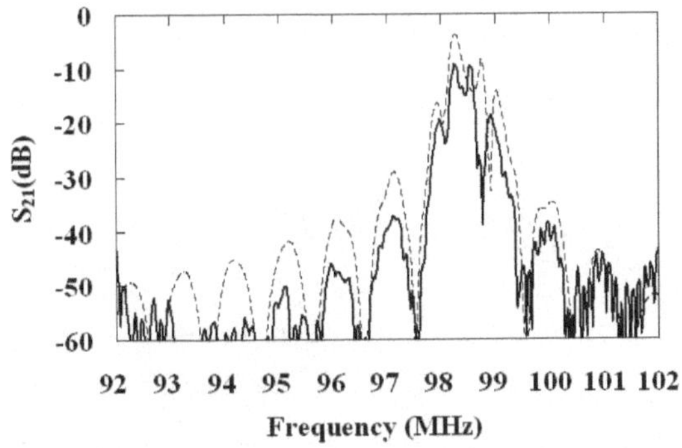

Fig. 3.1 Simulation (dotted line) and measurement (solid line) of frequency response of a two-port SAW resonator.

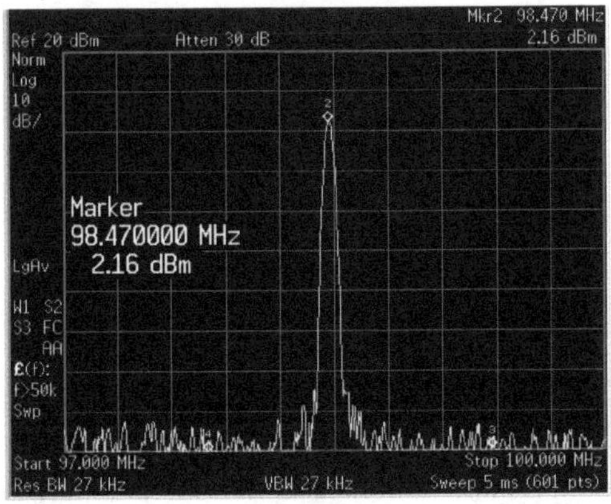

Fig. 3.2 Oscillation of two-port SAW resonator in measurement.

This study made use of a dual-device configuration shown in Fig. 3.3. A resonator coated with polyaniline/WO3 composites was used as a sensor, and a non-coated resonator was used as a reference. All detection signals were taken the output difference between the sensor and the reference to reduce interference from the environment such as pressure, flow rate, humidity, etc. The SAW resonators were connected to the RF electronic oscillator circuits to generate RF signals. Dry N_2 was the carrier gas. Mass flow controllers (Sierra, USA) were used to control flow rate at 110 ml/min. The dual-device configuration was put into a sealed 5 cm3 sensing chamber. All detections were performed at room temperature. A frequency counter connected to a computer system via a RS-232 interface board monitored the frequency differences between the sensor and the reference. Just before the gas detection process began, the dual-device configuration was exposed to dry nitrogen for 30min to stabilize the initial SH-SAW signals. The noise measurement was determined from data collected for 10min at 20 points per min, and noise was taken as the standard deviation of the residuals of the linear least squares fit through the data.

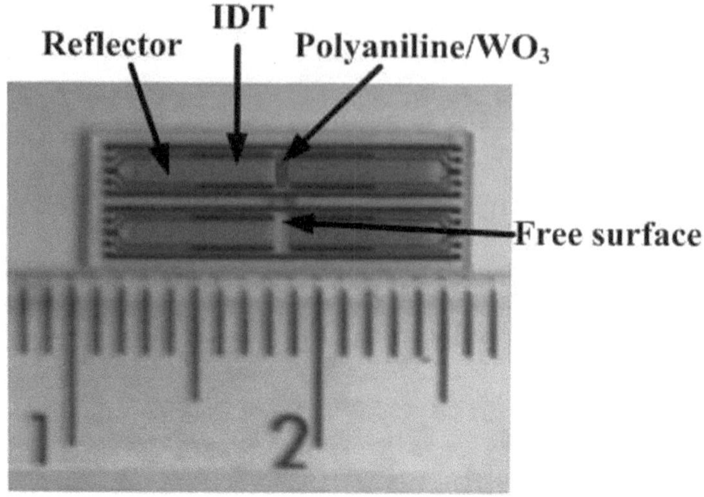

Fig. 3.3 Photograph of a dual-device configuration.

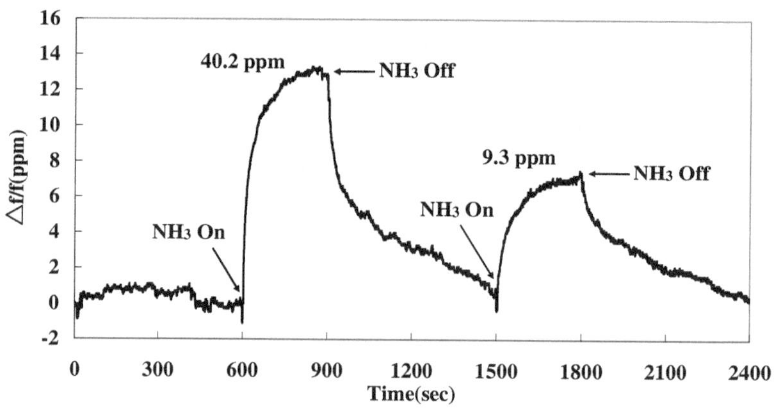

Fig. 3.4 Frequency responses of a SAW sensor to 40.2 and 9.3ppm ammonia.

Fig. 3.4 shows the SAW responses for a sensor upon exposure to pure N_2 and mixed NH_3/ N_2 gas streams. Several points are worth noting. First, the positive frequency shifts are approximately proportional to the concentration of NH_3. Second, the sensor has fast response time: for 40.2ppm NH_3, 90% of the maximum frequency shift is attained in 110s. Third, NH_3 reversibly binds to the chemically sensitive surface. Detecting to 40.2ppm ammonia, the frequency shift is 11.6ppm, the noise level is 0.16ppm, and a signal-to-noise ratio is 72.5. This sensor was also presented the sensitive response to 9.3ppm ammonia. The frequency shift is 5.9ppm, the noise level is 0.18 ppm, and a signal-to-noise ratio is 32.8.

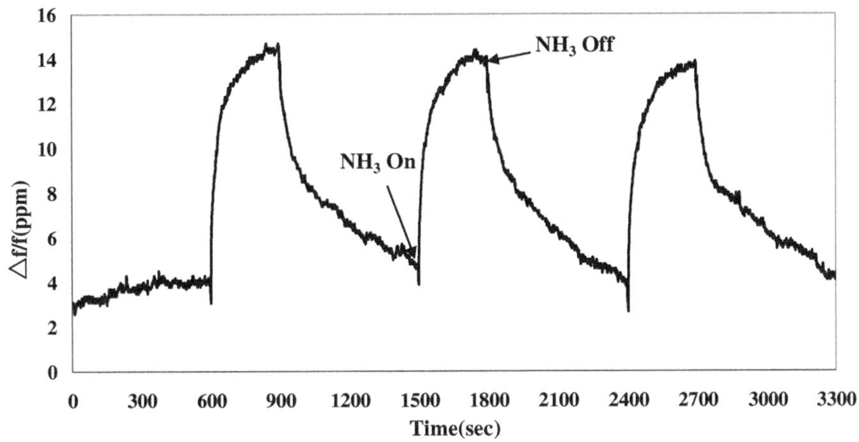

Fig. 3.5 Frequency responses of a SAW sensor to 26.2ppm ammonia.

The real-time responses of a SAW sensor to 26.2ppm ammonia in dry nitrogen are shown in Fig. 3.5. This SAW sensor detects the presence of ammonia gas after ammonia is on. The response returns to its original condition when dry nitrogen purges. The response is also repeatable becausee each gas on/off cycle presents similar response.

4. Conclusions

In this work, SHM was applied to analyze the Rayleigh SAW propagating under the periodic *Al* grating on the ST-cut quartz. The open grating shows the enhanced reflectivity. The thick *Al* strip thickness enlarges the attenuation. The SAW sensor coated with polyaniline /WO_3 composites exhibited responses to ammonia at room temperature. The response could be recovered by dry nitrogen easily and presented reversibility and repeatibility. This helps us achieve our goal of developing SAW-based gas sensors that are sensitive to ppm-level ammonia at room temperature.

Acknowledgments: The authors would like to thank the National Science Council, R.O.C, for partially supporting this research under Contract No. NSC96-2221-E-214-002.

References

[1] Timmer, B., Olthuis, W. A., Berg,van den: Ammonia sensors and their applications - a review. *Sens. Actuat. B* **107** 666-677 (2005).

[2] Wu, T. T., Chen, Y. Y., Chou, T. H.: A high sensitivity nanomaterial based SAW humidity sensor. *J. Phys. D: Appl. Phys.* **41** 085101 (2008).

[3] Huang, F. C., Chen, Y. Y., Wu, T. T.: A room temperature SAW hydrogen sensor with Pt coated ZnO nanorods. *Nanotechnology* **20** 065501 (2009).

[4] Benetti, M., Cannatà, D., Pietrantonio, F. Di, Fedosov, V. I., Verona, E.: Gigahertz-range electro-acoustic devices based on pseudosurface-acoustic waves in AlN/diamond/Si structures. *Appl. Phys. Let.* **87** 033504 (2005).

[5] Makkonen, T., Plessky, V. P.: Modeling longitudinal leaky saw propagation under periodic electrode arrays. *Proc. IEEE Ultrason. Symp.* 691-694 (2005).

[6] Hofer, M., Finger, N., Kovacs, G., Zaglmayr, S., Langer, U., Lerch, R.: Finite-element simulation of wave propagation in periodic piezoelectric SAW structures. *IEEE Trans. Ultrason., Ferroelect., Freq. Contr.* **53** 1192-1201 (2006).

[7] Laude, V., Reinhardt, A., Wilm, M., Khelif, A., Ballandras, S.: Fast FEM/BEM simulation of SAW devices via asymptotic waveform evaluation. *IEEE Trans. Ultrason., Ferroelect., Freq. Contr.* **51** 359-363 (2004).

[8] Duran, M., Nedelec, J. C., Ossandon, S.: Numerical study of the spectral 3-D Green's function singularities for piezoelectric SAW components. *IEEE Trans. Ultrason., Ferroelect., Freq. Contr.* **52** 2395-2402 (2005).

[9] Laude, V., Hanckes, C. F. J., Ballandras, S.: Surface Green's function of a piezoelectric half-space. *IEEE Trans. Ultrason., Ferroelect., Freq. Contr.* **53** 420-428 (2006).

[10] Kitabayashi, H., Smith, P. M.: Analysis of SAW propagation in gratings on ZnO/diamond substrates. *IEEE Trans. Ultrason., Ferroelect., Freq. Contr.* **51** 249-261 (2001).

[11] Sato, T., Abe, H.: Propagation of longitudinal leaky surface waves under periodic metal grating structure on lithium tetraborate. *IEEE Trans. Ultrason., Ferroelect., Freq. Contr.* **45** 394-408 (1998).

[12] Kovalenko, V. V., Zhukova,A. A., Rumyantseva, M. N., Gaskov, A. M., Yushchenko, V. V., Ivanova, I. I., Pagnier, T.: Surface chemistry of nanocrystalline SnO2: Effect of thermal treatment and additives. *Sens. Actuat. B* **126** 52-55 (2007).

[13] Karunagaran, B., Uthirakumar, P., Chung, S. J., Velumani, S., Suh, E. K.: TiO2 thin film gas sensor for monitoring ammonia, *Materials Characterization.* **58** 680-684 (2007).

[14] Sengupta, P. P., Barik, S., Adhikari, B.: Polyaniline as a gas-sensor material. *Materials and Manufacturing Processes* **21** 263-270 (2006).

[15] Koul, S., Chandra, R.: Mixed dopant conducting polyaniline reusable blend for the detection of aqueous ammonia, *Sens. Actuat. B* **104** 57-67 (2005).

[16] Pissoort, D., Michielssen, E., Olyslager, F., Zutter, D. D.: Fast analysis of 2D electromagnetic crystal structures using a periodic Green function approach, *IEEE J. Lightwave Technol.* **23** 2294-2308 (2005).

Three Dimensional Displacement Measurement of Transient Elastic Wave Propagation Using a Multidimensional Point-Wise Fiber Bragg Grating Displacement Sensor System

Kuo-Chih Chuang and Chien-Ching Ma

Department of Mechanical Engineering, National Taiwan University, Taipei 106, Taiwan

ccma@ntu.edu.tw

Abstract. A fiber Bragg grating (FBG) sensor setup method that can allow the FBG to detect out-of-plane or in-plane point-wise displacement is demonstrated in this study. Based on this method, we successfully establish a multidimensional sensing system which contains one out-of-plane displacement sensor and two in-plane displacement sensors to measure the transient elastic wave propagation in a square thick plate subjected to impact loadings. The proposed system employs wavelength-optical intensity modulation techniques based on a long-period fiber grating filter and two FBG filters. The transient responses of out-of-plane particle motions measured by the FBG displacement sensor are compared with those obtained simultaneously by a laser Doppler vibrometer (LDV). The experimental results show that the proposed FBG displacement sensor system is capable of performing real-time measurement of three dimensional dynamic displacements for transient elastic wave propagation.

1. Introduction

For the last decade, fiber Bragg gratings (FBGs) have been employed as important optical sensors to measure physical quantities such as strain and temperature [1, 2]. In most cases, since an FBG is completely mounted on the surface of a specimen, it is not used to measure the displacement of a specific point on an arbitrary surface. Traditionally, optical non-contact displacement sensors based on Doppler effect [3] or heterodyne grating interferometry [4] can provide high-precision out-of-plane or in-plane displacement measurement. However, they are high in cost

T.-T. Wu and C.-C. Ma (eds.), *IUTAM Symposium on Recent Advances of Acoustic Waves in Solids*, IUTAM Bookseries 26, DOI 10.1007/978-90-481-9893-1_28,

and often occupy much space. In an attempt to use a low-cost and small size FBG to measure the displacement, Chuang and Ma [5] recently proposed an FBG setup method and successfully used it to detect the out-of-plane or in-plane dynamic displacement.

To investigate behaviors of transient elastic waves, Ma and Lee [6] studied three-dimensional transient elastic waves in a multilayered medium subjected to dynamic loadings by theoretical analysis, numerical calculation and experimental measurement by a commercial NBS (National Bureau of Standards) conical transducer. However, the conical transducer can only be used to measure the out-of-plane motions. In this paper, we propose an FBG displacement sensor setup method to establish a multidimensional displacement sensor system [5] and use this system to measure the transient elastic wave propagation in a square thick plate subjected to impact loadings. In this work, the responses of out-of-plane FBG displacement sensor are demodulated by a long-period fiber grating (LPFG). For the in-plane FBG sensors, the FBG filter-based demodulation technique is employed to obtain larger sensitivity and signal-to-noise ratio (SNR). The measured out-of-plane displacement responses are compared simultaneously with those obtained by a laser Doppler vibrometer (LDV). The in-plane displacement measurement ability is examined by two orthogonal FBG in-plane sensors. Finally, the resonant frequencies of the thick plate determined from the transient time-domain responses are discussed with the corresponding mode shapes predicted by FEM computations.

2. Principle and Setup of the FBG Sensing System

The configuration of the FBG displacement sensor system used in this study is illustrated in Fig. 3.1. Light from a broadband source (BBS) is transmitted to an isolator, a splitter, an LPFG, and two FBG filters. Each light enters a directional circulator and is reflected back to the corresponding output channel by an FBG sensor. According to Bragg's law, the Bragg wavelength λ_{Bs} of an FBG sensor is given by [5]

$$\lambda_{Bs}(z,t) = 2n_{\mathrm{eff}}(z,t)\Lambda(z,t) \tag{2.1}$$

where $n_{\mathrm{eff}}(z,t)$ is the effective refractive index of the fiber core and $\Lambda(z,t)$ is the Bragg grating period. The shift in Bragg wavelength due to dynamic strain can be expressed as

$$\Delta\lambda_{Bs}(z,t) = (1 - p_e)\lambda_{Bs}(z,t)\varepsilon(z,t) \tag{2.2}$$

where $\varepsilon(z,t)$ is the dynamic strain of the fiber, and p_e is an effective photoelastic coefficient. By fixing one end of a length l_0 of fiber containing an FBG sensor to a stationary boundary and fixing the other end to a point on a movable object under measurement, the displacement of the object can be expressed by

$$D(t) = \int_0^{l_0} \varepsilon(z,t)dz\,. \tag{2.3}$$

In order to estimate demodulation behavior, the transmittance of the FBG filter can be approximated by

$$F(\lambda) = 1 - R_f \exp[-4\ln 2(\frac{\lambda - \lambda_{Bf}}{\sigma_f})^2] \tag{2.4}$$

where λ_{Bf} is the Bragg wavelength of the filter, R_f is the maximum reflectivity, and σ_f is the grating full-width at half maximum (FWHM) of the FBG filter. The spectrum reflectance of the FBG sensor can also be approximated by a Gaussian function, given by

$$S(\lambda,t) = R_s \exp[-4\ln 2(\frac{\lambda - \lambda_{Bs}(z,t)}{\sigma_s})^2] \tag{2.5}$$

where R_s is the maximum reflectivity and σ_s is the grating FWHM of the FBG sensor. When the intensity of the BBS transmitted into the FBG sensor is $I_i(\lambda)$, the total light power $P_d(t)$ detected on the photodetector (PD) can be expressed by

$$P_d(t) = \int_0^{\infty} I_d(\lambda)d\lambda = k\int_0^{\infty} I_i(\lambda)S(\lambda,t)F(\lambda)d\lambda \tag{2.6}$$

where $I_d(\lambda)$ is the light intensity into the PD and k is a parameter reflects the optical intensity attenuation through the light path, splitter or circulator. The demodulation behavior using the LPFG filter is similar to that of the FBG filter. The only difference lies on the spectrum of the LPFG, which has much larger linear portion than that of the FBG filter.

Details regarding the FBG displacement sensor setup method can be found in Chuang and Ma [5]. To attach the out-of-plane FBG sensor to the surface of the specimen, one end of it is glued to the translation stage with strain gauge cement and subsequently with a layer of epoxy glue to provide extra strength. The other end of the FBG sensor is cut at an angle to decrease return reflection loss. Then, a mix of epoxy resin and hardener is put on the sensing point. Within the first minute of the curing cycle of the epoxy glue, the FBG displacement sensor is moved slightly up and down by the translation stage in order to reduce light reflection and the end face of the FBG sensor is finally glued to a point on the surface to be measured. If λ_{Bf} of the FBG or the transmittance peak of the LPFG filter attached to a translation stage is adjusted to the left of $\lambda_{Bs}(z,0)$ before the measurement, then $\lambda_{Bs}(z,t)$ will lie in the linearized positive slope spectrum of the FBG or LPFG filter. The PD output electrical signal will linearly increase as the FBG sensor is elongated and decrease as the FBG sensor is compressed under the measurement on the order of sub-micrometers. Moreover, the setup of the FBG sensor system also allows the FBG displacement sensor to detect the dynamic responses like acoustic vibration.

3. Experimental Results and Discussion

The experimental setup is shown in Fig. 3.1. Light from a broadband source (BBS) with a wavelength ranging from 1520 to 1570 nm is transmitted to an isolator, coupled by a 1×3 splitter and divided into three paths. Each light beam enters a corresponding LPFG or FBG filter (FBG filters 1, 2, and LPFG filter) attached to a translation stage and a directional circulator and is reflected back to the output channel by the FBG sensors, including one out-of-plane FBG displacement sensor and two in-plane FBG displacement sensors. A photodetector (PDA10CS, InGaAs amplified detector, Thorlabs, Inc) then receives the optical intensity signal and transforms it into an electrical signal, which is recorded by an oscilloscope (LeCroy 64Xs). The original Bragg wavelengths of the FBG sensors as well as the LPFG filter and FBG filters are also depicted in Fig. 3.1. All the FBG grating lengths are 10 mm.

The proposed FBG displacement sensor system is employed to measure the transient elastic wave propagation on an aluminum square thick plate (made of 6061-T6). The aluminum thick plate is loosely placed on top of four silicon rubber pads to simulate the traction-free boundary. The dimension of the thick plate and sensing locations of the three orthogonal FBG displacement sensors (consisting of one out-of-plane displacement sensor (FBG1), and two in-plane displacement sensors (FBG2 and FBG3)) and LDV on the thick plate are shown in Fig. 3.2. The thickness of the thick plate is 6.65 mm. Elastic transient waves are generated by a steel ball impacted vertically from a height onto point A (i. e., center of the surface of the thick plate) or B (as shown in Fig. 3.2) on the top surface.

The out-of-plane transient displacement responses measured by the proposed FBG displacement sensor can be verified by the LDV. Hence, the out-of-plane FBG displacement sensor (FBG1) is performed first. A 5/32 in. steel ball is dropped onto the square aluminum thick plate on location A. FBG1 as well as LDV are employed simultaneously to measure the transient displacement responses of the wave propagation. Fig. 3.3 shows the experimental results of the transient displacement responses. Five wave groups can be clearly observed in the measured waveforms, which represent five impacts of the steel ball. The maximum peak to peak value of LDV is 2238.29 nm. Fig. 3.4 focuses on the time period of 20 ms of the initial displacement response. The agreement between results measured from FBG1 and LDV is excellent.

Since the dynamic measurement ability of the FBG is already demonstrated by the out-of-plane FBG displacement sensor, we now perform multidimensional measurement simultaneously using four sensors (i. e., FBG1, FBG2, FBG3, and LDV). Fig. 3.5 shows parts of the first wave group of the multidimensional displacement responses of the square thick plate after a 3/16 in. steel ball has impacted at location A. Again we can see that the out-of-plane response obtained from FBG1 agrees well with that obtained from LDV. The maximum peak to peak value of LDV is 3135.52 nm. Examining the out-of-plane transient responses in FIG. 3.5 and FIG. 3.4, we can see that the repeatability of the experimental results is excellent. It should be noted that the responses of the FBG in-plane displacement sensors are normalized for comparison.

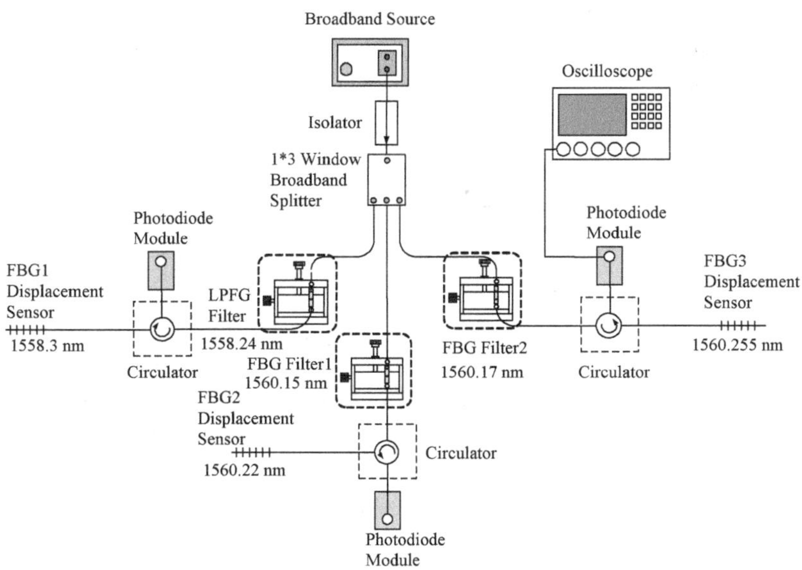

Fig. 3.1 Experimental setup of the multidimensional FBG displacement sensor system.

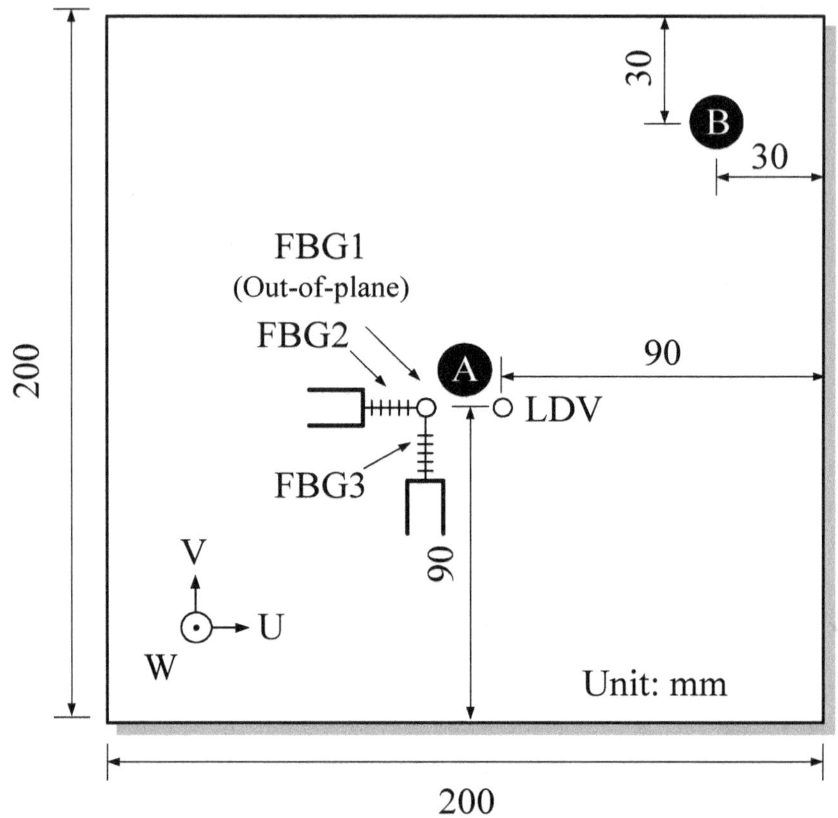

Fig. 3.2 Illustration of sensing points on the square thick plate.

The frequency spectrum can be constructed from taking the fast Fourier transform of the time-domain transient responses obtained by all the displacement sensors. Fig. 3.6 shows the frequency spectra of the out-of-plane transient responses presented in Fig. 3.5 to 20 kHz. It is clearly shown that the resonant frequencies determined from the transient responses obtained from LDV and FBG1 sensor are nearly the same. Thus, the measurement ability of the FBG sensor can be demonstrated by comparison with the LDV in either the time domain or the frequency domain. The frequency spectra of the in-plane transient responses presented in Fig. 3.5 are shown in Fig. 3.7. It is noted that the resonant frequencies determined from the in-plane transient responses for FBG2 and FBG3 sensors are also nearly the same.

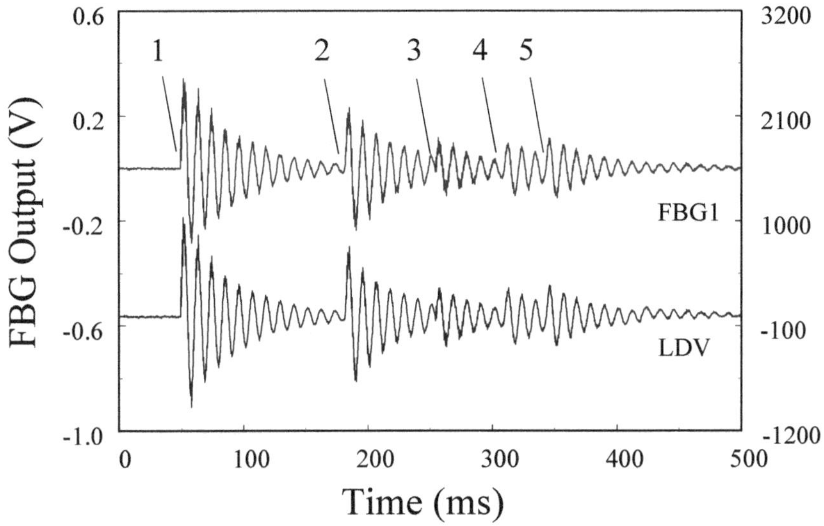

Fig. 3.3 Transient out-of-plane displacement response of the square thick plate for the impact location at A.

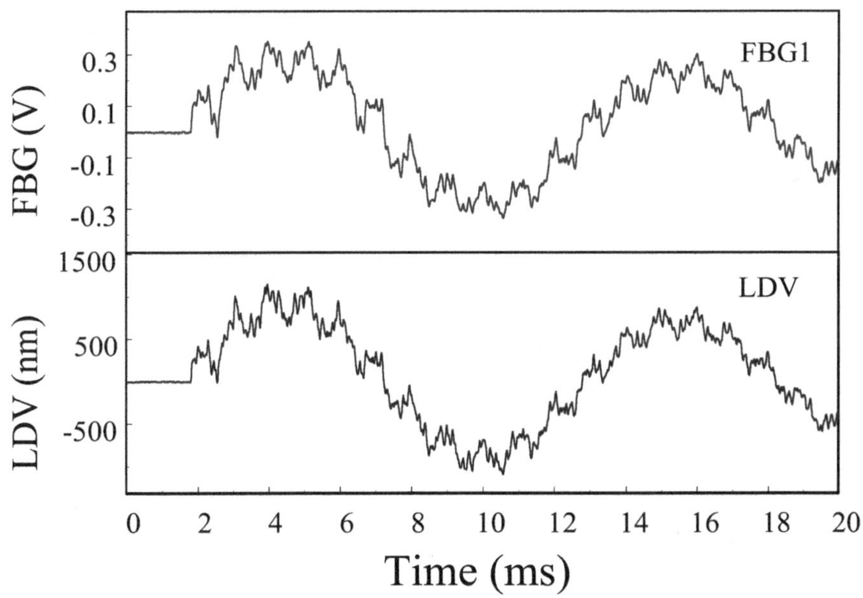

Fig. 3.4 Transient out-of-plane displacement response of the square thick plate within 20 ms.

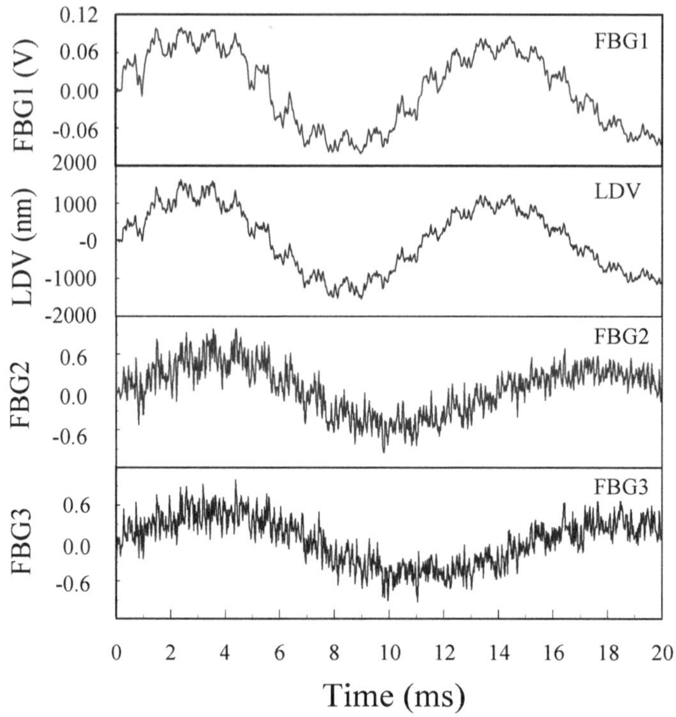

Fig. 3.5 Transient multidimensional displacement responses of the square thick plate.

In order to acquire complete information on the dynamic characteristics of the aluminum square thick plate, FEM computations are employed to provide three-dimensional vibration mode shapes of the thick plate in resonance. The boundary condition of the thick plate for computations is assumed to be traction-free. For FEM calculations, the mass density of 6061-T6 aluminum is 2710 kg/m^3, the Young's modulus is 73.5 GPa, and the Poisson's ratio is 0.34. It should be noted that a total of 46 modes can be obtained from FEM computations within 20 kHz. Fig. 3.8 presents the first 21 mode shapes. The definition of the three orthogonal directions (i.e., U, V, and W) is indicated in Fig. 3.2. The dashed lines and solid lines indicate the displacement in different phases. The nodal line is represented by a bold line. To discuss the relationship between the impact location, sensing locations, and nodal line, frequency spectra obtained from out-of-plane sensors (as shown in Figs. 3.6(a) and 3.6(b)) are considered. In fact, the first mode is not detectable due to the fact that the impact location is directly on the nodal line (see W column). Other than the reason due to impact location, the second mode is not detectable because the sensing locations of FBG1 and LDV are both on the nodal line. Since impact location A is far away from the nodal line of mode 3, the amplitude of mode 3 is relatively large compared to other detectable modes.

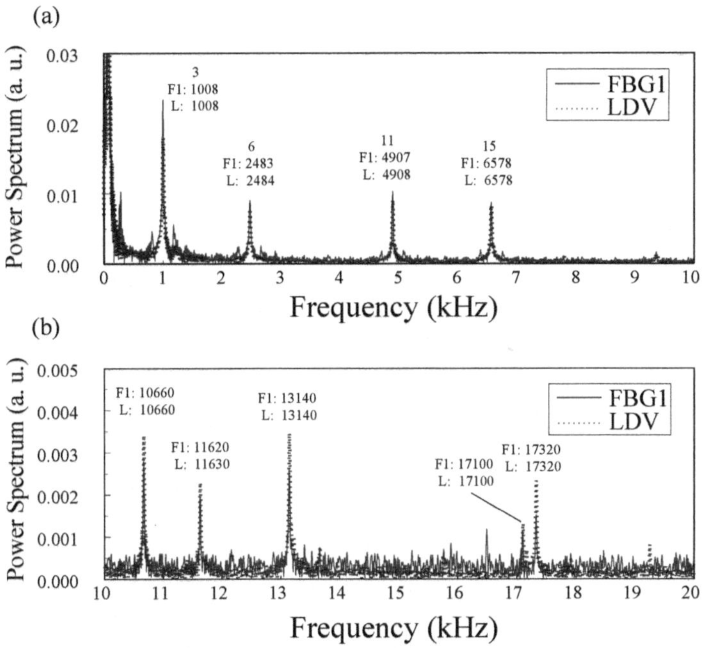

Fig. 3.6 Frequency spectrum of the out-of-plane responses for location A.

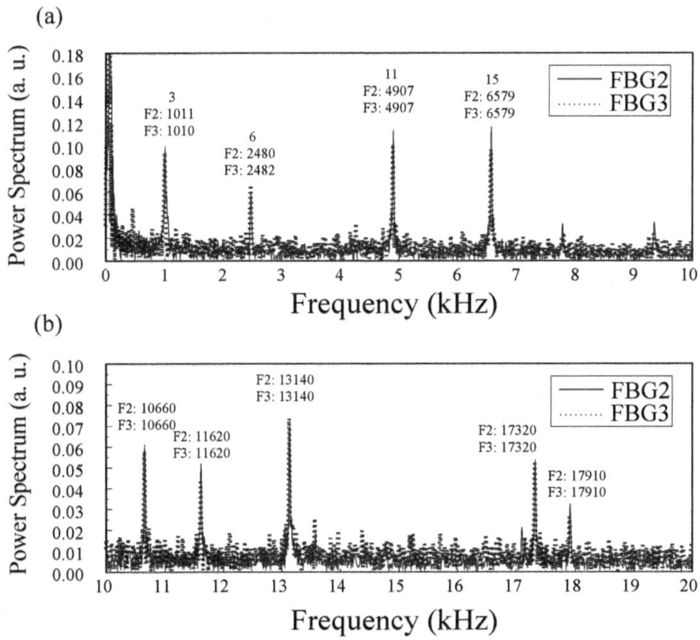

Fig. 3.7 Frequency spectrum of the in-plane responses for location A.

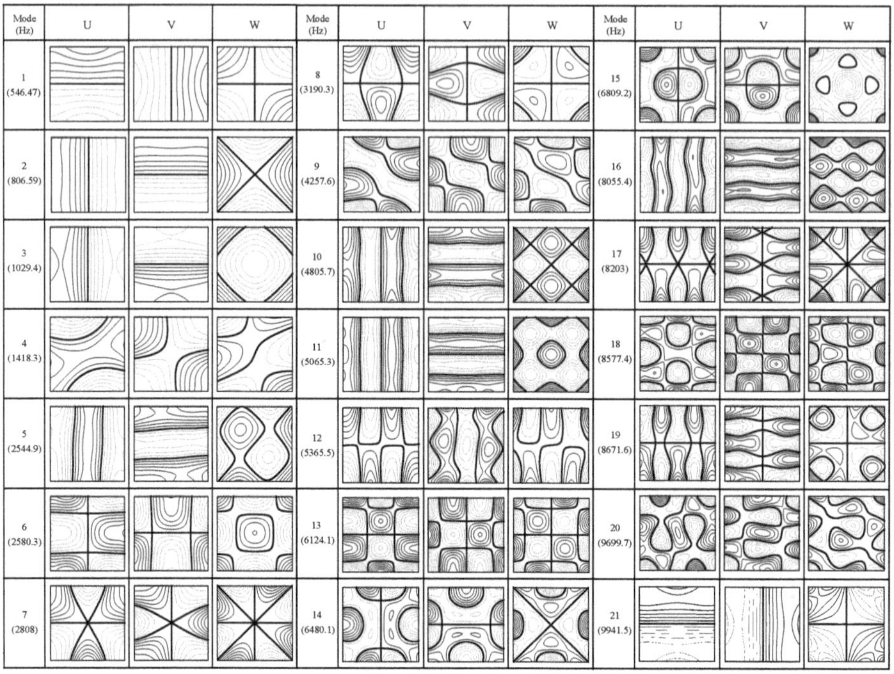

Fig. 3.8 Parts of out-of-plane and in-plane vibration mode shapes.

Fig. 3.9 shows parts of the wave group of the multidimensional displacement responses of the square thick plate after the 3/16 in. steel ball has impacted at location B (as shown in Fig. 3.2). The maximum peak to peak value of LDV is 2780.64 nm.

Fig. 3.10 shows the frequency spectra of the out-of-plane transient responses presented in Fig.3.9 to 20 kHz. The resonant frequencies determined from the transient responses obtained from LDV and FBG1 sensor are still nearly the same. Out-of-plane modes 16 and 19, which are not detectable when the impact location is at A, can be detected due to the fact that impact location B is not on the nodal line. The frequency spectra of the in-plane transient responses presented in Fig. 3.9 are shown in Fig. 3.11.

4. Conclusions

In this study, we propose a setup method for a multidimensional FBG displacement sensor system and investigate its transient measurement ability by measuring the dynamic responses of a square thick plate impacted by steel balls. The out-of-plane point-wise transient responses are measured by an FBG displacement sensor and LDV simultaneously and excellent agreement between these two measurements is obtained. The resonant frequencies determined from the frequency spectrum obtained from taking the fast Fourier transform of the transient responses obtained by the proposed sensor system agree well with the relationship between the impact locations, sensing locations, and the nodal line of the mode shapes predicted by FEM computations. The experimental results show that the proposed FBG displacement sensor system is capable of performing real-time measurement of three dimensional displacement of transient elastic wave propagation.

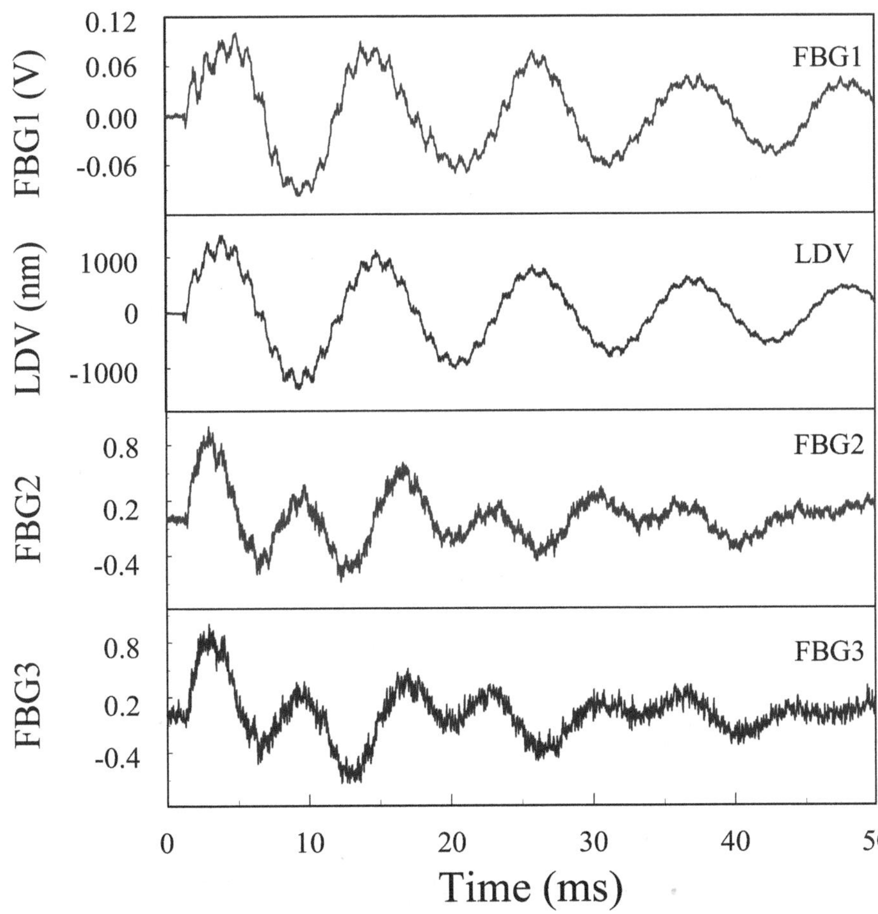

Fig. 3.9 Transient multidimensional displacement responses of the square thick plate.

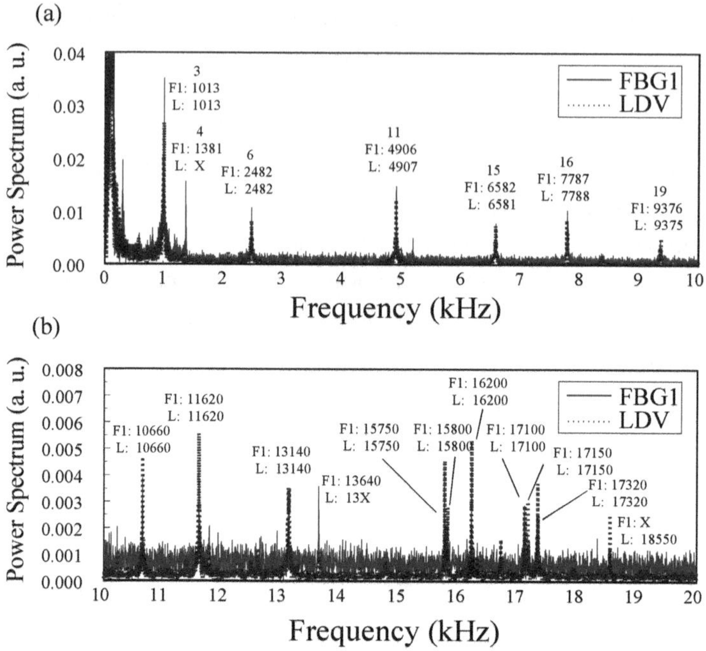

Fig. 3.10 Frequency spectrum of the out-of-plane responses for location B.

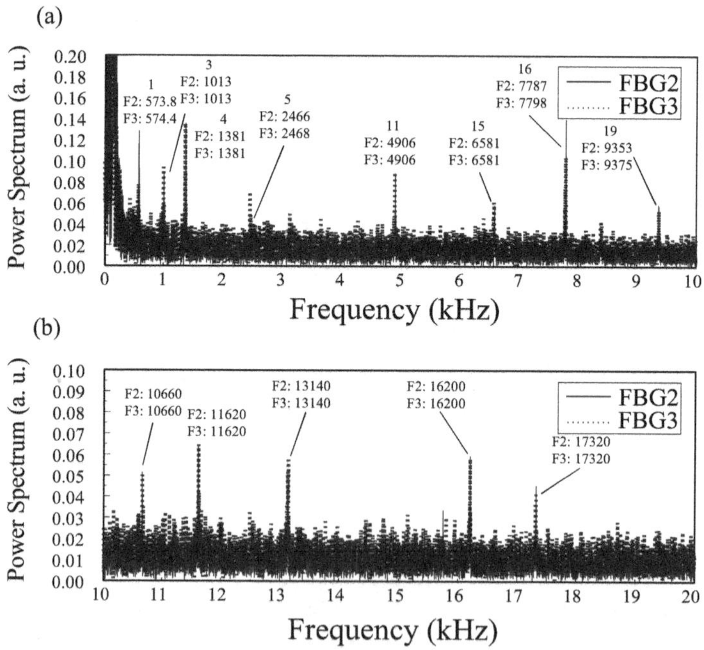

Fig. 3.11 Frequency spectrum of the in-plane responses for location B.

Acknowledgements: The authors would like to thank Professor W. F. Liu of the Department of Electrical Engineering, Feng Chia University for useful help in FBG fabrication. The authors also gratefully acknowledge the financial support of this research by the National Science Council (Republic of China) under Grant NSC96-2221-E-002-210-MY3.

References

[1] Kersey, A. D., Davis, M. A., Patrick, H. J., LeBlanc, M., Koo, K. P., Askin,s C. G., Putnam, M. A. , Friebele E. J.: Fiber grating sensors. *J. Lightwave Technol.* **15** 1442-1463 (1997).

[2] Rao, Y. J.: In-fibre Bragg grating sensors. *Meas. Sci. Technol.* **8** 355-375 (1997).

[3] Lee, C. K., Wu, G. Y., Teng, C. T., Wu, W. J., Lin, T., Hsiao, W. H., Shih, H. C., Wang, J. S., Lin, S. C., Lin, C. C., Lee, C. F., Lin, Y. C.: A high performance Doppler interferometer for advanced optical storage systems. *Jpn. J. Appl. Phys.* **38** 1730-1741 (1999).

[4] Hsu, C. C., Wu, C. C., Lee, J. Y., Chen, H. Y., Weng, H. F.:Reflection type heterodyne grating interferometry for in-plane displacement measurement. *Opt. Commun.* **281** 2582-2589 (2008).

[5] Chuang, K .C., Ma, C. C.: Pointwise fiber Bragg grating displacement sensor system for dynamic measurements. *Appl. Opt.* **47** 3561-3567 (2008).

[6] Ma, C. C., Lee, G. S.: General three-dimensional analysis of transient elastic waves in a multilayered medium. *ASME J. Appl. Mech.* **73** 490-504 (2006).

Part V
Seminar Section

Band Structure Calculations by Modal Analysis

Mahmoud I. Hussein

Department of Aerospace Engineering Sciences, University of Colorado at Boulder, Colorado 80309, USA

mih@colorado.edu

Abstract. In this paper we present a new paradigm by which modal analysis – which is well established in engineering structural dynamics – is applied to band structure calculations for phononic crystals, or periodic media in general. Our method, which we refer to as *reduced Bloch mode expansion (RBME)*, is essentially an expansion employing a natural basis composed of a selected reduced set of Bloch eigenfunctions. This reduced basis is selected within the Irreducible Brillouin Zone at high symmetry points determined by the crystal structure and group theory (and possibly at additional related points). At each of these high symmetry points, a number of Bloch eigenfunctions are selected up to the frequency range of interest for the band structure calculations. Since it is common to initially discretize the problem at hand using some choice of basis, reduced Bloch mode expansion constitutes a secondary expansion using a set of Bloch eigenvectors, and hence keeps and builds on any favorable attributes a primary expansion approach might exhibit. We report phonon band structure calculations by the proposed method showing up to two orders of magnitude reduction in computational effort with negligible loss in accuracy.

1. Introduction

The study of wave propagation in phononic crystals, or periodic media in general, utilizes Bloch theory, which allows for the calculation of dispersion curves (frequency band structure) and density of states. Due to crystallographic symmetry, the Bloch wave solution needs to be applied only to a single unit cell in the reciprocal lattice space covering the first Brillouin zone (BZ) [1]. Further utilization of symmetry reduces the solution domain, even more, to the irreducible Brillouin zone (IBZ). There are several techniques for band structure calculations for phononic crystals (which are also applicable to photonic crystals).These include the planewave method [2], the transfer matrix method [3], the finite difference method [4], the finite element (FE) method [5-6], among others (see [7] for a review). Some of the methods involve expanding the periodic domain and the wave field

T.-T. Wu and C.-C. Ma (eds.), *IUTAM Symposium on Recent Advances of Acoustic Waves in Solids*, IUTAM Bookseries 26, DOI 10.1007/978-90-481-9893-1_29,

using a truncated basis. This provides a means of classification in terms of the type of basis, e.g., the planewave method involves a Fourier basis expansion and the FE method involves a real space basis expansion. The pros and cons of the various methods are discussed in depth in the literature [7].

Regardless of the type of system and type of method used for band structure calculations, the computational effort is usually high because it involves solving a complex eigenvalue problem and doing so numerous times as the value of the wave vector, k, is varied. The size of the problem, and hence the computational load, is particularly high for the following cases: (a) when the unit cell configuration requires a large number of degrees of freedom to be adequately described, (b) when the presence of defects is incorporated in the calculations, thus requiring to model large supercells, and (c) when a large number of calculations are needed such as in band structure optimization [8]. All these cases suggest that a fast technique for band structure calculation would be very beneficial.

Some techniques have been developed to expedite band structure calculations, examples include utilization of the multigrid concept [9], development of fast iterative solvers for the Bloch eigenvalue problem [5-6, 10], and extension of homogenization methods to capture dispersion [11–14]. In this paper we provide a fundamentally different approach for fast band structure calculations. We present *reduced Bloch mode expansion*, which is an expansion employing a natural basis composed of a selected reduced set of Bloch eigenfunctions[1]. This reduced basis is selected within the IBZ at high symmetry points determined by the crystal structure and group theory (and possibly at additional related points). At each of these high symmetry points, a number of Bloch eigenfunctions are selected up to the frequency range of interest for the band structure calculations. As mentioned above, it is common to initially discretize the problem at hand using some choice of basis. In this manner, reduced Bloch mode expansion constitutes a secondary expansion using a set of Bloch eigenvectors, and hence keeps and builds on any favorable attributes a primary expansion approach might exhibit. The proposed method is in line with the well known concept of modal analysis, which is widely used in various fields in the physical sciences and engineering[2].

In the next section, a description of the reduced Bloch mode expansion process and its application in a discrete setting using finite elements is given for a phononic crystal problem, followed by results and conclusions.

---

[1] The same mode selection concept, but in the context of a multiscale two-field variational method, was presented in [13, 14].

[2] The concept of modal analysis is rooted in the idea of extracting a reduced set of representative information on the dynamical nature of a complex system. This practice is believed to have originated by the Egyptians in around 4700 B.C. in their quest to find effective ways to track the flooding of the Nile and predict celestial events [15].

2. Reduced Bloch Mode Expansion

2.1 Method

The starting point for the RBME method is the discrete generalized eigenvalue problem emerging from Bloch theory applied to a standard periodic unit cell model. This yields an equation of the form

$$(\mathbf{K}(\mathbf{k}) - \omega^2\mathbf{M})\tilde{\mathbf{U}} = \mathbf{0} \tag{2.1.1}$$

where $\mathbf{M}$ and $\mathbf{K}$ are the global mass and stiffness matrices, respectively, $\tilde{\mathbf{U}}$ is the discrete Bloch vector which is periodic in the unit cell domain, $\mathbf{k}$ is the wave vector, and ω is the frequency. Equation (2.1) is then solved at a reduced set of selected wave vector points (i.e., reduced set of **k**-points), providing the eigenvectors from which a reduced Bloch modal matrix, denoted $\mathbf{\Psi}$, is formed. Several schemes are proposed for **k**-point selection, the simplest of which is the set of eigenvectors corresponding to the first few branches at the high symmetry points Γ, X, M for 2D and Γ, X, M, R for 3D, as illustrated in Fig. 2.1 for simple cubic cells (a detailed description of selection schemes is given in [16]). The matrix $\mathbf{\Psi}$ is then used to expand the eigenvectors $\tilde{\mathbf{U}}$, i.e.,

$$\tilde{\mathbf{U}}_{(n\times 1)} = \mathbf{\Psi}_{(n\times m)}\tilde{\mathbf{V}}_{(m\times 1)}, \tag{2.1.2}$$

where $\tilde{\mathbf{V}}$ is a vector of modal coordinates for the unit cell Bloch mode shapes. In (2.1.2), n and m refer to the number of rows and number of columns for the matrix equation. For the **k**-point selection schemes proposed, $m \ll n$. Substituting (2.1.2) into (2.1.2), and premultiplying by the complex transpose of $\mathbf{\Psi}$,

$$\mathbf{\Psi}^*\mathbf{K}(\mathbf{k})\mathbf{\Psi}\tilde{\mathbf{V}} - \omega^2\mathbf{\Psi}^{\mathrm{T}}\mathbf{M}\mathbf{\Psi}\tilde{\mathbf{V}} = \mathbf{0}, \tag{2.1.3}$$

yields a reduced eigenvalue problem of size $m\times m$,

$$\bar{\mathbf{K}}(\mathbf{k})\tilde{\mathbf{V}} - \omega^2\bar{\mathbf{M}}\tilde{\mathbf{V}} = \mathbf{0}, \tag{2.1.4}$$

where $\bar{\mathbf{M}}$ and $\bar{\mathbf{K}}(\mathbf{k})$ are the generalized mass and stiffness matrices. The reduced eigenvalue problem given in (2.1.4) can then be solved for the entire region of interest within the IBZ at a significantly lower cost compared to using the full model given in (2.1.1).

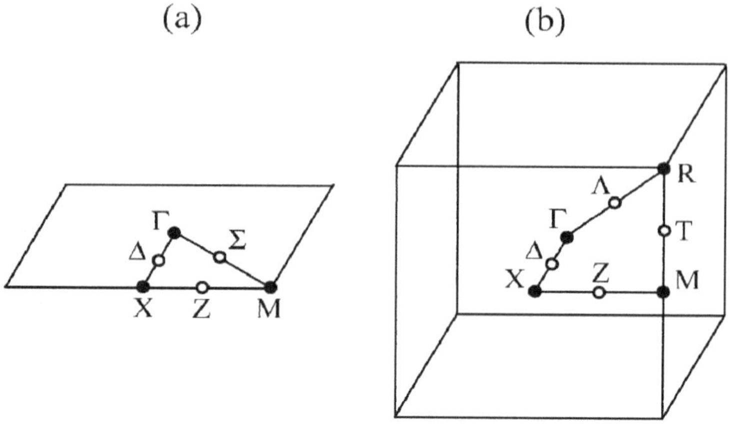

Fig 2.1 Unit cell in reciprocal lattice space with the irreducible Brillouin Zone, high symmetry **k**-points (solid circles) and intermediate **k**-points (hollow circles) shown. (a) 2D square unit cell, (b) 3D simple cubic unit cell.

2.2 Results

To demonstrate the reduced Bloch mode expansion approach, we consider a linear elastic, isotropic, continuum model of a 2D phononic crystal under plain strain conditions. As an example, a square lattice is considered with a bi-material unit cell. One material phase is chosen to be stiff and dense and the other compliant and light. In particular, a ratio of Young's moduli of E_2/E_1 = 16 and a ratio of densities of ρ_2/ρ_1 = 8 are chosen. The topology of the material phase distribution in the unit cell is shown in the inset of Fig. 2.2. The unit cell is discretized into 45 × 45 uniformly sized 4-node bilinear quadrilateral finite elements, i.e., 2025 elements. With the application of periodic boundary conditions, the number of degrees of freedom is n = 4050. Fig. 2.2 shows the calculated band structure and density of states using 2-point expansion, that is, the selection is carried out at the Γ, X, M points in **k**-space. In the calculations, eight modes were selected at each of these selection points. As such, a total of 24 eigenvectors (m = 24) were used to form the Bloch modal matrix. The results for the full model are overlaid for comparison indicating *excellent agreement*, despite a reduction of model size from 4050 to 24 degrees of freedom. For models with a larger number of degrees of freedom, and a calculation with high **k**-point sampling, two orders of magnitude or more reduction in computational cost will be achieved (as shown in Fig. 2.3).

3. Conclusions

Reduced Bloch mode expansion was presented as an approach for efficient and accurate calculation of band structures for periodic media such as phononic crystals. This modal analysis approach involves expanding the Bloch solution at all calculation **k**-points using, in its discrete form, a selected reduced set of Bloch eigenvectors to form the expansion basis. This basis is selected within the irreducible Brillouin zone at high symmetry points determined by the crystal's structure and group theory, i.e., the Γ, X, M, R points for the 3D simple cubic lattice. At each of the reciprocal lattice selection points, a number of Bloch eigenvectors are selected up to the frequency range (or dispersion branch number) of interest for the band structure calculations. Since it is common to initially discretize the periodic unit cell and solution field using some choice of basis, e.g., using finite elements, reduced Bloch mode expansion is practically a secondary expansion that keeps, and builds on, any favorable attributes a primary expansion approach might exhibit.

Results presented for 2D plain strain phononic band structure calculations are in excellent agreement with those obtained from the full model. Bloch mode shapes (not shown) and density of state predictions also agree. Two orders of magnitude in reduction of computation time has been recorded for the proposed approach, and more savings are expected for models of larger size and for calculations based on more refined sampling of the reciprocal lattice space (or IBZ circuit lines). The proposed approach is also applicable to discrete lattice dynamics calculations, and to photonic and electronic band structure calculations.

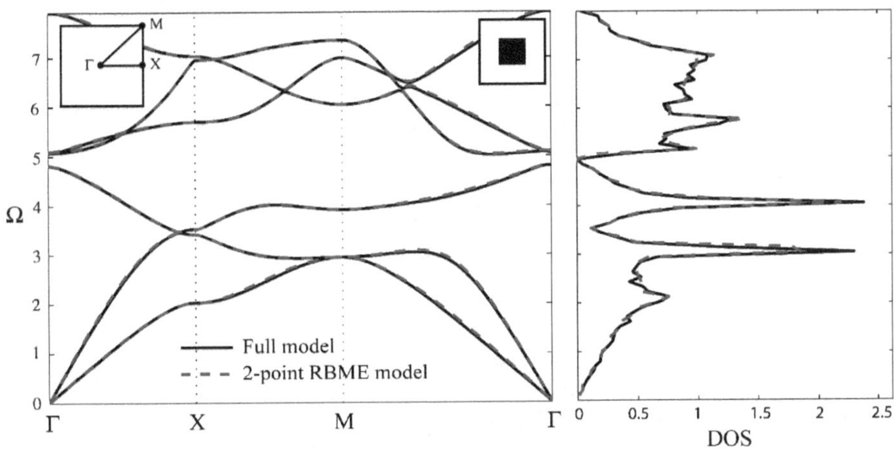

Fig 2.2 Phononic band structure and density of states (DOS) calculated using full model (matrix size: 4050×4050) and reduced Bloch mode expansion model (matrix size: 24×24). The IBZ and eigenvector selection points are shown in the left inset. The 2D unit cell is shown in the right inset; the stiff/dense material phase is in black, and the complaint/light material phase is in white. The finite element method was used for the primary expansion.

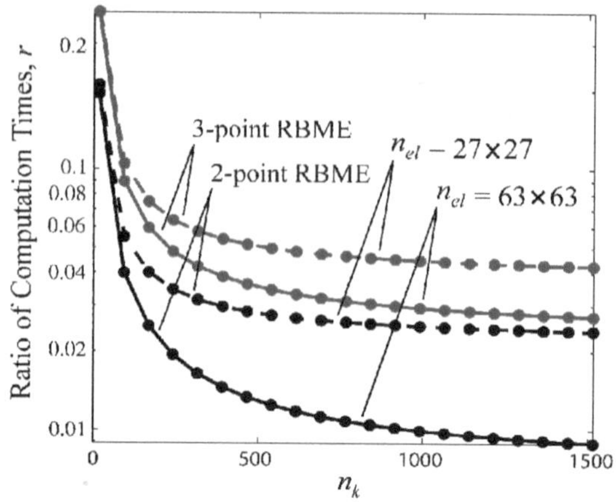

Fig 2.3 Computational efficiency: ratio of reduced Bloch mode expansion model to full model calculation times, r, versus number of sampled k-points along the border of the IBZ, n_k (for two 2D finite element meshes).

References

[1] Brillouin, L. *Wave Propagation in Periodic Structures*. Dover, New York (1953).

[2] Ho, K.M. et al.: Existence of a photonic gap in periodic dielectric structures. *Phys. Rev. Lett.* **65** 3152-3155 (1990).

[3] Pendry, J.B., MacKinnon, A.: Calculation of photon dispersion-relations. *Phys. Rev. Lett.* **69** 2772-2775 (1992).

[4] Yang, H.Y.D.: Finite difference analysis of 2-D photonic crystals. *IEEE T. Microw. Theory* **44** 2688-2695 (1996).

[5] Dobson, D.C.: An efficient method for band structure calculations in 2D photonic crystals. *J. Comput. Phys.* **149** 363-76 (1999).

[6] Axmann, W., Kuchment, P.: An efficient finite element method for computing spectra of photonic and acoustic band-gap materials - I. Scalar case. *J. Comput. Phys.* **150** 468-481 (1999).

[7] Busch, K. et al.: Periodic nanostructures for photonics. *Phys Reports* **444** 101-202 (2007).

[8] Burger, M. et al.: Inverse problem techniques for the design of photonic crystals. *IEICE T. Electron.* **E87C** 258-265 (2004).

[9] Chern, R.L., Chang, C.C. and Hwang, R.R.: Large full band gaps for photonic crystals in two dimensions computed by an inverse method with multigrid acceleration. *Phys. Rev. E* **68** 026704 (2003).

[10] Johnson, S.G. and Joannopoulos, J.D.: Photonic crystals: Putting a new twist on light. *Opt. Express* **8** 173 (2001).

[11] McDevitt, T.W., Hulbert, G.M. and Kikuchi, N.: An assumed strain method for the dispersive global-local modeling of periodic structures. *Comput. Method. Appl. M.* **190** 6425 (2001).

[12] Nagai, G., Fish, J. and Watanabe, K.: Stabilized nonlocal model for dispersive wave propagation in heterogeneous media. *Comput. Mech.* **33** 144 (2004).

[13] Hussein, M.I. *Dynamics of Banded Materials and Structures: Analysis, Design and Computation in Multiple Scales*. Ph.D. Thesis, University of Michigan–Ann Arbor, USA (2004).

[14] Hussein, M.I., Hulbert, G.M.: Mode-enriched dispersion models of periodic materials within a multiscale mixed finite element framework. *Finite Elem. Anal. Des.* **42** 602-612 (2006).

[15] Døssing, O.: IMAC-XIII keynote address: Going beyond modal analysis, or IMAC in a new key. Modal Analysis. *Int. J. Anal. Exp. Modal Analysis* **10** 69 (1995).

[16] Hussein, M.I., Reduced Bloch mode expansion for periodic media band structure calculations. arXiv:0807.2612v4 (2008).

Band Gap in Phononic Crystal Thin Plate with/without Mirror Plane

Zhilin Hou[1] and Badreddine M. Assouar[2]

[1]Department of Physics, South China University of Technology, Wushan, Guangzhou, 510640, China

[2]Institut Jean Lamour, CNRS - Nancy University, BP: 239, Bd des Aiguillettes, 54506 Vnadoeuvre les Nancy

Badreddine.Assouar@lpmi.uhp-nancy.fr

Abstract. The mechanism of opening a band gap in the free phononic crystal (PC) thin plate with or without a mirror plane is investigated. It is found that, in a PC plate with a mirror plane, the permitted modes can be separated into symmetric and antisymmetric modes, and the band gap in such a system can be opened by the interaction between the modes of the same kind and/or the breaking of the degeneracy of the mode at the edge of the Brillouin zone. However, for a PC plate without a mirror plane, mode separation can no longer be performed, and interaction can occur between any two permitted modes. As a result, a new kind of band gap can be opened.

1. Introduction

In this study, we report on the opening of band gap of 1D PC plate with and without mirror plane. In PC plate, because of the finite size in the thickness direction, elastic waves in a PC plate can be scattered not only by periodically arranged scatterers but also by the surface of the plate, which makes the mechanism for opening the Acoustic band gaps (ABGs) in a PC plate more complicated than that in a bulk-wave PC. As a consequence, the width of ABGs is sensitive both on the periodic structure and on the thickness of the plate [1–3]. The relationship between the ABGs and the PC plate's thickness and structure has been investigated by many researchers, and has shown that ABGs usually appear in a relatively thin PC plate with a large elastic contrast between the different components [3,4]. However, the understanding of how the structure changes the width of the ABGs in a PC plate is

T.-T. Wu and C.-C. Ma (eds.), *IUTAM Symposium on Recent Advances of Acoustic Waves in Solids*, IUTAM Bookseries 26, DOI 10.1007/978-90-481-9893-1_30,

still incomplete. We find that, as far as the symmetry is concerned, only two kinds of PC plates were studied in previous works. One of them is uniform along the thickness direction, which means the system is mirror symmetric around its middle plane. The other kind of system, which has been studied recently is the PC plate covered with a uniform substrate layer (PCS) [4–6]. This system is asymmetric in the thickness direction. It was pointed out that the width of the ABGs in such a system can be adjusted by the thickness and by the elastic parameters of the material of the substrate [6,7]. We have pointed out in [4,5] that the band structures of the symmetric PC and PCS plate are quite different. Some phenomena, such as the energy conversion between the symmetric and antisymmetric Lamb modes in the transmission spectrum, can only take place in the PCS system. We concluded in [5] that the appearance of the ABGs in a PCS system can be seen as a result of the strong interaction between the different modes.

2. Numerical Study

We know that one kind of technique for opening a new band gap in bulk-wave PC is to change the configuration of the periodic unit cell. It is shown in [8,9] that ABGs can be adjusted by adding extra scatters. The main idea of this technique is to alter the symmetry of the PC structure. Expanding on this idea, we will give an understanding of the ABGs in a PC plate from a symmetric point of view in this paper. We will show that the difference in the band structure between the symmetric PC and the PCS plates mentioned above results from the break in the mirror symmetry of the plates around their middle planes. The special features of the band structures in a PCS system can also appear in other asymmetric PC plates.

The numerical calculation is performed by the Eigen-mode-matching-theory (EMMT) presented in [5], by which the band structure of the infinite plate, and the corresponding transmission coefficient of the system with finite periods can also be obtained at the same time. To use this method, the system is first cut into layers, waves in each layer are expressed as a superposition of plane wave function set, and then the boundary condition is used to connect the nearest layers.

The schematic of studied PC plate is given in the figure 1, in which two different materials shown by dark grey (labelled A) and light grey (labelled B) are, respectively, Pb and epoxy in our calculation. We consider a symmetric structure as shown schematically in figure 1(a).

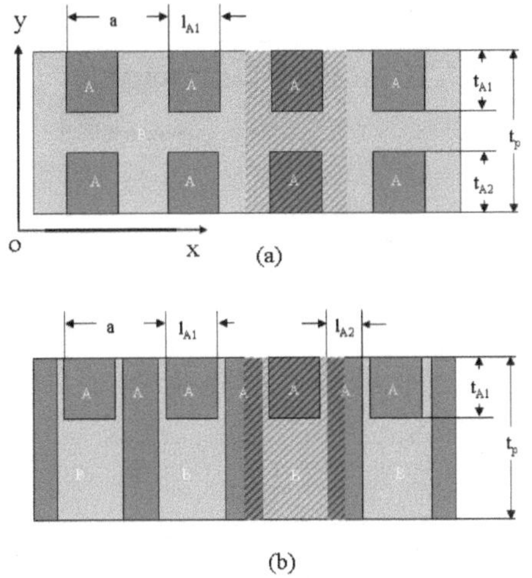

Fig. 1 One-dimensional PC plate constructed by elastic material A and B.

The results for the system with $t_p = 0.5a$ and $l_{A1} = 0.4a$, where a is the lattice constant, are shown in figure 2, in which figures 2(a) and (b) are for the systems with $t_{A1} = t_{A2} = 0.2a$ and $t_{A1} = 0.1a$, $t_{A2} = 0.2a$, respectively, which are the symmetric and asymmetric plates. The three panels in figures 2(a) or (b) show the band structure (middle panel) and the transmission coefficient (energy flux) stimulated by the symmetric (left panel) and antisymmetric (right panel) 0th Lamb modes, respectively. For the transmission coefficient, a finite system with 16 periods in the x direction connected by two uniform half-infinite plate (material B) is considered. Similarly to the results given in [7], from figure 2, we can find two intrinsic differences. Firstly, the crossover points between the different bands in the middle panel of figure 2(a) can no longer be found in figure 2(b). As a result, new band gaps, especially the lowest gap between the second and the third bands, appear. Secondly, the transmission coefficient shows that the bands in figure 2(a) can be separated into two independent parts, which can only be excited by the symmetric and antisymmetric 0th Lamb waves separately. But, in figure 2(b), such a separation can no longer be performed.

It is known that the waves propagating in the freestanding uniform plate with a lossless boundary must satisfy the *transverse resonance principle* [10], which means the permitted modes are travelling waves along the x direction and resonant standing waves in the thickness direction. This suggests that the transverse component of the modes must experience a phase shift of some integral multiple of 2π during each round trip in the thickness direction. So, if the plate is symmetric around its middle plane, the reflection of the modes from each surface should be the same. This means that the permitted modes can be separated into two independent sets: the symmetric and the antisymmetric modes for a mixed mode or odd and even modes for the SH mode.

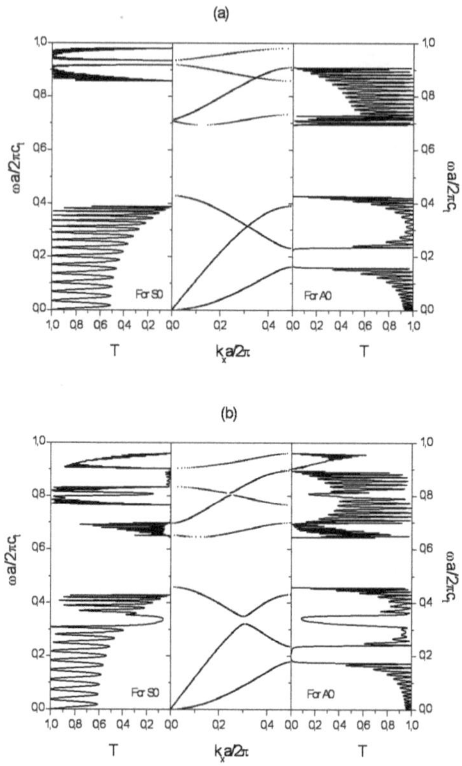

Fig. 2 Band structure and the transmission coefficient of the (a) symmetric and (b) asymmetric PC plate.

For the periodic composite structure of the PC, the statement above is still true if the PC plate is mirror symmetric around its middle plane, which means that the permitted modes can also be separated into the symmetric/antisymmetric modes for a mixed mode (or odd/even modes for SH mode). In this kind of system, the effect of the Bragg scattering by the periodic structure can lead to two changes in the band structure. The first one is the opening of the folding points of the modes at the edge of the Brillouin zone (BZ) and the second is the coupling effect between the modes that are of the same kind (symmetric or antisymmetric mode). To show this, we will still consider a symmetric structure as shown schematically on figure 1(*a*), but with a very small filling fraction so that the band structure is only slightly different from that of the uniform plate. The parameters of the system are chosen to be $t_p = 1a$, $t_{Al} = t_{A2} = 0.05a$, and $l_{Al} = 0.1a$, respectively. The band structure of such a PC system is shown in figure 3(*b*), and as a comparison, a band structure of a uniform plate is also shown in figure 3(*a*). From the figure, we can clearly see that, because of the opening of the folding points of the mode at the edge of BZ (labelled B) and the coupling effect between the A0 and A1 modes (labelled C), a band gap appears. But the crossover point between A0 and S0 (shown as D1) and the point between S0 and A1 (shown as D2) still exists in figure 3(*b*). This means the no interaction between the symmetric and antisymmetric modes can be induced by Bragg scattering. However, the situation is quite different for an asymmetric plate. In this kind of plate, the separation of the modes can no longer be performed. As a consequence, a coupling effect can occur between any two modes, so that Bragg scattering can not only open the folding point of the modes at the edge of BZ but also cause a coupling effect between different modes. The band structure of an asymmetric plate with a small filling fraction is shown in figure 3(*c*). From figure 3(*c*) we can observe that, besides the points B and C, the crossover points D1 and D2 in figure 3(*b*) have also opened. In fact, no crossover point can be found in figure 3(*c*), which means that the coupling between symmetric and antisymmetric modes can be caused by breaking the mirror symmetry of the plate. But here, no new band gap is opened by this effect.

From the understanding mentioned above, we can conclude the following about the mechanism of opening the ABGs in a PC plate. For a symmetric PC plate, if the thickness of the plate is relatively large ($t_p = 1a$, for example), as shown in figure 3(*b*), the lowest band gap can be obtained by opening the degeneracy of points B2 and C (which are the folding points of S0 at the edge of BZ and the crossover point of mode A1 and A0, respectively) at the same time. When the thickness of the plate is thin enough, the lowest band gap can also be opened by just opening the folding points of the modes at the edge of BZ. An example of such an ABG is shown in figure 4(*b*). The structure of the system is the same as the one shown in figure 1(*a*) with parameters $t_p = 0.5a$, $t_{A1} = t_{A2} = 0.1a$, and $l_{Al} = 0.1a$, respectively. To get the lowest ABG of this kind, the plate must be thin enough so that the beginning frequency of the mode A1 is higher than the folding points of A0 and S0 (B2 and B3 in figure 4(*a*)). Note that the small width of the band gap in figures 3(*b*) and 4(*b*) results from the small filling fraction of the PC plate. It can be enlarged by increasing the size of the scatterers.

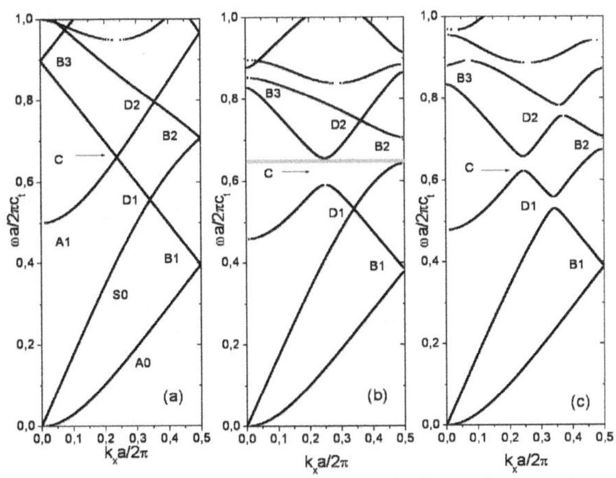

Fig. 3 Band structure of the system shown schematically in figure 1(*a*) with $tp = 1.0a$.

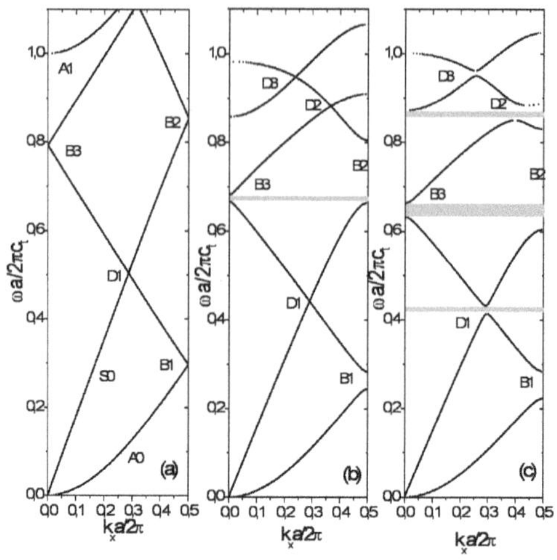

Fig. 4 Band structure of the system shown schematically in figure 1(*a*) with *t*p = 0.5*a*.

For an asymmetric plate, because the interaction can occur between any two modes, a new kind of band gap can be opened. In figure 4(*c*), the lowest ABG between the second and third bands is an example of this kind of ABG. To obtain this gap, the mirror symmetry of the PC plate for figure 4(*b*) is broken by increasing t_{A1} from 0.1*a* to 0.2*a*. By comparing figure 4(*c*) with figure 4(*b*), we can see that, as in figure 3(*c*), all of the crossover points marked by D1, D2 and D3 can no longer be found. This also shows that the coupling effect between modes is indeed a common feature in the asymmetric PC plate. We have to note that, to get the lowest ABG in such an asymmetric system, the PC plate should also be thin enough, so that the beginning frequency of mode A1 is higher than the crossover point D1.

We have also studied a 2D asymmetric PC plate. Figure 5 shows a band structure for 2D honeycomb PC based on empty holes in silicon substrate (tp=0.5*a*, t_{A1}=0.1*a* and *r*=0.48*a*, where *a* and *r* are the distance between the nearest holes and the radii of the holes). From which we can see that, because of the introduction of the structural asymmetry, a lowest band gap is opened in such high filled honeycomb structure.

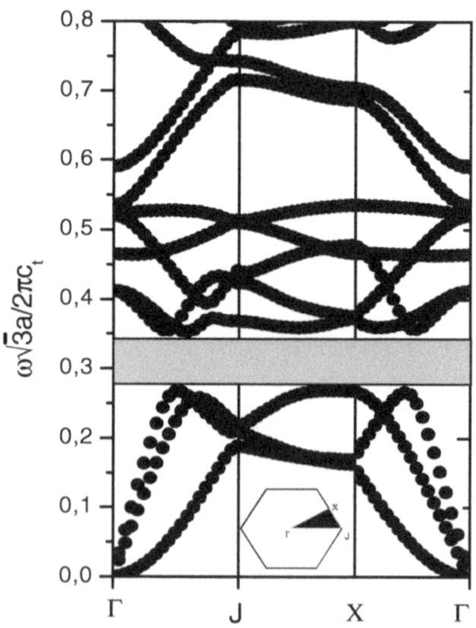

Fig. 5 Band structure of a 2D honeycomb PC plate with a thin layer uniform substrate.

3. Conclusion

In conclusion, the mechanism for opening a band gap in a 1D PC plate is investigated. It is shown that, for a PC plate with a mirror plane, the permitted modes can be separated into two independent parts, which are symmetric and antisymmetric modes. In such a system, the ABGs can be opened by both the interaction between modes of the same kind (symmetric or antisymmetric) and the opening of the folding points of the modes at the edge of BZ, or just by the latter mechanism if the plate is thin enough so that the interval between the two nearest isolated modes is large enough. However, for a PC plate without a mirror plane, the permitted modes can no longer be separated into two independent parts, and the interaction can occur between any two modes. As a result, a new band gap that is quite different from the one in the symmetric plate can be opened. The width of this kind of ABG is influenced by the degree that the system is away from the mirror symmetry. We have also presented the first result concerning the opening of lowest band gap in 2D PC asymmetric plate.

Acknowledgments: We gratefully acknowledge support in this work from the Agence Nationale de la Recherche ANR 2005 (projet No. ANR-05-JCJC-0075-01) and « La Région Lorraine ».

References

[1] Hsu, J-C., Wu, T T.: Efficient formulation for band-structure calculations of two-dimensional phononic-crystal plates. *Phys. Rev. B.* **74** 144303 (2006).
[2] Chen, J-J., Zhang, K-W., Gao, J., Cheng, J-C.: Stopbands for lower-order Lamb waves in one-dimensional composite thin plates. *Phys. Rev. B.* **73** 094307 (2006).
[3] Khelif, A., Aoubiza, B., Mohammadi, S., Adibi, A., Laude, V.: Complete band gaps in two-dimensional phononic crystal slabs. *Phys. Rev. E.* **74** 046610 (2006).
[4] Hou, Z., Assouar, M. B.: Modeling of Lamb wave propagation in plate with two-dimensional phononic crystal layer coated on uniform substrate using Plane-wave-expansion method. *Phys. Lett. A.* **372** 2091 (2008).
[5] Hou, Z., Assouar, M. B.: Transmission property of the one-dimensional phononic crystal thin plate. *J. Phys. D Appl. Phys.* **41** 095103 (2008).
[6] Gao, J., Zou, X –Y., Cheng, J-C., Li, B.: Band gaps of lower-order Lamb wave in thin plate with one-dimensional phononic crystal layer: Effect of substrate. *Appl. Phys. Lett.* **92** 023510 (2008).
[7] Vasseur, J O., Hladky-Hennion. A-C., Djafari-Rouhani, B., Duval, F., Dubus. B., Pennec, Y., Deymier, P A.: Waveguiding in two-dimensional piezoelectric phononic crystal plates. *J. Appl. Phys.* **101** 114904 (2007).
[8] Zhang, X., Liu, Z., Liu, Y., Wu, F.: Elastic wave band gaps for three-dimensional phononic crystals with two structural units. *Phys. Lett. A.* **313** 455 (2003).
[9] Lai, Y., Zhang, Z-Q.: Large band gaps in elastic phononic crystals with air inclusions. *Appl. Phys. Lett.* **83** 3900 (2003).
[10] Auld, B A.: *Acoustic Fields and Waves in Solids vol II,* 2nd edn, Malabar, FL: Krieger, (1990).

Radial Sonic Crystals

Daniel Torrent and José Sánchez-Dehesa

Wave Phenomena Group, Department of Electronics Engineering, Polytechnic University of Valencia, Valencia, Spain

jsdehesa@upvnet.upv.es

Abstract. This work introduces the so named radial sonic crystals, which are a new type of acoustic materials characterized by the fact that their acoustical properties are periodic in the radial direction, either in two dimensions or in three dimensions. These periodic conditions allow the application of Bloch theorem to the corresponding wave equation for sound propagation in spherical (3D) and cylindrical (2D) coordinates, respectively. The mentioned conditions leave the radial wave equation invariant under translation along r and can be achieved by using anisotropic fluid-like materials that can be engineered by using structures already proposed. We will report the rich variety of properties and potential applications of radial sonic crystals.

1. Introduction

In a general eigenvalue problem of the form $H\phi=\lambda\,\phi$ Bloch's theorem applies when the operator $H=H(r)$ is a differential operator invariant under certain set of translations, that is

$$H(\mathbf{r}+\mathbf{R})=H(\mathbf{r})$$

for a given set of vectors **R**, called lattice vectors.

In such case it can be shown that the eigenvectors ψ can be expressed as

$$\psi=exp(\boldsymbol{k}\,\boldsymbol{r})u$$

where u being a periodic function with the periodicity of the lattice **R**. The vector **k** is the so called Bloch's vector, and the eigenvalues are functions of this vector, that is, $\lambda=\lambda(\mathbf{k})$.

T.-T. Wu and C.-C. Ma (eds.), *IUTAM Symposium on Recent Advances of Acoustic Waves in Solids*, IUTAM Bookseries 26, DOI 10.1007/978-90-481-9893-1_31,

When the operator H corresponds to a differential wave equation (like the Schrödinger or Maxwell's equations) the eigenvalue λ is a function of the wave frequency ω, and then the solutions of the eigenvalue problem are of the form $\omega = \omega(\mathbf{k})$, which constitutes the dispersion relation of the medium also called the band structure. This band structure gives information about how the propagation of waves inside the crystal is.

Band structures have been studied for a wide variety of physical waves, like matter waves in semiconductors, electromagnetic waves in periodic dielectric or magnetic media (photonic crystals) and acoustic waves in periodic distributions of sound scatterers (sonic crystals). In all these systems the periodicity of the material was defined by the periodic spatial distribution of some material property, but always with a lattice vector R defined in rectangular coordinates.

Here a new type of periodic medium will be defined: A radially periodic medium. Though radially periodic systems have been widely studied in electromagnetic systems, it has not been done in the framework of Bloch's theorem, and then nothing similar to the band structure has been obtained for these systems. This is due to the fact that, in general, wave equations in polar or spherical coordinates are not invariant under translations. But when the medium is anisotropic this invariance can be obtained. The following discussion will be focused in two-dimensional acoustic waves, and other fields and dimensions will be considered at the end of the paper.

2. Invariance of the Wave Equation

The inhomogeneous acoustic wave equation in planar polar coordinates is

$$\frac{B}{r}\frac{\partial}{\partial r}\frac{r}{\rho}\frac{\partial P_q}{\partial r}+\left(\omega^2-q^2\frac{B}{r^2\rho}\right)P_q=0$$

where an angular dependence of the form $exp(iq\theta)$ is assumed. In order to make the differential operator invariant under translations under the form $r+nd$, the coefficients of the operator should be periodic with periodicity d. It is clear that it cannot be done with any choice of the acoustic parameters B and ρ, because the terms r/ρ and $r\rho$ cannot be periodic at the same time.

However a more general acoustic wave equation has to be considered when studying acoustic cloaking devices [1]. These materials are fluid like materials with anisotropic mass density, and the equation describing waves in such medium is, in polar coordinates

$$\frac{B}{r}\frac{\partial}{\partial r}\frac{r}{\rho}\frac{\partial P_q}{\partial r}+\left(\omega^2-q^2\frac{B}{r^2\rho}\right)P_q=0$$

note that now the density ρ is a tensor with a radial and an angular component, and that the differential operator can be made invariant under translations. This invariance can be achieved because the quantities

$$\frac{B}{r}\frac{\partial}{\partial r}\frac{r}{\rho}\frac{\partial P_q}{\partial r}+\left(\omega^2-q^2\frac{B}{r^2\rho}\right)P_q=0$$

can be periodic at the same time, thus Bloch's theorem can be applied to obtain some kind of radial band structure. A medium satisfying these periodic conditions is called a "Radial Sonic Crystal" (RSC).

In one dimensional periodic media the simpler case to analyze is the binary periodic system. Such a medium is made of two alternate materials, say type A and B, with homogeneous physical parameters. An example of the equivalent medium in RSC is shown in Fig. 1, where it can be seen that here the acoustic parameters increase with the radial coordinate.

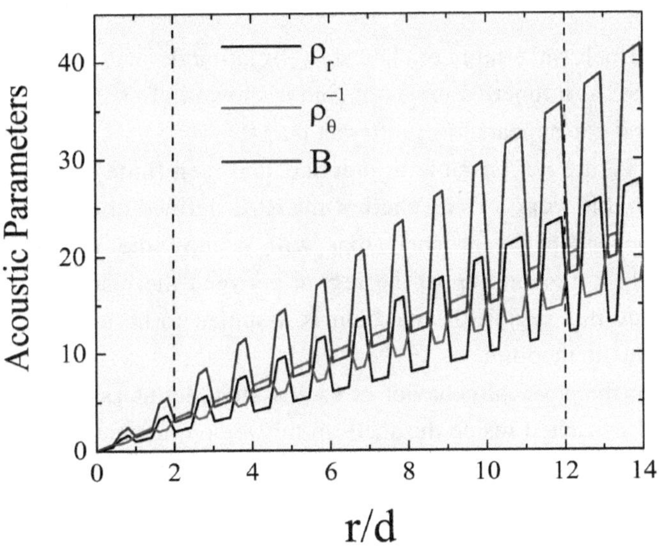

Fig. 1 Radial dependence of the acoustic parameters in a Radial Sonic Crystal

3. Band Structure

The anisotropic acoustic wave equation with the material parameters defined in Fig. 1 becomes a homogeneous wave equation with constant coefficients

$$\frac{\partial^2 P_q}{\partial r^2} + \left(\omega^2 \alpha - q^2 \beta\right) P_q = 0$$

This wave equation has plane-wave solutions with a dispersion relation given by $k^2 = \omega^2 \alpha - q^2 \beta$. This dispersion relation is similar to that of a waveguide with a cutoff frequency given by $\omega = q\sqrt{\beta / \alpha}$. Then, it is expected that the periodic medium of alternating layers of type A and B presents some kind of low frequency band gaps. Note also that for high frequency the dispersion relation becomes linear and independent of q and the system behaves like the one-dimensional periodic multilayer

The band structure is obtained in the same way as that of the one dimensional periodic medium, arriving to a dispersion relation given by

$$\cos Kd = \cos k_1 d_1 \cos k_2 d_2 + \gamma \sin k_1 d_1 \sin k_2 d_2$$

K being the Bloch wave number. The above relation defines the dispersion relation of the RSC as a function $\omega=\omega(K)$ and is shown in Fig. 2. Note how the low frequency band gap appears for q different than 0.

Infinite RSC are not possible in practice, and then finite slabs of them have to be used. In this work a finite shell of the RSC defined in Fig. 1 will be used. This RSC is a ten layers shell that will occupy the region defined by $2d<r<<12d$, corresponding to the region between the black vertical lines in Fig 1. Outside this region the medium is assumed to be a homogeneous and isotropic acoustic medium.

To analyze the physical behavior of such system, let us assume that a general acoustic field is excited inside the shell. In this region, the medium is homogeneous and isotropic, and then the response to the field will be

$$P(r, \vartheta) = \sum_q A_q^0 H_q(kr) e^{iq\vartheta}$$

where H_q are the Hankel functions, $k=\omega/c$ and c is the speed of sound in the homogeneous medium. As a response to this field, inside the cavity there will be a "reflected" field

$$P(r,\vartheta)=\sum_q R_q A_q^0 J_q(kr)e^{iq\vartheta},$$

where J_q are the Bessel functions. Outside the RSC there will be a "transmitted" field

$$P(r,\vartheta)=\sum_q T_q A_q^0 H_q(kr)e^{iq\vartheta}$$

The magnitude of the coefficients **T** will be indicative of the nature of the system. For the mentioned slab these coefficients have been plotted in the right panel of Fig. 2 in logarithmic scale together with the band structure. Note that these coefficients, for each q, are very small in the regions of frequency where there is a band gap. Also, there is some kind of Fabry-Perot like oscillations in the regions where the band structure is almost linear. Note also that there are some peaks in the transmission coefficient inside the band gaps. These peaks can be considered as localized states due to the presence of defects, where the defect here is the homogeneous cavity inside the RSC.

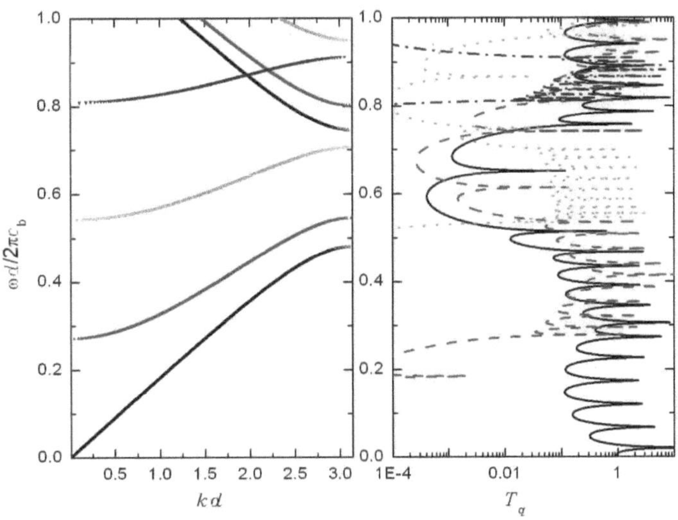

Fig. 2 Left: Band structure of the RSC considered in the text. Right: Transmission coefficient of a finite slab made of RSC.

4. Resonances

In the right panel of Fig. 2 can be seen some resonance peaks due to the cavity formed by the RSC. Note that these resonances are different for different q, what makes these systems useful for building some kind of artificial q- poles. To see this effect, in Fig. 3 a point source oscillating at a frequency $\omega d / 2\pi c = 0.408$ has been placed outside two identical RSC. The frequency corresponds to a resonance of the q=1 mode and it is seen how this mode is excited from outside the cavity. Note that the pole is oriented in the direction of the point source, which makes these systems potentially useful as antennas for sound source detectors.

RSC are also possible in three dimensions, where the acoustic wave equation for radially symmetric problems is in this case.

$$\frac{B}{r^2}\frac{\partial}{\partial r}\frac{r^2}{\rho_r}\frac{\partial P_q}{\partial r}+\left(\omega^2-l(l+1)\frac{B}{r^2\rho_\vartheta}\right)P_q=0$$

Therefore the periodicity in the coefficients implies that the quantities

$$\frac{B}{r^2},\frac{r^2}{\rho_r},\rho_\vartheta,$$

have to be periodic. The same procedure to obtain the band structure for two dimensional RSC applies here.

In principle the material properties of the RSC looks like impossible to build. However, recently [3, 4, 5] it has been shown that anisotropic and inhomogeneous fluid-like materials are possible to build in a wide range of frequencies.

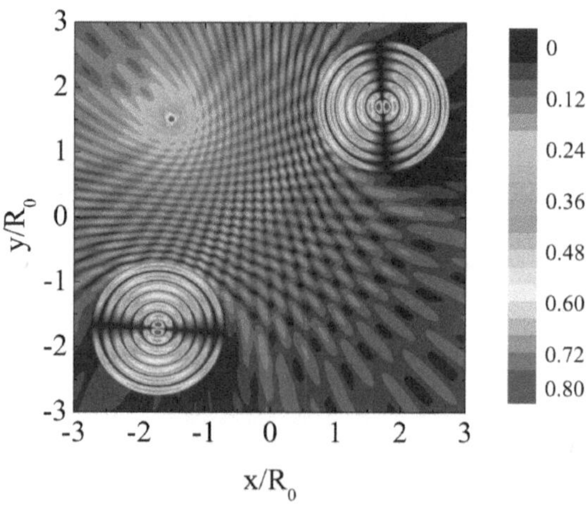

Fig. 3 Interaction between a punctual sound source and two RSCs at the resonance frequency 0.408.

5. Conclusions

This work has introduced a new type of crystals called Radial Sonic Crystal, which are radially periodic systems in which Bloch theorem can be applied thanks to the used of acoustic metamaterials with mass anisotropy. These RSC are possible to build by using sonic crystals based on non-isotropic lattices. These new type of acoustic materials have a potential use as waveguides, sound isolators, q-pole generators and dynamically orientated acoustic antennas, between others. Electromagnetic systems radially periodic are also possible by using EM metamaterials with anisotropic parameters. They can be designed by using the same procedure employed here and can be defined in 2D as well in 3D.

Acknowledgements: This work has been partially supported by U.S. Office of Naval Research under Grant No. N000140910554 and by the Spanish Ministry of Science and Innovation under Grants No. TEC2007-67239 and No. CSD2008-00066 (CONSOLIDER Program).

References

[1] Cummer S A and Schurig D.: One path to acoustic cloaking. *New J. Phys.* **9** 45 (2007).
[2] Tretyakov S.: *Analytical Modeling in Applied Electromagnetism*, Norwood, MA, Artech House, (2000).
[3] Torrent D and Sánchez-Dehesa J.: Anisotropic mass density by two dimensional acoustic metamaterials, *New J. Phys.* **10** 023004 (2008).
[4] Torrent D and Sánchez-Dehesa J.: Acoustic metamaterials for new two dimensional sonic devices, *New J. Phys.* **9** 323 (2007).
[5] Torrent D and Sánchez-Dehesa J.: Acoustic cloaking in two dimensions: A feasible approach. *New J. Phys.* **10** 063015 (2008).

Polarization States in 2D Phononic Crystals and Phononic Crystal Waveguides

Younes Achaoui[1], Abdelkrim Khelif[2], Sarah Benchabane[1] and Vincent Laude[1]

[1]Institut FEMTO-ST, Université de Franche-Comté, CNRS, ENSMM, UTBM
32 avenue de l'observatoire F-25044 Besançon Cedex, France

[2]GeorgiaTech-CNRS, Atlanta, GA, USA

younes.achaoui@femto-st.fr

Abstract. We investigate, using the Plane Wave Expansion method and an energy balance criterion, the polarization states of elastic waves in a two dimensional phononic crystal made of vacuum holes in a silicon host matrix. Pure modes can be identified for the out of plane polarization while a continuous variation of the in-plane polarization with the Bloch wave vector can be observed. The study is then extended to two-dimensional phononic crystal waveguides by using the super-cell technique. The dependence of the elastic wave polarization state on the wave guide width is investigated and shows as well a continuous variation with the Bloch wave vector. Moreover, the variation of the width of the waveguide can be used to tune both the dispersion and the polarization of guided waves.

1. Introduction

Phononic crystals are inhomogeneous two or three dimensional periodic structures made of at least two materials exhibiting different elastic properties [1,2]. The propagation of elastic waves inside such a medium can give rise to directional or complete band gaps providing that the contrast between the mass density and elastic constants of the two materials is large, that the filling fraction is high enough, and that the host material is arranged following an adequate lattice topology and specific geometrical shapes. Within such a frequency band gap, a phononic crystal can be seen as a mirror for incident waves, because of destructive interferences between waves scattered on the periodic inclusions. Phononic band gap effects have been reported for bulk [2, 3, 4] and surface [5, 6, 7] acoustic waves as well as

T.-T. Wu and C.-C. Ma (eds.), *IUTAM Symposium on Recent Advances of Acoustic Waves in Solids*, IUTAM Bookseries 26, DOI 10.1007/978-90-481-9893-1_32,

for slabs [8, 9]. Phononic crystals also allow for elastic wave confinement or guiding through the introduction of point or linear defects. Phononic waveguides, resonators and stubs have been proposed as possible ways to create complex devices based on coupling, filtering and wave guiding phenomena [10–13].

For acoustic waves propagating in a fluid, where no transverse polarisation is possible, the calculation of the wave dispersion (and of the transmission coefficient in the case of a waveguide) is usually sufficient to characterize phononic structures. But in the case of waves propagating in a solid, both transverse and longitudinal polarizations can exist and are usually coupled. Taking the polarization states into account in the band diagram is hence a very relevant way to complete the available information on elastic wave propagation. This has for example been shown by a previous study dedicated to the polarization effects in a perfect 2D phononic crystal made of air inclusions in an epoxy matrix [14]. This work highlighted the influence of the filling fraction on the coupling between transverse and longitudinal polarizations in the sagittal plane. The continuity of elastic displacement fields when the wave vector sweeps the Brillouin zone has also been investigated in order to study the repulsion level between different branches in the band diagram in a 2D PMMA/Ni phononic crystal [15]. No equivalent works have however been reported for phononic waveguides.

In this paper, we report on the evolution of the polarization states of a two-dimensional phononic crystal made of a square lattice of void inclusions in a silicon matrix. The Plane Wave Expansion method along with the energy balance criterion are used to compute band structures displaying additional information related to the weighting of each polarization. The study is then extended to a phononic waveguide obtained by inserting a line defect in the initial phononic crystal thanks to the super-cell technique. The influence of the wave guide width on the guided modes dispersion relations as well as on their polarization properties is also investigated.

2. Polarization States inside a Two Dimensional Phononic Crystal

We consider a two dimensional phononic crystal made of a square lattice of cylindrical holes in (010) silicon with a 58% filling fraction. Band structures as well as modal distributions are computed using the Plane Wave Expansion method. The PWE method is based on the decomposition of the strain and displacement fields using the Bloch-Floquet theorem.

The elastic waves are then represented as the product of a periodic function given by a discrete sum over Fourier harmonics in the reciprocal-lattice space, with a time-harmonic exponential function with frequency and Bloch wave vector. The displacement field in the x direction then reads

$$u_x = \sum_n U_{nx} \exp(-i\mathbf{G}_\mathrm{n}.\mathbf{r})\exp(-i\mathbf{k}\cdot\mathbf{r}) \tag{1}$$

where $\mathbf{G}_\mathrm{n}$ are the reciprocal lattices vectors, and $\mathbf{k}$ is the wave vector.

The two dimensional Fourier series are here truncated to the 36 first harmonics to allow for a reasonable compromise between computation time and convergence. In order to avoid possible confusions, we will use throughout this paper the notations u_x, u_y and u_z the displacement fields instead of referring to transverse horizontal and shear and longitudinal polarizations that are dependent on the direction of propagation. The weighting of the polarization along the x direction is given by:

$$p_x^2 = \frac{\int |u_x|^2\, dx}{\int \left(|u_x|^2 + |u_y|^2 + |u_z|^2\right) dr} \tag{2}$$

with the integral taken over the unit-cell. Similar expressions hold for the amounts of polarization along the y and the z axes.

Figures 1-(b), (c) and (d) display the dispersion curves of elastic waves that propagates inside the phononic crystal schematized in Figure 1-(a). The crystal shows a complete band gap with a 30% fractional bandwidth. The three band structures actually display the same dispersion relations but with additional information on the contribution of the three polarizations.

It can be inferred from Figure 1 that the u_z component is completely decoupled from the two in-plane components u_x and u_y. This is expected in a cubic material as silicon, where in-plane and out-of-plane components are decoupled. The in-plane components u_x and u_y are conversely strongly coupled, due to the scattering introduced by the two-dimensional periodic structuration of the propagation medium. For the same reason, the polarization varies as the Bloch wave vector sweeps the Brillouin zone for the in-plane components of the field only, while the polarization of u_z remains constant for a particular branch over the whole Brillouin zone.

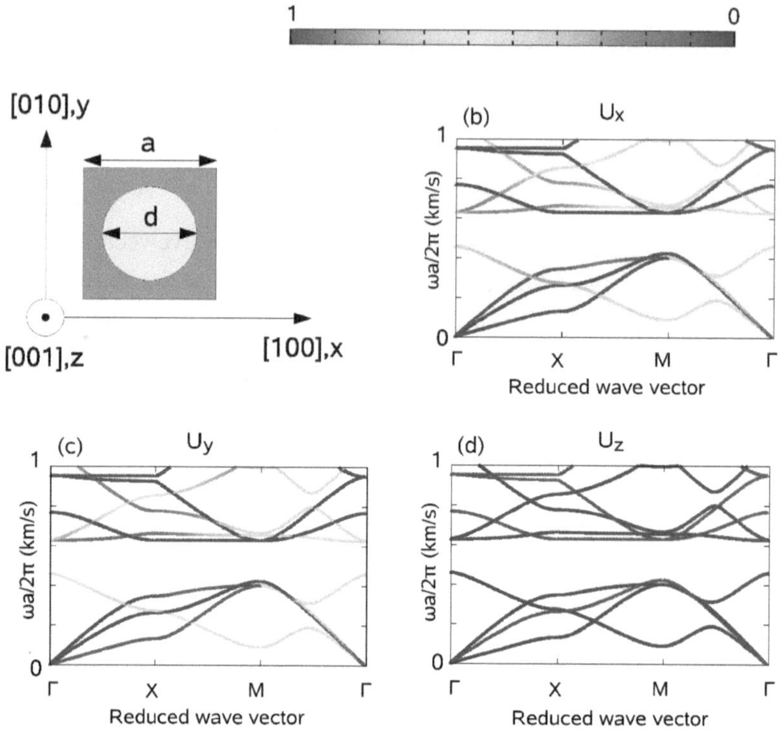

Fig. 1 (a) 2D phononic crystal consisting of a square lattice array of cylindrical holes in silicon. The three band diagrams depict the same dispersion relations, but the colouring of the bands shows the amount of polarization along the (b) x axis, (c) the y axis and (d) the z axis.

3. Polarization States inside a Two Dimensional Phononic Crystal Waveguide

We now consider a phononic crystal waveguides managed by removing a row of holes along the x direction from the initially perfect periodic structure presented in the previous section. Here again, the polarization properties of the obtained structure are investigated. The waveguide width is varied from $0.2a$ to a, with increments of $0.2a$, where a is the lattice period to evaluate the influence of this parameter on both the wave guiding conditions and on the defect mode polarization properties. The calculations are performed using the PWE method and the Super Cell technique. The considered unit-cell is in this case seven times longer along the *y* direction than along the *x* direction compared to the cell used for the perfect phononic structure. The computed structure then consists of a guiding section surrounded by three holes on each side. The resulting 6 holes separation between two wave guiding sections in the computation ensures that neighboring waveguides are isolated in the y-direction, hence avoiding artificial branches that could be intro duced by adjacent waveguide mode coupling. The number of Fourier harmonics is multiplied by a factor of six in the *y*-direction to achieve similar conditions of convergence to the previous section.

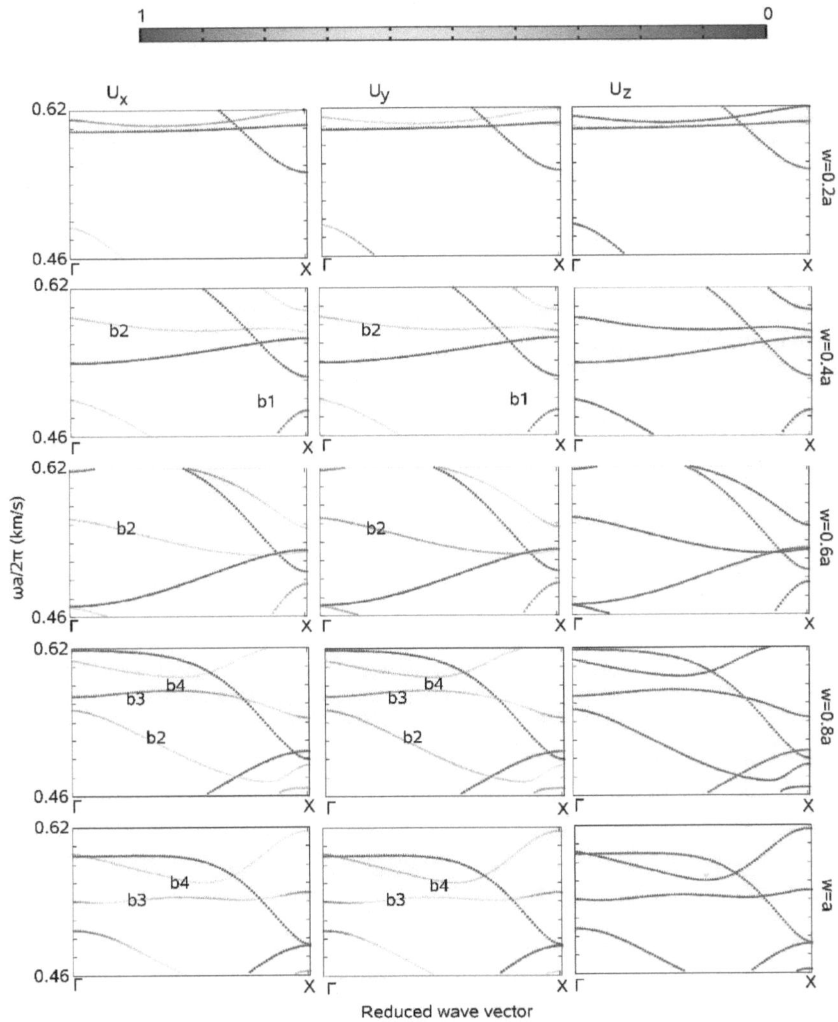

Fig. 2 Band structure along the ΓX direction for different line defect widths, *w*=0.2*a*, 0.4*a*, 0.6*a*, 0.8*a* and *a*, for a waveguide created inside the 2D phononic crystal. For each width, the band structure is repeated three times to highlight the influence of the polarization.

Figure 2 displays the dispersion relations of waves guided along the x-direction for frequencies falling within the complete band gap. The band structure is here again represented for the three polarizations and for each waveguide width. The most striking point at first sight is that these phononic waveguides are multimode even for very low waveguide width (e.g. for example for a width of $0.2a$) and present at least two out-of-plane and two in-plane polarized branches. The overall number of guided modes obviously increases with the defect width. As a general rule, we can notice that if the polarization of the out-of-plane component u_z remains constant, as no coupling with any other type of wave can happen due to the crystallographic symmetry of silicon, the repartition between u_x and u_y varies as the Bloch wave vector sweeps the ΓX axis.

The coupling between the two in-plane components is here very strong and very few quasi-pure shear modes can be found, if we except for example for the branch labeled b1 for a waveguide width of $0.4a$, or for modes close to the X point on branch labeled b2, amongst others. Computations of the displacement field mode shape confirmed that the polarization varies continuously, though the individual behavior of each branch can be quite different and is highly dependent on the line defect width. A branch can for instance start as predominantly longitudinal at point Γ and reach point X with a longi-transverse polarization (e.g. branch b4 for w=$0.8a$) or can enter and exit the Brillouin zone with the same polarization after having gone through a mixed polarization state, as in the case of branch b3 for a guide of width a.

4. Conclusion

In summary, we calculated the polarization states in a two dimensional silicon phononic crystal and in phononic crystal waveguides with different widths. The in-plane and out-of-plane polarizations are decoupled. The two in-plane polarizations are strongly coupled and the variation of the coupling is continuous with regard to the wave vector. For the waveguides, this last property is also verified. However, by varying the width of the guide, branches fall into the band gap, and the polarization of the modes change or not with regard to the width of the guide depending on the frequency and the wave vector.

References

[1] Sigalas, M.M. and Economou, E.N.: Band structure of elastic waves in two dimensional systems. *Solid Stat Comm.* **86** 141 (1993).

[2] Kushwaha, M. S., Halevi, P., Dobrzynski, L., and Djafari-Rouhani, B.: Acoustic band structure of periodic elastic composites. *Phys. Rev. Lett.* **71** 2022-2025 (1993).

[3] Kushwaha, M.S., Halevi, P., Martinez, G., Dobzynski, L., and Djafari-Rouhani, B.: Theory of acoustic band structure of periodic elastic composites. *Phys. Rev B* **49**2313 (1994).

[4] Kushwaha, M.S., Halevi, P.: Band-gap engineering in periodic elastic composites. *Appl. Phys. Lett.* **64** 1085 (1994).

[5] Tanaka, Y. and Tamura, S.: Surface acoustic waves in two-dimensional periodic elastic structures. *Phys. Rev. B* **58** 7958 (1998).

[6] Tanaka, Y. and Tamura, S.: Acoustic stop bands of surface and bulk modes in two-dimensional phononic lattices consisting of aluminum and a polymer. *Phys. Rev. B* **60** 13294 (1999).

[7] Wu, T., Huang, Z. and Lin, S.: Surface and bulk acoustic waves in two-dimensional phononic crystal consisting of materials with general anisotropy. *Phys. Rev. B* **69** 094301 (2004).

[8] Khelif, A., Aoubiza, B., Mohammadi, S., Adibi, A., and Laude, V.: Complete band gaps in two-dimensional phononic crystal slabs. *Phys. Rev. E* **74** 046610 (2006).

[9] Sun, J. H. and Wu, T. T.: Efficient formulation for band-structure calculations of two-dimensional phononic-crystal plates. *Phys. Rev. B* **74** 144303 (2006).

[10] Sigalas, M. M.: Defect states of acoustic waves in a two-dimensional lattice of solid cylinders. *J. Appl. Phys.* **84** 3026 (1998).

[11] Torres, M., Montero de Espinosa, F.R, Garcia-Pablos, D. and Garcia, N.: Sonic Band Gaps in Finite Elastic Media: Surface States and Localization Phenomena in Linear and Point Defects. *Phys. Rev. Lett* **82** 3054 (1999).

[12] Khelif, A., Choujaa, A., Djafari-Rouhani, B., Wilm, M., Ballandras, S., and Laude, V.: Trapping and guiding of acoustic waves by defect modes in a full-band-gap ultrasonic crystal. *Phys. Rev.* B **68** 214301 (2003)

[13] Pennec, Y., Djafari-Rouhani, B., Vasseur, J. O., Larabi, A., Khelif, A., Choujaa, A., Benchabane, S., and laude, V.: Acoustic channel drop tunneling in a phononic crystal. *Appl. Phys. Lett.* **87** 261912 (2005).

[14] Manzanares-Martinez, B. and Ramos-Mendieta, F.: Sagittal acoustic waves in phononic crystals: k-dependent polarization. *Phys. Rev. B* **76** 134303 (2007).

[15] T.-T. Wu and Z. G Huang, Level repulsions of bulk acoustic waves in composite materials. *Phys. Rev. B* **70** 214304 (2004).

Dispersion Analysis of Wave Motion in Microstructured Solids

Tanel Peets and Kert Tamm

Centre for Nonlinear Studies, Institute of Cybernetics at Tallinn University of Technology, Akadeemia tee 21, 12618 Tallinn, Estonia

tanelp@cens.ioc.ee

Abstract. The Mindlin-type model is used for describing longitudinal waves in microstructured solids. This model involves explicitly the internal parameters and therefore tends to be rather complicated. An hierarchical approximation is derived, which is able to grasp the main effects of dispersion with wide variety of parameters. Attention is paid to the internal degrees of freedom of the microstructure and their influence on the dispersion effects. It is shown how the internal degrees of freedom can change the effects of dispersion.

1. Introduction

It is well recognized by modern science, that matter is not continuous but has an internal structure. Clearly this microstructure plays a significant role when modelling wave propagation – waves that have a wavelength shorter than a certain threshold value, "feel" the microstructure.

There are two approaches in modelling the microstructure - one group of models are based on lattice theory [1-3], another on continuum theory [4-6].

In the discrete approach the volume elements of the matter are treated as point masses with a defined distribution and some interaction between the discrete masses. The governing equations are then deduced following the Newton's law.

In the microcontinuum theory, the macro- and microstructure of the continua are separated. Then the conservation laws for both structures should either be separately formulated [4,5], or the microstructural quantities (cells) are separately taken into account in one set of conservation laws. Engelbrecht *et al.* [6] have derived the one-dimensional model for longitudinal waves in microstructured materials based on Mindlin model [5]. This model will be the basis of our analysis. These governing equations of wave motion tend to be rather complicated and therefore there is a need for simplification. A slaving principle is used in order to derive a hierarchical asymptotic Whitham-type model.

T.-T. Wu and C.-C. Ma (eds.), *IUTAM Symposium on Recent Advances of Acoustic Waves in Solids*, IUTAM Bookseries 26, DOI 10.1007/978-90-481-9893-1_33,

An important effect caused by microstructure is dispersion. A wave packet can be viewed as a collection of harmonic waves. If such a wave travels through a microstructured material, then different harmonics "feel" the microstructure according to their wavelength and travel with different speeds. The variation of phase velocity with wavenumber is the hallmark of dispersion [7,8].

Generally if there is N particles per unit cell in discrete model, then N dispersion curves appear (3N in case of 3D model). The lower curve is called an acoustic branch, the upper curves are called optical branches and they only appear when there are at least 2 particles per unit cell. Optical branches are said to reflect the internal degrees of freedom [1,9].

Because of the inclusion of the microstructure, the dispersion curves derived from the 1D microcontinuum model, also give two distinct curves [4,6,9]. As in discrete model these curves are acoustical and optical modes where an optical modes are interpreted as internal degrees of freedom or "internal modes"[5,9]. The dispersion curve derived from the Whitham-type approximate model has only an acoustical branch. It means that the approximate model does not account directly for internal degrees of freedom. The authors have shown that this approximation is acceptable with wide variety of parameters. However the question that remains is when the internal modes can be ignored.

2. The Basic Model

The basic model is that of Mindlin [5] and we follow the presentation of its ideas in [6]. The main idea is to distinguish between macro- and microdisplacements $u_i(x_i,t)$ and $u_j'(x_i',t)$, respectively. Assuming that microdisplacement is defined in coordinates x_k', moving with a microvolume (cell), we define $u_j' = x_k'\varphi_{kj}(x_i,t)$, where φ_{kj} is an arbitrary function. It is clear that actually it is the microdeformation while $\partial u_j^{\prime} / \partial u_i^{\prime} = \partial_i^{\prime} u_j^{\prime} = \varphi_{ij}$. Further we consider the simplest 1D case and drop the indices *i* and *j*.

Now the fundamental balance laws can be formulated separately for macroscopic and microscopic scales. Introducing the Lagrangian $L = K - W$, formed from the kinetic and potential energies

$$K = \frac{1}{2}\rho u_t^2 + \frac{1}{2}I\varphi_t^2, \quad W = W(u_x, \varphi, \varphi_x), \tag{2.1}$$

where ρ and I denote the macroscopic density and the microinertia, respectively, we can use the corresponding Euler-Lagrange equations:

$$\left(\frac{\partial L}{\partial u_t}\right)_t + \left(\frac{\partial L}{\partial u_x}\right)_x - \left(\frac{\partial L}{\partial u}\right) = 0, \quad \left(\frac{\partial L}{\partial \varphi_t}\right)_t + \left(\frac{\partial L}{\partial \varphi_x}\right)_x - \left(\frac{\partial L}{\partial \varphi}\right) = 0. \tag{2.2}$$

Here and further, the indices x and t denote differentiation.

The partial derivatives

$$\sigma = \partial W / \partial u_x, \quad \eta = \partial W / \partial \varphi_x, \quad F = \partial W / \partial \varphi, \tag{2.3}$$

are recognized as the macrostress, the microstress and the interactive force, respectively.

The simplest potential energy function describing the influence of a microstructure is a quadratic function

$$W = \frac{1}{2} a u_x^2 + A\varphi u_x + \frac{1}{2} B\varphi^2 + \frac{1}{2} C\varphi_x^2, \tag{2.4}$$

where a,A,B,C denote material constants. Introducing Eq. (2.4) into Eq. (2.3) we get finally

$$\rho u_{tt} = a u_{xx} + A\varphi_x, \quad I\varphi_{tt} = C\varphi_{xx} - Au_x - B\varphi. \tag{2.5}$$

This is the governing system of two second-order equations that can also be represented in the form of one fourth-order equation

$$u_{tt} = \left(c_0^2 - c_A^2\right) u_{xx} - p^2 \left(u_{tt} - c_0^2 u_{xx}\right)_{tt} + p^2 c_1^2 \left(u_{tt} - c_0^2 u_{xx}\right)_{xx}, \tag{2.6}$$

where material parameters $c_0^2 = a/\rho, \quad c_1^2 = C/I, \quad c_A^2 = A^2/\rho B, \quad p^2 = I/B$, are introduced. The parameters c_0, c_1, c_A are velocities while p is a time parameter. This is the basic linear equation governing 1D longitudinal waves in microstructured solids. It has been shown by Sun *et al.* that Mindlin type model can also be used for modeling wave dispersion in layered media [10].

An approximation of Eq. (2.6) can be obtained by using the slaving principle. It is supposed that the inherent length-scale l is small compared with the wavelength L of the excitation. The following dimensionless variables and parameters are introduced $U=u/U_0$, $X=x/L$, $T=c_0t/L$, $\delta=(l/L)^2$, $\varepsilon=U_0/L$, where U_0 is the amplitude of the excitation. In addition it is assumed that $I=\rho l^2 I^*$ and $C=l^2C^*$, where I^* is dimensionless and C^* has the dimensions of stress.

Next the system (2.5) is rewritten in its dimensionless form and the slaving principle [11] is applied. Then we get finally

$$U_{TT} = \left(1 - \frac{c_A^2}{c_0^2}\right) U_{XX} + \frac{c_A^2}{c_B^2}\left(U_{TT} - \frac{c_1^2}{c_0^2} U_{XX}\right)_{XX}, \tag{2.7}$$

where $c_B^2=L^2/p^2=BL^2/I$. Note that c_B involves the scales L and l and c_A includes the interaction effects through parameter A. Restoring dimensions, Eq. (2.7) yields

$$u_{tt} = \left(c_0^2 - c_A^2\right) u_{xx} + p^2 c_A^2 \left(u_{tt} - c_1^2 u_{xx}\right)_{xx}. \tag{2.8}$$

This is an example of the Whitham-type hierarchical equation.

3. Dispersion Analysis

The dispersion relations for Eqs. (2.6) and (2.8) are

$$\begin{aligned}\omega^2 &= \left(c_0^2 - c_A^2\right)k^2 + p^2\left(\omega^2 - c_0^2 k^2\right)\left(\omega^2 - c_1^2 k^2\right),\\ \omega^2 &= \left(c_0^2 - c_A^2\right)k^2 - p^2 c_A^2\left(\omega^2 - c_1^2 k^2\right)k^2.\end{aligned} \tag{3.1}$$

In order to reduce the number of independent variables, the wave number, the frequency and the propagation speeds are normalized defining $\xi = pc_0 k$, $\eta = p\omega$, $\gamma_A = c_A/c_0$, $\gamma_1 = c_1/c_0$. Using these new quantities the dispersion relations (3.1) assume the forms

$$\begin{aligned}\eta^2 &= \left(1 - \gamma_A^2\right)\xi^2 + \left(\eta^2 - \xi^2\right)\left(\eta^2 - \gamma_1^2 \xi^2\right),\\ \eta^2 &= \left(1 - \gamma_A^2\right)\xi^2 - \gamma_A^2\left(\eta^2 - \gamma_1^2 \xi^2\right)\xi^2,\end{aligned} \tag{3.2}$$

where the parameters γ_A and γ_1 have the values $0 < \gamma_A < 1$ and $0 < \gamma_1 < 1$ respectively.

The characteristic dispersion curves are shown in Fig. 3.1. The full dispersion relation (3.2a), which is represented by the continuous line, represents two distinct branches – acoustical and optical. The acoustical branch is analogous to the case of elastic vibrations where all the cells move in unison. These are external modes. The optical branch reflects the role of the internal modes, which involve the distortion of the cells [5,9].

The optical branch is always concave, the acoustical branch can be either concave or convex or linear, which represents anomalous, normal or no dispersion respectively. This concavity and convexity of the acoustic dispersion curve shows explicitly the influence of basic material properties [12].

The full model (2.6) and approximate model (2.7) can be compared using numerical analysis. The initial value problem in dimensionless form under periodic boundary conditions is solved using the pseudospectral method [13]. The initial profile is chosen $U(X,0) = sech^2(\kappa X/2)$, where κ is the width of the profile.

Figures 3.2 and 3.3 show the results of the numerical analysis. Figure 3.2 represents the case when acoustical branch is concave (anomalous dispersion). It is clear from the numerical experiment that although there are small differences between the full model (2.6) and approximation (2.7), the approximation is able to display the main effects of dispersion i.e. the type of the dispersion.

Figure 3.3 shows a numerical experiment in case when there is no dispersion in approximate dispersion relation (3.2b) and in the acoustic branch of the full dispersion relation (3.2a). The approximate model indeed shows no dispersion effects – the initial profile moves with constant speed and shape. The full model (2.6) however displays a small effect of dispersion, which is due to the optical bra nch or internal modes. The dispersion effects do not appear immediately, but may take some time to appear.

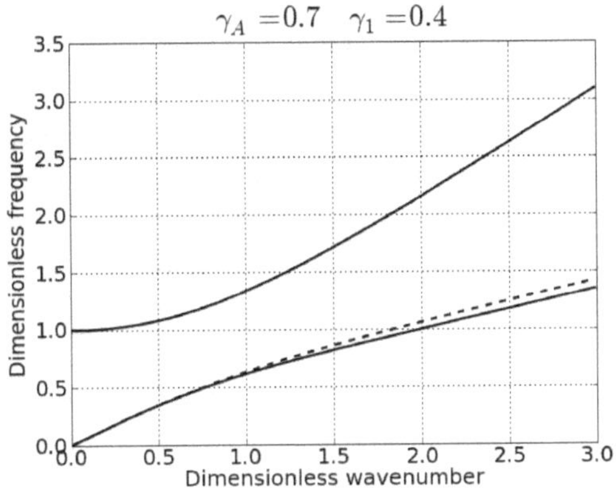

Fig. 3.1 The characteristic dispersion curves. Solid lines represent full dispersion relation, dashed line represents approximate dispersion relation.

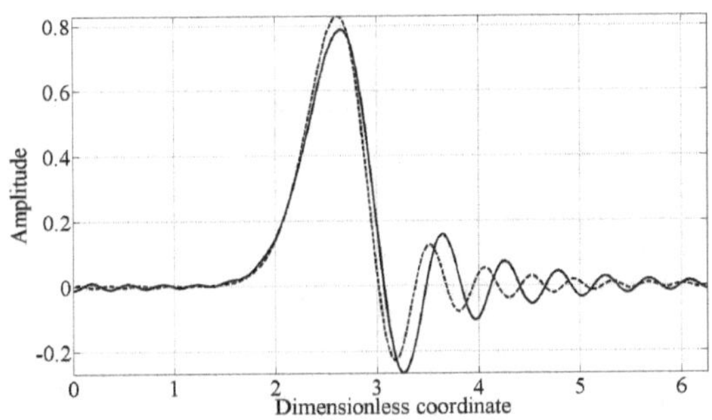

Fig. 3.2 The solutions of full model (solid line) and approximation (dashed line), in case of γ_A=0.9 and γ_1=0.7.

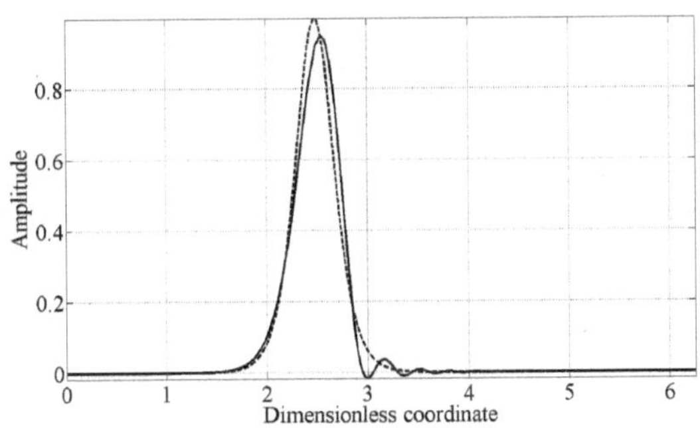

Fig. 3.3 The solutions of full model (solid line) and approximation (dashed line), in case of γ_A=0.9 and γ_1=0.7.

4. Final Remarks

The numerical analysis demonstrates that the full and approximate models give in most cases similar results. There are however conditions when approximate dispersion curve (3.2b) coincides well with the acoustic branch of the full dispersion curve (3.2a), but the numerical experiment gives different types of dispersion for the full model (2.6) and for the approximation (2.7). This is likely to be present when acoustic curve displays normal dispersion.

These effects need further investigations and will be presented in further publications.

Acknowledgements: The authors gratefully acknowledge the financial support from Estonian Science Foundation.

References

[1] Brillouin, L.: *Wave Propagation in Periodic Structures. Dover Publications*, INC, NY (1953).

[2] Metrikine, A.V., Askes, H.: One-dimensional dynamically consistent gradient elasticity models derived from a discrete microstructure. Part 1: Generic formulation. *Eur. J. Mech. A. Solids* **21** (2002) 555-572.

[3] Maugin, G.A.: *Nonlinear Waves in Elastic Crystals.* Oxford University Press (1999).

[4] Eringen, A.C.: Linear theory of micropolar elasticity. *J. Math. Mech.* **15** 909-923 (1966).

[5] Mindlin, R.D.: Microstructure in linear elasticity. *Arch. Rat. Mech. Anal.* **16** 51-78 (1964).

[6] Engelbrecht, J., Berezovski, A., Pastrone, F., Braun, M.: Waves in microstructured materials and dispersion. *Philos. Mag.* **85**(33-35) 4127-4141(2005).

[7] Nettel, S.: *Wave physics. Oscillations - Solitons - Chaos.* New York: Springer (1995).

[8] Main, I.G.: *Vibrations and Waves in Physics.* Cambridge University Press (1984).

[9] Chen, Y., Lee, J.D.: Analysis of photon dispersion relations from atomic model to continuum theory. *Nanotech* **1** 396-399 (2002).

[10] Sun, C.-T., Achenbach, J.D., Herrmann, G.: Continuum theory for a laminated medium. *J. Appl. Mech. Trans. ASME* **35**(3) 467-475 (1968).

[11] Christiansen, P.L., Muto, V., Rionero, S.: Solitary wave solution to a system of Boussinesq-like equations. *Chaos Solitons Fractals* **2** 45-50 (1992).

[12] Peets, T., Randrüüt, M., Engelbrecht, J.: On modelling dispersion in microstructured solids. *Wave Motion* **45** 471-480 (2008).

[13] Salupere, A., Tamm, K., Engelbrecht, J.: Numerical simulation of interaction of solitary deformation waves in microstructured solids. *Int. J. Non Linear Mech.* **43** 201-208 (2008).

On the Role that Scholte Waves Play in Acoustic Propagation along a Fluid-Solid Interface

Piotr Borejko

Department of Civil Engineering, Vienna University of Technology, A-1040 Vienna, Austria

piotr.borejko@tuwien.ac.at

Abstract. The reflection of sound from a point source in a fluid at a plane interface with an elastic solid, and sound propagation from a point source in a wedge of fluid supported by an elastic solid are investigated using the method of generalized ray. Time records of the acoustic pressure received at large ranges are evaluated both for a fluid-solid interface alone and for a wedge of fluid overlaying an elastic solid when the shear wave speed in the solid is lower than the sound speed in the fluid (so-called "slow-speed elastic bottom"). The most prominent phase (large in amplitude and wide in time) is that constituted by Scholte waves.

1. Introduction

An important element of many problems in physical acoustics is the phenomenon of acoustic reflection and transmission at an interface between two media, one (upper) fluid and the other (lower) an elastic solid. The general problem of determining the reflected and transmitted acoustic fields due to a spherical wave that radiates from a point source and impinges on a plane fluid-solid interface is surprisingly difficult and has occupied several researches over the years[1-7].

The acoustic fields may be determined from the asymptotic solutions of wave equations at high frequencies, the acoustical theory known as geometrical acoustics or ray acoustics[8]. It should be emphasized that the acoustic fields may differ significantly from the geometrical-acoustics approximation if either the source or receiver is near the interface (in acoustic wavelengths) or if the stationary phase path (the ray path) is near the critical angle. It is for this reason that the exact wave-theory solutions of wave equations with appropriate

T.-T. Wu and C.-C. Ma (eds.), *IUTAM Symposium on Recent Advances of Acoustic Waves in Solids*, IUTAM Bookseries 26, DOI 10.1007/978-90-481-9893-1_34,

boundary conditions (the general solutions) must be considered, constituting a valid representation of the entire acoustic field (i.e., including all wave-forms comprising the field[9]).

The general solutions for the acoustic fields may be derived by applying the theory of generalized ray[10] which involves direct construction of integral representations for such solutions from the known source functions, the reflection and transmission coefficients for plane waves, the receiver functions, and the phase functions. These integrals can then be evaluated exactly by applying the Cagniard method[11]. In the literature they are known as ray integrals, but they are quite different from ray integrals of the theory of geometrical acoustics[8].

Specifically, the generalized-ray method may be used to treat the fluid-solid interface problem. If both a point source and a point receiver are placed in a fluid, the general solution is then set as a sum of two ray integrals in which the first integral represents the entire field of the point source and the second one represents the entire reflected field (excited in a fluid by the source field which impinges on the interface). The latter integral thus represents not only the regularly reflected (according to Snell's law) wave but also (depending on the wave speed ratios of the adjacent media, the distance of the source and receiver from the interface, and the distance between them measured along the interface[10]) the critically refracted longitudinal and shear waves (also called the head[11] (or lateral[12]) waves or the diffracted components of the reflected wave[9,13]) and the pseudo-Rayleigh[14,15] and Scholte[11,16] (also called Stoneley[14]) interface waves.

Although the problem of acoustic propagation in a wedge of fluid overlaying a liquid or elastic bottom has already been tackled[17,18], the effect of a sloping slow-speed elastic bottom on the acoustic field within the wedge was not considered in detail. In this paper, we thus concentrate on the interpretation of this effect in terms of the generalized-ray theory, and in particular our emphasis is on the role that Scholte waves play in the field as observed by a submerged distant receiver.

In Sec. 2, we review the generalized-ray theory of acoustic propagation from a point source in a wedge of fluid supported by an elastic solid.

In Sec. 3, we present results of pressure-field calculations both for a fluid-solid interface alone and for a wedge of fluid supported by an elastic solid. In the two cases, a fluid overlaying a slow-speed elastic bottom with discontinuities at the interface both in the velocity and in the density is considered.

2. Theory

Referring to Fig. 2.1, consider a wedge-shaped layer of homogeneous fluid of density ρ and of sound speed c, bounded above by a horizontal pressure-release plane and below by a sloping (acoustically penetrable) plane interface with a semi-infinite homogeneous elastic solid of density ρ_2 and of P and S wave speeds c_P and c_S, respectively. These boundary planes intersect along the line of apex, and the apex angle is α. The origin O of a Cartesian coordinate system (x, y, z) is placed at the horizontal boundary at distance d from the apex, and x, y and z are the range (normal to the apex), cross-range (parallel to the apex) and depth coordinates, respectively. An omnidirectional point source is located in the layer at depth z_0 directly below O, and the layer thickness at the source location is h.

The acoustic field in the layer is represented by the velocity potential ϕ satisfying the inhomogeneous wave equation

$$c^2\nabla^2\phi - \frac{\partial^2\phi}{\partial t^2} = -f(t)\delta(x)\delta(y)\delta(z - z_0), \tag{2.1}$$

where $f(t)$ is a causal function of time t. The problem is to determine the time record of the pressure at a receiver placed in the layer at a range r from a point source. A solution for this problem is obtained by applying geometric-acoustic arguments of the method of images and a modified generalized-ray method.

First, the acoustic field in the layer is expanded into a finite series of wave fields (the first is that radiated from a point source, the next two are those reflected once at either boundary, etc.) plus the wave diffracted at the apex. Thus, omitting the effect of diffraction at the apex, which in certain cases is negligible or entirely absent[19,20], the potential for the total field is then represented by

$$\phi = \phi_0 + \sum_{k=1}^{N} \phi_{\pm k}. \tag{2.2}$$

In the above ϕ_0 is the potential for the incident (source) field; $\phi_{\pm k}$ is that for the multi-reflected field undergoing k reflections, where the +/– sign meaning that the first reflection is at the sloping/horizontal boundary; and $N = \pi/\alpha$, where α is an integral submultiple of π.

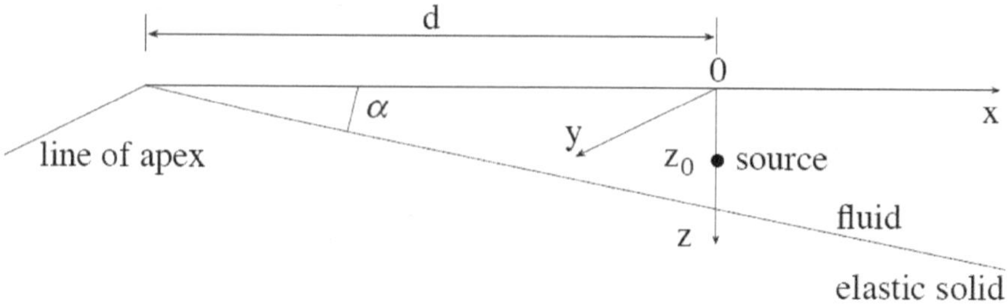

Fig. 2.1 Geometry for a wedge of fluid having a sloping interface with an elastic solid.

Next, the incident potential is expanded into an integral over plane waves, and the plane-wave reflection coefficients are used to synthesize an integral representation of the multi-reflected potential. The time-reduced (i.e., Laplace time-transformed) potentials then are

$$\begin{aligned}\bar{\phi}_0 &= \frac{s\bar{f}(s)}{8\pi^2 c^2}\int_{-\infty}^{+\infty}\int_{-\infty}^{+\infty} S\exp(sg_0)d\xi d\kappa,\\ \bar{\phi}_{\pm k} &= \frac{s\bar{f}(s)}{8\pi^2 c^2}\int_{-\infty}^{+\infty}\int_{-\infty}^{+\infty} S\Pi_{\pm k}\exp(sg_{\pm k})d\xi_{\pm l}d\kappa_{\pm l}.\end{aligned} \tag{2.3}$$

In the above S is the source function; $\Pi_{\pm k}$ is a product of the reflection coefficients; ξ,κ and are the wave slownesses of the 0-th field; $\xi_{\pm l},\kappa_{\pm l}$ are the local wave slownesses of the $\pm k$-th field; and $g_0, g_{\pm k}$ are the phase functions of the 0-th, $\pm k$-th fields, respectively. In Eqs. (2.3); $\bar{\phi}_0$ is the particular solution of time-reduced inhomogeneous wave equation (2.1); and $\bar{\phi}_{\pm k}$ is the solution of time-reduced homogeneous wave equation (2.1, $f(t)=0$), satisfying the boundary conditions for the layer. Substituting these solutions into time-reduced Eq. (2.2), we thus obtain the general solution of the Laplace transformed potential $\bar{\phi}$.

Finally, the inverse Laplace transforms of $\bar{\phi}_0$ and $\bar{\phi}_{\pm k}$ are evaluated exactly by applying the Cagniard method. Since $p=-\rho\partial\phi/\partial t$, where p is the acoustic pressure, the solutions for the pulses p_0 and $p_{\pm k}$ thus obtained are

$$\begin{aligned}p_0 &= \frac{p_c}{R}\dot{f}(t-R/c) = p_c H(t-t_0)\int_{t_0}^{t}\ddot{f}(t-\tau)I_0(\tau)d\tau,\\ p_{\pm k} &= p_c H(t-t_{\pm k})\int_{t_{\pm k}}^{t}\ddot{f}(t-\tau)I_{\pm k}(\tau)d\tau,\end{aligned} \tag{2.4}$$

where $p_c=-\rho/(4\pi c^2)$. The superposition of these solutions

$$p = p_0 + \sum_{k=1}^{N} p_{\pm k} \tag{2.5}$$

yields the general solution of p for a point source in a non-parallel layer of fluid, which is exact other than the omission of wave diffracted at the apex. The R is the source-to-receiver distance; $H(t)$ is the Heaviside step function; $t_0, t_{\pm k}$ are the minimum arrival times of the 0-th, $\pm k$-th pulses, respectively; and

$$I_0(\tau) = 2\,\mathrm{Re}\int_0^{q(\tau)} S\frac{dg_0^{-1}}{d\tau}dq, \qquad I_{\pm k}(\tau) = 2\,\mathrm{Re}\int_0^{q(\tau)} S\Pi_{\pm k}\frac{dg_{\pm k}^{-1}}{d\tau}dq$$

are the ray-integral representations of the entire 0-th and $\pm k$-th fields, respectively. The Re stands for "real part"; $g_0^{-1}, g_{\pm k}^{-1}$ are the inverse phase functions; and q is the wave slowness.

Note that the pressure due to the incident and bottom-reflected pulses is

$$p = p_0 + p_{+1}. \tag{2.6}$$

This solution ignores entirely the existence of the upper boundary, and thus represents the exact solution of p for a point source in a semi-infinite fluid $z < h$ bounded below by a horizontal interface $z = h$ with a semi-infinite elastic solid $z > h$. Here h is the height of O above the interface, and a point source is placed at $z = z_0$ below O. The heights of the source and receiver above the interface then are $z_S = h - z_0$ and $z_R = h - z$, respectively.

3. Examples

Solutions (2.5) and (2.6) were used to compute the received pressure records for $\rho = 1.0$, $c = 1.0$, $\rho_2 = 2.2$, $c_P = 1.78$, $c_S = 0.73$. A fluid overlaying a slow-speed elastic bottom with discontinuities at the interface both in the velocity and in the density is thus considered. The $f(t)$ in Eqs. (2.4) used for the computed examples is a parabolic ramp function[10] so that the source pulse p_0 has the form of an isosceles triangle, where Δ is the rise time of the pulse, 2Δ is the pulse duration and $(p_c/R)\Delta$ is the peak value of the pulse.

The first calculations were carried out for the fluid-solid interface problem alone [solution (2.6)], and the time records of the pressure are shown in Fig. 3.1 for two large-range locations (in terms of the source height z_S): $r/z_S = 200$ and $r/z_S = 400$. Each record exhibits four distinct wave-forms, enumerated in ascending order according to their arrival times at the receiver: the critically refracted longitudinal wave, the wave emitted from the source (the source pulse), the totally reflected wave, and the Scholte wave which is a very prominent wave-form on the response curve. (In the case of a slow-speed elastic bottom, both the critically refracted shear wave and the pseudo-Rayleigh wave are absent.)

This problem was treated by de Hoop and van der Hijden[16] who applied a modified Cagniard's method to derive integral representations for the incident and reflected acoustic fields, and computed the pressure records due to a source pulse of Heaviside form, but their computations were restricted to small values of r/z_S.

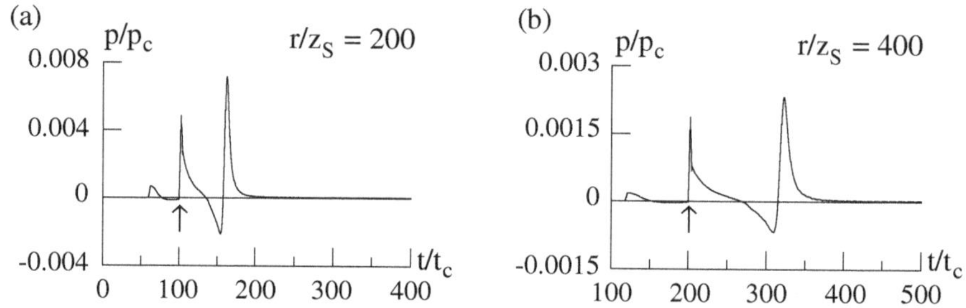

Fig. 3.1 Pressure records due to a source pulse of triangular form ($t_c = 2z_S/c$, $\Delta/t_c = 2.0$, and the vertical arrow indicates the arrival time of the source pulse); (a) at $r/z_S = 200$ ($z_R/z_S = 6.24$); (b) at $r/z_S = 400$ ($z_R/z_S = 11.48$).

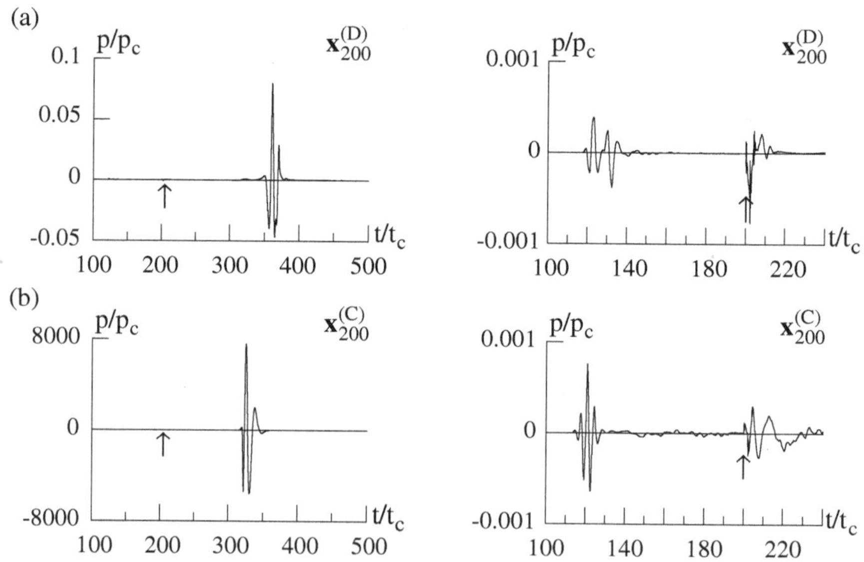

Fig. 3.2 Pressure records due to a source pulse of triangular form ($t_c = h/c$, $\Delta/t_c = 2.0$, and the vertical arrow indicates the arrival time of the source pulse); (a) at down-slope location $\mathbf{x}_{200}^{(D)}$ ($r/h = 200$); (b) at cross-slope location $\mathbf{x}_{200}^{(C)}$ ($r/h = 200$).

The second calculations were done for a 3 degree wedge [solution (2.5)]. The point source and the two receivers were located on the bisecting plane of the wedge. Figure 3.2 shows the time records of the pressure received at two large-range locations (in terms of the wedge depth at the source location h): $\mathbf{x}_{200}^{(D)}$ and $\mathbf{x}_{200}^{(C)}$, where the D and C letters in parentheses attached to $\mathbf{x}$ stand for down-slope and cross-slope of the source, respectively, and the subscript 200 of $\mathbf{x}$ is the value of r/h . The records exhibit three phases: the early time response composed of critically refracted longitudinal waves arriving along refracted ray paths, the intermediate time response composed of multi-reflected spherical waves arriving along ray paths, and the late time response composed of Scholte waves which is the most prominent phase in the records. Note that the method of normal modes cannot be extended to treat acoustic propagation in the present wedge because "penetrable" boundary conditions at the sloping interface preclude the separability of the Helmholtz equation[21].

References

[1] Brekhovskikh, L. M.: *Waves in Layered Media. Academic Press*, New York (1980).

[2] Gerjuoy, E.: Refraction of waves from a point source into a medium of higher velocity. *Phys. Rev.* **73** 1442-1449 (1948).

[3] Paul, D. I.: Acoustical radiation from a point source in the presence of two media. *J. Acoust. Soc. Am.* **29** 1102-1109 (1957).

[4] Friedricks, K. O. and Keller, J. B.: Geometric acoustics II. Diffraction, reflection, refraction of a weak spherical or cylindrical shock at a plane interface. *J. Appl. Phys.* **26** 961-966 (1955).

[5] Ott, V. H.: Reflexion und Brechung von Kugelwellen; Effekte 2. Ordnung. *Ann. Phys.* **41** 443-466 (1942).

[6] Paul, D. I.: Wave propagation in acoustics using the saddle point method. *J. Math. Phys.* **38** 1-15 (1959).

[7] Stickler, D. C.: Reflected and lateral waves for the Sommerfeld model. *J. Acoust. Soc. Am.* **60** 1061-1070 (1976).

[8] Pierce, A. D.: Acoustics: *An Introduction to Its Physical Principles and Applications. Acoustical Society of America*, New York (1989).

[9] Chapman, D. M. F.: Using streamlines to visualize acoustic energy flow across boundaries. *J. Acoust. Soc. Am.* **124** 48-56 (2008).

[10] Pao, Y. H. and Gajewski, R.: *The generalized ray theory and transient responses of layered elastic solids. In Physical Acoustics, ed. by Mason W. P. and Thurston R. N., Academic Press*, New York, 183-265 (1977).

[11] Cagniard, L.: *Reflection and Refraction of Progressive Seismic Waves*. Mc-Graw Hill, New York (1962).

[12] Chapman, D. M. F.: The normal-mode theory of air-to-water sound transmission in the ocean. *J. Acoust. Soc. Am.* **87** 601-618 (1990).

[13] Godin, O. A.: Low-frequency sound transmission through a gas-liquid interface. *J. Acoust. Soc. Am.* **123** 1866-1879 (2008).

[14] Roever, W. L., Vining, T. F., and Strick, E.: Propagation of elastic wave motion from an impulsive source along fluid/solid interface I. Experimental pressure response II. Theoretical pressure response III. The pseudo-Rayleigh wave. *Phil. Trans. Roy. Soc.* **A251** 455-523 (1959).

[15] van der Hijden, J. H. M. T.: Quantitative analysis of the pseudo-Rayleigh phenomenon. *J. Acoust. Soc. Am.* **75** 1041-1047 (1984).

[16] de Hoop, A. T. and van der Hijden, J. H. M. T.: Generation of acoustic waves by an impulsive point source in a fluid/solid configuration with a plane boundary. *J. Acoust. Soc. Am.* **75** 1709-1715 (1984).

[17] Pao, Y. H., Ziegler, F., and Wang, Y. S.: Acoustic waves generated by a point source in a sloping fluid layer. *J. Acoust. Soc. Am.* **85** 1414-1426 (1989).

[18] Borejko, P., Chen, C. F., and Pao, Y. H.: Application of the method of generalized ray to acoustic waves in a liquid wedge over elastic bottom. *J. Comput. Acoust.* **9** 41-68 (2001).

[19] Buckingham, M. J.: Theory of acoustic radiation in corners with homogeneous and mixed perfectly reflecting boundaries. *J. Acoust. Soc. Am.* **86** 2273-2291 (1989).

[20] Borejko, P. and Dong, H.: On acoustic propagation in a perfect wedge. In Theoretical and Computational Acoustics 2007, ed. by Taroudakis M. and Papadakis P., *E-Media University of Crete,* Heraklion, 247-256 (2008).

[21] Buckingham, M. J.: Theory of three-dimensional acoustic propagation in a wedgelike ocean with a penetrable bottom. *J. Acoust. Soc. Am.* **82** 198-210 (1987).

Forced Motions in Rectangular Elastic Waveguide

Anastasiya A. Bondarenko[1]**, Viatcheslav V. Meleshko**[2] **and**
Alexander N. Trofimchuk[1]

[1]Department of Natural Resources, Institute of Telecommunications and Global Information Space of the NAS Ukraine, Kiev 03186, Ukraine

[2]Department of Theoretical and Applied Mechanics, Kiev National Taras Shevchenko University, Kiev 01601, Ukraine

anastasiya.bondarenko@gmail.com

Abstract. Based on the superposition method, an analytic technique is developed for evaluation of the energy division among various modes in rectangular isotropic waveguide excited by loadings distributed on a portion of bounding faces. Results calculated for a square waveguide with uniform loading of constant magnitude are presented graphically. At each frequency there is more than one dominant mode. Backward modes are dominant for most feasible frequencies. Thus the loading considered is recommended for experimental investigations of these waves.

1. Introduction

Nowadays guided waves in infinitely wide plates, circular and hollow cylinders are well studied and successfully applied in long-range non-destructive evaluation, surface cracks determination, material properties measurement, etc. In order to develop NDE for waveguides of more complex cross section, it is necessary to find dispersion characteristics of a large number of propagating modes and to separate dominant modes that transfer the most of the energy depending on frequency. The problem is considerably complicated by high modal density and intricate structure of the wave field due to the interference of the bounding faces of finite cross section. To investigate dispersion characteristics of rectangular bars [1], rails [2], and L-shaped waveguides [3], computer simulations, as well as finite and boundary element methods are usually applied. In spite of a great number of numerical and experimental results for wave dispersion in waveguides of finite cross section, a qualitative comprehension of the energy propagation process and effective mode excitation has been not achieved yet.

T.-T. Wu and C.-C. Ma (eds.), *IUTAM Symposium on Recent Advances of Acoustic Waves in Solids*, IUTAM Bookseries 26, DOI 10.1007/978-90-481-9893-1_35,

The purpose of this paper is to analyze dispersion characteristics of rectangular elastic waveguide, wave motions in which are excited by harmonic loadings on a portion of the bounding faces. The set of propagating modes at each frequency is determined analytically by means of the superposition method. To evaluate the energy division among various modes, an efficient technique is proposed. The paper is organized as follows. Sec. 2 presents statement of the problem and a brief description of the solution process. In Sec. 3, the method developed is applied for numerical evaluation of amount of the energy going into each longitudinal mode excited in a square waveguide under the uniform normal loading. Main conclusions are given in Sec. 4.

2. Theoretical Analysis

2.1 Statement of the Problem

Consider an elastic waveguide of rectangular cross section made of homogeneous isotropic material with density ρ, shear modulus G and Poisson's ratio ν. The waveguide is referred to rectangular Cartesian coordinates with z-axis directed along the waveguide axis and bounding faces $x = \pm a$, $y = \pm b$. Time-harmonic waves of frequency ω are excited by loadings distributed uniformly on portion $-d \le z \le d$ of faces $y = \pm b$.

Displacements (u_x, u_y, u_z) are expressed in terms of scalar ϕ and vector $\vec{\psi}$ potentials satisfying Helmholtz equations and a condition that $\vec{\psi}$ is solenoidal

$$\nabla^2\phi + \frac{\omega^2}{c_1^2}\phi = 0\,, \quad \nabla^2\vec{\psi} + \frac{\omega^2}{c_2^2}\vec{\psi} = 0\,, \quad \frac{\partial\psi_x}{\partial x} + \frac{\partial\psi_y}{\partial y} + \frac{\partial\psi_z}{\partial z} = 0\,. \tag{2.1.1}$$

Here ∇^2 is the three-dimensional Laplace operator, c_1 and c_2 are velocities of compressional and shear waves, respectively. Harmonic factor $e^{-i\omega t}$ is omitted.

Boundary conditions on the waveguide faces are written as follows

$$\sigma_x = \tau_{xy} = \tau_{xz} = 0 \text{ at } x = \pm a\,;$$

$$\sigma_y = \begin{cases} 2GF(x,z), & |z| \le d, \\ 0, & |z| > d, \end{cases} \quad \tau_{yx} = \tau_{yz} = 0 \text{ at } y = \pm b\,. \tag{2.1.2}$$

Function $F(x,z)$ can be represented as Fourier transform with respect to propagation constant γ

$$F(x,z)=\frac{1}{2\pi}\int_{-\infty}^{+\infty}\overline{f}(x,\gamma)e^{i\gamma z}d\gamma,\ \overline{f}(x,\gamma)=\int_{-d}^{d}F(x,z)\cos\gamma z\,dz\,. \tag{2.1.3}$$

A period-average power through the waveguide cross section is determined as

$$W_y=\int_{-d}^{d}\int_{-a}^{a}P_y\,dxdz\,,\ P_y=-\frac{i\omega}{4}(\sigma_y u_y^*-\sigma_y^* u_y)=-\omega GF\,\mathrm{Im}\,u_y(x,b,z)\,, \tag{2.1.4}$$

where * denotes complex conjugate.

2.2 Method of Solution

Solution to the problem on forced motions in the rectangular waveguide consists of two steps: determination of a set of propagating normal modes in the waveguide with free faces, and evaluation of period-average power for each mode. Under the assumption that function *F(x,z)* is even in both coordinates, further theoretical analysis is limited to longitudinal wave motions, for which mode shapes are symmetric relative to middle planes of the waveguide.

2.2.1 Normal modes in rectangular waveguide

The superposition method is applied to construct an analytical solution to the problem on normal modes in a rectangular waveguide. The main idea of the method consists in using two ordinary Fourier series, each term of which identically satisfies Eqs. (2.1.1) and zero shear stresses on bounding faces. Thus displacement, for instance $u_y(x,y,z)=U_y(x,y)e^{i\gamma z}$ has the form of

$$U_y=-a\sum_{k=0}^{\infty}(-1)^k Y_k f_k\beta_k\left\{\frac{\beta_k^2+q_2^2+\gamma^2}{2q_1}\cdot\frac{\cosh q_1x}{\sinh q_1a}-q_2\cdot\frac{\cosh q_2x}{\sinh q_2a}\right\}\sin\beta_k y$$
$$-b\sum_{n=0}^{\infty}(-1)^n X_n b_n\left\{\frac{\alpha_n^2+p_2^2+\gamma^2}{2}\cdot\frac{\sinh p_1y}{\sinh p_1b}-\left(\alpha_n^2+\gamma^2\right)\cdot\frac{\sinh p_2y}{\sinh p_2b}\right\}\cos\alpha_n x, \tag{2.2.1}$$

where following notations are introduced

$$\alpha_n = \frac{n\pi}{a},\ \beta_k = \frac{k\pi}{b},\ q_i^2 = \beta_k^2 + \gamma^2 - \Omega_i^2,\ p_i^2 = \alpha_n^2 + \gamma^2 - \Omega_i^2,\ i = 1,2;$$

$$,\ \Omega_2 = \frac{\omega}{c_1},\ f_k = \begin{cases} 1, & k = 0, \\ 1/\beta_k^2, & k > 0; \end{cases}\ b_n = \begin{cases} 1, & n = 0, \\ 1/\alpha_n^2, & n > 0; \end{cases} \tag{2.2.2}$$

Two sets of Fourier coefficients Y_k, X_n $(k,n = 0,1,...)$ are determined such that normal stresses vanish on boundaries. Since the expressions for a coefficient of a term in one series will depend on all coefficients of the other series and vice versa, final solution requires solving an infinite system of linear algebraic equations

$$Y_k a\Delta_k(q) + \varepsilon_k \sum_{n=0}^{\infty} X_n b_n \left[\frac{2\alpha_n^2\beta_k^2}{\alpha_n^2 + q_1^2} - \frac{2\alpha_n^2\beta_k^2}{\alpha_n^2 + q_2^2} - \frac{\Omega_0^2\left(2\gamma^2 - \Omega_2^2\right)}{\alpha_n^2 + q_1^2} \right] = 0,$$

$$X_n b\Delta_n(p) + \varepsilon_n \sum_{k=0}^{\infty} Y_k f_k \left[\frac{2\alpha_n^2\beta_k^2}{\beta_k^2 + p_1^2} - \frac{2\alpha_n^2\beta_k^2}{\beta_k^2 + p_2^2} - \frac{\Omega_0^2\left(2\gamma^2 - \Omega_2^2\right)}{\beta_k^2 + p_1^2} \right] = 0. \tag{2.2.3}$$

Here $\varepsilon_i = 1/2$ for $i = 0$, and $\varepsilon_i = 1$ for $i > 0$; $\Omega_0^2 = \nu\Omega_1^2/(1-2\nu)$; $\Delta_k(q)$, $\Delta_n(p)$ correspond to Rayleigh-Lamb dispersion equations for infinite plates of thickness $2a$ and $2b$, respectively. The determinant of system (2.2.3) governs dispersion relation between frequencies and propagation constants of normal modes for given values of ν and a/b. Approaches fo correct reduction of the infinite system based on the asymptotic behavior of unknowns with large indices, as well as further details on the solution process and related references are given in [4].

2.2.2 *Efficiency of mode excitation*

To evaluate the period-average power supplied with loading (2.1.2), let represent the displacement at faces $y = \pm b$ by the Fourier integral. Taking into account (2.2.1), we have

$$u_y(x,b,z) = \frac{1}{2\pi}\int_{-\infty}^{\infty} \overline{U}_y(x,b)e^{i\gamma z}d\gamma = b\frac{\Omega_2^2}{2}\sum_{n=0}^{\infty}(-1)^n b_n \cos\alpha_n x \left[\frac{1}{2\pi}\int_{-\infty}^{\infty} \overline{X}_n(\gamma)e^{i\gamma z}d\gamma \right]. \tag{2.2.4}$$

Here a bar over the integrand is introduced with the purpose of uniformity of designations for functions being transformed. For given frequency ω, $\overline{X}_n(\gamma)$ have poles at points $\pm\gamma_m$ $(m=1,2,...,M)$, which are roots of the dispersion equation and correspond to propagating normal modes. Since $\overline{X}_n(\gamma)$ are analytic functions except for these points, the residue theorem can be applied to compute the integral in (2.2.4). Then

$$\operatorname{Im} u_y(x,b,z) = b\frac{\Omega_2^2}{2}\sum_{n=0}^{\infty}(-1)^n b_n \cos\alpha_n x \sum_{m=1}^{M} \operatorname*{Res}_{\gamma=\gamma_m} X_n(\gamma)\cos\gamma_m z\,. \tag{2.2.5}$$

Since period-average power W_y is a superposition of period-average powers $W_y^{(m)}$ for each propagating m mode, according to (2.1.4), it is easily to obtain that

$$W_y^{(m)} = -\omega G b\frac{\Omega_2^2}{2}\sum_{n=0}^{\infty}(-1)^n b_n \int_{-a}^{a}\overline{f}(x,\gamma)\cos\alpha_n x \operatorname*{Res}_{\gamma=\gamma_m} X_n(\gamma)dx\,. \tag{2.2.6}$$

Functions $X_n(\gamma)$ can be expressed as $d_n(\gamma)/D(\gamma)$, where $d_n(\gamma)$ are smooth functions with no singularities at points $\pm\gamma_m$, $D(\gamma)$ is the determinant of system (2.2.3). Then $\operatorname*{Res}_{\gamma=\gamma_m} X_n(\gamma) = d_n(\gamma_m)/D'(\gamma_m)$. On the other hand,

$$D'(\gamma_m) = d_n(\gamma_m)\cdot\frac{d}{d\gamma}\left(\frac{1}{X_n(\gamma)}\right)\Bigg|_{\gamma=\gamma_m}\,. \tag{2.2.7}$$

Finally, the period-average power for m mode is calculated as follows

$$W_y^{(m)} = -\omega G b\frac{\Omega_2^2}{2}\sum_{n=0}^{\infty}(-1)^n b_n \left(\frac{d}{d\gamma}\left(\frac{1}{X_n(\gamma)}\right)\Bigg|_{\gamma=\gamma_m}\right)^{-1}\int_{-a}^{a}\overline{f}(x,\gamma)\cos\alpha_n x dx\,. \tag{2.2.8}$$

Expression (2.2.8) does not contain functions $d_n(\gamma)$, $D(\gamma)$ and permits to evaluate the power simply by solving the finite system of equations corresponding to (2.2.3).

3. Numerical Results

The method developed is applied for evaluation of the energy division among various modes excited by uniform normal loading of constant magnitude $F(x,z)=1$. Fourier representation of this function can be easily obtained, and according to (2.2.7) the period-average power for m mode is calculated as follows

$$W_y^{(m)} = -2\omega Gab\Omega_2^2 \frac{\sin \gamma_m d}{\gamma_m} \left(\frac{d}{d\gamma} \left(\frac{1}{X_0(\gamma)} \right) \Bigg|_{\gamma=\gamma_m} \right)^{-1} . \quad (3.1.1)$$

Let consider a square waveguide ($\nu = 0.25$), where the loading is distributed on portion $-1 \le z \le 1$ of faces $y = \pm 1$. Owing to diagonal symmetry of the cross-section, longitudinal modes separate into breathing *L*-modes, $U_x(x,y)=U_y(y,x)$, and screw S_1-modes, $U_x(x,y)=-U_y(y,x)$. Calculations were performed for each mode type retaining 10 equations in the finite system corresponding to (2.2.3). Comparatively small number of equations in the reduced system provides correct values for frequencies and propagation constants that qualitatively agree with experimental data [4]. Results obtained are shown in Fig. 2.1, $\Gamma = 2\gamma b / \pi$ vs $\Omega = 2\omega b / \pi c_2$. The dominant mode at each frequency is indicated by the heavier line superimposed on the dispersion branch. A bar over a symbol indicates that the reflection of a segment in the $\mathrm{Re}\,\Gamma = 0$ plane is shown; the segments correspond to backward waves with phase and group velocities of opposite sign.

Fig. 2.1 shows that most of the energy goes into one mode, but the dominant mode is not the same at all frequencies. So, in the frequency range $0 < \Omega \le 1.378$, all of the energy does into $L^{(1)}$-mode. At $\Omega = 1.379$, there appears $S_1^{(1)}$-mode (45% of the energy) and backward $\overline{S}_2^{(1)}$-mode (54% of the energy). At frequency $\Omega = \sqrt{2} = 1.414$ of the Lamé mode, 96% of all the energy goes into the backward wave, while $S_1^{(1)}$-mode transfers 3.3% and $L^{(1)}$-mode – 0.7% of all the energy, respectively. For $\Omega > \sqrt{2}$, the backward wave disappears, and most of the energy is transferred by $S_1^{(1)}$-mode. As the frequency increases, $L^{(1)}$-mode becomes more powerful; for $\Omega \ge 1.686$, it transfers more than a half of the energy, until $\overline{S}_3^{(1)}$- and $S_2^{(1)}$-modes appear. At higher frequencies, $L^{(1)}$-mode never dominates. The mode behavior described is similar for all other modes. Therefore, based on these calculations some general conclusions can be made. Each propagating modes becomes dominant shortly after it appears, that is most of the energy goes into modes with small propagation constants. Redistribution of the energy in the vicinity of cutoff frequencies is similar to that observed for Rayleigh-Lamb waves in the infinite plate under the same conditions of excitation [5].

Backward waves are dominant for most feasible frequencies. Therefore, uniform normal loading can be recommended for experimental investigations of these waves. However, this loading is inappropriate for effective excitation of an edge mode (the $S_1^{(1)}$-mode at high frequencies). Intensive motions in this mode are concentrated near the edges of the waveguide [4]. In a square waveguide excited by the uniform loading, the edge mode receives less than 1% of the energy.

The same conclusions can be drawn from results calculated by means of the semi-analytical finite element method and experimental data presented in [2] for a point loading on the face of a square waveguide.

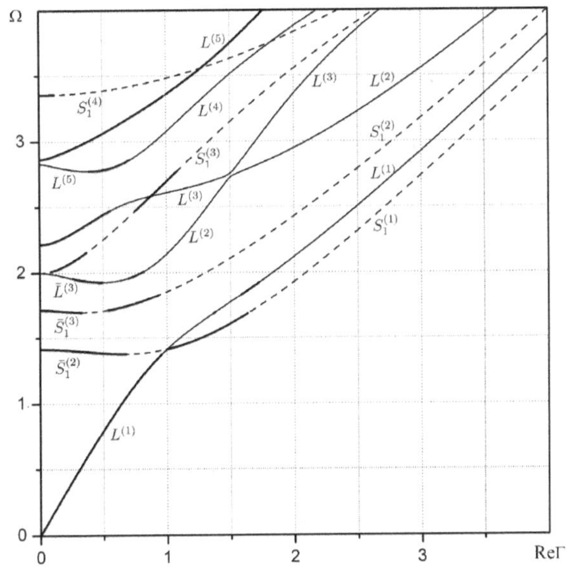

Fig. 2.1 Frequency spectrum for breathing L (—) and screw S_1 (---) modes in a square waveguide with dominant mode indicated, $\nu = 0.25$.

References

[1] Cegla, F.B.: Energy concentration at the center of large aspect ratio rectangular waveguides at high frequencies. *J. Acoust. Soc. Amer.* **123** 4218-4226 (2008).

[2] Hayashi, T., Song, W.J., Rose, J.L.: Guided wave dispersion curves for a bar with an arbitrary cross-section, a rod and rail example. *Ultrasonics* **41** 175–183 (2003).

[3] Gunawan, A., Hirose, S.: Boundary element analysis of guided waves in a bar with an arbitrary cross-section. *Eng. Anal. Bound. Elem.* **29** 913-924 (2005).

[4] Bondarenko, A.A.: Normal wave propagation in a rectangular elastic waveguide. In: *CD-ROM Proceedings of the ICTAM 2008* No. **10932** (2008).

[5] Torvik, P.J., McClatchey, J.J.: Response of an elastic plate to a cyclic longitudinal force. *J. Acoust. Soc. Amer.* **44** 59-64 (1968).

Normal Waves in Anisotropic Cylinders of Sector Cross-Section

Storozhev Valeriy, Troyan Renata and Puzyrev Vladimir

Mathematical Department, Donetsk National University, Universitetskaya 24, Donetsk, 83001, Ukraine

renata_shev@rambler.ru

Abstract. The work is devoted to theoretical research of normal elastic and electroelastic waves in anisotropic cylindrical waveguides class C_{6v} hexagonal systems of not typical geometry cross-section from piezoactive and nonpiezoactive materials. Waveguides of the circular sections with sector cuts of any sizes and waveguides of the concentric ring sections with sector cuts are considered. Possibilities of using section geometry for changing structure and properties of spectrum of traveling and edge stationary normal waves are investigated. The approach is based on exact analytical integration of the wave dynamics equations for linear elastic and electroelastic medium of hexagonal system with the generalized wave potentials introduction and reducing the considered problem to a spectrum boundary problem in the section of a waveguide. The potentials expressed through special cylindrical functions of different types with fractional indexes are used for satisfaction the edge conditions on boundary parts of section.

1. Introduction

The problem of normal elastic and electroelastic waves propagation in anisotropic cylinders from materials of hexagonal systems with the canonical circular or ring form of section is investigated in works [1-3] on many aspects. The dispersion spectrums of normal elastic and electroelastic waves are investigated for cylindrical waveguides with non classic sections to a lesser degree. In particular, this problem is considered in works [4-7]. In this work the method and the analysis of the normal waves in cylindrical waveguides with not canonical section, particularly in circular and ring waveguides with sector cut of any angular measure in section are presented. The carried out researches are of interest for search of new fundamental laws and specific effects in waveguide properties of cylinders anisotropic in an axial direction. The section geometry variation is considered to be the mechanism of dispersion spectrum properties controlling, such as values of critical frequencies and distribution of different free traveling waves modes, forms of wave displacement, types of high-frequency short-wave localization. The laws found out as a result of this research can be used at hypersonic techniques designing.

T.-T. Wu and C.-C. Ma (eds.), *IUTAM Symposium on Recent Advances of Acoustic Waves in Solids*, IUTAM Bookseries 26, DOI 10.1007/978-90-481-9893-1_36,

2. Formulation

2.1 Geometrical and Material Description of the Waveguides

Extended transverse-isotropic cylindrical waveguides with longitudinal axis of elastic symmetry are considered. The sector cut of any angular measure is contained in circular or concentric ring section. The waveguides in dimensionless cylindrical coordinates system (r,θ,z) occupy areas $V_1=\{0\le r\le R,-\alpha\le\theta\le\alpha,-\infty<z<\infty\}$ or $V_2=\{R_1\le r\le R_2,-\alpha\le\theta\le\alpha,-\infty<z<\infty\}$. The complicate contour $\Gamma=\Gamma_-\cup\Gamma_+\cup\Gamma_0$ of waveguide V_1 section is the superposition of lines $\Gamma_0=\{ r=R, -\alpha\le\theta\le\alpha\}$, $\Gamma_+=\{ 0\le r\le R_1, \theta=\alpha\}$, $\Gamma_-=\{ 0\le r\le R_1, \ \theta=-\alpha\}$. The contour of waveguide V_2 section is $\tilde{\Gamma}=\tilde{\Gamma}_-\cup\tilde{\Gamma}_+\cup\tilde{\Gamma}_{01}\cup\tilde{\Gamma}_{02}$, where $\tilde{\Gamma}_{0i}= \ =\{ r=R_i, -\alpha\le\theta\le\alpha\}$, $\tilde{\Gamma}_+=\{ R_1\le r\le R_2, \theta=\alpha\}$, $\tilde{\Gamma}_-=\{ R_1\le r\le R_2, \ \theta=-\alpha\}$.

The analysis of full dispersion spectrum for considered piezoactive waveguides is based on exact analytical integration of the wave dynamics equations of linear electroelastic hexagonal system medium. They give amplitude functions $u_j^{(0)}$, $\varphi^{(0)}$ in elastic displacements vector components $u_j=\mathrm{Re}\left(u_j^{(0)}\cdot\exp\left(-i\left(\omega t-\tilde{k}x_3\right)\right)\right)$ and quasistatic electric field potential $\varphi=\mathrm{Re}(\varphi^{(0)}\cdot\exp\left(-i\left(\omega t-\tilde{k}x_3\right)\right))$ for harmonical electroelastic waves. In the matrix-operational form the stationary motion equations are given by

$$\left\|L_{ij}\right\|\overline{\Phi^{(0)}}=0\ \left(i,j=\overline{1,4}\right), \tag{2.1.1}$$

where $\overline{\Phi^{(0)}}=(u_1^{(0)},u_2^{(0)},u_3^{(0)},\varphi^{(0)})^T$, $L_{11}=c_{11}\partial_1^2+c_{66}\partial_2^2-c_{44}k^2+\Omega^2$,
$L_{12}=L_{21}=(c_{12}+c_{66})\partial_1\partial_2$, $L_{13}=L_{31}=ik(c_{13}+c_{44})\partial_1$, $L_{14}=-L_{41}=-ik(e_{31}+e_{15})\partial_1$,
$L_{22}=c_{66}\partial_1^2+c_{11}\partial_2^2-c_{44}k^2+\Omega^2$, $L_{23}=L_{32}=ik(c_{13}+c_{44})\partial_2$,
$L_{24}=-L_{42}=-ik(e_{31}+e_{15})\partial_2$, $L_{33}=c_{44}\partial_1^2+c_{44}\partial_2^2-c_{33}k^2+\Omega^2$,
$L_{34}=-L_{43}=-e_{15}(\partial_1^2+\partial_2^2)+e_{33}k^2$, $L_{44}=\varepsilon_{11}(\partial_1^2+\partial_2^2)-\varepsilon_{33}k^2$, $\partial_j=\partial/\partial x_j$, $k=\tilde{k}R_*$, $\Omega=\omega R_*(\rho/c_*)^{1/2}$, c_{pj} – dimensionless related to c_* elastic constants, ρ – density, e_{pj} – dimensionless related to c_*/E_* piezoelectric constants, ε_{pj} – dimensionless related to $\varepsilon_*=c_*/E_*^2$ dielectric constants of piezoactive waveguide material, R_* - normalizing parameter for u_j, x_j.

The system of the wave amplitude equations for nonpiezoactive waveguide is given by

$$\left\|L_{ij}\right\|\overline{\Psi^{(0)}}=0\ \left(i,j=\overline{1,3}\right),\ \overline{\Psi^{(0)}}=(u_1^{(0)},u_2^{(0)},u_3^{(0)})^T. \tag{2.1.2}$$

Some variants of boundary conditions are considered. The boundary conditions on the free surface Γ_0 with electrodes coverings for piezoactive waveguides of circular section are given by:

$$(u_r)_{\Gamma_0}=(u_\theta)_{\Gamma_0}=(u_z)_{\Gamma_0}=(\varphi)_{\Gamma_0}=0. \tag{2.1.3}$$

The conditions on the fixed surfaces Γ_0 with electrodes coverings are of the form

$$(\sigma_{rr})_{\Gamma_0} = (\sigma_{r\theta})_{\Gamma_0} = (\sigma_{rz})_{\Gamma_0} = (\varphi)_{\Gamma_0} = 0. \tag{2.1.4}$$

The surfaces of sector cuts $\Gamma_\pm$ have thin flexible not extensible electroconductive coverings and corresponding conditions are given by

$$(\sigma_{\theta\theta})_{\Gamma_\pm} = (u_r)_{\Gamma_\pm} = (u_z)_{\Gamma_\pm} = (\varphi)_{\Gamma_\pm} = 0. \tag{2.1.5}$$

The conditions on the free internal and external cylindrical parts of boundary with electrodes coverings for piezoactive waveguides of ring section are of the form:

$$\begin{aligned}&(\sigma_{rr})_{\Gamma_j} = (\sigma_{r\theta})_{\Gamma_j} = (\sigma_{rz})_{\Gamma_j} = (\varphi)_{\Gamma_j} = 0,\\ &(\sigma_{\theta\theta})_{\Gamma_\pm} = (u_r)_{\Gamma_\pm} = (u_z)_{\Gamma_\pm} = (\varphi)_{\Gamma_\pm} = 0.\end{aligned} \tag{2.1.6}$$

2.2 Obtaining an Analytic Form of Dispersing Relations

By investigations waves conventionally are shared on the symmetric (S-type), when $u_r(r,-\theta) = u_r(r,\theta)$, $u_\theta(r,-\theta) = -u_\theta(r,\theta)$, $u_z(r,-\theta) = u_z(r,\theta)$, $\varphi(r,\theta) = \varphi(r,-\theta)$ and antisymmetric (A-type) with $u_r(r,-\theta) = -u_r(r,\theta)$, $u_\theta(r,-\theta) = u_\theta(r,\theta)$, $u_z(r,-\theta) = -u_z(r,\theta)$, $\varphi(r,\theta) = -\varphi(r,-\theta)$. At a stage of construction the basic dispersion relation for full spectrum of the normal elastic and electroelastic waves, connected with the systems of the wave equations (2.1.1) (2.1.2) integration, the technique of generalized wave potentials φ_j introduction is used. In piezoactive waveguides [3] generalized potentials φ_j are related to complex amplitude functions of wave displacements $u_j^0(x_1,x_2,x_3)$ with formulas

$$u_1^{(0)} = \partial_1\varphi_1 + \partial_2\varphi_4, \quad u_2^{(0)} = \partial_2\varphi_1 - \partial_1\varphi_4, \quad u_3^{(0)} \equiv \varphi_2, \quad \varphi^{(0)} \equiv \varphi_3, \tag{2.2.1}$$

where $\varphi_p = \sum_{j=1}^{3} \beta_{pj} F_j(x_1,x_2)$ $(p=\overline{1,3})$, $\varphi_4 \equiv F_4(x_1,x_2)$; functions F_j are defined from the metaharmonic equations $D^2 F_j + \gamma_j^2 F_j = 0$ $\left(j=\overline{1,4}\right)$.

At research of S-waves in piezoactive waveguides of circular section metaharmonical functions are introduced in the form:

$$\begin{aligned}&F_j(r,\theta) = \sum_{n=0}^{\infty} A_{nj}^{+} J_{v_n^+}(\gamma_j r)\cos(v_n^+\theta) \ (j=\overline{1,3}),\\ &F_4(r,\theta) = \sum_{n=0}^{\infty} A_{n4}^{+} J_{v_n^+}(\gamma_4 r)\sin(v_n^+\theta),\end{aligned} \tag{2.2.2}$$

where $v_n^+ = (2n+1)\pi/2\alpha$; $J_{v_n^+}(\gamma_j r)$ - cylindrical Bessel functions of the first kind. They provide exact satisfaction to edge conditions on parts of a boundary surface $\Gamma_\pm$ of sector cut. For S-waves in waveguides of ring section functions F_j are of the form

$$F_j(r,\theta)=\sum_{n=0}^{\infty}(A_{nj}^{+}J_{\nu_n^+}(\gamma_j r)+B_{nj}^{+}Y_{\nu_n^+}(\gamma_j r))\cos(\nu_n^+\theta) \quad (j=\overline{1,3}),$$

$$F_4(r,\theta)=\sum_{n=0}^{\infty}(A_{n4}^{+}J_{\nu_n^+}(\gamma_4 r)+B_{n4}^{+}Y_{\nu_n^+}(\gamma_4 r))\sin(\nu_n^+\theta). \qquad (2.2.3)$$

It is shown that for expressions of stresses amplitude functions the singularities occur in a point $r=0$ - top of sector cut due to the presence of a terms kind $r^{-\alpha}J_{\nu_n^\pm}(\gamma_j r)$ if $\nu_n^\pm-\alpha<0$ in corresponding expressions.

The equations for obtaining critical frequencies of normal waves and the dispersion equations are received from edge conditions (2.1.3-2.1.6) by using method of orthogonal series. In particular, the sequence of dispersion equations for waveguides of circular section in case of S-waves are given by

$$\Theta_n^{(u)}(\Omega,k)=\begin{vmatrix}\Delta_r^{(u,1,n)} & \Delta_r^{(u,2,n)} & \Delta_r^{(u,3,n)} & \Delta_r^{(u,4,n)}\\ \Delta_\theta^{(u,1,n)} & \Delta_\theta^{(u,2,n)} & \Delta_\theta^{(u,3,n)} & \Delta_\theta^{(u,4,n)}\\ \Delta_z^{(u,1,n)} & \Delta_z^{(u,2,n)} & \Delta_z^{(u,3,n)} & \Delta_z^{(u,4,n)}\\ \Delta_\varphi^{(u,1,n)} & \Delta_\varphi^{(u,2,n)} & \Delta_\varphi^{(u,3,n)} & \Delta_\varphi^{(u,4,n)}\end{vmatrix}=0 \quad (n=\overline{0,\infty}), \qquad (2.2.4)$$

where $\Delta_r^{(u,j,n)}=\beta_{1j}\left(\upsilon_n^+R^{-1}J_{\upsilon_n^+}(\gamma_j R)-\gamma_j J_{\upsilon_n^++1}(\gamma_j R)\right),\ldots,\ \Delta_\varphi^{(u,j,n)}=\beta_{3j}J_{\upsilon_n^+}(\gamma_j R)$.

Each of these dispersion equations describes independent partial spectrum of normal elastic or electroelastic waves with certain circle wave number, and the general spectrum is formed as superposition of all partial spectrums. This result has analogy to structure of the general dispersion spectrum of normal waves in waveguides with circular and concentric ring sections.

3. Numerical Results

In this section numerical results of the received dispersion equations, and also expressions for traveling and edge standing elastic and electroelastic normal waves in waveguides from ceramics $BaCaTiO_3$ with normalized material constants c_{11}=15.8, c_{12}=6.9, c_{13}=6.75, c_{33}=15.0, c_{44}=4.5, $e_{31}=-3.1$, $e_{33}=13.5$, $e_{15}=10.9$, $\varepsilon_{11}=88.5$, $\varepsilon_{33}=80.5$ and parameters $c_*=10^{10}\ N/m^2$, $E_*=10^{10}\ N/K$, $R_*=1.0$ m are presented. The basic properties of critical frequencies distributions are investigated. Established that it is possible to receive various alternation orders and concentration of critical frequencies distribution for traveling normal waves by angular sizes variation of circular waveguides section.

The researches of high-frequency short-wave localization of normal elastic and electroelastic traveling waves are held. An asymptotic for phase velocities of all modes from various partial spectrums is a volume elastic shear waves velocity in an axis of mechanical properties symmetry direction in case of the fixed cylindrical boundary surface of a circular nonpiezoactive waveguide.

An asymptotic for phase velocity of lowest mode waves is the phase velocity of the generalized surface Rayleigh wave along an axis of elastic properties symmetry in anisotropic semispace of class C_{6v} hexagonal systems with a free surface in case of a free cylindrical boundary surface (Fig 3.3).

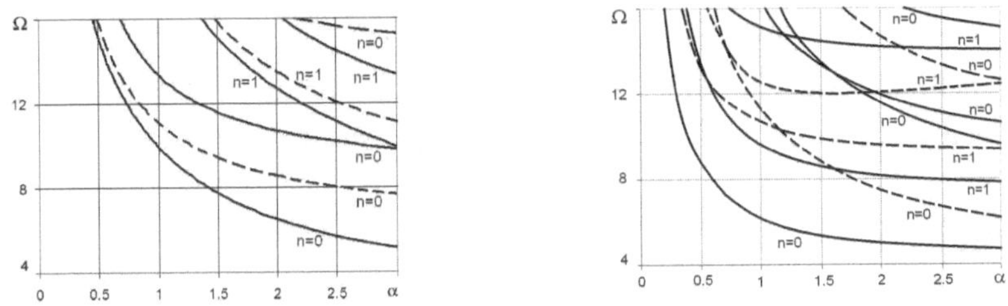

(a) piezoactive circular waveguide. Solid lines represent critical frequencies for waves of lateral type, dotted lines represent critical frequencies for electroelastic waves of longitudinal type.

(b) nonpiezoactive ring waveguide. Solid lines represent critical frequencies for waves of lateral type, dotted lines represent critical frequencies for waves of longitudinal type.

Fig. 3.1 S-waves critical frequencies distributions in waveguides with a fixed cylindrical surface.

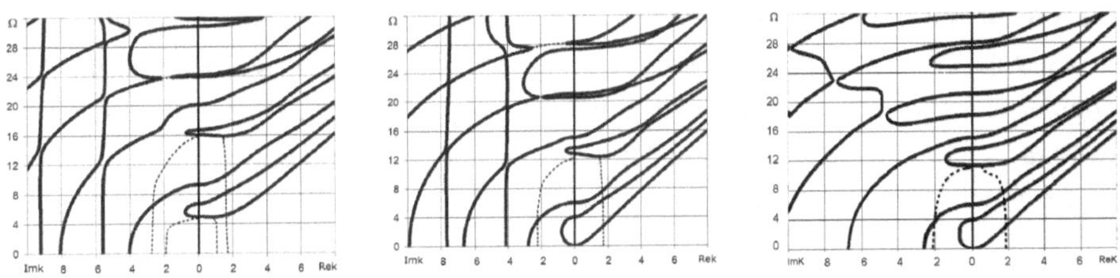

(a) piezoactive, waveguide, $\alpha = \pi / 4$

(b) piezoactive, waveguide, $\alpha = \pi / 2$

(c) nonpiezoactive, waveguide, $\alpha = \pi / 2$

Fig. 3.2 S-waves dispersion spectrums in circular waveguides with a free cylindrical surface, $n = 0$.

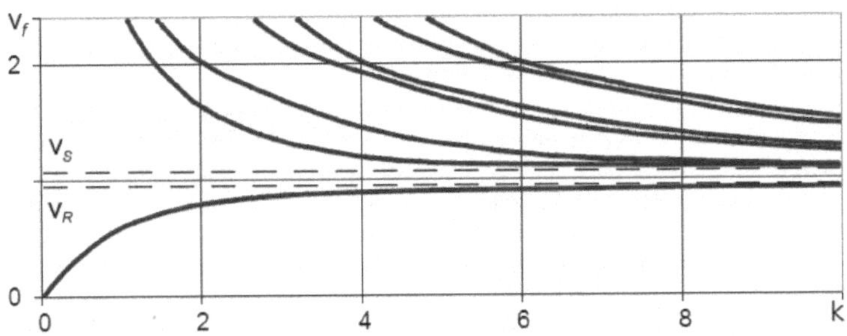

Fig. 3.3 S-waves normalized phase velocities in a waveguide of circular section with a free cylindrical surface, $\alpha = \pi/2$, $n = 0$.

The change in the described laws is in tending phase velocity of lowest modes waves from all partial spectrum to velocity of an electroelastic Rayleigh wave along a corresponding direction in piezoactive semispace for piezoactive waveguides. Asymptotic velocities for waves of all other modes are the velocities of purely elastic shear waves in the aforementioned direction. The asymptotic phase velocity tending to the elastic or electroelastic surface Rayleigh waves velocity is observed at two lowest modes from partial dispersion spectrum, and asymptotical velocity of other modes waves is the velocity of a volume elastic shear wave for ring waveguides.

The obtained results can find application by working out methods of optimum excitation and reception the signals transferred by investigated waves.

4. Conclusion

Theoretical research of normal elastic and electroelastic waves in anisotropic cylindrical waveguides class C_{6v} hexagonal systems of not typical geometry cross-section from piezoactive and nonpiezoactive materials had been conducted. The approach was based on exact analytical integration of the wave dynamics equations with the generalized wave potentials introduction. Numerical results were investigated for the waveguides from ceramics $BaCaTiO_3$. It was shown, that the angular sizes variation is the mechanism of dispersion spectrum properties controlling, such as values of critical frequencies and distribution of free traveling waves modes, types of high-frequency short-wave localization.

References

[1] Mirsky, J.: Wave propagation in transversely isotropic circular cylinder. *J. Acoust. Soc. Amer.* **37** 1022-1026 (1965).

[2] Suncheleyev, R.Ya.: Dynamic problems for transverse-isotropic elastic cylinder. *Prikl. Math. and Mech.* **30** 38-44 (1966) (in russian).

[3] Shulga, N.A.: Propagation of harmonic waves in anisotropic piezoelectric cylinders. Homogeneous piezoceramics waveguides. *Int. Appl. Mech.* **38** 933-953 (2002).

[4] Storozhev, V.I.: Numerically-analytical method of elastic waves investigation in orthotropic cylinder of nonlinear section. *Theoret. and Applied Mech.* **25** 97-102 (1995) (in russian).

[5] Shulga, N.A.: Propagation of harmonic waves in anisotropic piezoelectric cylinders. Waveguides with complicated properties. *Int. Appl. Mech.* **38** 1440-1463 (2002).

[6] Damljanovic, V., Weaver, R.L.: Propagating and evanescent elastic waves in cylindrical waveguides of arbitrary cross-section. *J. Acoust. Soc. Amer.* **115** 1572-1581 (2004).

[7] Bartoli, I., Marzania, A., Scalea, F. L., Viola, E.: Modeling wave propagation in damped waveguides of arbitrary cross-section. *J. Sound Vib.* **295** 658-707 (2006).

Dispersion Behaviors of ASF Modes Propagating along Wedges Tips with Coatings

Po-Shien Tung[1], Sheng-Wei Tang[2] and Che-Hua Yang[1]

[1]Graduate Institute of Manufacturing Technology, National Taipei University of Technology, Taipei 106, Taiwan.

[2]Graduate Institute of Mechanical and Electrical Engineering, National Taipei University of Technology, Taipei 106, Taiwan.

chyang@ntut.edu.tw

Abstract. Acoustic sensors frequently rely on the detection of small mass changes that results from binding of a coated layer coupled to the active sensor surface. Anti-symmetric flexural (ASF) modes are anti-symmetric type of guided waves propagating along the tip of wedge-shaped waveguides. Like surface acoustic waves, ASF can be a potential candidate for the purpose of acting as acoustic sensors. This paper employs a combined numerical and experimental study for the dispersion behaviors of ASF modes propagating along wedge tips with a layer of coating. The numerical study is based on finite element method while the experimental study employs a laser ultrasound technique. Velocity for the ASF mode propagating along the wedge tip with a slower coating is found out to start at the ASF velocity of the matrix wedge at the low frequency regime, and gradually influenced by the coating as the frequency increases. Enhanced loaded effects from the coating are founded due to the wedge tip geometry. In general, the numerical results show good agreement with the measurements.

1. Introduction

Surface acoustic wave (SAW) sensors are frequently used as physical, chemical or biological sensors. The sensor principle is based on the detection of small mass changes that result from the binding of a coated layer coupled to the active sensor surface. The propagation velocity of acoustic waves traveling through a layer at the sensor surface is very sensitive to additional mass loading. Thus, the change in the velocity is a measure of small mass changes at the sensor surface [1].

T.-T. Wu and C.-C. Ma (eds.), *IUTAM Symposium on Recent Advances of Acoustic Waves in Solids*, IUTAM Bookseries 26, DOI 10.1007/978-90-481-9893-1_37,

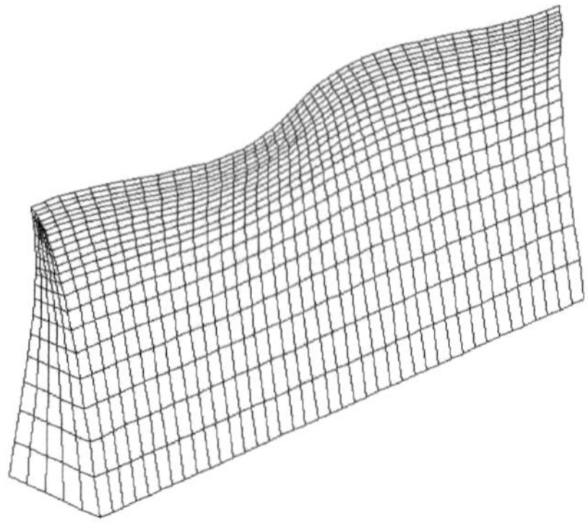

Fig. 1 Wave motion pattern of an ASF mode.

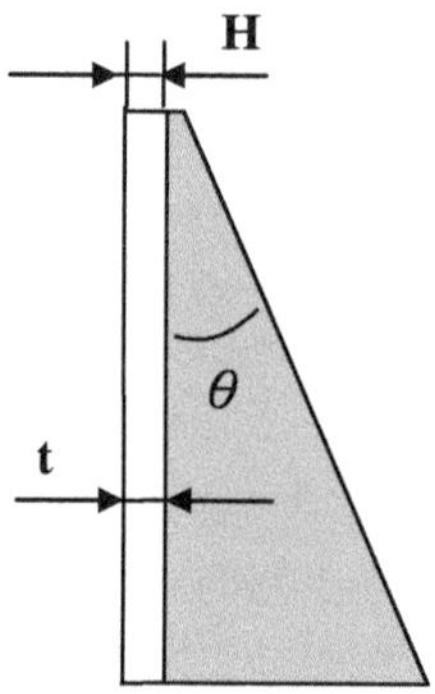

Fig. 2 A schematic showing the geometry of a coated wedge tip.

Discovered by Lagasse and coworkers [2, 3] in the early 1970's through a numerical study, wedge waves (WWs) are guided acoustic waves that propagate along the tip of a wedge, and their energy is tightly confined near the apex. As shown in Fig. 1, like Lamb waves, WWs with a displacement field that is anti-symmetric about the mid-apex plane are called anti-symmetric flexural (ASF) modes. To date, no elasticity-based exact solution has been available for WW problems. By assuming the wedge to be a thin plate of variable thickness, McKenna et al. obtained a theoretical approximation for the prediction of the dispersion relation of a truncated wedge. [4] Also, Krylov and coworkers used a geometric acoustic approximation [5] and [6] to obtain the phase velocity of ASF modes. Experimental works employing piezoelectric transducers, miniature non-contact electromagnetic acoustic transducers (EMAT) [7] and [8] and laser ultrasound techniques [9] – [11] have been conducted to investigate different aspects of ASF modes, including the influences of apex angles, apex truncation [10], fluid loading effects, and the effect of curvature of the wedge apex [11].

For a flat matrix, the effects of a thin film coating on the propagating behaviors of guided waves are well known [12], however, not for the case of ASF modes. This paper describes a combined numerical and experimental investigation for the dispersion behaviors of ASF modes propagating along wedges with a layer of thin coating.

2. Materials and Samples

Two wedge samples are studied in this research; one without coating and one with a layer of coating on one side of the wedge tip. The wedge sample without coating is called a general wedge (GW), and the wedge sample with coating is called coated wedge (CW). Fig. 2 shows the geometry of the wedge tip with a layer of coating. Here, the apex angle is denoted as θ, the apex truncation is denoted as H, and the thickness of the coating is denoted as t. The aluminum GW sample with an apex angle of 30° is designated as Al30. The CW sample, with aluminum matrix coated with copper film, is designated as AlCu30. Apex truncations (H) for the wedge samples are 8.4 μm for the Al30 sample and 11.2 μm for the AlCu30 sample. These truncations are small compared with typical wavelengths of the investigated ASF modes.

3. Laser Ultrasound Measurements

A laser-generation/laser-detection ultrasound technique (LUT) is used for the measurements of ASF dispersion spectra. As shown in Fig. 3, the experimental configuration consists of a pulsed laser for ultrasonic wave generation and a laser optical probe to detect the acoustic waves. The excitation source is a Nd:YAG laser with a power of approximately 100 mJ, a wavelength of 532 nm, and a pulse duration of 6.6 ns. A commercial heterodyne Doppler laser optical receiver (OFV 511 and OFV 2700, Polytec, Germany) is applied to detect the ASF modes propagating along the tips of wedges. A B-scan scheme is used for the measurement of the dispersion behaviors of WWs. During scanning, the optical detector is located at a fixed point, while the generation laser beam is scanned along the wedge tip. Along the wedge tip, there are 200 scanning steps with a step size of 10 mm. A two-dimensional fast Fourier transform [13], first taken with respect to time and then with respect to the scanned position, is used to obtain the dispersion relation from the B-scan data. A more detailed description for the experimental technique and signal processing scheme can be found in [11].

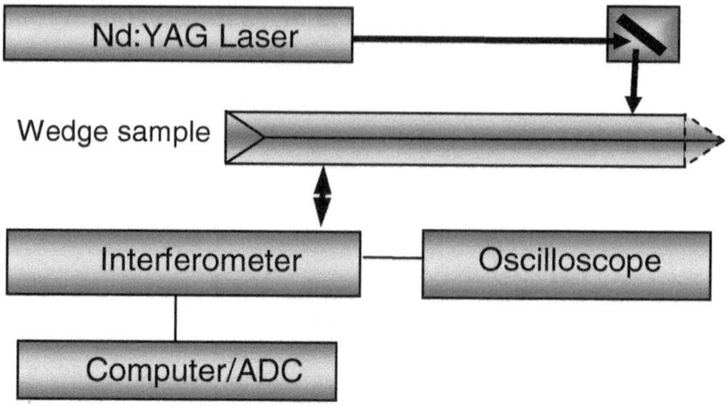

Fig. 3 A schematic showing the experimental configuration of LUT.

4. Results and Discussions

Dispersion curves of ASF modes are first presented for the Al30 sample, an aluminum wedge without coating. Fig. 4 shows the FE-simulated and LUT-measured ASF dispersion spectra for the Al30 sample. The three ASF modes are labeled as A_1, A_2, and A_3, respectively. For the Al30 sample with a small truncation, an FE mesh without apex truncation is used as a close approximation. Without apex truncation, the ASF modes are known to be free of dispersion, which is also observed for the FE results in Fig. 4. Since these ASF modes are dispersion free, velocities for the three ASF modes are constant values over the investigated frequency range. Limited bandwidths without whole coverage of the whole investigated frequency range are observed for the three ASF modes. For example, the measured A_1 mode covers a bandwidth from 0.2 MHz to 1.8 MHz, and 2.0 MHz to 4.2 MHz for the A_2 mode. The limited bandwidths can due to the laser excitation mechanism and also from the filtering effect of the signal processing scheme. Also, the measured ASF modes are nearly dispersion free, however, show minor dispersion with phase velocities slightly increase as the frequency increases due to the 8.4 μm truncation. In general, the FE-modeled ASF velocities agree well with the LUT measurements while the FE-simulated ASF modes are extended in the investigated frequency range.

For a wedge with an apex truncation of 11.2 μm as the AlCu30 sample, a slight positive dispersion, or increasing velocity as frequency increases, is expected if it were the case of a common general wedge. However, while the wedge is coated a layer of film with slower velocity, the ASF dispersion becomes obviously negative or decreasing velocity as frequency increases. Fig. 5 shows the measured and modeled dispersion curves for the AlCu30, a faster aluminum wedge matrix coated with a layer of slower copper film on a side of its wedge tip. The three ASF modes for the AlCu30 are labeled CA_1 CA_2 and CA_3. ASF velocities for the coated wedge start at the ASF mode of the same order for the matrix wedge at the low frequency regime, and gradually influenced by the coating while frequency increases. Like the case of surface acoustic waves propagating along a flat surface with thin coating, loaded phenomenon are observed for the ASF modes in coated wedges. Moreover, enhanced loaded effects due to the wedge tip geometry are founded in the frequency range where the ASF wavelength is much larger than the film thickness. While the negative dispersion trends are considered for each of the CW/ASF modes, the FE-modeled dispersion spectra show good agreement with the measurement. The measured CW/ASF modes show limited bandwidths just like the CW case.

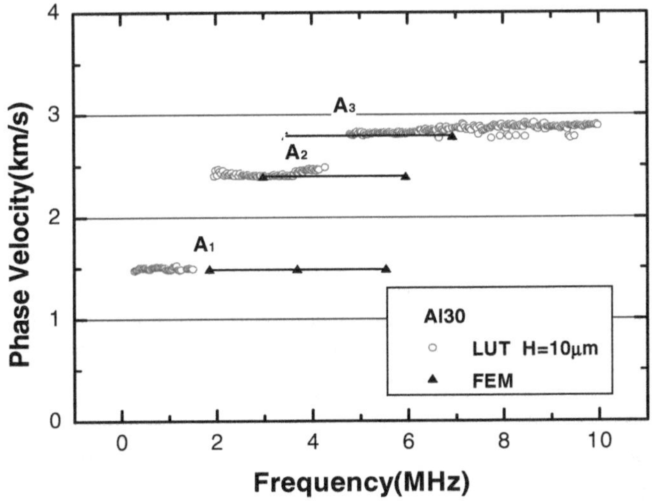

Fig. 4 Simulated and measured dispersion spectra for the Al30 sample.

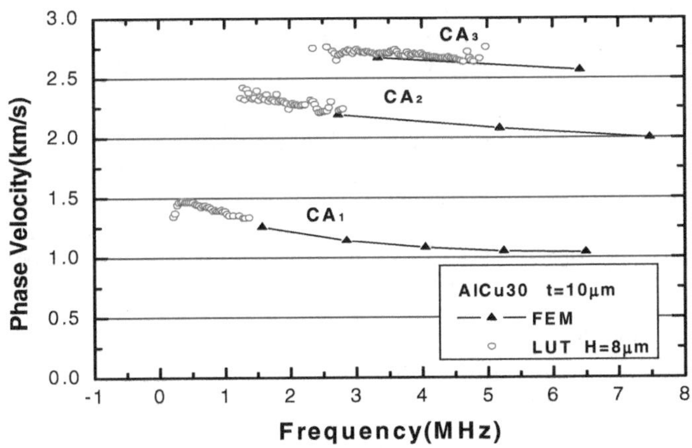

Fig. 5 Simulated and measured dispersion spectra for the AlCu30 sample.

5. Conclusions

Dispersion behaviors of anti-symmetric flexural (ASF) modes traveling in coating wedges (CW), and their relation with those of general wedges (GW), were investigated with a combined numerical and experimental study based on a laser ultrasound technique. Dispersion behaviors were observed for ASF modes propagating along the tips of a small truncation wedge with one of its surfaces coated with a layer. Transition behaviors of the ASF modes from those of a GW to those of a CW were characterized. Similar to the case of a flat matrix coated with a thin film, the phase velocity of the CW/ASF mode gradually approaches the velocity of the coated layer as frequency increase. Enhanced loaded effects due to the wedge tip geometry are founded in the frequency range where the ASF wavelength is much larger than the film thickness.

References

[1] Talbi, A., Sarry, F., Elhakiki, M., Brizoual, L.Le.: ZnO/quartz structure potentiality for surface acoustic wave pressure sensor. *Sens. Actuators A* **128** 78-83 (2006).

[2] Lagasse, P.E.: Analysis of dispersion free guide for elastic waves. *Electron. Lett.* **8** 372 (1972).

[3] Lagasse, P.E., Mason I.M., Ash E.A.: Microwave Theory and Technique, *IEEE Trans. Sonics Ultrason.* **20** 143-154 (1973).

[4] McKenna, J., Boyd, G.D., Thurston R.N.: Plate theory solution for guided flexural acoustic waves along the tip of a wedge, *IEEE Trans. Sonics and Ultrasonics.* **21** 178-186 (1974).

[5] Krylov, V.V., Shanin, A.V.: Scattering of an acoustic wedge wave by a shallow notch. *Sov. Phys. Acoust.* **37** 65 (1991).

[6] Krylov, V.V., Parker, D.F.: Harmonic generation and parametric mixing in wedge acoustic waves. *Wave Motion.* **15** 185 (1992).

[7] Chamuel, J.R.: Edge waves along immersed elastic elliptical wedge with range dependent apex angle. *J. Acoust. Soc. Am.* **95** 2893 (1994).

[8] Chamuel, J.R.: Flexural edge waves along free and immersed elastic wave guides. *Review of Progress in Quantitative Nondestructive Evaluation* **16** 129-136 (1997).

[9] Jia, X., De Billy, M.: Observation of the dispersion behavior of surface acoustic waves in a wedge waveguide by laser ultrasonics. *Appl. Phys. Lett.* **61** 2970-2972 (1992).

[10] Yang, C.-H., Tsai, K.-Y.: Assessment of anisotropy and Apex Truncation Effects of the Piezoelectric Wedges. *Jpn. J. Appl. Phys.* **43** 4392 (2004).

[11] Yang, C-.H., Liaw, J.-S.: A comparison laser ultrasound measurements and finite-element simulations for dispersion behavior of anti-symmetric flexural modes propagating along wedge tips with coatings. *Jpn. J. Appl. Phys.* **39** 2741 (2000).

[12] Farnell, G.W., Adler, E.L. *Physical Acoustics*, Mason, W.P., Thurston R.N., (eds.), 9, 35 Academic Press, New York (1972).

[13] Alleyne, D.N., Cawley, P.: A two-dimensional fourier transform method for the measurement of propagating multimode signals. *J. Acoust. Soc. Am.* **89** 159 (1991).

Scattering from a Rectangular Crack in a Cladding

Per-Åke Jansson

Department of Applied Mechanics, Chalmers University of Technology, Gothenburg, Sweden

per-ake.jansson@chalmers.se

Abstract. Ultrasonic testing of a thick plate with a rectangular crack in a cladding is modelled analytically for a fully three-dimensional case. The effects of a corrugated interface and anisotropy are both taken into account. The wave scattering problem is solved by a hypersingular integral equation approach. A model for the ultrasonic transmitter and receiver is also included. Some preliminary numerical results for the change in signal response due to the presence of the crack are given.

1. Introduction

A situation that frequently occurs in the nuclear power industry is that of a thick plate or pipe with an austenitic cladding to reduce or prevent corrosion. A common method to apply the cladding is to use a welding process. From an ultrasonic testing point of view this leads to some complications. The cladding material is anisotropic, and the interface between the cladding and the base material is usually corrugated.

For several reasons it is desirable to develop a mathematical model for ultrasonic testing of a component with a crack in a cladding. Such a model can be useful for planning and qualification of testing procedures, and to increase the understanding of the influence of the cladding. Besides, using a good mathematical model is considerably cheaper than experiments, in particular when parameter studies are needed.

The aim of this paper is to develop a fully three-dimensional analytical model for a thick plate with a rectangular crack in a cladding taking the effects of a periodic interface as well as anisotropy into account. Previously, wave propagation in similar structures, with and without cracks, has been studied in [1-2].

T.-T. Wu and C.-C. Ma (eds.), *IUTAM Symposium on Recent Advances of Acoustic Waves in Solids*, IUTAM Bookseries 26, DOI 10.1007/978-90-481-9893-1_38,

In this paper the fully three-dimensional problem is solved exactly in the sense that no restrictions are imposed on the frequency or the shape of the periodic interface, although numerical results are only given for a sinusoidal surface. The results are believed to be valid, at least qualitatively, not only for a plate but also for a thick-walled pipe, as long as the radius of the pipe is sufficiently large.

2. Problem Formulation

The geometry of the wave propagation problem is depicted in Fig.2.1. A thick plate is composed of two layers of different generally anisotropic materials in welded contact. The crystal axes may be arbitrarily oriented. The interface S_0 is assumed to be periodic with a period a. This is believed to be a reasonable approximation, since real interfaces are more or less periodic. The layer of thickness d_1 is the base material. The cladding of thickness d_2 contains a rectangular crack S_C (sides $2c_1$ and $2\ c_2$), which may be tilted arbitrarily with respect to the back wall S_2. An ultrasonic transmitter is placed on the free surface S_1 of the base material. All numerical examples given here are for a transducer working in pulse-echo mode, even though the analysis is valid for the case of tandem inspection as well. Only the case of a transducer working at a fixed angular frequency ω will be considered, although it is straightforward, but somewhat time-consuming, to obtain results in the time domain.

For time-harmonic conditions the displacement fields u_j^i in the two materials i = 1, 2 satisfy

$$\frac{\partial}{\partial x_m}\sigma_{mj}^i + \rho^i \omega^2 u_j^i = 0, \tag{2.1}$$

where ρ^i is the corresponding density, and σ_{mj}^i is the stress tensor, which is related to the displacement by the constitutive relation

$$\sigma_{mj}^i = c_{mjm'j'}^i \frac{\partial}{\partial x_{j'}} u_{m'}^i. \tag{2.2}$$

The boundary conditions to be satisfied are $t_j = \sigma_{kj} n_k = 0$ on S_1 (except directly below the transducer), S_2, and S_C. Furthermore, u_j and t_j are continuous on the interface S_0, $x3 = s(x1)$, which is assumed to be periodic, in fact, it is taken as sinusoidal in the following. Actually, real interfaces are more or less sinusoidal. It should be pointed out, however, that there is no fundamental difficulty in choosing some other periodic function, as long as it is differentiable.

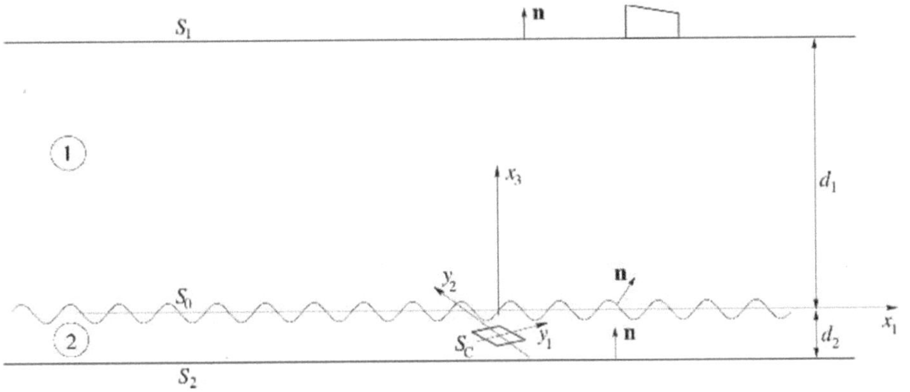

Fig. 2.1 The geometry of a plate with a crack inside a cladding.

3. The Green's Tensor for the Structure without a Crack

The first step of the solution is to determine the Green's tensor for the layered plate without a crack with a source in the cladding. The following integral representations are used:

$$\int_{S_1-S_0}\left[\Sigma_{mjl}(\mathbf{x};\mathbf{x}')G^1_{jl'}(\mathbf{x};\mathbf{x}'')-G_{jl}(\mathbf{x};\mathbf{x}')\Sigma^1_{mjl'}(\mathbf{x};\mathbf{x}'')\right]n_m\mathrm{d}S=\begin{cases}G_{l'l}(\mathbf{x}'';\mathbf{x}'), & \mathbf{x}''\text{in material 1,}\\ 0, & \mathbf{x}''\text{in material 2.}\end{cases} \tag{3.1}$$

$$\int_{S_0-S_2}\left[\Sigma_{mjl}(\mathbf{x};\mathbf{x}')G^2_{jl'}(\mathbf{x};\mathbf{x}'')-G_{jl}(\mathbf{x};\mathbf{x}')\Sigma^2_{mjl'}(\mathbf{x};\mathbf{x}'')\right]n_m\mathrm{d}S$$
$$=\begin{cases}-G^2_{l'l}(\mathbf{x}'';\mathbf{x}'), & \mathbf{x}''\text{in material 1,}\\ -G^2_{l'l}(\mathbf{x}'';\mathbf{x}')+G_{l'l}(\mathbf{x}'';\mathbf{x}'). & \mathbf{x}''\text{in material 2.}\end{cases} \tag{3.2}$$

Here $G_{jl}(\mathbf{x};\mathbf{x}')$ is the Green's tensor to be determined, $G^i_{jl}(\mathbf{x};\mathbf{x}')$ are halfspace Green's tensors for material $i=1,2$, and Σ^i_{mjl} are the corresponding stress tensors. Following Krasnova [1] the halfspace Green's tensors as well as the scattered field in material 2, *i.e.* $G_{l'l}-G^2_{l'l}$, are expanded as Fourier transforms in x_1 and x_2. In a similar way the surface fields G_{jl} and $\Sigma_{mjl}n_m$ at the interface can be expanded as Fourier transforms in x_1. Inserting the Fourier expansions for the Green's tensors and the surface fields into the integral representations, Eqs. (3.1)-(3.2), yields after some tedious algebra a set of simultaneous equations for the Fourier coefficients. Thereby the Green's tensor $G_{jl}(\mathbf{x};\mathbf{x}')$ is determined.

4. The Integral Equation for the Crack Opening Displacement

Starting from the Green's tensor derived in the previous section an integral representation for the crack opening displacement $\Delta u_j(\mathbf{x})$ can easily be derived. The result is

$$\int_{S_c} \Delta u_j(\mathbf{x})\Sigma_{mjl}(\mathbf{x};\mathbf{x}')n_m \mathrm{d}S_c = -u_l^{\mathrm{inc},2}(\mathbf{x}') + u_l^2(\mathbf{x}'). \tag{4.1}$$

Here $\Sigma_{mjl}(\mathbf{x};\mathbf{x}')$ is the Green's stress tensor for the structure without a crack, and $u_l^{\mathrm{inc},2}(\mathbf{x}')$ is the displacement field that would have existed in material 2 if the crack had not been there. This field was determined by Krasnova [1]

Transforming the Green's stress tensor to a coordinate system $y_1 y_2 y_3$, see Fig.1, operating with the traction operator with respect to the field point $\mathbf{x}'$ and taking the limit as y_3' approaches zero yield an integral equation for the crack opening displacement.

Next the crack opening displacement is expanded in Chebyshev functions ψ_m (see [2] for details) which exhibit, like the displacement field, a square root behavior at the crack tips:

$$\Delta u_j(y_1, y_2) = \sum_{m,m'=1}^{\infty} \gamma_{jmm'}\psi_m(y_1 / c_1)\psi_{m'}(y_2 / c_2), \tag{4.2}$$

Inserting the expression for Δu_j into the integral equation and projecting onto the Chebyshev functions yield a linear system of equations for the unknowns $\gamma_{jmm'}$. Hence, the crack opening displacement is determined.

5. The Signal Response

Next, the reciprocity result of Auld [3] is used to relate the crack opening displacement (COD) to the output voltage from the receiving probe:

$$\delta\Gamma = \frac{\mathrm{i}\omega}{4P}\int_{-c_1}^{c_1}\int_{-c_2}^{c_2} \Delta u_j(y_1, y_2)\sigma_{3j}^{\mathrm{re}}(y_1, y_2)\mathrm{d}y_1\mathrm{d}y_2, \tag{5.1}$$

where P is the incident electric power to the probe, Δu_j is the COD due to the incoming field, and $\sigma_{3j}^{\mathrm{re}}$ is the traction on the crack with the receiving probe acting as a transmitter in the absence of the crack. The quantity $\delta\Gamma$ denotes the extra electric reflection coefficient from the receiving probe due to the presence of the crack; this is essentially the quantity measured in practice.

Using the expansion of the COD, Eq. (4.2), and projecting onto the Chebyshev functions, Eq. (5.1) becomes

$$\delta\Gamma = \frac{\mathrm{i}\omega}{4P}\sum_{j=1}^{3}\sum_{mm'=1}^{\infty}\gamma_{jmm'}S^{\mathrm{re}}_{jmm'}, \tag{5.2}$$

where $S^{\mathrm{re}}_{jmm'}$ is a matrix determined from $\sigma^{\mathrm{re}}_{3j}(y_1, y_2)$.

6. Numerical Results

In this section some numerical results that illustrate the effect of the interface will be given. The base material 1 has a thickness $d_1 = 30$ mm and is regarded as an isotropic steel with density $\rho_1 = 8.4$ g/cm^3, and with logitudinal wave velocity 5.9 mm/μs and shear velocity 3.2 mm/μs. The cladding has a thickness $d_2 = 5$ mm and is assumed to be an austenitic transversely isotropic steel with density $\rho_2 = 8.5$ g/cm^3 and stiffness constants (in GPa) c^2_{11}, $c^2_{12} = 115$, $c^2_{33} = 250$, $c^2_{13} = 115$, and $c^2_{44} = 100$. The crystal axes coincide with the coordinate axes.

A 2 by 2 mm quadratic crack with sides parallel to the $x2x3$-axes and with its centre 2.5 mm above the back surface is used. The interface has a fixed period a = 5 mm. Ultrasonic vertically polarized shear waves are emitted in the $x1x3$-plane at 45° relative to the normal of the free surface from a 10 by 10 mm transducer. The model of the transmitter was developed by Boström and Wirdelius [4]. Multiple reflection towards the free surface $S1$ is not taken into account. Only results for a fixed frequency of 1 MHz are given. This is computationally efficient and is believed to give a good idea of the influence of the parameters to be varied.

Figure 6.1 shows a few results for the signal response (in dB) as a function of the position of the transmitter relative to the crack (in mm) for different values of the amplitude (b) of the corrugation. Obviously, the corrugation has only a minor influence on the quantitative response in this case, but it should be kept in mind that the contrast between the two materials is relatively small. It is difficult, however, to draw any conclusions about any systematic behavior.

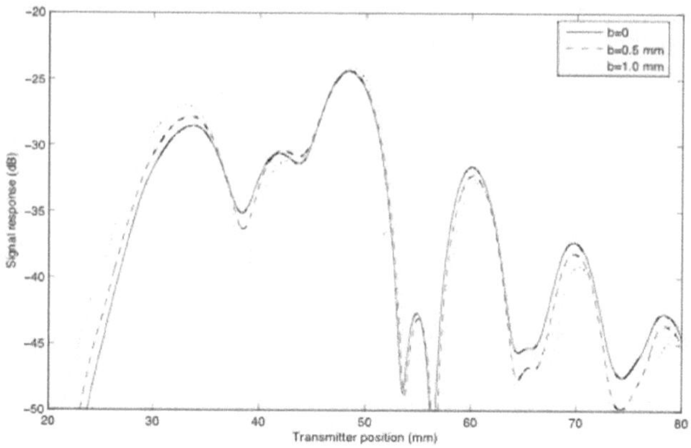

Fig. 6.1 The signal response from a crack as a function of the position of the transmitter for different values of the amplitude (b) of the corrugation.

7. Concluding Remarks

In this paper the effect of a corrugated interface on 3D wave scattering by a crack in a cladding has been investigated. A mathematical model has been developed, where the elastic wave propagation problem is solved exactly. It is believed that the model will provide a useful tool for planning and qualification of ultrasonic testing procedures. From systematic studies with varying values of the governing parameters it should also be possible to acquire a deeper understanding of the influence of the properties of the interface, the size, location, and orientation of the crack, the material parameters, etc. Obviously, much more numerical work is needed before any definite conclusions can be drawn. It is also desirable to obtain results in the time domain. This is, however, a straightforward extension.

Acknowledgements: The present work is sponsored by.the Swedish Radiation Safety Authority, and this is gratefully acknowledged.

References

[1] Krasnova, T.: *Elastic wave scattering from corrugated surfaces in anisotropic media.* PhD thesis, Chalmers University of Technology, Sweden. (2005).

[2] Jansson, P-Å., Zagbai, T.: 2D P-SV wave scattering by a crack in a cladding. In: Thompson, D.O., Chimenti, D.E. (eds.) *Review of Progress in Quantitative Nondestructive Evaluation* 26A, pp. 71–78. American Institute of Physics, Melville, NY (2007).

[3] Auld, B.A.: General electromechanical reciprocity relations applied to the calculations of elastic wave scattering coefficients. *Wave Motion* **1** 3-10 (1979).

[4] Boström, A., Wirdelius, H.: Ultrasonic probe modeling and nondestructive crack detection. *J. Acoust. Soc. Am.* **98** 2836-2848 (1995).

Defect Inspection of Complex Structure in Pipes by Guided Waves

Ping-Hung Lee and Shiuh-Kuang Yang

Department of Mechanical and Electro-Mechanical Engineering, National Sun Yat-Sen University, Kaohsiung 804, Taiwan

d933020006@student.nsysu.edu.tw

Abstract. The pipeline systems are widely used in refinery and petro-chemical industries to carry high pressure, high temperature or even highly corrosive fluids. Leaks or sudden failures of pipes can cause injuries, fatalities, and environmental damage. Discover in recent years, the ultrasonic guided wave can propagate considerable distances in pipe, especially the torsional mode. But some complex pipelines features will distort the signals; even attenuate the transmitted energy substantially. To enable the torsional mode have the ability to detect the defects hided in the welded support. In this study, finite element method was adopted to establish FEM model of the complex features with defect firstly. The transient analysis and mode extraction technique were also used to analyze the characteristics, like mode conversion and dispersion, of the echo when torsional mode impinges onto the pipe features. Secondly, we use the wavelet transform as a signal processing tool to extract the signal of defect from the complex structures and then discuss the feasibility of defect detection of complex structures. The results show that guided wave technique with the wavelet transform performs the ability of defect inspection of complex structures in pipes.

1. Introduction

In many processing industries such as chemical, power plants, oil and gas refineries, etc., thousands of pipes have a wide distribution. Corrosions in pipes exposed to aggressive corrosive environment can lead to serious injuries, fatalities and environmental damage and these needs to be monitored at regular intervals. For the long length of pipe rested on the welded support bracket, the welded region is a candidate for corrosion section due to the water trapping. Wide area corrosion that leads to

T.-T. Wu and C.-C. Ma (eds.), *IUTAM Symposium on Recent Advances of Acoustic Waves in Solids*, IUTAM Bookseries 26, DOI 10.1007/978-90-481-9893-1_39,

wall thickness reduction at pipe supports is the most common cause of external piping corrosion failure. Consequently, pipe inspection plays an important role for avoiding a catastrophic situation of the pipeline system. Some inspection methods of measuring wall thickness, such as conventional ultrasonic, radiography and eddy currents require many measurements and access to the entire testing surface directly. It causes that the condition of the pipe in the hidden region, i.e., insulated, buried, coated or other inaccessible regions cannot be easily inspected without expensive preparatory work. Contrary to the localized inspection methods, guided wave technique can be used to inspect the hidden region of pipeline and allow reliable, rapid and low cost inspection to be performed. Many researches used guided wave for corrosion inspection in pipes. In 1997, site trials of the portable, dry-coupled piezoelectric transducer system showed the ability of the L(0,2) mode of propagating over distances approaching 50 m in a 6 inch steel pipe [1]. In 2001, the T(0,1) mode had been employed at lower frequencies to perform the rapid screening of long sections of pipe for different discontinuity types and other pipe features [2]. In 2003, Cawley *et al.* [3] have discussed the progress of guided wave testing design for long range inspection of pipe from research work to a commercial testing system at frequencies below 100 kHz. To improve the guided wave pipe inspection technique, the advanced signal processing method and the interaction of guided waves with the complex geometries are the major issues to study. The circumferential and longitudinal welded supports are one of the common examples of the complex geometry in pipes. The pipe supports are used to line the pipelines and integrate the manufacturing process for refinery, chemical and petro-chemical industrials. The bottom portion of pipe surface at the support is difficult to be evaluated by conventional ultrasonic methods or visual inspection. Unfortunately, very little work effort is reported on the effect of guided wave interacting with the pipe supports. In 2007, Cheng *et al.* [4] indicated that the reflection of the T(0,1) mode increased as the tightness of the clamp support increased. Moreover, the resonance effect was observed at a frequency band of 18 to 22 kHz in the case of a clamp support with a rubber gasket at a specific contact condition.

In this study, the interaction of torsional guided wave with longitudinal welded supports has been investigated by finite element method, firstly. Wave propagation can be visualized and the echoes from the welded supports have been collected. Secondly, the model of a defect on the pipe surface was built up to collect the signals related to defect at the welded support in a pipe for the purpose of understanding the difference between the reflected signals of a normal support and a corrosive one. Finally, the time-frequency analyses of the reflection coming from different cases were compared each other by continuous wavelet transformed method.

3. Finite Element Simulation

3.1 Finite Element Model

The geometry of the finite element models is shown in Fig. 3.1 and the total length of the 2 sets of pipe is 4 m. The excitation signal and reflected signal from the support are monitored at line C. The distance between the end A and line C is 1 m. From Fig. 3.1(a) and 3.1(b), a longitudinal welded support and a longitudinal welded support with a notch are located on the position 2.5 m away from the end A in turn. The size of the longitudinal welded support is 260 mm in length, 100 mm in high, and 7-mm in width; the circumferential distribution of the through-thickness notch is 1/12 of the full circumference and the axial length is 25 mm.

The FE models produced to investigate the propagation of T(0,1) mode through the longitudinal welded supports were built up by the commercial program ANSYS. A membrane finite element, SHELL 63, with 6 degrees of freedom at each node defined in ANSYS has adopted to simulate the vibration and propagation of torsional modes in pipes. In addition, there are 76 elements chosen around the circumferential section of the pipe. The longitudinal welded support is modeled by SOLID 45 elements and the axial length of SHELL and SOLID elements is 5 mm. All the size of elements matched the convergence rules for wave propagation simulation [5] and gave reliable results.

As for the mode shape of T(0,1) mode, the excitation of T(0,1) mode was achieved by applied the same sequence at all of the nodes around the circumference of the end A of the pipe. A four-cycle 26 kHz tone burst in Hanning window was applied as a sequence of prescribed displacements in the circumferential dire ction of the pipe. The received signals at line C are separated into single-mode waveforms, such as T(0,1) or F(1,2) mode, by using a mode extraction technique.

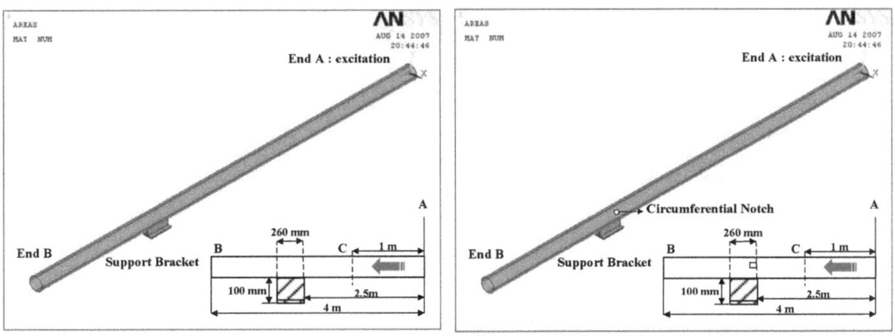

(a) (b)

Fig. 3.1 The geometry of the finite element models (a) for a normal support and (b) a corrosive support.

3.2 The Reflection of the Support Bracket

The time records from the simulations on a 6 inch steel pipe are shown in Fig. 3.2, 3.3 and 3.4. The reflection signals of guided wave modes from the support, the circumferential notch and the end of B are received at line C. For the model of a welded support without notch on the pipe, the signal processing to extract the 0-order mode T(0,1) is shown in Fig. 3.2(a). As for the longitudinal welded support is asymmetric, the reflected echoes are composed of the axisymmetric mode and the non-axisymmetric modes. Fig. 3.2(b) shows the signal when the same raw data are processed to extract the order 1 flexural F(1,2) and F(1,3) modes.

Fig. 3.3(a) and 3.3(b) are the T(0,1) mode and the order 1 flexural modes for the model of a welded support with circumferential notch on a pipe, respectively. By comparing the reflection in Fig. 3.2(a) and 3.3(a), the effect of the notch can be seen. Owing to the existence of the one-twelve circumferential notch, there is a bigger signal around 1.2 ms to 1.3 ms in the time domain. The amplitude of the torsional mode reflected from the welded support reduced and the amplitude of the flexural modes increased on the contrary.

Moreover, the echoes of torsional mode and flexural modes are summed together and shown in Fig. 3.4(a) and 3.4(b) for a normal support and a corrosive support, respectively. In Fig. 3.4(a), the first wave packet E1 is the incident T(0,1) mode and the last wave packet E2 is the reflection of the pipe end B. The wave packets W1, W2, and W3 are the reflection signals of the longitudinal welded support. The packet W1 is the direct echo, the packet W2 is delayed echo, and the packet W3 is the trailing echo. Similarly, Fig. 3.4(b) shows the echoes reflected from the circumferential notch and the welded support at the same time. The P1sign indicates the directed echoes reflected from the notch and the support. The P2 sign is the delayed echoes reflected from the bottom of the support and the P3 sign is the echoes included the trailing echoes and the reverberation between the end A and the notch.

To sum up, the time histories of the T(0,1) waves in the longitudinal welded support are related to the direct echo, the delayed echo, the trailing echo, and mode conversion. Both the delayed and trailing signals are leakages from the support and larger than the direct echo. When there is a notch at support on a pipe, the incident T(0,1) mode will convert more energy to the flexural modes. Consequently, the flexural modes reflected from a normal support and a corrosive support is compared with each other by the following time-frequency analysis.

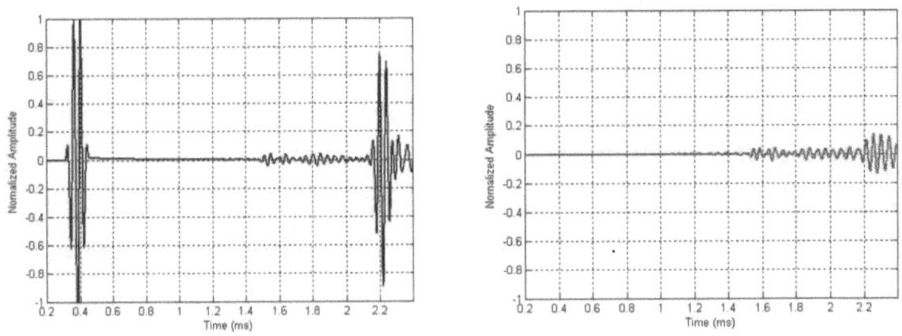

(a) (b)

Fig. 3.2 The time records of (a) T(0,1) mode and (b) order 1 flexural modes in the normal support model.

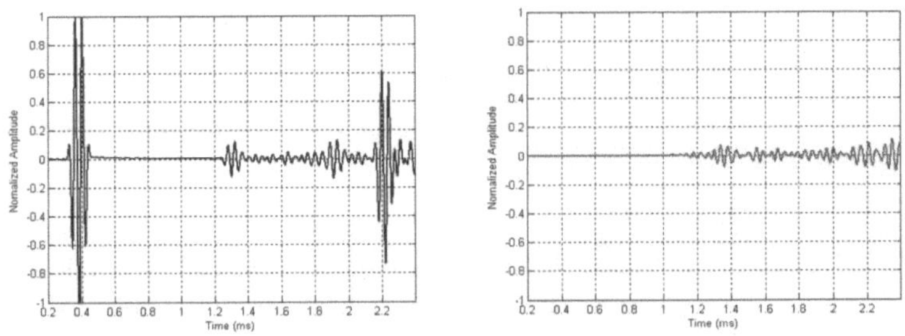

(a) (b)

Fig. 3.3 The time records of (a) T(0,1) mode and (b) order 1 flexural modes in the corrosive support model.

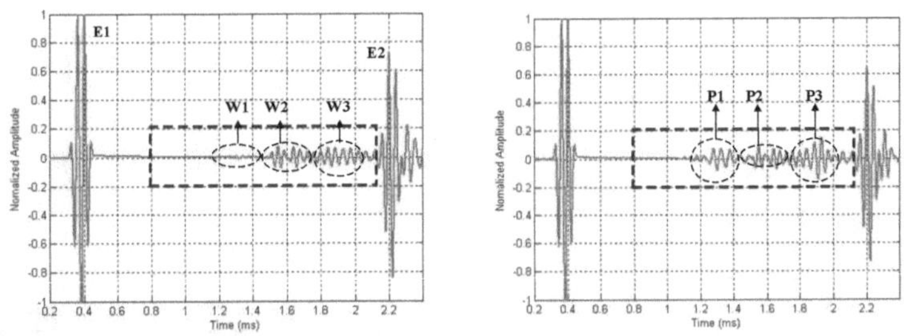

(a) (b)

Fig. 3.4 The summation of time records for (a) a normal support model and (b) a corrosive support model.

4. Time-Frequency Analysis

An advanced signal processing is needed to interpret the signals of guided wave reflected from the complex structure in pipes. Features extracted in a spectrum of time-scale vs. amplitude are being considered to apply to the non-stationary signal analysis and continuous wavelet transform (CWT) is one of the time-frequency analyses.

4.1 Wavelet Transform [6]

The wavelet transform analysis is implemented using the software Matlab. The continuous wavelet transform breaks down a signal x(t) into components (wavelet function Φ). The continuous wavelet transform is defined as the sum over the whole signal duration time multiplied by the wavelet function which related to the two parameters-scale and shift.

4.2 The Spectrum Analysis of Continuous Wavelet Transform

The motivation for time-frequency analysis by continuous wavelet transform is to see the different behavior of the reflected signal between a normal welded support and a corrosive one. The two windowed time records in Fig. 3.4(a) and 3.4(b) are processing by the mother wavelet 'dB4' to get the spectrum of time-scale vs. amplitude. As shown in Fig. 4.1(a), the signal after 1.4 ms is larger than the signal distributed around 1.2 to 1.4 ms. And all the reflection show less dispersive behavior in the corresponding spectrum of time-frequency analysis. The reflection of notch at support is composed of a great deal of torsional mode and flexural modes. From the results in Fig. 4.1(b), there is a big echo around 1.2 to 1.4 ms and the echo shows distinct dispersive behavior in the corresponding spectrum.

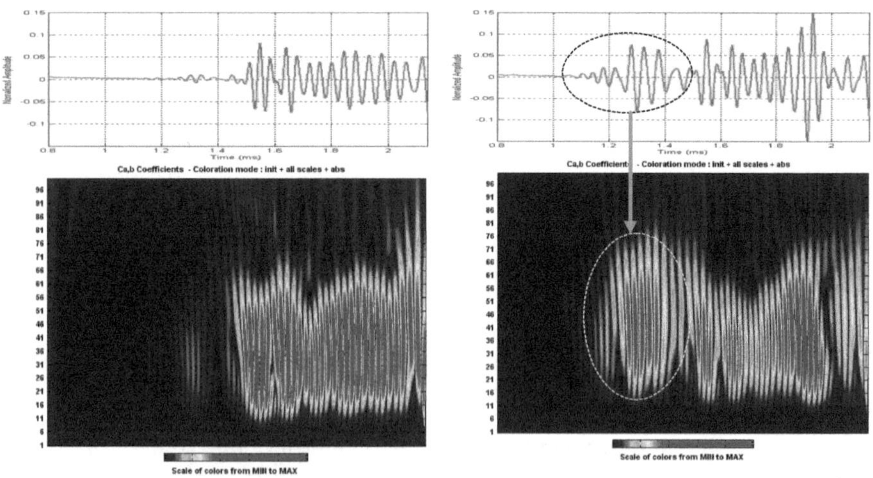

(a) (b)

Fig. 4.1 The spectrum of time-scale vs. amplitude, (a) a normal support model and (b) a corrosive support model.

5. Conclusions

Using finite element method, a technique for guided wave simulation was developed to see the effect of the welded support on the pipe for T(0,1) mode propagation. The reflection of the welded support has been identified as three parts formed by the direct echo, the delayed echo and the trailing echo. Then to improve the ability of guided wave inspection technique, an advanced signal processing tool named continuous wavelet transform was adopted to analyze the signal reflected from the corrosive support. By comparing the spectrum of time-scale vs. amplitude, the degree of dispersive behavior is different between the reflection of a normal support and a corrosive one.

References

[1] Alleyne, D. N., and Cawley, P.: Long range propagation of lamb waves in chemical plant pipework. *Mater. Eval.* **45** 504-508 (1997).

[2] Alleyne, D. N., Pavlakovic, B., Lowe, M. J. S., and Cawley, P.: Rapid long range inspection of chemical plant pipework using guided waves. *Insight.* **43** 93-96 (2001).

[3] Cawley, P., Lowe, M. J. S., Alleyne, D. N., Pavlakovic, B., and Wilcox, P.: Practical long range guided wave inspection - applications to pipes and rail. *Mater. Eval.* **61** 66-74 (2003).

[4] Cheng, J. W., Yang, S. K., and Li, B. H.: Guided wave attenuation in clamp support mounted pipelines. *Mater. Eval.* **65** 317-322 (2007).

[5] Alleyne, D. N., Cawley, P., and Lowe, M. J. S.: The reflection of guided waves from circumferential notches in pipes. *J. Appl. Mech.* **65** 635-641 (1998).

[6] Misiti, M., Misiti, Y., Oppenheim, G., and Poggi, J. M.: *Wavelet toolbox for use with MATLAB*, Version 1, The MathWorks Inc. (1996).

SAW Gas Sensor with Nanostructured Sensing Materials

Yung-Yu Chen[1], Tsung-Tsong Wu[2], Tai-Hsu Chou[2] and Fu-Chun Huang[2]

[1]Department of Mechanical Engineering, Tatung University, Taipei, Taiwan

[2]Institute of Applied Mechanics, National Taiwan University, Taipei, Taiwan

yychen@ttu.edu.tw

Abstract. Surface acoustic wave (SAW) sensors with nanostructured sensing materials for gas detection are reported in this paper. The SAW sensors were fabricated based on a 128°YX-$LiNbO_3$ substrate with an operating frequency of 145 MHz. A dual delay line configuration was adopted to eliminate external environmental fluctuations. The camphor sulfonic acid doped polyaniline nanofibres and Pt coated ZnO nanorods were employed for the detections of humidity and hydrogen due to high surface-to-volume ratio, large penetration depth and fast charge diffusion rate. The nanostructured sensing materials were synthesized by the interfacial polymerization method. Finally, the SAW sensors were tested toward humidity and hydrogen. Results show our proposed SAW sensors exhibit fast response, good sensitivity and short-term repeatability while operating at room temperature.

1. Introduction

Over the past few decades, a number of research works focused on surface acoustic wave (SAW) devices because of small size, low cost and compatibility with the integrated circuit process. Moreover, SAWs are sensitive to surface perturbation such as mass loading, viscoelastic change, or electrical alternation since their acoustic energy is confined within one or two wavelengths near the surface. This feature contributes to make SAW devices be a real-time sensor with high stability, short response time and good reproducibility.

For gas detection, a sensing material must be coated on SAW sensors. Recently, there are many investigations focused on the sensing characteristics of various sensitive materials, mainly conductive polymers [1-4] and metal oxide films [5-11]. Although the selective films offer a feasible way to grab gas molecules in the environment, there is still a

T.-T. Wu and C.-C. Ma (eds.), *IUTAM Symposium on Recent Advances of Acoustic Waves in Solids*, IUTAM Bookseries 26, DOI 10.1007/978-90-481-9893-1_40,

detection limitation on account of less interaction area. Nanostructured materials such as nanorod, nanobelts, and nanofibers have received much attention due to their high surface-to-volume ratio, large penetration depth, and fast charge diffusion rate. Therefore, the gas sensors with nanostructured sensing materials not only possess high sensitivity but also accelerate response and recovery time.

In this paper, surface acoustic wave (SAW) sensors with nanostructured sensing materials for gas detection are reported. The SAW sensors were fabricated based on a 128°YX-LiNbO3 substrate with an operating frequency of 145 MHz. A dual delay line configuration was adopted to eliminate external environmental fluctuations. The camphor sulfonic acid (CSA) doped polyaniline (PANI) nanofibers [12] and Pt coated ZnO nanorods [13] were employed for the detections of humidity and hydrogen respectively. The nanostructured sensing materials were synthesized by the interfacial polymerization method due to simple fabrication and low cost. Finally, several experiments were carried out for evaluating the sensing performances of the fabricated sensors operated at room temperature, including temperature effect, the short-term repeatability and sensitivity.

2. Experimental Details

A two-port SAW resonator based on a 128°YX-$LiNbO_3$ substrate was employed as gas sensors. The coupling-of-modes model was utilized to predict its device performances prior to fabrication. The SAW resonators were fabricated by the micro-electro-mechanical system process: interdigital transducers (IDTs) and reflectors were patterned by exposure and development after an aluminium film was deposited on the substrate. The wavelength, centre frequency and the delay line were designed to be 27 μm, 145 MHz and 4320 μm respectively. The number of electrode pairs of IDTs is 15, while the electrode number of the reflectors is 50.

To eliminate external environmental fluctuations, the SAW resonators were mounted on a PCB and wire bonded with an amplifier to configure a dual delay line system shown in Fig. 1. The gain of the amplifier must be larger than the insertion loss of the SAW resonator and the total phase delay in the oscillator loop must be a multiply of 360°. The supplied voltage for the amplifier is 4 V. The dual delay line system which consisted of two counterparts in oscillator where one is coated with selecting material and the other is bare to execute common mode rejection, was realized to eliminate external environmental fluctuations. To function as an active element, the coated one contributes to a frequency shift by the interaction between sensing material and the target gas. By comparison, the reference one which has bare surface offers the signal of the environmental effects.

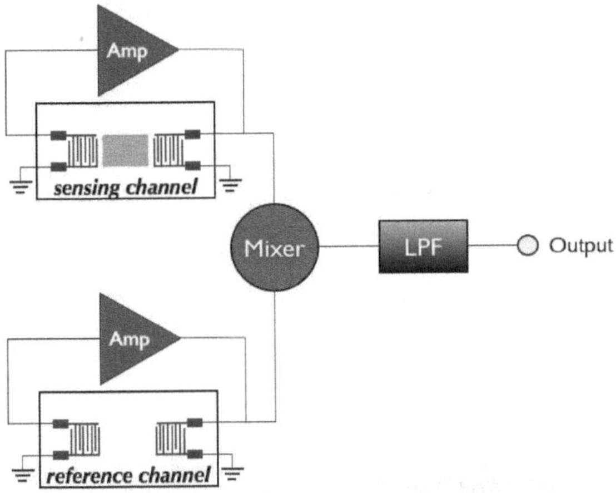

Fig. 1 Schematic diagram of a dual delay line configuration.

In this study, two nanostructured sensing materials were synthesized by the interfacial polymerization method due to simple fabrication and low cost. The camphor sulfonic acid (CSA) doped PANI nanofibers and Pt coated ZnO nanorods were chosen for the detections of humidity and H2 respectively. The selective materials were deposited selectively on the delay line between the two IDTs and then evaluated with the scanning electron microscope and the x-ray diffraction. The results are shown in Fig. 2. A small PDMS hood with the content of about 0.34 cm3 was covered on the SAW resonators. Adhesives and silicon sealant were applied to seal any possible crevice to prevent gas leakage. The chamber was drilled to provide the gas entrance and exit. The frequency of the dual delay line system was acquired by a frequency counter.

3. Measurement Results

The fabricated dual delay line configuration was evaluated under the thermal and humidity testing prior to the gas detection. Results show that the configuration indeed suppresses effectively the noise of temperature and humidity perturbations and enhances sensing stability. Then, some measurements were implemented for investigating the short-term repeatability of the sensors. At the initial stage, the steady state of the baseline frequency was reached, and afterwards nitrogen or mixed water vapour flowed into the chamber to alter the relative humidity (RH) inside the acrylic chamber. Testing cycles were implemented with constant exposure time and purge time to reach a new steady state or re-establish the baseline. The results shown in Fig. 3(a) indicate that the frequency shifts are 60, 80 and 110 kHz corresponding to the RH variation of 20 %, 30 % and 45 %. The result also demonstrates that the SAW humidity sensor possesses good reaction response and recovery response.

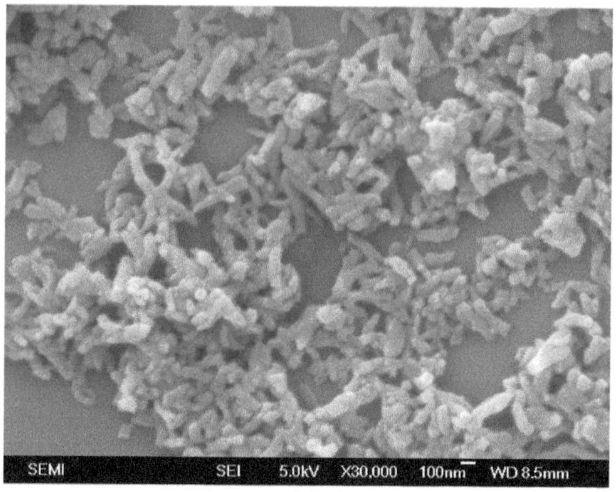

(a)

(b)

Fig. 2 SEM images of (a) CSA-doped polyaniline nanofiber [12] and (b) ZnO nanorod [13].

The real-time responses of the dual-channel sensor to different H2 concentrations are shown in Fig. 3(b). At the initial stage, the steady state of the base frequency was reached, and then nitrogen or hydrogen was led into the PDMS chamber. Testing cycles were implemented with constant exposure time and purge time to reach a new steady state or return to the baseline. The sensor was then exposed to different concentrations of hydrogen: 200, 500, 1500, 2500, and 6000 ppm at room temperature. The responses are 8.36, 12.66, 17.47, 20, and 26.2 kHz respectively. It takes less than 15s to reach about 90% of the steady state, and the recovery time is about 2 minute. The result shows that the frequency shift of the SAW hydrogen sensor increases quickly at low concentrations and the response tends to saturate at large concentrations.

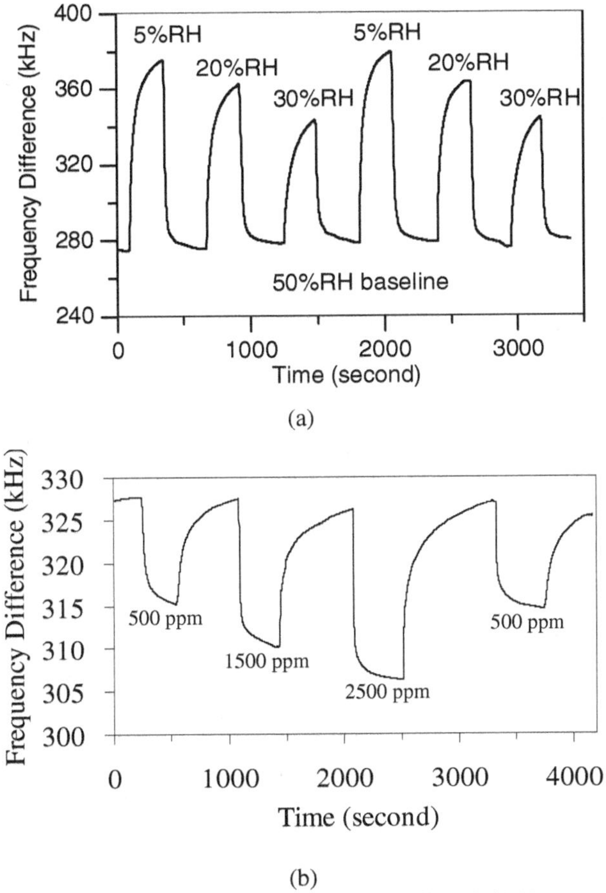

Fig. 3 Short-term repeatability toward different (a) RH [12] and (b) H_2 concentrations [13].

4. Conclusions

SAW sensors with nanostructured sensing materials like the camphor sulfonic acid (CSA) doped polyaniline nanofibres and the Pt coated ZnO nanorods have been successfully developed for humidity and hydrogen detections. Some experiments are implemented to evaluate the performances of the sensors. Results show that temperature perturbation can be well suppressed by utilizing a dual delay line configuration. In addition, the frequency shifts are 60, 80 and 110 kHz corresponding to the RH variation of 20 %, 30 % and 45 %, while the frequency shifts toward hydrogen concentrations of 200 ppm and 6000 ppm are 8.36 kHz and 26.2 kHz. The fabricated SAW sensors do provide high sensitivity, fast response and good repeatability while operating at room temperature.

Acknowledgments: The authors gratefully acknowledge the financial support of this research from the National Science Council of Taiwan through the grant NSC 95-2221-E-002-052-MY2.

References

[1] Jakubik, W. P.: Investigations of thin film structures of WO3 and WO3 with Pd for hydrogen detection in a surface acoustic wave sensor system. *Thin Solid Films* **515** 8345-8350 (2007).

[2] Ippolito, S. J., Kandasamy, S., Kalantar-zadeh, K., Wlodarski, W., and Holland, A.: Comparison between conductometric and layered surface acoustic wave hydrogen gas sensors. *Smart Materials and Structures* **15** S131-S136 (2006).

[3] Ippolito, S. J., Kandasamy, S., Kalantar-zadeh, K., Wlodarski, W., Galatsis, K., Kiriakidis, G., Katsarakis, N., and Suchea, M.: Highly sensitive layered ZnO/LiNbO3 SAW device with InOx selective layer for NO2 and H2 gas sensing. *Sensors and Actuators B: Chemical* **111-112** 207-212 (2005).

[4] Fechete, A. C., Wlodarski, W., Kalantar-Zadeh, K., Holland, A. S., Antoszewski, J., Kaciulis, S., and Pandolfi, L.: SAW-based gas sensors with rf sputtered InOx and PECVD SiNx films: Response to H2 and O3 gases. *Sensors and Actuators B: Chemical* **118** 362-367 (2006).

[5] Nomura, T., Oofuchi, K., Yasuda, T., Furukawa, S.: SAW humidity sensor using dielectric hygroscopic polymer film. *IEEE Ultrasonics Symposium* 503-506 (1994).

[6] Braga, E. R., Nakano, A. Y., da Cunha, M. P.: A SAW resonator sensor system employed in humidity measurements. *SBMOEEE MIT-SIMOC'99 Proceedings* 342-345 (1999).

[7] Penza, M., Anisimkin, V.I.: Surface acoustic wave humidity sensor using polyvinyl-alcohol film. *Sensor and Actuators A: Physical* **76** 162–166 (1999).

[8] Penza, M., Cassano, G.: Relative humidity sensing by PVA-coated dual resonator SAW oscillator. *Sensors and Actuators B: Chemical* **68** 300–306 (2000).

[9] Tashtoush, N.M., Cheeke, J.DN., Eddy, N.: Surface acoustic wave humidity sensor based on a thin PolyXIO film, *Sensors and Actuators B: Chemical* **49** 218–225 (1998).

[10] Nieuwenhuizen, M. S., Nederlof, A. J.: A SAW gas sensor for carbon dioxide and water: preliminary experiments. *Sensors and Actuators B: Chemical* **2** 97-101 (1990).

[11] Korsah, Kofi, Ma, C. L., Dress, Bill.: Harmonic frequency analysis of SAW resonator chemical sensors: application to the detection of carbon dioxide and humidity. *Sensors and Actuators B: Chemical* **50** 110-116 (1998).

[12] Wu, T. T., Chen, Y. Y., Chou, T. H.: A high sensitivity nanomaterial based SAW humidity sensor. *J. Phys. D: Appl. Phys.* **41** 085101 (2008).

[13] Huang, F. C., Chen, Y. Y., Wu, T. T.: A room temperature SAW hydrogen sensor with Pt coated ZnO nanorods. *Nanotechnology* **20** 065501 (2009).

Design of IF Two-Track Surface Acoustic Wave Filters Using (100) AlN/Diamond Structures

Ruyen Ro[1], Chia-Chi Sung[2], Ruyue Lee[1] and Yuan-Feng Chiang[2]

[1]Department of Electrical Engineering, I-Shou University, Kaohsiung 840, Taiwan

[2]Department of Engineering Science and Ocean Engineering, National Taiwan University, Taipei 104, Taiwan

ryro@isu.edu.tw

Abstract. In this paper, characteristics of two-track surface acoustic wave (SAW) filters based on Cu electrode/(100) AlN/diamond structures were analyzed using the finite element method (FEM) and the transmission matrix method. To achieve optimum performance, the Chebyshev window function was used to modify the reflectivity distributions of reflectors. The SAW parameters for weighted reflectors with various electrode thicknesses and different metallization ratios are retrieved using the FEM. The transmission matrix method was then employed to calculate the scattering parameters of weighted reflectors and two-track SAW filters. Simulation results show a two-track SAW filter with low insertion loss, good shape factor, good sidelobe suppression, small group delay, and flat passband can be tailored and applicable to the code division multiple access (CDMA) system.

1. Introduction

The increasing popularity in the global market of wireless and mobile communications systems has created a vast demand for electronic devices with designed specifications such as duplexers, IF and RF filters. The surface acoustic wave (SAW) device due to its miniature size, rugged structure and ease of mass production is a promising candidate for the design of IF and RF filters for wireless applications. [1–4] In this study, two-track SAW filters are proposed to design IF filters applicable to the code division multiple access (CDMA) system. Traditional transverse SAW filters cannot meet the specifications of CDMA IF filters because of the triple-transit interference (TTI) and other second-order effects. In the two-track SAW filter, if the input interdigital transducers (IDTs) in both tracks are identical but out of phase, then the output IDTs are

T.-T. Wu and C.-C. Ma (eds.), *IUTAM Symposium on Recent Advances of Acoustic Waves in Solids*, IUTAM Bookseries 26, DOI 10.1007/978-90-481-9893-1_41,

identical and in phase, and vice versa, which indicates that if there is no reflectors in each track the transmission coefficient will vanish, and the TTI effect can be eliminated. This implies that the frequency responses of a two-track SAW filter are dominated mainly by the reflectors. The detailed information about the two-track SAW filters can be found in elsewhere. [5] For a uniform reflector, the reflection characteristics, 3 dB bandwidth and sidelobe suppression are not sufficient to be employed for designing a CDMA IF filter. To overcome this problem, some researchers employed the window function to modify the reflectivity or line width distributions of reflectors to gain an improved suppression capability. [6–8] In this study, the Chebyshev window function, which has been widely used in the optimum design of antennas and filters, is proposed to modify the reflectivity distributions of reflectors.

Recently, SAW devices in layered structures including a diamond layer have been employed for super high frequency applications. For example, the SAW device based on the (100) AlN/diamond structure can possess a high phase velocity of 10780 m/s and a relatively large electromechanical coupling coefficient (K^2) of 2.4% in mode 1 (Sezawa mode). [9] The increase of the phase velocity will increase the wavelength and correspondently the line width of the IDT. This indicates that without an improvement in the line width resolution limit technology of the IDTs the use of diamond based structures can realize the SAW devices with width-modulated configurations, such as single phase unidirectional transducers (SPUDTs) and width-controlled reflectors, which are applicable for double mode SAW (DMS) filters, two-track SAW filters, ladder filters, resonators, and sensors. [10–12]

In this study, we attempt to design the two-track SAW filter with width-weighted reflectors in the (100) AlN/diamond structure. Cu IDT is applied due to its reflectivity is two times larger than conventional Al IDT. The finite element method (FEM) model is established to calculate the mutual-coupling coefficients and phase velocities of each electrode region with various metallization ratios (MRs) and different electrode thicknesses. The transmission matrix method is then applied to evaluate frequency responses of SAW filters. [5, 6]

2. SAW Parameters Calculation

In this study, the FEM is employed to compute effective phase velocity (V_{eff}) and mutual-coupling coefficient (k) of each electrode region with various MRs and different electrode thicknesses. Two vibration modes of SAWs, a symmetric mode and an asymmetric mode, can be observed using the FEM model. The eigenfrequencies of these two modes can be employed to evaluate k and V_{eff} as given by [2]

$$V_{\mathrm{eff}} = p(f_{\mathrm{sym}} + f_{\mathrm{asym}}), \tag{1}$$

$$k = \frac{\pi\left(f_{\mathrm{asym}} - f_{\mathrm{sym}}\right)}{p\left(f_{\mathrm{asym}} + f_{\mathrm{sym}}\right)}, \tag{2}$$

where p is the period of an electrode and f_{sym} and f_{asym} are the resonance frequencies of the symmetric and asymmetric modes, respectively. For a free or metalized surface, both modes have the same resonance frequency; i.e., $f_{sym} = f_{asym}$. The constructed FEM model can be employed to calculate the parameters of coupling-of-mode (COM) theory such as phase velocity, electromechanical coupling coefficient, mutual-coupling coefficient, and static capacitance of layered media.

With given those COM parameters, reflection and transmission characteristics of reflection gratings and two-track SAW filters are calculated using the transmission matrix method. [5, 6] Detailed information of the transmission matrix and the associated procedure for the calculation of the S-parameters of the two-track SAW filter can be found in references. [3, 5, 6]

3. Results and Discussions

In this study, Cu IDT/(100) AlN/diamond structure was adopted to tailor the two-track SAW filters applicable for the CDMA system. The corresponding material constants were presented in references. [13, 14] Simulation results indicated that the sidelobe suppression level of reflectors with uniform width can not meet the requirements for CDMA IF filters. Accordingly, it is necessary to modulate the line width or reflectivity distributions of reflectors. Figure 1 shows the calculated effective phase velocity and reflectivity of an electrode with various MRs and different electrode thicknesses. As the electrode thickness increases, the maximum value of the reflectivity increases while the effective phase velocity decreases. For the 0.2, 0.6, 0.8, and 1.5% curves, the maximum reflectivity values are 0.0129, 0.0152, 0.0177, and 0.04 with MRs being equal to 0.2, 0.325, 0.4, and 0.6, respectively.

To enhance the sidelobe suppression of grating reflectors, the Chebyshev window function is employed to modulate the reflectivity distributions of reflectors. With the sidelobe level 60dB and the number of electrode 240, the generated Chebyshev distributions of the reflectivity for the gratings with different electrode thicknesses are presented in Fig. 2. The maximum reflectivity values shown in Fig. 2 are obtained according to the data presented in Fig. 1. By using the numerical interpolation method, the corresponding MR distributions of the gratings with different electrode thicknesses are presented in Fig. 3. The corresponding effective phase velocities calculated using the FEM are also presented in Fig. 3.

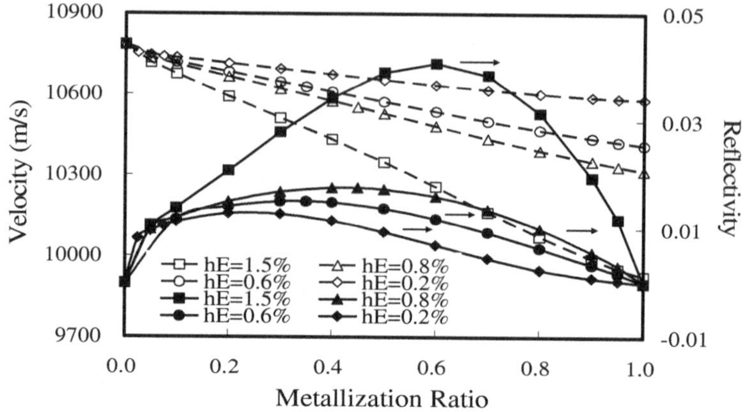

Fig. 1 Reflectivity and effective phase velocity of an electrode versus MR with different electrode thicknesses, hEs.

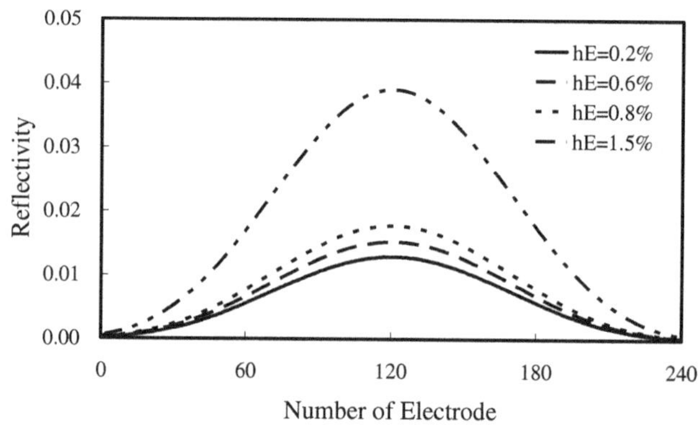

Fig. 2 Chebyshev distribution of the reflectivity for the gratings with different electrode thickness ratios.

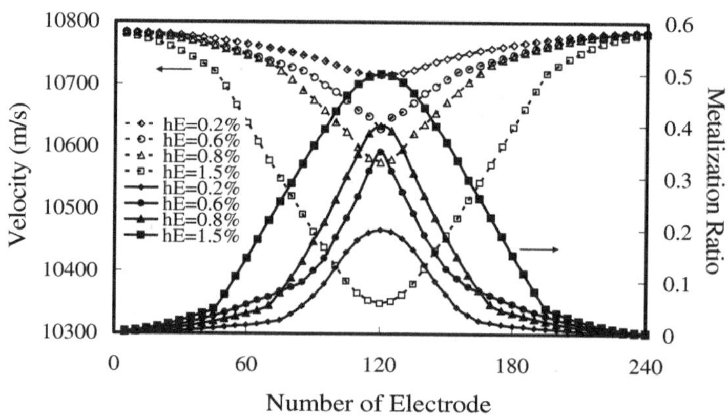

Fig. 3 The retrieved MR and corresponding effective velocity for the weighted reflectors with Chebyshev distributions as presented in Fig. 2.

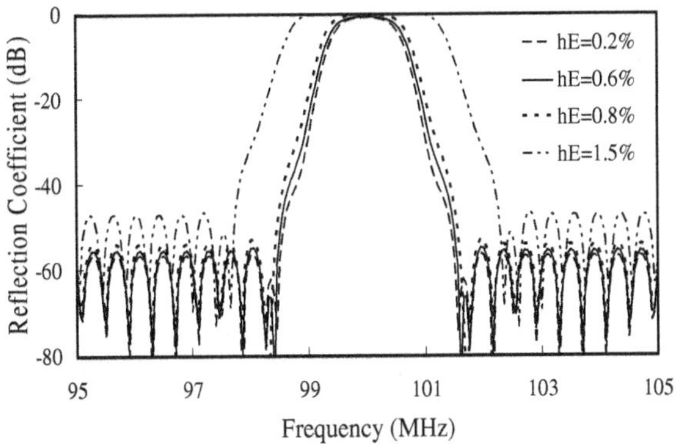

Fig. 4 Magnitude of reflection coefficients of weighted reflectors with various electrode thickness ratios.

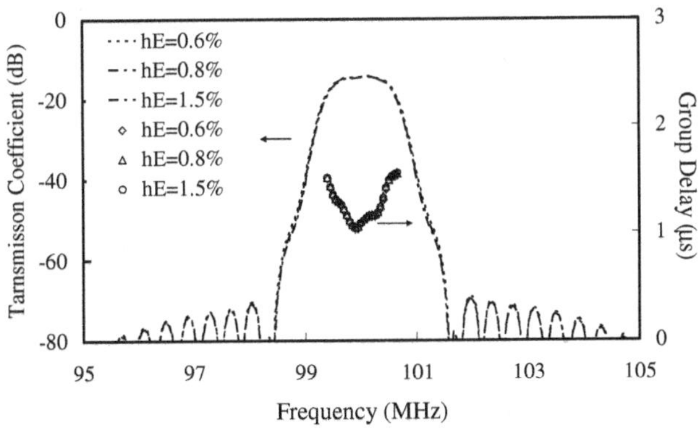

Fig. 5 Transmission coefficient and group delay of two-track SAW filters with weighted reflectors for various electrode thickness ratios.

According to Fig. 3, the reflection coefficients of weighted gratings calculated using the transmission matrix method are shown in Fig. 4. It can be seen that the 1.5% curve has the largest bandwidth among these four cases in Fig. 4, however with the poorest sidelobe suppression level. This implies that the larger the reflectivity, the larger the bandwidth and the poorer the sideband suppression level. The reflection characteristics presented in Fig. 4 can not meet the requirements for the CDMA system.

By multiplying the reflectivity distributions of gratings by different scaling factors, the transmission coefficients of the two-track SAW filters applicable for the CDMA IF filters are represented in Fig. 5. The corresponding filter parameters such as shape factor, insertion loss, sidelobe rejection, and modulated reflectivity range for different electrode thickness ratios are summarized in Table 1. It is shown that the specifications of CDMA IF filters can be satisfied if the maximum reflectivity of reflectors is around 0.014. Thus, for the electrode thickness being equal to 0.2% the reflectivity is not large enough for the design of CDMA IF filters.

Table 1 The corresponding parameters of two-track SAW filters applicable for the CDMA system.

hE (%)	Shape factor	Insertion loss (dB)	Sidelobe suppression (dB)	Group dalay(μs)
0.6	1.48	14.38	54.85	1.04–1.57
0.8	1.48	14.55	55.47	1.06–1.56
1.5	1.48	14.47	56.82	1.05–1.56

4. Conclusions

In this paper, the Chebyshev functions are employed to design the grating reflectors with different reflectivity distributions to achieve a high sidelobe suppression response. The SAW parameters such as phase velocity, electromechanical coupling coefficient, mutual-coupling coefficient, and static capacitance are calculated using the FEM and then substituted into the transmission matrix method to evaluate the corresponding scattering parameters. Simulation results indicate that two-track filters with weighted reflectors can be used to design the CDMA IF filters, provided that the maximum value of reflectivity is around 0.014.

References

[1] Hashimoto, K.: *Surface Acoustic Wave Devices in Telecommunications: Modeling and Simulation*. Springer Verlag, Berlin (2000).

[2] Plessky, V., Koskela, J.: Coupling-of-modes analysis of SAW devices. *Int. J. of High Speed Electron. and Syst.* **10** 867–947 (2000).

[3] Campbell, C. K.: *Surface Acoustic Wave Devices for Mobile and Wireless Communications*. Academic Press, New York (1998).

[4] Morgan, D. P.: *Surface Acoustic Wave Filters*. Academic Press, Amsterdam (2007).

[5] Ro, R., Tung, H., Wu, S.: Design of two-track surface acoustic wave filters with width-controlled reflectors. *Jpn. J. of Appl. Phys.* **43** 688–694 (2004).

[6] Tung, H., Ro, R.: Frequency responses of surface-acoustic-wave filters with two-track slanted finger interdigital transducers. *Jpn. J. of Appl. Phys.* **44** 4025–4031 (2005).

[7] Omori, T., Tajima, M., Akasaka, J., Hashimoto, K., Yamaguchi, M.: Design of weighted reflectors and their application to surface acoustic wave bandpass filter. *Jpn. J. of Appl. Phys.* **39** 3024–3027 (2000).

[8] Omori, T., Akasaka, J., Arai, M., Hashimoto, K., Yamaguchi, M.: Optimisation of weighted SAW grating reflectors with minimized time delay deviation. *IEEE Int. Freq. Control Symp. & PDA Exhib.* 666–670 (2001).

[9] Wu, S., Ro, R., Lin, Z. X., Lee, M. S.: Rayleigh surface acoustic wave modes of interdigital transducer/(100) AlN/(111) diamond. *J. Appl. Phys.* **104** 064919 (2008).

[10] Hartmann, C. S., Abbott, B. P.: Overview of design challenges for single phase unidirectional SAW filters. *Proc. of the IEEE Ultrason. Symp.* **1** 79–89 (1994).

[11] Bergmann, A., Machui, J., Wagner, K., Weigel, R.: Two-track-reflector-filters for CDMA mobile telephones. *Proc. of the IEEE Ultrason. Symp.* **1** 57–60 (1996).

[12] Yamada, T., Nakamura, H., Tsuzuki, S.: A miniaturized SAW reflectors filter for IF stage of CDMA cellular phone applications. *Proc. of the IEEE Ultrason. Symp.* **1** 405–408 (1999).

[13] Benetti, M., Cannat'a, D., Pietrantonio, F. D., Verona, E.: Growth of piezoelectric film on diamond for high-frequency surface acoustic wave devices. *IEEE Trans. Ultrason. Ferroelectr. Freq. Control* **52** 1806–1811 (2005).

[14] Gualtieri, J. G., Kosinski, J. A., Ballato, A.: Piezoelectric materials for acoustic wave applications. *IEEE Trans. Ultrason. Ferroelectr. Freq. Control* **41** 53–59 (1994).

Acoustic Wave Properties of (100) AlN Films

Sean Wu[1, 2]**, Zhi-Xun Lin**[2] **and Ruyen Ro**[3]

[1]Department of Electronics Engineering and Computer Science, Tung Fang Institute of Technology, No. 110, Tung-Fung Rd., Hunei Shiang, Kaohsiung County, 829,Taiwan.

[2]Advance Design Technology INC., Tainan County 710, Taiwan

[3]Department of Electrical Engineering, I-Shou University, No. 1, Sec. 1, Syuecheng Rd., Dashu Township, Kaohsiung County 840, Taiwan.

wusean.tw@gmail.com

Abstract. AlN films have been widely investigated for the application in surface acoustic wave (SAW) and films bulk acoustic wave (FBAW) devices. (100) AlN films have been deposited, but there is few research to study their acoustic properties. Different orientation piezoelectric films will form different acoustic properties. Bulk acoustic wave (BAW) and SAW properties of (100) AlN films were theoretically analyzed in this research. As regards the BAW properties, (100) AlN films can excite a pure shear mode (velocity=5867 m/s, K^2=2.45%) and exhibited a great potential for FBAR liquid biosensors. As regards the Rayleigh SAW modes of (100) AlN films on diamond, especial for mode 1, the phase velocity is 10474 m/s and the K^2 is 2.31 % at the films thickness ratio (h/λ) is 0.3. (100) AlN films on diamond also can excite high velocity shear horizontal (SH) SAW modes. Especial for mode 0, the K^2 curve shows a maximum value (1.27%) at h/λ=0.28 and the velocity is 7496 m/s. Those research results provide an important theoretical basis for further application on FBAW and SAW devices.

1. Introduction

Aluminum nitride (AlN) thin films are widely used in the fabrication of SAW and FBAW devices because of its high acoustic velocity and suitable piezoelectric coupling factor. The crystalline orientation of piezoelectric thin films is a key issue in optimizing the piezoelectric response. It usually shows (002) texture with c-axis perpendicular to the substrate due to the lower surface free energy for (002) plane. Different orientations of piezoelectric films will form different acoustic

T.-T. Wu and C.-C. Ma (eds.), *IUTAM Symposium on Recent Advances of Acoustic Waves in Solids*, IUTAM Bookseries 26, DOI 10.1007/978-90-481-9893-1_42,

properties. For a FBAW RF-filter based on AlN, it is required to grow films with a perfect (002) orientation in order to excite optimally the longitudinal thickness mode. But in a liquid medium, longitudinally polarized wave resonators show significant acoustic leakage into the liquid, which results in a substantial loss of resolution. Shear mode wave does not produce any compressional motion in the liquid; thus, no energy leakage occurs. Therefore, what kind of piezoelectric films can excite shear mode is very important for FBAW liquid or bio sensors.

Diamond is a non-piezoelectric material for high velocity SAW devices due to the highest SAW velocity among all materials and it needs to add a piezoelectric layer on the top to excite SAWs. AlN is the highest SAW velocity among current piezoelectric materials. Therefore, AlN films on diamond will form a very attractive high velocity SAW substrate. Recently, (100) AlN films have been deposited, but there is few research to study its acoustic properties. In our research, we study the BAW properties (100) AlN films. Rayleigh SAW modes and shear horizontal (SH) SAW modes of (100) AlN films on diamond also were studied in this research.

2. Bulk Acoustic Wave Properties of (100) AlN Films

The three-dimensional bulk acoustic wave equation, in general, is referred to as the Christoffel equation. The elements in the 3×3 Christoffel matrix are functions of material properties of piezoelectric materials and the propagation direction of acoustic waves. With the knowledge of material parameters and the propagation direction of acoustic wave, the stiffen Christoffel matrix can be solved numerically. Solving the Christoffel matrix gives us three eigenvalues and associated eigenvectors; the eigenvalue is the phase velocity and piezoelectric coupling constant (K^2) of the acoustic wave and the eigenvector tells us the corresponding vibration direction of the particle (mode). For more details of the calculations, please refer to reference [1-5].

By solving the Christoffel matrix of the (100) AlN films, we can obtain one pure longitudinal mode with phase velocity $(c_{11}/\rho)^{1/2}$, one pure shear mode with phase velocity $(c_{66}/\rho)^{1/2}$, and one stiffen shear mode with phase velocity $[(c_{44} + e_{15}^2/\varepsilon_{11})/\rho)^{1/2}$ and K^2 being equal to $e_{15}^2/c_{44}\varepsilon_{11}$. With the AlN material properties given in Table 1, the pure longitudinal mode has the phase velocity 9911 m/s, the pure slow shear mode has the phase velocity 5597 m/s, and the fast shear mode has the phase velocity 5867 m/s with K^2 2.45%. Therefore, the (100) AlN films provided a pure fast shear mode.

3. Surface Acoustic Wave Properties of (100) AlN Films on Diamond

Following the similar approach as developed by Campbell [6], the matrix method is effectively employed here to calculate the Rayleigh and SH wave velocity and electromechanical coupling coefficients (K^2) in a layered piezoelectric structure. For more details of the calculations and material constants, please refer to reference [7-9].

3.1 Rayleigh SAW Modes of (100)AlN/Diamond

The phase velocity dispersion curves of first five Rayleigh SAW modes propagation in the IDT/(100)AlN/diamond structure are shown in Fig. 1. The phase velocity of each mode decreases as the films thickness ratio increased. For mode 0, the value of phase velocity is from the SAW velocity of diamond (10934 m/s) at $h/\lambda=0$. As the films thickness ratio (h/λ) is increasing, the phase velocity curve rapidly decreases. At $h/\lambda=3$, the velocity of the (100)AlN/diamond is 5438 m/s. Mode 1, mode 2, mode 3, and mode 4 show cutoff at the critical point, where the phase velocity is equal to the shear bulk wave velocity in diamond (12323 m/s). Mode 1 occurs at $h/\lambda>0.175$, mode 2 occurs at $h/\lambda>0.298$, mode 3 occurs at $h/\lambda>0.585$, and mode 4 occurs at $h/\lambda>0.7101$. The K^2 dispersion curves of first five Rayleigh-type modes propagation in the (100)AlN/diamond structure are shown in Fig. 2. For mode 0, the K^2 curve shows the optimal value (0.76%) at $h/\lambda=0.1467$ and the velocity is 9282 m/s. For mode 1, the K^2 curve shows the optimal value (2.31%) at $h/\lambda=0.3$ and the velocity is 10474 m/s. For mode 2, the K^2 curve shows the optimal value (0.85%) at $h/\lambda=0.5495$ and the velocity is 11049 m/s. For mode 3, the K^2 curve shows the optimal value (1.59%) at $h/\lambda=0.848$ and the velocity is 10916 m/s. For mode 4, the K^2 curve shows the optimal value (1.24%) at $h/\lambda=1.1965$ and the velocity is 10893 m/s.

3.2 Shear Horizontal SAW Modes of (100)AlN/Diamond

The phase velocity dispersion curves of SH SAW modes propagation in the IDT/(100)AlN/diamond structure are shown in Fig. 3. The phase velocity of each mode decreases as the films thickness ratio increased. Mode 0, mode 1, mode 2, mode 3 and mode 4 show cutoff at the critical point, where the phase velocity is equal to the shear bulk wave velocity in diamond (12323 m/s). Mode 1 occurs at h/λ>0.37, mode 2 occurs at h/λ>0.64, mode 3 occurs at h/λ>0.91 and mode 4 occurs at h/λ>1.12. As h/λ is increasing, the phase velocity curve decreases. The K^2 dispersion curves of SH modes propagation in the IDT/(100)AlN/diamond structure are shown in Fig. 4. Those curves become smoother and smaller as the mode increases. For mode 0, the K^2 curve shows the optimal value (1.27%) at h/λ=0.28 and the velocity is 7496 m/s. For mode 1, the K^2 curve shows the optimal value (0.46%) at h/λ=0.7 and the velocity is 8400 m/s. For mode 2, the K^2 curve shows the optimal value (0.27%) at h/λ=1.21 and the velocity is 8320 m/s. For mode 3, the K^2 curve shows the optimal value (0.196%) at h/λ=1.7 and the velocity is 8331 m/s. For mode 4, the K^2 curve shows the optimal value (0.153%) at h/λ=2.19 and the velocity is 8346 m/s.

4. Conclusion

BAW and SAW properties of (100) AlN films will be theoretically analyzed in this research. As regards the BAW properties, (100) AlN films can excite a pure shear mode (velocity=5867 m/s, K^2=2.45%) and exhibited a great potential for FBAR liquid biosensors. Rayleigh SAW properties of (100) AlN films on diamond exhibited smaller films thickness ratios, higher phase velocities and larger K^2 than the ones of (002) AlN on diamond. Especial for mode 1, the phase velocity is 10474 m/s and the K^2 is 2.31 % at the films thickness ratio is 0.3. This structure also can excite high velocity SH SAWs. Especial for mode 0, the K^2 curve shows a maximum value (1.27%) at h/λ=0.28 and the velocity is 7496 m/s. Those research results provide an important theoretical basis for further application on FBAW and SAW devices.

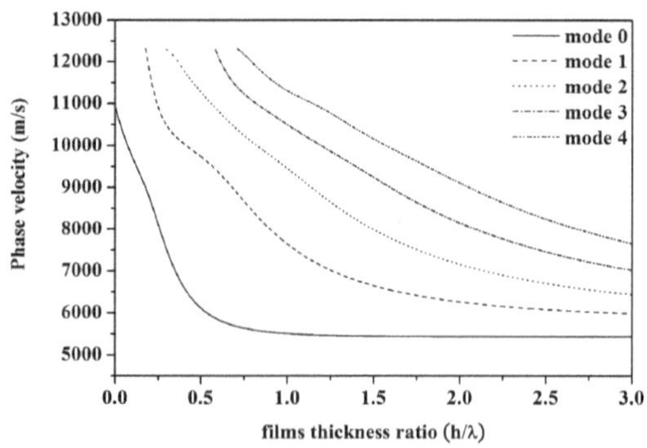

Fig. 1 Calculated phase velocities dispersion curves of Rayleigh SAW modes propagation in the IDT/(100)AlN/diamond.

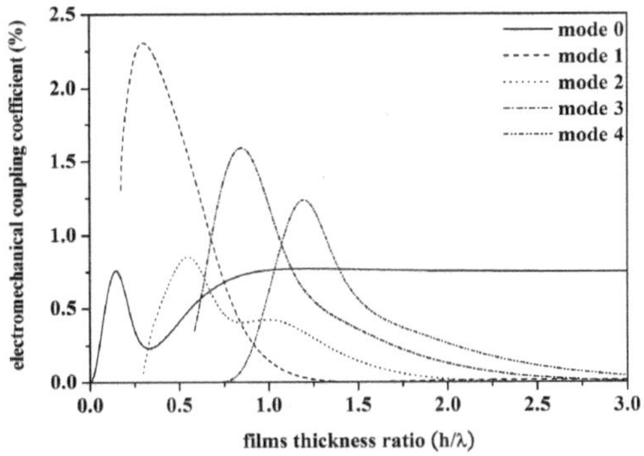

Fig. 2 Calculated electromechanical coupling coefficients (K^2) dispersion curves of Rayleigh SAW modes propagation in the IDT/(100)AlN/diamond.

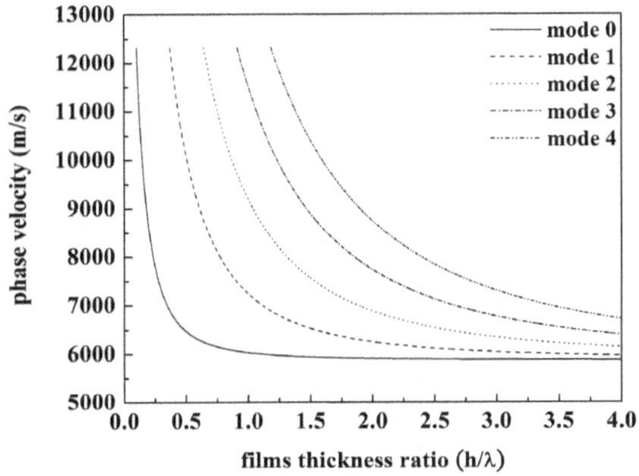

Fig. 3 Calculated phase velocities dispersion curves of SH SAW modes propagation in the IDT/(100)AlN/diamond.

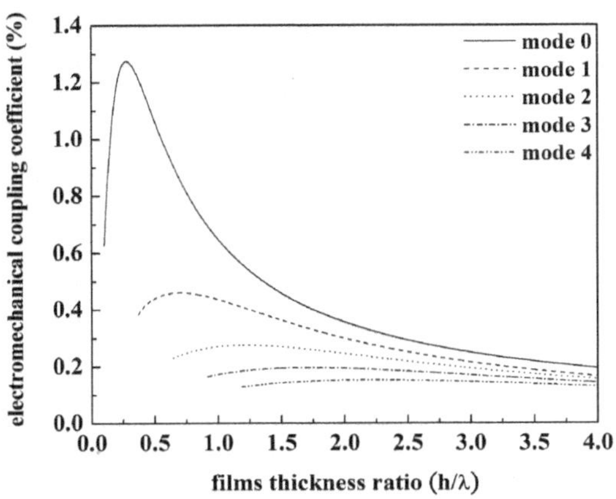

Fig. 4 Calculated electromechanical coupling coefficients (K^2) dispersion curves of SH SAW modes propagation in the IDT/(100)AlN/diamond.

Table 1 Material parameters of AlN films.

		AlN
Density (kg/m^3)	ρ	3512
Elastic stiffness (G Pa)	c_{11}	345
	c_{12}	125
	c_{13}	120
	c_{33}	395
	c_{44}	118
	c_{66}	110
Piezoelectric stress constant (C/m^2)	e_{15}	-0.48
	e_{31}	-0.45
	e_{33}	1.55
Dielectric permittivity (ε_0)	ε_{11}	9
	ε_{33}	11

References

[1] Wu, S., Lin, Z. X., Lee, M. S., Ro, R.: Theoretical BAW Properties of the (103) AlN Films. *Jpn. J. Appl. Phys.* **46** 3471 (2007).

[2] Lee, M. S., Wu, S., Lin, Z. X., Ro, R.: Bulk Acoustic Wave Analysis of Crystalline-Plane-Oriented Aluminum Nitride Films. *Jpn. J. Appl. Phys.* **46** 6719 (2007).

[3] Wu, S., Lin, Z. X., Lee, M. S., Ro, R.: Bulk Acoustic Wave Analysis of Crystalline-Plane-Oriented ZnO Films. *J. Appl. Phys.* **102** 084908 (2007).

[4] Wu, S., Lin, Z. X., Lee, M. S., Ro, R.: Pure Shear Bulk Acoustic Wave of the (100) oriented ZnO. *16th IEEE International Symposium on the Application of Ferroelectrics*, Nara city, Japan, 27-31 (2007).

[5] Wu, S., Lin, Z. X., Lee, M. S., Ro, R., Chen, K. I.: Pure Shear Bulk Acoustic Wave of the Oriented AlN. *16th IEEE International Symposium on the Application of Ferroelectrics*, Nara city, Japan, 27-31 (2007).

[6] Campbell, J. J., Jones, W. R.: A Method for Estimating Optimal Crystal Cuts and PropaGation Direction for Excitation of Piezoelectric Surface waves. *IEEE Trans. Sonics Ultrason.* **15** 209 (1968).

[7] Wu, S., Ro, R., Lin, Z. X., Lee, M. S.: Rayleigh SAW modes of IDT/(100) AlN/(111) diamond. *J. Appl. Phys.* **104** 064919 (2008).

[8] Wu, S., Ro, R., Lin, Z.X.: Rayleigh SAW modes of IDT/(100)ZnO/(111)diamond. *Appl. Phys. Lett.* **94** 032908 (2009).

[9] Wu, S., Ro, R., Lin, Z. X., Lee, M. S.: High Velocity Shear Horizontal Surface Acoustic Wave Modes of Interdigital Transducer/(100)AlN/(111)Diamond, *Appl. Phys. Lett.* **94** 092903 (2009).

Probing Viscoelastic Properties of Polymer Solution Boundary Layers Using Quartz Crystal Resonator

Ping Wang[1], Jiajie Fang[1], Yihong Kang[2], Sheng Qin[1], Osung Kwan[2] and Da-Ming Zhu[2,1]

[1]Department of Modern Physics, Univ. of Sci. & Tech. of China, Hefei, China 230027, China

[2]Department of Physics, University of Missouri, Kansas City, Missouri 64110, USA

zhud@umkc.edu

Abstract. Viscoelastic behaviors of polyethylene glycol (PEG) solution boundary layers were studied using a quartz crystal resonator technique. The technique probes the solution boundary layer adjacent to the quartz crystal surface by detecting resonant acoustic excitations that decay in the solution. Two different types of the electrode surfaces on the quartz crystal were used in the study; one is bare gold and another is gold plated with thiol-group (SH) attached polyethylene glycol (SH-PEG). The results show that different surfaces have little effect on the measured viscoelastic properties of the solution boundary layer over a wide range of concentration. Near the semidilute concentration of the solution, the viscosity of the boundary layer increases rapidly, following a power law with an exponent of 1.5 in its concentration dependence. The dynamic shear modulus of the boundary solution layer is nearly zero at the low concentration but rises rapidly as the concentration of the solution is approaching the semidilute concentration. More detailed experimental studies and theoretical modeling are needed in order to understand these intriguing results.

1. Introduction

Although the properties of polymer solutions have been studied for quite a long time, the subject still draws considerable interests due to the important roles played by polymer solutions in many different fields [1-5]. A number of import issues that remain to be resolved include how the properties of polymer solutions are affected by the presence of solid boundaries. The results from recent studies of boundary layers of simple molecules

T.-T. Wu and C.-C. Ma (eds.), *IUTAM Symposium on Recent Advances of Acoustic Waves in Solids*, IUTAM Bookseries 26, DOI 10.1007/978-90-481-9893-1_43,

such as water indicate that boundary effect can be significant; in some cases the presence of the boundary can completely change the structures or alter the phase of the boundary layers [4-8]. However, a quantitative description of how the viscoelastic properties of polymer solution boundary layers are affected by the boundary has not yet been achieved. The aim of our work presented here is to investigate the viscoelastic behaviors of boundary layers of a commonly used polymer solution near a solid-liquid interface to elucidate the boundary effects to the properties of polymer solutions.

2. The Technique

To probe directly the viscoelastic properties of polymer boundary layers in solutions can be a challenge task since the presence of the liquid makes many techniques developed for studying solid-vapor interfaces difficult to use. In this study we used a quartz crystal resonator, more commonly known as quartz crystal microbalance (QCM), technique which replies on the inverse piezoelectric effect of a quartz crystal [9-15]. By applying an AC electric field across a quartz crystal, a shear deformation (strain) oscillation is excited in the crystal. The resonance of such an oscillation is extremely sensitive to the media in contact with the surface of the crystal. Assuming that an overlayer on the top of a quartz crystal can be described as a viscoelastic element which consists of a spring and a dashpot in parallel and setting the normal direction of the overlayer to be in y-direction while the direction parallel to the interface to be in x-direction, the shear stress σ_{xy} applied to the overlayer and its elastic response contribute to the stress/strain relation follow a relation [9-12]:

$$\sigma_{xy} = \mu \frac{\partial u_x(y,t)}{\partial y} + \eta \frac{\partial v_x(y,t)}{\partial y} \tag{1}$$

where u_x, v_x are the shear displacement and the corresponding velocity of a small element in the overlayer; μ and η are the elastic shear modulus and the viscosity of the overlayer, respectively. The wave equation governing the shear waves propagating in the overlayer is

$$\mu^* \frac{\partial^2 u_x(y,t)}{\partial y^2} = -\rho\omega^2 u_x(y,t) \tag{2}$$

where $\mu^* = \mu + i\omega\eta$ is a complex shear modulus of the overlayer. Assuming a "no-slip" boundary condition between the overlayer and the quartz crystal, a general solution of the wave equation (2) can be obtained. From the general solution, the shifts in resonance frequency Δf and dissipation factor ΔD are obtained [10-12]. The general behaviors of Δf and ΔD depend on a viscous penetration depth $\delta = (2\eta/\rho\omega)^{1/2}$. If the overlayer thickness is much smaller than δ, the dissipation factor vanishes, while $\Delta f/f$ is proportional to the layer thickness, a result widely known as Sauerbrey relation [9]. In this case, a mass deposited on the order of a few nano-grams or less can be detected, and the technique has been widely used for monitoring thin film depositions over the past several decades [9]. On the other hand, if the overlayer thickness is much larger than δ, the acoustic response of the QCM due to the contact with the overlayer can be described by

$$\Delta f \approx -\frac{1}{2\pi\rho_q h_q}\sqrt{\frac{\rho}{2}}(\eta\omega\sqrt{\frac{\sqrt{\mu^2+\eta^2\omega^2}+\mu}{\mu^2+\eta^2\omega^2}}-\mu\sqrt{\frac{\sqrt{\mu^2+\eta^2\omega^2}-\mu}{\mu^2+\eta^2\omega^2}}) \quad (3)$$

$$\Delta D \approx \frac{1}{\pi f \rho_q h_q}\sqrt{\frac{\rho}{2}}(\eta\omega\sqrt{\frac{\sqrt{\mu^2+\eta^2\omega^2}-\mu}{\mu^2+\eta^2\omega^2}}+\mu\sqrt{\frac{\sqrt{\mu^2+\eta^2\omega^2}+\mu}{\mu^2+\eta^2\omega^2}}) \quad (4)$$

where ρ_q and h_q are the density and thickness of quartz crystal electrode, respectively [9-12]. But the contributions to Δf and ΔD are from the layer with thickness δ adjacent to the quartz crystal in the overlayer. If the overlayer's density and viscosity are similar to that of water and the resonant frequency of the quartz crystal is 10 MHz, the viscous penetration depth δ is about a few hundred nanometers. Thus, if a quartz crystal operated at such frequency is submerged in a solution with viscosity similar to that of water, the measured Δf and ΔD essentially probe the viscoelastic properties of the solution boundary layer with a thickness of about a few hundred nanometers adjacent to the crystal-solution interface.

Thus, by measuring Δf and ΔD simultaneously on a thick solution overlayer, the viscosity and shear modulus of the solution boundary layer adjacent to the crystal surface can be determined using Eqs. (3) and (4). Fig. 1 is a plot of $-\Delta f/\Delta D$ as a function of η and μ. A close look of Fig. 1 as well as Eqs. (3) and (4) finds that in the viscous limit, $\mu \cong 0$, the ratio $-\Delta f/\Delta D$ is a constant that depends only on the measuring frequency ($\omega/4\pi$). Such a characteristic has been tested in the determination of the viscosities of a number of simple molecular solutions, and good agreements have been found between the viscosity of the solution layer using QCM and that obtained using other techniques [9-15]. If the measured $-\Delta f/\Delta D$ shows a systematic variation as a function of the solution conditions (e.g. the concentration), that would be an indication that the shear modulus of the solution layer cannot be neglected. Thus, the variation of $-\Delta f/\Delta D$ as a function of solution condition can be used as a criterion for determining whether a solution boundary layer can be treated as purely viscous or not. This approach allows experimentally determining the viscosity and shear modulus of solution boundary layers.

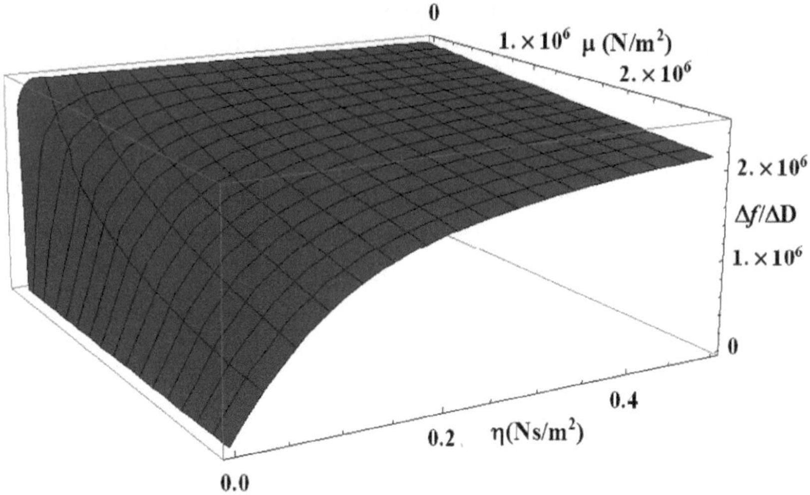

Fig. 1 The ratio $-\Delta f/\Delta D$ as a function of η and μ calculated using Eqs. (3) and (4). Notice that in the viscous limit ($\mu \cong 0$) the ratio is a constant. Beyond that limit, the ratio varies with the viscosity as well and the shear modulus.

3. Experimental

In this work, we used the quartz crystal resonator technique to investigate the properties and behaviors of boundary layers polyethylene glycol (PEG) solutions. PEG is a commercially important polymer which has a wide range of application in many different fields including biomedical and clinic research [16]. In the study, a freshly cleaned quartz crystal with a fundamental resonant frequency of 5 MHz operated at its third harmonics (15 MHz) was used. The details about the crystal surface preparation and experimental procedure can be found in Ref. 17. The electrodes on the quartz crystal were connected to a network analyzer (Agilent technologies E5061A) for measuring the acoustic responses from the solution; the resonant peaks at the third harmonics of the fundamental frequency of the crystal were monitored to obtain Δf and ΔD as a function of PEG concentration in solution. The resonant frequency ($f_{resonant}$) is measured by sweeping the driving frequency and monitoring the value of the frequency at which the response amplitude is the largest. The dissipation factor (D) is determined by dividing the fall width at half maximum ($FWHM$) of the resonant peak with the resonant frequency ($D = FWHW/f_{resonant}$). A schematic illustration of the experimental setup used for measuring Δf and ΔD and a general behavior Δf and ΔD as a function of solution concentration are illustrated in Fig.2.

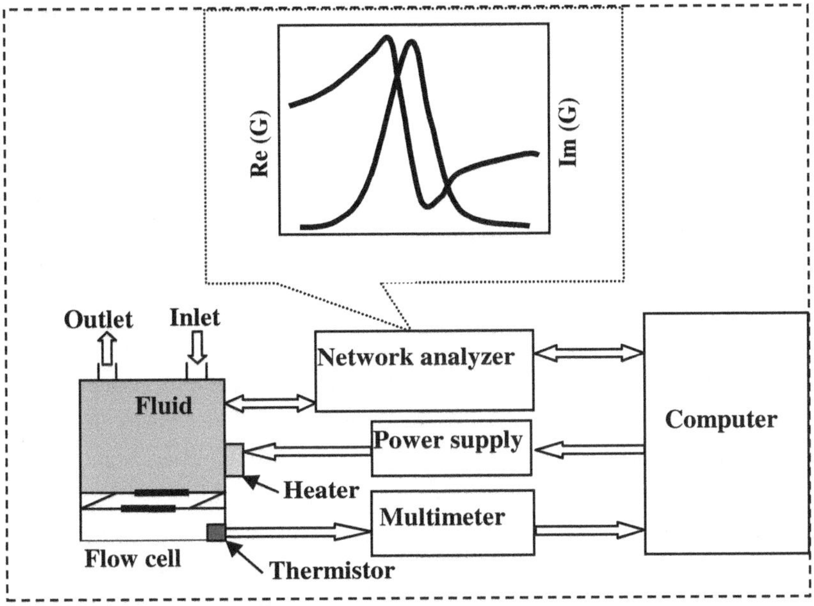

Fig. 2 An illustration of the experimental setup.

The polyethylene glycol (PEG) samples used have molecular weight M_w=1 x 10^4 g/mol and polydispersity index M_w/M_n = 1.06 were used. In order to check if the measured results are sensitive to the molecule-substrate interaction, we used quartz crystal with bare Au electrodes and then platted Au surface with a layer of thoil-attached PEG (SH-PEG) repeating the experiment [16]. By comparing the results obtained using the quartz crystal with bare Au electrode surface and with SH-PEG plated Au surface, any significant effect of the surface interaction to the measured viscoelastic properties of the boundary solution layer can be addressed.

4. Experimental Results and Discussion

In Figs. 3 and 4 we plot the resonant frequency shift (Δf_w) and the dissipation factor change (ΔD_w) of the quartz crystals with the bare gold electrode and with the SH-PEG plated gold electrode as a function of the PEG concentration scaled by a semidilute concentration c^*. Here, Δf_w and ΔD_w are in reference with the resonant frequency f_w and the dissipation factor D_w measured after crystals were submerged in pure water solution (before PEG was added); c^* is defined as $c^*=3M/(4\pi<R^2>^{1.5}N_{av})$, where M is the molecular weight, R is the gyration radius of the polymer coil, and N_{av} is Avogadro's number [18,19]. At c^* the chains of

PEG molecules are supposedly in contact but without significant entanglements. The measured Δf_w and ΔD_w can be related to Δf and ΔD by adding the differences of the resonance frequency and the dissipation factor in pure water and in air. Δf_w measured with SH-PEG plated quartz crystal has a constant shift comparing to that measured by un-plated crystal. The shift is due to chemisorption of the SH-PEG layer directly onto the gold electrode surface [17]. This shift can be subtracted off from Δf_w by measuring the acoustic responses of the quartz crystal resonator right after the gold electrode surface being cleaned and then again after the SH-PEG was chemisorbed onto the gold surface [17].

Figs. 3 and 4 show that acoustic responses of the quartz resonator as a function of the PEG concentration change very little in solutions at the low concentration, and rise rapidly as the concentration is above about 0.1 c^*. The acoustic responses for the two different surfaces move together, except a small and constant shift in Δf_w. The constant shift was identified as due to the chemisorpted first layer of SH-PEG. The results indicate that beyond the adsorbed first layer, adsorptions of additional PEG and the concentration profiles of PEG layers for the two different surfaces are very close to each other. Thus, different surfaces on a quartz crystal have little effect on the viscoelastic properties of the solution boundary layer with a thickness of about a few hundred nanometers.

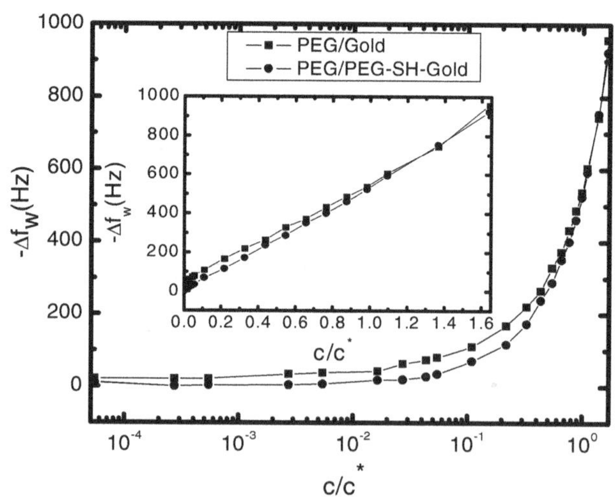

Fig. 3 Measured resonant frequency shift of the quartz crystal resonator as a function of concentration of PEG in solution. The frequency shift is in reference with that measured in pure water. (■) The surface of electrode is pure gold; (●) The surface of electrode is plated with a layer of SH-PEG. SH-groups chemically bond to the gold electrode surface. The lines are guides to the eye. The inlet is a linear plot.

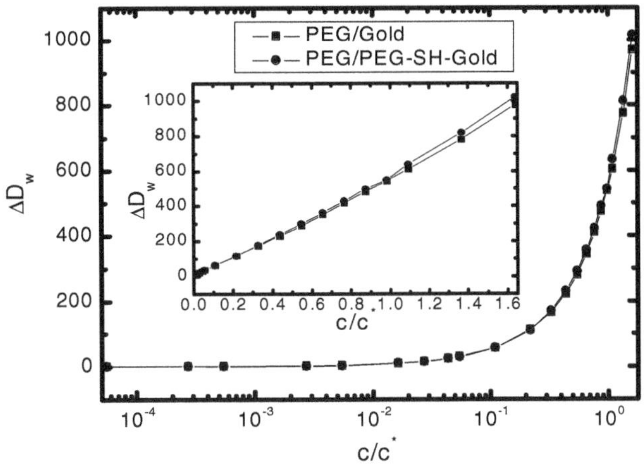

Fig. 4 Measured dissipation factor changes of the quartz crystal resonator as a function of concentration of PEG in solution. The changes are in reference with that measured in pure water. (■) The surface of electrode is pure gold; (•) The surface of electrode is plated with a layer of SH-PEG. The lines are guides to the eye. The inlet is a linear plot.

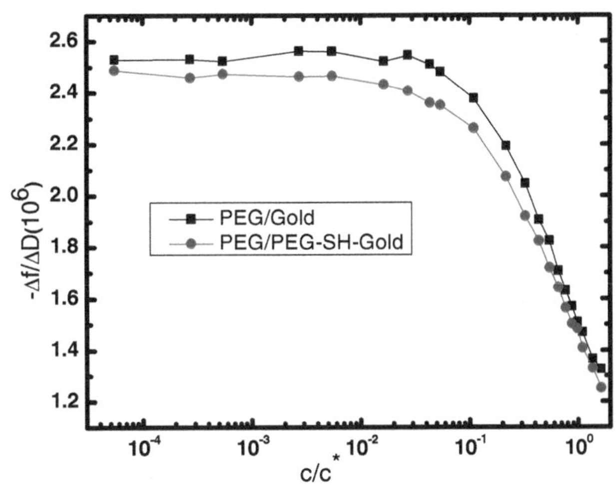

Fig. 5 The ratio $-\Delta f/\Delta D$ as a function of the scaled PEG concentration. The lines are the guide to the eye.

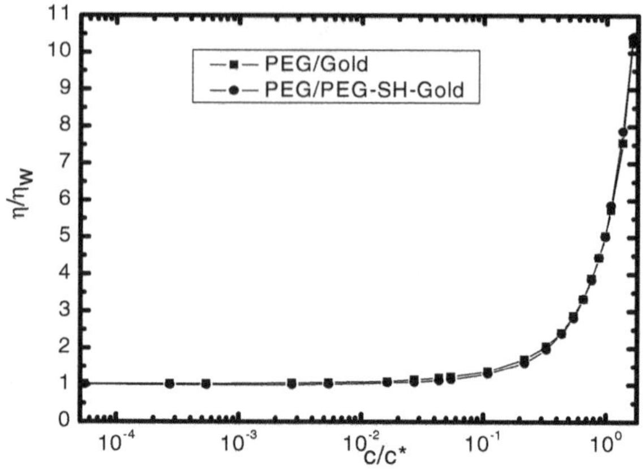

Fig. 6 Viscosity of PEG solution boundary layer as a function of PEG concentration, derived using Eqs. (3) and (4).

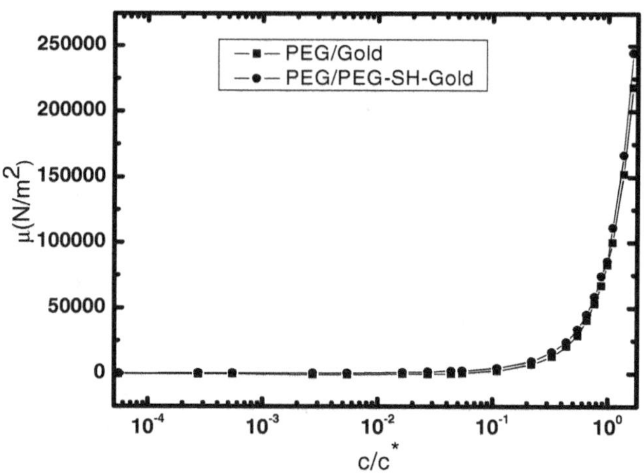

Fig. 7 Shear modulus of PEG solution boundary layer as a function of PEG concentration, derived using Eqs. (3) and (4).

Fig. 5 plots the ratio $-\Delta f/\Delta D$ as a function of the scaled PEG concentration. The ratio is nearly a constant at the low concentration, but decreases noticeably as the concentration is above 0.1 c^*. Such a decrease indicates that the acoustic response of the boundary solution layer is not purely viscous; the shear modulus of the boundary layer contributes to the response as the concentration is close and above the semidilute concentration c^*. We thus use Eqs. (3) and (4) to derive the viscosity η and the shear modulus μ of the solution boundary layer from the measured

Δf and ΔD. The results are plotted in Figs. 6 and 7. The viscosity of the solution boundary layer is very close to that of pure water as the concentration is below 0.1 c^*. This is not surprising since at such a dilute concentration, PEG molecules move nearly independently of each other, resulting in a very week concentration dependence of the viscosity.

Comparing to the viscosity of bulk PEG solutions of about the molecular weight, the viscosity of the boundary layer display similar concentration dependence. At the concentration close to and above the semidilute concentration, the dependence of viscosity can be fitted with a power law with an exponent approximately equal to 1.5 [17]. Such an exponent is smaller than that obtained for bulk PEG solution, and is also smaller than that predicted by theory [17]. Our recent studies further confirm that such a deviation exists for higher molecular weight PEG solutions.

The shear modulus of the PEG solution boundary layer in contact to the quartz crystal surface derived from the measured Δf and ΔD using Eqs. (3) and (4) show a similar concentration dependence as that for viscosity. The shear modulus is nearly zero at the low concentration, but rises rapidly as the concentration is close to and above the semidilute concentration. It should be noted that the shear modulus measured using the quartz crystal resonator technique is the dynamic elastic modulus of the boundary layer at the measuring frequency. The measured rise in shear modulus for the boundary solution layer could be simply due to an increase in relaxation time in the boundary layer as the concentration increases. However, it is also possible that such a rise is due to the changes in the rigidity in the solution boundary layer because of the interaction of the layer with the boundary and the geometric confinement to the layer from the boundary. Several recent studies demonstrated that a thin boundary layer of water could behaves like a solid layer at the room temperature [6-8]. The results of the PEG solution boundary layers presented here suggest that similar phenomena might exist in polymer systems. We are currently conducting studies using combined quartz crystal resonator and atomic force microscopy techniques to explore further the nature of the elastic behaviors of PEG solution boundary layers.

In summary, we present the results of a recent study of viscosity and shear modulus of polyethylene glycol (PEG) solutions using a quartz crystal resonator technique. The results show that near the semidilute concentration of the solution, both the viscosity and shear modulus of the solution boundary layer rises rapidly. A more detailed quantitative study of such an intriguing behavior is apparently called for in order to understand the viscoelastic nature of polymer solution boundary layers in general and PEG solution boundary layers in specific.

Ackledgements: D. Zhu is supported in part by grants from University of Missouri Research Board and Research Corporation.

References

[1] Teraoka, I.: *Polymer solutions.* John Wiley & Sons, New York (2002).

[2] Tanford, C.: *Physical chemistry of macromolecules.* Wiley, New York (1961).

[3] Morawetz, H.: *High polymers*, 2nd ed., Macromolecules in solution. Wiley: New York (1975).

[4] Dutta, A. K., Nayak, A., Belfort, G.: Viscoelastic properties of adsorbed and cross-linked polypeptide and protein layers at a solid–liquid interface. *J. Colloids and Interface Science* **324** 55-60 (2008).

[5] Hu, H., Granick, S.: Viscoelastic dynamics of confined polymer melts. Science 258,1339-1342(1992); Zhu, Y., Granick, S.: Viscosity of interfacial water. *Phys. Rev. Lett.* **87** 096104 (2001).

[6] Raviv, U., Laurat, P., Klein, J.: Fluidity of water confined to subnanometre films. Nature **413** 51-54(2001); Jagla, E. A.: Boundary lubrication properties of materials with expansive freezing. *Phys. Rev. Lett.* **88** 245504 (2002).

[7] Choi, E., Yoon, Y., Lee, S., Kang, H.: Freezing transition of interfacial water at room temperature under electric fields. *Phys. Rev. Lett.* **95** 085701 (2005).

[8] Jinesh, K. B., Frenken, J. W. M.: Capillary condensation in atomic scale friction: How water acts like a glue. *Phys. Rev. Lett.* **96** 166103(2006); Jinesh, K. B., Frenken, J. W. M.: Experimental evidence for ice formation at room temperature. *Phys. Rev. Lett.* **101** 036101(2008).

[9] Sauerbrey,G. Z.: Use of quartz crystal vibrator for weighting thin films on a microbalance. *Phys.* **155** 206-222 (1959); Bottom, V. E.: *Introduction to quartz crystal unit design.* Van Nostrand Reinhold Co.: New York (1982).

[10] Rodahl, M., Hook, F., Krozer, A., Brzezinski, P., Kasemo, B.: Quartz crystal microbalance setup for frequency and Q-factor measurements in gaseous and liquid environments. *Rev. Sci. Instr.* **66** 3924 (1995).

[11] Kanazawa, K. K., Gordon, J. G.: Frequency of a quartz microbalance in contact with liquid. *Anal. Chem.* **57** 1770-1771 (1985).

[12] Rodahl, M., Kasemo, B.: On the measurement of thin liquid overlayers with the quartz-crystal microbalance. *Sensors and Actuators A* **54** 448-456 (1996).

[13] Wu, B., Wu, K., Wang, P., Zhu, D. M.: Adsorption kinetics and adsorption isotherm of poly(N-isopropylacrylamide) on gold surfaces studied using QCM-D. *J. Phys. Chem. C* **111** 1131-1135 (2007).

[14] Wu, K., Wu, B., Wang, P., Hou, Y., Zhang, G., Zhu, D. M.: Adsorption isotherms and dissipation of adsorbed poly(N-isopropylacrylamide) in its swelling and collapsed states. *J. Phys. Chem B* **111** 8723-8727 (2007).

[15] Zhu, D. M., Wu, K., Wu, B., Wang, P., Fang, J. J.: Physisorption of poly(N-isopropylacrylamide) in its swollen and collapsed states: effects of molecular conformation and substrate interaction. *J. Phys. Chem. C* **111** 18679-18686 (2007).

[16] Kozer, N., Kuttner, Y. Y., Haran, G., Schreiber, G.: Protein-protein association in polymer solutions: from dilute to semidilute to concentrated. *Biophysical Journal* **92** 2139-2149 (2007).

[17] Wang, P., Fang, J., Hou, Y., Du, X., Zhu, Da. M.: Viscelastic properties of polyethylene glycol (PEG) boundary layers near a solid subtrate, *J. Phys. Chem. C* **113** 729-735 (2009).

[18] De Gennes, P. G.: Dynamics of entangled polymer solutions. II. Inclusion of hydrodynamic interactions. *Macromolecules* **9** 594-598 (1976).

[19] De Gennes, P. G.: *Scaling concepts in polymer physics.* Cornell University Press: Ithaca, New York (1979).

Appendix I: Symposium Program

Tuesday, May 26	
8:20-8:30	***Opening Ceremony***

Section 1 Waves in General

Chair: Jan D. Achenbach

Center for Quality Engineering and Failure Prevention, Northwestern University, USA.

08:30-08:55 **Time-Reversed Waves and Super-Resolution**

Mathias Fink

Institut Langevin, Laboratoire Ondes et Acoustique, ESPCI, 10 rue Vauquelin, 75005, Paris, France.

08:55-09:20 **Existence of Exceptional Body Waves and Subsonic Surface Waves in Monoclinic and Orthotropic Materials**

T. C. T. Ting

Division of Mechanics and Computation, Durand 262, Stanford University, Stanford, CA 94305, USA.

09:20-09:45 **Deformation Waves in Microstructured Materials: Theory and Numerics**

Jüri Engelbrecht, Arkadi Berezovski, and Mihhail Berezovski

Centre for Nonlinear Studies, Institute of Cybernetics at Tallinn University of Technology, Tallinn, Estonia.

10:10-10:30 ***Coffee Break***

Section 2 Waves and NDE (I)

Chair: Arthur G. Every

School of Physics, University of the Witwatersrand, South Africa.

10:30-10:55 **The Velocity of Anti-Plane Surface Waves on a Body with Depth-Dependent Properties**

Jan D. Achenbach

Center for Quality Engineering and Failure Prevention, Northwestern University, Evanston, IL 60208, USA.

10:55-11:20 **Resonance Ultrasound Microscopy for Imaging Young's Modulus of Solids**

Masahiko Hirao and Hirotsugu Ogi

Graduate School of Engineering Science, Osaka University, Toyonaka, Osaka 560-8531, Japan.

11:20-11:45 **Nonclassical Nonlinearity in Solids for Defect-Selective Imaging and NDE**

Igor Solodov

Department of Non-Destructive Testing,Institute for Polymer Technology, University of Stuttgart,Pfaffenwaldring 32, 70569 Stuttgart, Germany.

11:45-13:30 ***Picture Taken & Lunch***

Seminar Section

Chair: Chin-Teh Sun

School of Aeronautics and Astronautics, Purdue University, USA.

13:30-13:35 **Band Structure Calculations by Modal Analysis**

Mahmoud I. Hussein

Department of Aerospace Engineering Sciences, University of Colorado at Boulder, Boulder, CO 80309, USA.

13:35-13:40 **Band Gap in Phononic Crystal Thin Plate with/without Mirror Plane**

Zhilin Hou[1] and Badreddine Assouar

[1]Department of Physics, South China University of Technology, Wushan, Guangzhou, 510640, China.

[2]LPMIA, CNRS - Nancy University, BP: 239, Bd des Aiguillettes, 54506 Vnadoeuvre les Nancy, France.

13:40-13:45 **Radial Sonic Crystals**

Daniel Torrent and José Sánchez-Dehesa

Wave Phenomena Group, Department of Electronics Engineering, Polytechnic University of Valencia, C/Camino de Vera s/n, Valencia, 46022, Spain.

13:45-13:50 **Polarization States in 2D Phononic Crystals and Waveguides**

Y. Achaoui[1], A. Khelif[2], S. Benchabane[1] and V. Laude[1]

[1]*Institut FEMTO-ST, Université de Franche-Comté, CNRS, ENSMM, UTBM, 32 avenue de l'Observatoire, F-25044 Besançon Cedex, France.*

[2]*Georgia Tech Lorraine, UMI Georgia Tech -CNRS 2958, Metz Technopôle 2-3, rue Marconi, 57070 Metz, France.*

13:50-13:55 **Dispersion Analysis of Wave Motion in Microstructured Solids**

Tanel Peets

Centre for Nonlinear Studies, Institute of Cybernetics at Tallinn University of Technology, Akadeemia tee 21, 12618 Tallinn, Estonia.

13:55-14:00 **On the Role that Scholte Waves Play in Acoustic Propagation along a Fluid-Solid Interface**

Piotr Borejko

Department of Civil Engineering, Vienna University of Technology, Karlsplatz 13/E206/3, Vienna, A-1040, Austria.

14:00-14:05 **Forced Motions in Rectangular Elastic Waveguide**

A. A. Bondarenko[1], V. V. Meleshko[2] and A. N. Trofimchuk[1]

[1]*Department of Natural Resources, Institute of Telecommunications and Global Information Space of the NAS Ukraine, Chokolivs'ka 13, Kiev, 03186, Ukraine.*

[2]*Department of Theoretical and Applied Mechanics, Kiev National Taras Shevchenko University, Volodymyrs'ka 60, Kiev, 01601, Ukraine.*

14:05-14:10 **Normal Waves in Anisotropic Cylinders of Sector Cross-Section**

Storozhev Valeriy, Troyan Renata and Puzyrev Vladimir

Mathematical Department, Donetsk National University, Universitetskaya 24, Donetsk, 83117, Ukraine.

14:10-14:15 **Dispersion Behaviors of ASF Modes Propagating along Wedges Tips with Coatings**

Po-Shien Tung[1], Sheng-Wei Tang[2] and Che-Hua Yang[1]

[1]*Graduate Institute of Manufacturing Technology, National Taipei university of Technology, 1, Sec. 3, Chung-Hsiao E. Rd. Taipei 106, Taiwan, R.O.C.*

[2]*Graduate Institute of Mechanical and Electrical Engineering, National Taipei university of Technology,1, Sec. 3, Chung-Hsiao E. Rd. Taipei 106, Taiwan, R.O.C.*

14:15-14:20 **Scattering from a Rectangular Crack in a Cladding**

Per-Åke Jansson

Department of Applied Mechanics, Chalmers University of Technology, SE-412 96 Gothenburg, Sweden.

14:20-14:25 **Defect Inspection of Complex Structure in Pipes by Guided Waves**

Ping-Hung Lee, and Shiuh-Kuang Yang

Department of Mechanical and Electro-Mechanical Engineering, University of National Sun Yat-Sen, No. 70, Lienhai Rd, Kaohsiung, 804, Taiwan R.O.C.

14:25-14:30 **SAW Gas Sensor with Nanostructured Sensing Materials**

Yung-Yu Chen[1], Tsung-Tsong Wu[2], Tai-Hsu Chou[2] and Fu-Chun Huang[2]

[1]*Department of Mechanical Engineering, Tatung University, No.40, Sec. 3, Zhongshan N. Rd., Taipei, Taiwan.*

[2]*Institute of Applied Mechanics, National Taiwan University, No. 1, Sec. 4, Roosevelt Road, Taipei, Taiwan.*

14:30-14:35 **Acousto-Optic Response of Nematic Liquid Crystals to Interface Acoustic Waves**

Ching-Chung Yin, Kang-Che Huang and I-Han Chang

Department of Mechanical Engineering, National Chiao Tong University, 1001 Ta Hsueh Road, Hsinchu, 300, Taiwan, Republic of China.

14:35-14:40 **Design of IF Two-Track Surface Acoustic Wave Filters Using (100) AlN/Diamond Structures**

Ruyen Ro[1], Chia-Chi Sung[2], Ruyue Lee[1] and Yuan-Feng Chiang[2]

[1]*Department of Electrical Engineering, I-Shou University, 1, Section 1, Hsueh-Cheng Rd. ,Ta-Hsu Hsiang, Kaohsiung County 840, Taiwan, Republic of China.*

[2]*Department of Engineering Science and Ocean Engineering, National Taiwan University, 1, Section 4, Roosevelt Rd. Da-an District, Taipei County 106, Taiwan, Republic of China.*

14:40-14:45 **Acoustic Wave Properties of (100) AlN and ZnO Films**

Sean Wu[1], Zhi-Xun Lin[2], Ruyen Ro[3] and Maw-Shung Lee[2]

[1]*Department of Electronics Engineering and Computer Science, Tung Fang Institute of Technology, No. 110, Tung-Fung Rd., Hunei Shiang, Kaohsiung County, 829, Taiwan.*

[2]*Department of Electronic Engineering, National Kaohsiung University of Applied Sciences,No. 415, Chien Kung Rd., Kaohsiung City 807, Taiwan.*

[3]*Department of Electrical Engineering, I-Shou University, No. 1, Sec. 1, Syuecheng Rd., Dashu Township, Kaohsiung County 840, Taiwan.*

14:45-14:50 **Study Mechanical and Viscoelastic Properties of Polymer Solutions near a Solid-Fluid Interfaces Using Combined Quartz Crystal Microbalance and Atomic Force Microscopy**

Ping Wang[1], Jiajie Fang[1], Yihong Kang2, Sheng Qin[1], Osung Kwan[2] and Da-Ming Zhu[2,1]

[1]*Department of Modern Physics, Univ. of Sci. & Tech. of China,Hefei, China 230027, China.*

[2]*Department of Physics, University of Missouri, Kansas City, Missouri 64110, USA.*

15:00-15:40 ***Coffee Break***

Section 3 Phononic Crystals (I)

Chair: Vincent Laude

FEMTO-ST, Université de Franche-Comté, CNRS, ENSMM, UTBM, France.

15:40-16:05 **Guided Elastic Waves at Periodically Structured Surfaces and Interfaces**

A. G. Every and A.A. Maznev

School of Physics, University of the Witwatersrand, PO Wits 2050, Johannesburg, South Africa.

16:05-16:30 **Metafluids Based on Sonic Crystals**

José Sánchez-Dehesa and Daniel Torrent

Wave Phenomena Group, Department of Electronics Engineering, Polytechnic University of Valencia, Valencia, Spain.

16:30-16:55 **Band Structure and Phonon Transport in a Phononic Crystal Made of a Periodic Array of Dots on a Membrane**

B. Djafari Rouhani, Y. Pennec and H. Larabi

Institut d'Electronique, Microélectronique et Nanotechnologie, UMR CNRS 8520, Avenue Poincaré, Université de Lille1 Sciences et Technologies, 59652 Villeneuve d'Ascq, France.

16:55-17:20 **Material Parameters Determining the Band Gaps of Solid-Solid Phononic Crystals**

Xiao-Zhou Zhou[1], Yue-Sheng Wang[1] and Chuanzeng Zhang[2]

[1]Institute of Engineering Mechanics, Beijing Jiaotong University, Beijing, 100044, China.

[2]Department of Civil Engineering, University of Siegen, Siegen, D-57076, Germany.

18:00-21:00 ***Visit Taipei 101***

Wednesday, May 27

Session 4 Wave Electronics (I)

Chair: Yook-Kong Yong

Civil and Environmental Engineering,Rutgers,the State University of New Jersey, USA.

08:30-08:55 **Phase-Sensitive and Fast-Scanning Laser Probe System for RF SAW/BAW Devices**

Ken-ya Hashimoto[1], Nan Wu[1], Keisuke Kashiwa[1], Tatsuya Omori[1], Masatsune Yamaguchi[1], Osamu Takano[2], Sakae Meguro[2], Naoki Kasai[2] and Koichi Akahane[2]

[1]Graduate School of Engineering, Chiba University, Chiba 263-8522, Japan.

[2]Neoark Co. Ltd, Hachi-Ohji, Tokyo 192-0015, Japan.

08:55-09:20 **SAW-Tags - New Generation**

Victor Plessky[1] and Sanna Härmä[2]

[1]GVR Trade SA, Ch. du Vignoble 31C, Bevaix, 2022, Switzerland.

[2]Department of Applied Physics, Helsinki Univ. of Technology, P.O. Box 3500, FI-02015 TKK, Espoo, Finland.

09:20-09:45 **A Two-Dimensional Analysis of Surface Acoustic Waves in Finite Piezoelectric Plates**

Ji Wang, Rongxing Wu and Jianke Du

Piezoelectric Device Laboratory, Department of Mechanics and Engineering Science,School of Engineering, Ningbo University, 818 Fenghua Road, Ningbo, Zhejiang 315211, China.

09:45-10:10 **Wireless Drive of Piezoelectric Components**

Junhui Hu and Satyanarayan Bhuyan

School of Electrical & Electronic Engineering, Division of Microelectronics, Nanayang Technological University, Singapore 639798, Singapore.

10:10-10:30 ***Coffee Break***

Session 5 Waves and NDE (II)

Chair: Igor Solodov

Department of Non-Destructive Testing, Institute for Polymer Technology, University of Stuttgart, Germany.

10:30-10:55 **Counterpropagating Ultrasonic Waves for Inhomogeneous Materials Characterization**

Arvi Ravasoo

Centre for Nonlinear Studies, Institute of Cybernetics at Tallinn University of Technology, Akadeemia tee 21, 12618 Tallinn, Estonia.

10:55-11:20 **Ultrasonic Characterization of the Mechanical Properties of Thin Films and Coatings**

Sridhar Krishnaswamy

Department of Mechanical Engineering, Northwestern University, Evanston, IL 60208, USA.

11:20-11:45 **Evaluation of Corrosion in Carbon Steel Pipes by Laser-Generated Guided Wave**

Do-Youn Kim[1], Joon-Hyun Lee[2], Younho Cho[2], Jaesun Lee[1] and Jan D. Achenbach[3]

[1]*Graduate School of Mechanical Engineering, Pusan National University, Jangjeon-dong, Gumjeong-gu, Busan, 609-735, Korea.*

[2]*School of Mechanical Engineering, Pusan National University, Jangjeon-dong, Gumjeong-gu, Busan, 609-735, Korea.*

[3]*Walter P. Murphy and Distinguished McCormick School Professor, Center for Quality Prevention and Failure Prevention, Northwestern University, USA.*

11:45-12:10 **Measurement of Surface and Lamb Waves with Application on Acoustic Sensing and Non-Destructive Evaluation**

Yung-Chun Lee, Shi Hoa Kuo and Cheng-Hsien Chung

Department of Mechanical Engineering, National Cheng-Kung University, #1 University Road, Tainan, 70101, Taiwan

12:10-17:30 ***National Palace Museum***

18:00-21:00 ***Banquet***

Thursday, May 28

Session 6 Phononic Crystals (II)

Chair: Thomas C. T. Ting

Division of Mechanics and Computation, Stanford University, Stanford, USA.

08:30-08:55 **Behavior of Wave Motion in an Acoustic Metamaterial with Anisotropic Mass Density**

C. T. Sun and H. H. Huang

School of Aeronautics and Astronautics, Purdue University, W. Lafayette, IN 47907, USA

08:55-09:20 **Complex Band Structure of Phononic Crystals and the Diffraction problem**

Vincent Laude[1], Younes Achaoui[1], Sarah Benchabane[1] and Abdelkrim Khelif[2]

[1]*Institut FEMTO-ST, Université de Franche-Comté, CNRS, ENSMM, UTBM; 32 avenue de l'Observatoire, F-25044 Besançon, France.*

[2]*Georgia Tech Lorraine, UMI Georgia Tech -CNRS 2958, Metz Technopôle, 2-3, rue Marconi, 57070 Metz, France.*

09:20-09:45 **Versatile phononic slabs**

I. E. Psarobas

Section of Solid State Physics, The University of Athens, Panepistimioupolis, 157 84 Athens, Greece.

09:45-10:10 **Omnidirectional Band Gap Mirror for Surface Acoustic Wave**

Abdelkrim Khelif[1,2], Abdelkrim Choujaa[1],
Sarah Benchabane[1] and Vincent Laude[1]

[1]*Institut FEMTO-ST, Université de Franche Comté, CNRS, ENSMM, UTBM, Besançon, France.*

[2]*GeorgiaTech-CNRS International Joint Laboratory UMI 2958, ECE, GT, Atlanta, USA.*

10:10-10:30 ***Coffee Break***

Session 7 Wave Electronics (II)

Chair: Ken-ya Hashimoto

Graduate School of Engineering, Chiba University, Japan

10:30-10:55 **Novel High Frequency, Temperature Stable, Quartz Gyroscopes**

Yook-Kong Yong[1], Mihir Patel[1] and Jianke Du[2]

[1]*Department of Civil and Environmental Engineering, Rutgers University, 623 Bowser Road, Piscataway, NJ 0885e, USA.*

[2]*Department of Engineering Mechanics, Ningbo University, Ningbo, China.*

10:55-11:20 **COM Model Characterization for RF SAW Devices**

Ben Abbott[1], Kevin Gamble[1], Natalya Naumenko[2],
Svetlana Malocha[1] and Marc Solal[1]

[1]*TriQuint Semiconductor, 1818 S. Hwy 441, Apopka, FL 32703, USA.*

[2]*Moscow Steel and Alloys Institute, Moscow, Russia.*

11:20-11:45 **Development of a Surface Acoustic Wave Sensor for Gas Detection**

Chi-Yen Shen, Cheng-Liang Hsu, Rume-Tze Tsai and Ming-Yau Su

Department of Electrical Engineering, I- Shou University, No. 1, Section 1, Syuecheng Rd., Dashu Township, Kaohsiung County, Taiwan.

11:45-12:10 **Three Dimensional Displacement Measurement of Transient Elastic Wave Propagation Using a Multidimensional Point-Wise Fiber Bragg Grating Displacement Sensor System**

Chien-Ching Ma and Kuo-Chih Chuang

Department of Mechanical Engineering, National Taiwan University, Taipei 106, Taiwan.

12:10-13:30 ***Lunch***

Session 8 Phononic Crystals (III)

Chair: José Sánchez-Dehesa

Department of Electronics Engineering, Polytechnic University of Valencia, Spain.

13:30-13:55 **Phononic Quasicrystals**

Walter Steurer, Sofia Deloudi and Daniel Sutter-Widmer

Laboratory of Crystallography, ETH Zurich, Wolfgang-Pauli-Strasse 10, 8006 Zurich, Switzerland.

13:55-14:20 **Band Structure of Three-Dimensional Phononic Crystals with an Opal structure**

Yukihiro Tanaka, Shin-ichiro Tamura and Takuro Okada

Department of Applied Physics, Hokkaido University, Sapporo 060-8628, Japan.

14:20-14:45 **Bandgap Characteristics of a 3D Phononic Meta Material Composed of Ordered Quantum Dots**

Chi-Kuang Sun

Department of Electrical Engineering,, National Taiwan University, Taipei 106, Taiwan.

14:45-15:10 **Analytical and Experimental Analysis of Bandgaps in Nonlinear One Dimensional Periodic Structures**

Nicholas Boechler[1], Chiara Daraio[1], Raj Narisetti[2], Massimo Ruzzene[2] and Michael J. Leamy[3]

[1]*Aeronautics & Applied Physics, California Institute of Technology, Pasadena CA, 91125, USA.*

[2]*School of Aerospace Engineering, Georgia Institute of Technology, Atlanta GA, 30332, USA.*

[3]*School of Mechanical Engineering, Georgia Institute of Technology, Atlanta GA, 30332, USA.*

15:10-15:35 **Lamb Waves in Phononic-Crystal Plates: Numerical Studies and Experiments**

Tsung-Tsong Wu[1], Jin-Chen Hsu[2] and Jia-Hong Sun[1]

[1]*Institute of Applied Mechanics, National Taiwan University, Taipei 106, Taiwan.*

[2]*Department of Mechanical Engineering, National Yunlin University of Science & Technology, Yunlin 640, Taiwan.*

15:35-16:00 ***Closing Ceremony***

Appendix II: List of Participants

Ben Abbott
TriQuint Semiconductor,
1818 S. Hwy 441, Apopka, FL 32703, USA.
bpabbott@tqs.com

Younes Achaoui
Institut FEMTO-ST, Universite de Franche-Comte, CNRS, ENSMM, UTBM,
32 avenue de l'Observatoire,
F-25044 Besancon, France.
younes.achaoui@femto-st.fr

Jan D. Achenbach
Department of Mechanical Engineering, Northwestern University,
Evanston IL 60208, USA.

Walter P. Murphy and Distinguished McCormick School Professor, Center for Quality, Prevention and Failure Prevention, Northwestern University,
Evanston IL 60208, USA.
achenbach@northwestern.edu

Badreddine M. Assouar
Institut Jean Lamour, CNRS - Nancy University,
BP, 239, Bd des Aiguillettes, 54506 Vnadoeuvre les Nancy, France.
Badreddine.Assouar@lpmi.uhp-nancy.fr

Anastasiya A. Bondarenko
Department of Natural Resources, Institute of Telecommunications and Global Information Space of the NAS Ukraine,
Kiev 03186, Ukraine.
anastasiya.bondarenko@gmail.com

Piotr Borejko
Department of Civil Engineering, Vienna University of Technology,
A-1040 Vienna, Austria.
piotr.borejko@tuwien.ac.at

Shuo-Hung Chang
Department of Mechanical Engineering, National Taiwan University,
Taipei 106, Taiwan.
shchang@ccms.ntu.edu.tw

Francis Chen
Tai-Saw Technology CO., LTD., No.3, Gongye 2nd Rd., Pingzhen City,
Taoyuan County 324, Taiwan.
francischen@mail.taisaw.com

Yung-Yu Chen
Department of Mechanical Engineering, Tatung University
No.40, Sec. 3, Zhongshan N. Rd., Taipei, Taiwan.
yychen@ttu.edu.tw

Younho Cho
School of Mechanical Engineering, Pusan National University,
Jangjeon-dong, Gumjeong-gu, Busan, 609-735, Korea.
mechcyh@pusan.ac.kr

Kuo-Chih Chuang
Department of Mechanical Engineering, National Taiwan University,
Taipei 106, Taiwan.
r89522806@ntu.edu.tw

Bahram Djafari-Rouhani
Institut d'Electronique, Microelectronique et Nanotechnologie, UMR CNRS 8520,
Avenue Poincare, Universite de Lille1 Sciences et Technologies,
59652 Villeneuve d'Ascq, France.
bahram.djafari-rouhani@univ-lille1.fr

Jüri Engelbrecht
Centre for Nonlinear Studies, Institute of Cybernetics at Tallinn University of Technology,
Tallinn, Estonia.
je@ioc.ee

Arthur G. Every
School of Physics, University of the Witwatersrand, PO Wits 2050,
Johannesburg, South Africa.
Arthur.every@wits.ac.za

Mathias Fink
Institut Langevin, Laboratoire Ondes et Acoustique, ESPCI, 10 rue Vauquelin,
75005, Paris, France.
mathias.fink@espci.fr

Ken-ya Hashimoto
Graduate School of Engineering, Chiba University,
Chiba 263-8522, Japan.
k.hashimoto@ieee.org

Masahiko Hirao
Graduate School of Engineering Science, Osaka University,
Toyonaka, Osaka 560-8531, Japan.
hirao@me.es.osaka-u.ac.jp

Jin-Chen Hsu
Department of Mechanical Engineering, National Yunlin University of Science &
Technology,
Yunlin 640, Taiwan.
hsujc@yuntech.edu.tw

Junhui Hu
School of Electrical & Electronic Engineering, Division of Microelectronics, Nanayang Technological University,
Singapore 639798, Singapore.
ejhhu@ntu.edu.sg

Mahmoud I. Hussein
Department of Aerospace Engineering Sciences, University of Colorado at Boulder, Colorado 80309, USA.
mih@colorado.edu

Chyanbin Hwu
Department of Aeronautics and Astronautics, National Cheng Kung University, Tainan, Taiwan.
chwu@mail.ncku.edu.tw

Per-Åke Jansson
Department of Applied Mechanics, Chalmers University of Technology,
Gothenburg, Sweden.
per-ake.jansson@chalmers.se

Abdelkrim Khelif
Institut FEMTO-ST, Universite de Franche-Comte, CNRS, ENSMM, UTBM,
32 avenue de l'Observatoire, F-25044 Besancon, France.

CNRS-Georgia Tech joint Laboratory, UMI 2958, Georgia Institute of Technology, Atlanta, GA 30332, USA.
abdelkrim.khelif@femto-st.fr

Sridhar Krishnaswamy
Department of Mechanical Engineering, Northwestern University,
Evanston, IL 60208, USA.
s-krishnaswamy@northwestern.edu

Mao-Kuen Kuo
Institute of Applied Mechanics, National Taiwan University,
No. 1, Sec. 4, Roosevelt Road, Taipei, Taiwan.
mkkuo@spring.iam.ntu.edu.tw

C. S. Lam
Epson Electronics America,
2580 Orchard Parkway, San Jose, California, 95131, USA.
cslam@eea.epson.com

Vincent Laude
Institut FEMTO-ST, Universite de Franche-Comte, CNRS, ENSMM, UTBM,
32 avenue de l'Observatoire, F-25044 Besancon, France.
vincent.laude@femto-st.fr

Ping-Hung Lee
Department of Mechanical and Electro-Mechanical Engineering,
National Sun Yat-Sen University,
Kaohsiung 804, Taiwan.
d933020006@student.nsysu.edu.tw

Yung-Chun Lee
Department of Mechanical Engineering, National Cheng Kung University,
Tainan 701, Taiwan.
yunglee@mail.ncku.edu.tw

Chien-Ching Ma
Department of Mechanical Engineering, National Taiwan University,
Taipei 106, Taiwan.
ccma@ntu.edu.tw

Hirotsugu Ogi
Graduate School of Engineering Science, Osaka University,
Toyonaka, Osaka 560-8531, Japan.
ogi@me.es.osaka-u.ac.jp

Shih-Yung Pao
TXC corporation, No.4, Gongye 6nd Rd., Pingzhen City,
Taoyuan County 324, Taiwan.
sypao@mems.iam.ntu.edu.tw

Yih-Hsing Pao
Institute of Applied Mechanics, National Taiwan University,
No. 1, Sec. 4, Roosevelt Road, Taipei, Taiwan.
yhpao@spring.iam.ntu.edu.tw

Tanel Peets
Centre for Nonlinear Studies, Institute of Cybernetics at Tallinn University of Technology,
Akadeemia tee 21, 12618 Tallinn, Estonia.
tanelp@cens.ioc.ee

Ioannis E. Psarobas
Section of Solid State Physics, The University of Athens, Panepistimioupolis,
157 84 Athens, Greece.
ipsarob@phys.uoa.gr

Arvi Ravasoo
Centre for Nonlinear Studies, Institute of Cybernetics at Tallinn University of Technology,
Akadeemia tee 21, 12618 Tallinn, Estonia.
arvi@ioc.ee

Troyan Renata
Mathematical Department, Donetsk National University,
Universitetskaya 24, Donetsk, 83001, Ukraine.
renata_shev@rambler.ru

Ruyen Ro
Department of Electrical Engineering, I-Shou University,
No. 1, Sec. 1, Syuecheng Rd., Dashu Township, Kaohsiung County 840, Taiwan.
ryro@isu.edu.tw

Massimo Ruzzene
Department of Aerospace Engineering, Georgia Institute of Technology,
Atlanta, GA, USA.
massimo.ruzzene@aerospace.gatech.edu

José Sánchez-Dehesa
Wave Phenomena Group, Department of Electronics Engineering, Polytechnic University of Valencia,
Valencia, Spain.
jsdehesa@upvnet.upv.es

Chi-Yen Shen
Department of Electrical Engineering, I-Shou University,
No. 1, Sec. 1, Syuecheng Rd., Dashu Township, Kaohsiung County 840, Taiwan.
cyshen@isu.edu.tw

Wen-Pin Shih
Department of Mechanical Engineering, National Taiwan University,
Taipei 106, Taiwan.
wpshih@ntu.edu.tw

Igor Solodov
Department of Nondestructive Testing, Institute of Polymer Technology (IKT-ZFP), University of Stuttgart,
70569 Stuttgart, Germany.
igor.solodov@ikt.uni-stuttgart.de

Chin-Teh Sun
School of Aeronautics and Astronautics, Purdue University,
W. Lafayette, IN 47907, USA.
sun@purdue.edu

Chi-Kuang Sun
Research Center for Applied Sciences, Academia Sinica,
Taipei 115, Taiwan.

Department of Electrical Engineering and Graduate Institute of Photonics and Optoelectronics, National Taiwan University,
Taipei 10617, Taiwan.
sun@ee.cc.ntu.edu.tw

Jia-Hong Sun
Institute of Applied Mechanics, National Taiwan University,
Taipei 106, Taiwan.
jiahongsun@ntu.edu.tw

Yukihiro Tanaka
Department of Applied Physics, Hokkaido University,
Sapporo 060-8628, Japan.
yuki@eng.hokudai.ac.jp

Thomas C. T. Ting
Division of Mechanics and Computation, Durand 262, Stanford University,
Stanford, CA 94305, USA.
tting@uic.edu

Daniel Torrent
Wave Phenomena Group, Department of Electronics Engineering, Polytechnic University of Valencia,
Valencia, Spain.
datorma1@upvnet.upv.es

Victor Plessky
GVR Trade SA, Ch. du Vignoble 31C, Bevaix, 2022, Switzerland.
victor.plessky@gmail.com

Walter Steurer
Laboratory of Crystallography,
ETH Zurich, Wolfgang-Pauli-Strasse 10, 8006 Zurich, Switzerland.
walter.steurer@mat.ethz.ch

Ji Wang
Piezoelectric Device Laboratory, Department of Mechanics and Engineering Science
School of Engineering, Ningbo University,
818 Fenghua Road, Ningbo, Zhejiang 315211, China.
wangji@nbu.edu.cn

Wei-Chung Wang
Department of Power Mechanical Engineering, National Tsing Hua University,
101, Section 2, Kuang-Fu Road, Hsinchu 30013, Taiwan.
wcwang@pme.nthu.edu.tw

Yue-Sheng Wang
Institute of Engineering Mechanics, Beijing Jiaotong University,
Beijing, 100044, China.
yswang@bjtu.edu.cn

Sean Wu
Advance Design Technology INC., Tainan County 710, Taiwan.
Department of Electronics Engineering and Computer Science,
Tung Fang Institute of Technology,
No. 110, Tung-Fung Rd., Hunei Shiang, Kaohsiung County, 829,Taiwan.
wusean.tw@gmail.com

Tsung-Tsong Wu
Institute of Applied Mechanics, National Taiwan University,
No. 1, Sec. 4, Roosevelt Road, Taipei, Taiwan.
wutt@ndt.iam.ntu.edu.tw

Wen-Fang Wu
Department of Mechanical Engineering, National Taiwan University,
Taipei 106, Taiwan.
wfwu@ntu.edu.tw

Che-Hua Yang
Graduate Institute of Manufacturing Technology, National Taipei University of Technology,
Taipei 106, Taiwan.
chyang@ntut.edu.tw

Shiuh-Kuang Yang
Department of Mechanical and Electro-Mechanical Engineering,
National Sun Yat-Sen University,
Kaohsiung 804, Taiwan.
skyang@mail.nsysu.edu.tw

Chau-Shioung Yeh
Institute of Applied Mechanics, National Taiwan University,
No. 1, Sec. 4, Roosevelt Road, Taipei, Taiwan.
csyeh@spring.iam.ntu.edu.tw

Ching-Chung Yin
Department of Mechanical Engineering, University of Chiao Tong University,
1001 Ta Hsueh Road, Hsinchu, 300, Taiwan, Republic of China.
ccyin@faculty.nctu.edu.tw

Yook-Kong Yong
Department of Civil and Environmental Engineering, Rutgers University,
623 Bowser Road, Piscataway, NJ 08854, USA.
yyong@rci.rutgers.edu

Da-Ming Zhu
Department of Modern Physics, Univ. of Sci. & Tech. of China, Hefei,
China 230027, China.

Department of Physics, University of Missouri, Kansas City,
Missouri 64110, USA.
zhud@umkc.edu

FSC
www.fsc.org
MIX
Papier | Fördert
gute Waldnutzung
FSC® C083411

Kolibri 360 GmbH
Ferdinand-Jühlke-Straße 7
99095 Erfurt, Deutschland
produktsicherheit@kolibri360.de